CROP PRODUCTION TECHNOLOGIES FOR SUSTAINABLE USE AND CONSERVATION

Physiological and Molecular Advances

CROP PRODUCTION TECHNOLOGIES FOR SUSTAINABLE USE AND CONSERVATION

Physiological and Molecular Advances

Edited by

Münir Öztürk, PhD, DSc
Khalid Rehman Hakeem, PhD
Muhammad Ashraf, PhD, DSc
Muhammad Sajid Aqeel Ahmad, PhD

Apple Academic Press Inc.
3333 Mistwell Crescent
Oakville, ON L6L 0A2, Canada

Apple Academic Press Inc.
1265 Goldenrod Circle NE
Palm Bay, Florida 32905, USA

Library and Archives Canada Cataloguing in Publication

Title: Crop production technologies for sustainable use and conservation : physiological and molecular advances / edited by Münir Öztürk, PhD, DSc, Khalid Rehman Hakeem, PhD, Muhammad Ashraf, PhD, DSc, Muhammad Sajid Aqeel Ahmad, PhD.

Names: Öztürk, Münir A. (Münir Ahmet), editor. | Hakeem, Khalid Rehman, editor. | Ashraf, Muhammad, 1963- editor. | Ahmad, M. S. A. (Muhammad Sajid Aqeel), editor.

Description: Includes bibliographical references and index.

Identifiers: Canadiana (print) 20189067896 | Canadiana (ebook) 2018906790X | ISBN 9781771887267 (hardcover) | ISBN 9780429469763 (PDF)

Subjects: LCSH: Agricultural innovations. | LCSH: Food crops—Physiology. | LCSH: Food crops—Molecular aspects. | LCSH: Sustainable agriculture. | LCSH: Agriculture—Environmental aspects.

Classification: LCC S494.5.I5 C76 2019 | DDC 338.1/6—dc23

Library of Congress Cataloging-in-Publication Data

Names: Öztürk, Münir A. (Münir Ahmet), editor.

Title: Crop production technologies for sustainable use and conservation : physiological and molecular advances / editors: Munir Ozturk, Khalid Rehman Hakeem, Muhammad Ashraf, Muhammad Sajid Aqeel Ahmad.

Description: Waretown, NJ : Apple Academic Press, 2019. | Includes bibliographical references and index.

Identifiers: LCCN 2018058766 (print) | LCCN 2019004937 (ebook) | ISBN 9780429469763 (ebook) | ISBN 9781771887267 (hardcover)

Subjects: LCSH: Crops. | Crop science. | Plant physiology. | Crop yields.

Classification: LCC SB185 (ebook) | LCC SB185 .C75 2019 (print) | DDC 631--dc23

LC record available at https://lccn.loc.gov/2018058766

ABOUT THE EDITORS

Münir Öztürk, PhD, DSc, has served at Ege University Izmir, Turkey, for 50 years in various positions. He is currently working as Vice President of the Islamic World Academy of Sciences. He has received fellowships from the Alexander von Humboldt Foundation, the Japanese Society for Promotion of Science, and the National Science Foundation (USA). Dr. Ozturk has served as Chairman of the Botany Department and Founding Director of the Centre for Environmental Studies at Ege University, Izmir, Turkey. He has served as Consultant Fellow at the Faculty of Forestry at Universiti Putra Malaysia, and as Distinguished Visiting Scientist at the International Center for Chemical and Biological Sciences (ICCBS) at Karachi University, Pakistan. His fields of scientific interest are plant eco-physiology, conservation of plant diversity, biosaline agriculture and crops, pollution, biomonitoring, and medicinal and aromatic plants. He has published 40 books, 70 book chapters, and 175 papers in journals with significant impact factors.

Khalid Rehman Hakeem, PhD, is Associate Professor at King Abdulaziz University, Jeddah, Saudi Arabia. He completed his PhD (Botany) at Jamia Hamdard, New Delhi, India in 2011. Dr. Hakeem previously worked as Postdoctorate Fellow in 2012 and Fellow Researcher (Associate Professor) from 2013 to 2016 at Universiti Putra Malaysia, Selangor, Malaysia. His speciality is plant eco-physiology, biotechnology and molecular biology, plant-microbe-soil interactions, and environmental sciences, and to date, he has edited and authored more than 30 books with various international publishers. He has also to his credit more than 110 research publications in peer-reviewed international journals as well as 55 book chapters in edited volumes. Dr. Hakeem is the recipient of many national and international awards and fellowships.

Muhammad Ashraf, PhD, DSc, is currently working at the Institute of Molecular Biology and Biotechnology at The University of Lahore, Lahore, Pakistan. He has served as Chairman of the Pakistan Science Foundation, Islamabad, Pakistan, as well as Visiting Professor at a number of international universities in the USA, UK, China, Korea, Egypt, Thailand, and Saudi Arabia. He has also worked as Vice Chancellor at the Muhammad Nawaz Shareef University of Agriculture, Multan, Pakistan. His primary area of research is plant stress physiology. His ISI citation index is now over 14,500, and his Google citations are over 30,000. He has more than 500 ISI-rated publications to his credit. He has also authored a number of books published by different international publishers. He earned his PhD from the University of Liverpool, UK, and was a Fulbright scholar at the University of Arizona, Tucson, USA. In 2011, he was awarded a DSc degree by the University of Liverpool, UK.

Muhammad Sajid Aqeel Ahmad, PhD, is Assistant Professor at the University of Agriculture, Faisalabad, Pakistan. His research areas include environmental biology and plant ecology. While working with many regional natural and artificial plant communities, particularly those located in the Cholistan Desert and Potohar Plateau, he has highlighted significant spatial and temporal variations occurring in these communities along the environmental gradients. Additionally, he determined the efficiency of various *in-situ* conservation practices such as fencing on preserving the flora of the natural reserves of Pakistan. These findings have provided efficient measures to support the endangered flora of the region. Out of his research work, he has published 59 papers in ISI-rated journals. Dr. Ahmad has an *h*-index of 15. He has published four books and has edited two issues of the *Pakistan Journal of Botany* as a Guest Editor. He is also an author of four chapters in international books.

CONTENTS

CONTRIBUTORS

Aqsa Abbasi
Atta-ur-Rahman School of Applied Biosciences, National University of Sciences and Technology, Islamabad, Pakistan

Habib Ahmad
Department of Genetics, Hazara University Mansehra–21300, Pakistan/Vice Chancellor, Islamia College University, Peshawar, Pakistan, Tel.: +92- (0)997–414131, E-mails: drhahmad@gmail.com; vc@icp.edu.pk

Muhammad Sajid Aqeel Ahmad
Department of Botany, University of Agriculture, Faisalabad (38040), Pakistan, E-mail: sajidakeel@yahoo.com

Parvaiz Ahmad
Department of Botany,S.P. College, Srinagar, Jammu and Kashmir, India

Waqar Ahmad
Department of Genetics, Hazara University Mansehra–21300, Pakistan, E-mail: wahmadhu@gmail.com

Eren Akcıcek
Ege University, Faculty of Medicine, Department of Gastroenterology, Izmir, Turkey

Ezed Ali
Department of Genetics, Hazara University Mansehra–21300, Pakistan, E-mail: ali.ezed@gmail.com

Volkan Altay
Biology Department, Faculty of Science and Arts, Hatay Mustafa Kemal University, Hatay, Turkey

Naser A. Anjum
Department of Botany, Aligarh Muslim University, Aligarh, 202002, India

Muhammad Ashraf
Institute of Molecular Biology and Biotechnology, The University of Lahore, Lahore, Pakistan

Shafia Zaffar Faktoo
ICAR- Central Institute of Temperate Horticulture, Rangreth, Air Field, Srinagar, Jammu and Kashmir, India

Sana Fatima
Department of Botany, University of Agriculture, Faisalabad (38040), Pakistan

Salih Gucel
Botany Department and Center for Environmental Studies, Ege University, Izmir, Turkey

Alvina Gul
Atta-ur-Rahman School of Applied Biosciences, National University of Sciences and Technology, Islamabad, Pakistan, E-mail: alvina_gul@yahoo.com

Khalid Rehman Hakeem
Department of Biological Sciences, Faculty of Science, King Abdulaziz University, Jeddah, Saudi Arabia, E-mail: kur.hakeem@gmail.com

Mansoor Hameed
Department of Botany, University of Agriculture, Faisalabad (38040), Pakistan

J. S. (Pat) Heslop-Harrison
Department of Biology, Molecular and Cytogenetic Lab, University of Leicester, United Kingdom

Sameen Ruqia Imadi
Atta-ur-Rahman School of Applied Biosciences, National University of Sciences and Technology, Islamabad, Pakistan

Javed Iqbal
School of Biological Sciences, University of the Punjab, Lahore Pakistan,
E-mail: javediqbal1942@yahoo.com

Muhammad Iqbal
Formerly Department of Botany, Jamia Hamdard (Hamdard University), New Delhi, 110062, India; Department of Plant Production, College of Food & Agricultural Sciences, King Saud University, PO Box 2460, Riyadh 11451, Saudi Arabia

Muhammad Shahid Iqbal
Cotton Research Institute, Faisalabad, Punjab, Pakistan

M. Mahbubul Islam
Chief Scientific Officer & Head, Agronomy Division, Bangladesh Jute Research Institute, Dhaka–1207, Bangladesh, Tel.: +8801552416537,
E-mails: mahbub_agronomy@yahoo.com; mahbubulislam63@gmail.com

Mohammad Islam
Department of Genetics, Hazara University, Mansehra, KPK, Pakistan

Cemil Islek
Nigde Omer Halisdemir University, Arts and Sciences Faculty, Biotechnology Department, 51240 Nigde, Turkey, E-mail: cislek@ohu.edu.tr

Sumira Jan
ICAR- Central Institute of Temperate Horticulture, Rangreth, Air Field, Srinagar, Jammu and Kashmir, India

Ihsan Khaliq
Department of Plant Breeding & Genetics, University of Agriculture, Faisalabad, Pakistan

H. Vakıf Mercımek
Tobacco (TAPDK) Izmir Office Specialist, Bornova, Izmir, Turkey

Javid Iqbal Mir
ICAR- Central Institute of Temperate Horticulture, Rangreth, Air Field, Srinagar, Jammu and Kashmir, India, E-mail: javidiqbal1234@gmail.com

Khushi Muhammad
Department of Genetics, Hazara University Mansehra–21300; School of Biological Sciences, University of the Punjab, Lahore, Pakistan, E-mail: khushisbs@yahoo.com

Maria Naqve
Department of Botany, University of Agriculture, Faisalabad, Pakistan

Ijaz Rasool Noorka
Department of Plant Breeding and Genetics, University College of Agriculture, University of Sargodha, Pakistan, Tel.: 0092300–6600301, E-mail: ijazphd@yahoo.com

Münir Öztürk
Botany Department and Center for Environmental Studies, Ege University Bornova, Izmir, Turkey,
E-mail: munirozturk@gmail.com

Yong- Bao Pan
USDA-ARS, Sugarcane Research Unit, 5883 USDA Road, Houma, Louisiana, 70360, USA,
E-mail: yongbao.pan@ars.usda.gov

Halil Polat
Transitional Zone Agricultural Research Institute, Soil and Water Resources Department,
Plant Nutrient and Soil Unit, Eskişehir, Turkey

Maham Saddique
Department of Botany, University of Agriculture, Faisalabad, Pakistan

Ayesha Sajid
Atta-ur-Rahman School of Applied Biosciences, National University of Sciences and Technology,
Islamabad, Pakistan

Wajida Shafi
ICAR-Central Institute of Temperate Horticulture, Rangreth, Air Field, Srinagar,
Jammu and Kashmir, India

Muhammad Shahbaz
Department of Botany, University of Agriculture, Faisalabad, Pakistan,
E-mail: shahbazmuaf@yahoo.com

Muhammad Rafiq Shahid
Cotton Research Institute, Faisalabad, Punjab, Pakistan

Kanwal Shahzadi
Department of Plant Sciences, Quaid-I-Azam University, Islamabad, Pakistan

Muneer Sheikh
ICAR- Central Institute of Temperate Horticulture, Rangreth, Air Field, Srinagar,
Jammu and Kashmir, India

Kadriye Taşpinar
Transitional Zone Agricultural Research Institute, Soil & Water Resources Department,
Plant Nutrient and Soil Unit, Eskişehir, Turkey

Shahid Umar
Department of Botany, Jamia Hamdard (Hamdard University), New Delhi, 110062, India,
E-mail: s_umar9@hotmail.com

Bengu Turkyilmaz Unal
Nigde Omer Halisdemir University, Arts and Sciences Faculty, Biotechnology Department,
51240 Nigde, Turkey, E-mail: bturkyilmaz@ohu.edu.tr

ABBREVIATIONS

AARI	Ayub Agricultural Research Institute
ABA	abscisic acid
ACC	1-aminocyclopropane–1-carboxylate
ACO	ACC oxidase
ACS	ACC synthase
AFLP	amplified fragment length polymorphism
AJK	Azad Jammu & Kashmir
APS	American Phytopathological Society
AVG	aminoethoxyvinylglycine
AVRDC	Asian Vegetable Research and Development Centre
BAC	bacterial artificial chromosome
BADC	Bangladesh Agricultural Development Corporation
BI	Biodiversity International
BJMC	Bangladesh Jute Mills Corporation
BSA	bovine serum albumin
BSA	bulked segregant analysis
BYDV	barley yellow dwarf virus
CAT	catalase
CCC	chlormequat chloride
CDK	cyclin-dependent kinases
CGIAR	Consultative Group on International Agricultural Research
CIAT	Center International de-Agricultural Tropical
CIFOR	Center for International Forestry Research
CIMMYT	Center International de-Mejoraments de maize Trigo
CIP	Center International de-Papa
CLCuD	cotton leaf curl disease
CO	Constans
CREs	cis-regulatory elements
CRISPR	clustered regularly interspaced short palindromic repeats
CTs	condensed tannins
CYCD	D-cyclins
DAE	day after emergence
DAF	DNA amplification fingerprinting
DArT	diversity arrays technology

DAS	days after sowing
DFl	drought at flowering
DPod	drought at pod development
DPre-fl	drought at pre-flowering
DSTC	digested sugar beet tailings
EBL	epibrassinolide
ECe	electric conductivity of the soil
EST	expressed sequence tags
ETC	electron transport chain
FAO	Food and Agriculture Organization
FC	flue-cured
FISH	fluorescence in situ hybridization
GA	gibberellins
GB	glycine betaine
GBSS	granule-bound starch synthase
GISH	genomic *in situ* hybridization
GM	genetically modified
GP	great plains
GPX	Glutathione peroxidase
GR	glutathione reductase
GSH	glutathione
HG	homology group
HNF	Hamdard National Foundation
IAA	indoleacetic acid
ICARDA	International Center for Agricultural Research in the Dry Areas
ICRAF	International Center for Research in Agroforestry
ICRISAT	International Crops Research Institute for the Semi-Arid Tropics
ICSB	International Consortium for Sugarcane Biotechnology
IFAD	International Fund Sponsor the CGIAR for Agricultural Development
IFPRI	International Food Policy Research Institute
IITA	International Institute of Tropical Agriculture
IMF	International Monetary Fund
IPGRI	International Plant Genetic Research Institute
IRGA	infra-red gas analyzer
IRRI	International Rice Research Institute
IWMI	International Water Management Institute
JA	jasmonates

JA	jasmonic acid
KH_2PO_4	potassium dihydrogen phosphate
KP	Khyber Pakhtunkhwa
KTIs	Kunitz protease inhibitor genes
LAI	leaf area index
LRK	lectin-like receptor kinase
LRR	leucine-rich-repeat
LSU	large subunit
MAB	marker-assisted breeding
MAS	marker-assisted selection
MASA	mutant allele-specific amplification
MDA	malondialdehyde
MG	monoploid genome
MLO	mycoplasma-like organisms
MSP	minimum support price
MT	metric tons
NAR	net assimilation rate
NARC	National Agricultural Research Council
NBS	nucleotide-binding site
OH^-	hydroxyl radicals
OP	osmotic pressure
PCR	polymerase chain reaction
PEG	polyethylene glycol
PG	polygalacturonase gene
PGSC	potato genome sequencing consortium
P_N	net photosynthetic rate
POD	peroxidase
PR	pathogenesis-related
PVX	potato virus X
QTL	quantitative trait loci
QTL	quantitative trait locus
RAPD	random amplified polymorphic DNA
RDP	ribosomal database project-II
RDW	root dry weight
RFLP	restriction fragment length polymorphism
RL	root length
RNA	ribonucleic acid
RNAi	RNA interference
RNA-Seq	RNA sequencing
RO	alkoxy radicals

ROL	radial oxygen loss
ROS	reactive oxygen species
RUE	radiation use efficiency
RWC	relative water content
SA	salicylic acid
SAGE	serial analysis of gene expression
SAM	S-adenosyl methionine
SBP	sugar beet pulp
SCAR	sequenced characterized amplified region
SCoT	start codon targeted markers
SDMs	single-dose markers
SE	standard error
SINE	short interspersed nuclear element
SMC	soil moisture content
SNP	single nucleotide polymorphism
SOD	superoxide dismutase
SSLP	simple sequence length polymorphism
SSN	sequence-specific nuclease
SSR	seven microsatellite loci
SSR	simple sequence repeats
SSU	small subunit
STMS	sequence tagged microsatellite site
STR	short tandem repeats
STS	sequence tagged site
TALENs	transcription activator-like effector nucleases
TDFs	transcripts derived fragments
TNL	TIR-NBS-LRR
TRV	tobacco rattle virus
TSP	total soluble proteins
UNDP	United Nations Development Programme
USDA	United States Department of Agriculture
VIGS	virus-induced gene silencing
VNTR	variable number tandem repeat
WARDA	West African Rice Development Association
WSMV	wheat streak mosaic virus

PREFACE

According to the United Nations' estimates, Earth's population achieved the landmark of 7.3 billion in 2015, a level of 7.5 billion in 2017, and is expected to go up to 7.7 billion in 2018. This means that there is an addition of nearly half a billion residents on the globe during only the last decade. The predictions are that the population will reach a number of 11.2 billion in the year 2100; this will mean that approximately 6 billion people have been added to the planet during the last century. This situation is alarming and raises the question of sustainability of nature and natural resources. Under these circumstances, will our natural resources continue to exist in a healthy state? Will our forest reserves sustain? Will our farmlands be able to feed the exponentially growing human population? Will our nonrenewable resources last enough to support our industry and economy? In what sort of environment will our future generations be living in? All these questions along with many others are stumbling our minds.

Human population relies on the cultivation of major food crops like wheat, sugarcane, sugar beet, rice, maize, sunflower, potato, tomato, peanuts, etc. Major fruits include grapes, citrus, banana, apples, etc. Major crops supporting our industry include cotton, which is the backbone of the textile industry. Some other major crops, like tobacco, jute, etc., are also among highly valuable commercial crops. All these crops are being extensively cultivated over large areas to fulfill the growing demand for food, fodder, and livelihood of the rapidly exploding human population.

During the past century, the crop varieties had poor yields and had little resistance to diseases and environmental stresses. The easiest and most economic approach was to clear the land for cultivation, land that had been previously occupied by forested systems. This resulted in the destruction of habitats and deterioration of the environment, particularly our soils and water resources. The condition has been much worse, particularly in Asia and Africa, having the largest share of the global population. After the agricultural revolution and the introduction of new high-yielding and disease-resistant crop varieties, it has been possible to partially achieve the goal of food security, with the least impact on environmental resources. However, scientists still question how long this revolution would be able to sustain the global food demand. There is a dire need to work harder toward the attainment of sustainable human development.

This book is an effort to put together the fundamentals of existing knowledge pertaining to crop production technologies of major crops used for *food* (wheat and mungbean), *sugar production* (sugarcane, sugar beet), *vegetables* (okra, capsicum, tomato and potato), *fruits* (pears) or in *commercial fiber production* (cotton and jute). Additionally, the centers of origin of crop plants and the sustainability of their use have also been discussed. There is a focus on future perspectives to enhance the yield of the major agricultural crops by using physiological and molecular enhancement. This book also focuses on the strategies to cope with the increasing demand of food supply. Thus, this volume will be a valuable addition to the efforts for addressing food security issues.

—Editors

CHAPTER 1

PLANT GENETIC RESOURCES OF MAJOR AND MINOR CROPS: ORIGIN, SUSTAINABLE USE, AND CONSERVATION

MANSOOR HAMEED[1], MUHAMMAD SAJID AQEEL AHMAD[1], MUHAMMAD ASHRAF[2], MÜNIR ÖZTÜRK[3], and SANA FATIMA[1]

[1]Department of Botany, University of Agriculture, Faisalabad (38040), Pakistan, E-mail: sajidakeel@yahoo.com

[2]Institute of Molecular Biology and Biotechnology, The University of Lahore, Lahore, Pakistan

[3]Botany Department and Center for Environmental Studies, Ege University Bornova, Izmir, Turkey

ABSTRACT

Plant genetic resources contribute to approximately 60% of human food and other socio-economic needs. However, the present situation of plant genetic resources is alarming because a number of traditional crops and varieties are out of cultivation and human dependence on crop plants is restricted to only a few cereals and legume cultivars. This situation is worst in some cases where germplasm of traditionally cultivated crops is being depleted or has been permanently lost. In this situation, exploration, and conservation of not only the genetic resources of all major and minor crops, but also of ancestral and wild and weedy species is crucial. The importance and use of genetic resources in crop improvement programs and strategies for their conservation have been discussed in detail in this chapter.

1.1   INTRODUCTION

Plant genetic resources (germplasm resources or gene pools) are the complete generative/vegetative or reproductive materials within or among the species, which are used mainly for economic, scientific or social purposes (Hammer et al., 1999; Frankel and Hawkes, 2011; Pautasso, 2012). Many fields of plant sciences are benefited significantly from genetic resources, e.g., ecology, and evolution, physiology, and biochemistry, cytogenetics, and molecular biology, agronomy, pathology, breeding, and many others. Conventional breeding programmes and genetic engineering largely depend upon the genetic resources (i.e., genetic diversity), not only for improving yield and nutritive quality (Evenson et al., 1998; Frankel and Hawkes, 2011), but also for improving resistance or adaptation potential against a variety of biotic and abiotic stresses (Cherry et al., 2000; Fritsche-Neto and Borém, 2012).

Genetic diversity of crop plants has been disappearing rapidly because of modern agricultural practices like preference of monocultures of high-yielding varieties and GM plants, and loss of primitive cultivars and land-races, mainly due to habitat conversion (Wolfenbarger and Phifer, 2000; Hyten et al., 2006; Schmidt and Wei, 2006: Kovach and McCouch, 2008). However, a large reservoir of genetic diversity still exists in the form of landraces, farmer's varieties and even wild relatives of crop plants with a tremendous amount of useful genetic variation (Harlan, 1975; Naeem et al., 2012). These gene pools are critically important in breeding for adaptation to environmental factors, resistance to biotic and abiotic stresses, or better nutritive value in crop plants, as these cannot be developed by artificial means (Ashraf and Akram, 2009; Ahmad et al., 2012). However, to a limited extent, these can be created using mutation breeding or genetic engineering techniques (Fridman and Zamir, 2012; Perez-de-Castro et al., 2012).

Plant genetic resources or gene pools can be categorized into three major groups on the basis of their degree of sexual compatibility: i) primary gene pool includes all crop plants and those species that can produce fertile hybrids, ii) secondary gene pool with plant species that contains certain barriers against crossing, and iii) tertiary gene pool with plant species that can be crossed only with the help of modern techniques (Gepts, 2000). However, (Gladis and Hammer, 2002) recognized the fourth gene pool, which includes synthetic strains that are developed through transgenic technologies and that do not occur in nature.

Evolution of species and domestication is largely dependent on genetic diversity in plants. Farmers have used this genetic diversity in developing promising and high-yielding crop species since the inception of agriculture

(Konig et al., 2004; Vaughan et al., 2007). Genetic variability can play a fundamental role in crop improvement and development of new varieties, as it provides basic material for breeding and genetic engineering, extremely helpful for desired traits in crop plants according to the human need (Konig et al., 2004; Vaughan et al., 2007). Today, only 12 species (mainly wheat, rice, maize, and potatoes) contribute to 70% of the total food supply of the world (Wilson, 1992). Food security in the future may be ensured through the exploitation of all available genetic resources, i.e., not only the consumption of traditional crops but also the utilization of other wild, minor, and under-utilized crops (Brussaard et al., 2010).

1.2 ANCESTORS OF CROP PLANTS

During the evolutionary history, crops have enlarged their geographical ranges, and therefore extended over large areas. Species continued evolving and picking up the germplasm from related species since the domestication started, and consequently further extended their range. Because of this, there was more complexity in the genetic makeup of the species. This is an important process, in which sometimes the ancestors of a species are completely absorbed and may become extinct, because the ancestors compete out by their hybrids. Its good example is maize, the traditional varieties of which are out of cultivation due to the extensive development of hybrids (Blixt, 1994; Glaszmann et al., 2010b).

Majority of the domesticated plants can still produce fertile hybrids by crossing them with their wild ancestors. In some crops like maize and date palm, such crossing is indispensable for maintaining genetic diversity (Gross and Olsen, 2010). It was predicted that the major cereal crops were wild ancestors at one time, and therefore in most cases, the domesticated plants and their wild ancestors are genetically closely related (Vaughan et al., 2008; Abbo et al., 2012).

Improvement in crop production and food security essentially depends on conservation of these ancestral species (Hawkes et al., 2000). Currently, wild ancestors and wild relatives of crop plants are under serious threat of becoming extinct, which are obligatory for maintaining useful genetic diversity in agricultural crops, as well as in natural species (Maxted et al., 2008; Maxted et al., 2011). Degradation and destruction of natural environments (habitat loss), conversion of natural habitats for other uses, loss of many fruits and industrial crop species (deforestation), wild relatives of cereal crops by overgrazing and desertification, and industrialization of agriculture radically reduced the occurrence of ancestral species of crop plants (Maxted et al., 1997; Heywood and Dulloo, 2006).

Bread wheat (*Triticum aestivum* L.) was the result of natural hybridization between wild species of einkorn wheat (*Triticum urartu* L.) and goat grass (*Aegilops speltoides* L.). The later is now extinct, which gave rise to emmer wheat (*T. diccocoides*). This species then went hybridization with another species of goat grass (*Aegilops tauschii*) to form *T. aestivum* (Eckardt, 2010; Peng et al., 2011).

Similarly, turnip (*Brassica rapa*) is a close relative of *B. oleracea* complex, which includes broccoli, Brussels sprout, and kale. Wild species of these species are still found on sea-cliffs and in semi-arid and arid regions. During evolutionary history, species with proliferate masses, longer flowering period and enlarged axillary buds have resulted. *Brassica napus* were resulted from the natural cross between *B. oleracea* and *B. rapa* (Prakash et al., 2010; Anjum et al., 2012).

Wild potatoes e.g., *Solanum jamesii, Solanum berthaultii*, etc. are bitter in taste due to the presence of many toxic alkaloids (Caruso et al., 2011). Ancestral species of potato (*Solanum tuberosum* L.) is not known, but species with low alkaloids like *S. brevicaule, S. canasense, S. leptophyes*, and *S. sparsipilum* may contribute a lot to the domestication of modern potato (Brown, 2011). However, more likely the ancient cultivated form *S. tuberosum* ssp. *andigena* is the ancestral species. Evolution has resulted in the development of a genotype with fewer flowers, shorter stolons and large nomogenous tubers with different day-length requirements (Hawkes, 1990; Cooper et al., 2001).

Lycopersicon esculentum and *L. pimpinellifolium* along with nine other closely related species are the ancestral species of cultivated tomato (*L. esculentum*), which are widespread in Central America. Disease resistance was the main focus in the domestication of modern cultivars (Jenkins, 1948; Breto et al., 1993). Ancestral species of the cultivated pea (*Pisum sativum*) is the wild relative of *P. elatius,* however, *P. fulvum* equally contributed to the domestication of modern pea cultivars. Non-shattering pods, large, and thin-coated seeds, and erect habit of plants were the main focus of domestication (Cupic et al., 2009; Jing et al., 2010; Jing et al., 2012).

Domestication of cultivated rice (*Oryza sativa*) is little controversial. Previously it was believed that rice was domesticated from wild Asian progenitors, *O. nivara* and *O. rufipogon* (Ge et al., 1999). More recently, it was investigated that introgression was the key factor controlling rice domestication. Critical domestication alleles e.g., (non- shattering) were fixed during the evolutionary history, and thereafter introgression frequently occur between early cultivar and its wild progenitors, and as a result, modern cultivars such as *indica* and *japonica* derived as hybrids (Sang and Ge, 2007).

It is now widely accepted that all pre-Columbian maize landraces arose from a specific teosinte (*Zea mays* ssp. *Parviglumis*), about 9000 years ago through a single domestication event in southern Mexico (Matsuoka, 2005). Teosinte can produce fertile hybrids with maize and other related species, such as *Tripsacum*, all geographically restricted to Mexico and Guatemala (Wllkes, 1977; de Wet et al., 1981; De Wet, 2009).

Ancestral species or wild relatives of the crop plants are now considered as an increasingly important resource for improving production and biotic/abiotic stress tolerance of the agricultural crops, and therefore, vital for ensuring food security for the future generations. These species contain a number of useful genes, which can be effectively incorporated in modern crops. Such practices have been successfully done in the past, and there are several examples of useful gene transfer from wild and ancestral species into major crops including cereals, legumes, oilseed, fiber, and fodder crops.

1.3 DOMESTICATION OF CROP PLANTS

Crop plants were domesticated when people started cultivation and management of the wild ancestral species. Historical evidence indicates that crop plants should have been domesticated between 3,000 to 10,000 years ago, i.e., soon after the end of the Pleistocene era. In the old world (e.g., Near East and Asia), domestication started during the early Holocene era, whereas in the New World (North America, South America, and Mexico) by the middle of Holocene (Diamond, 2002; Abbo et al., 2009). In early days, ecological, and physiological conditions of plants were rather critical due to the expansion of forests, the evolution of river systems, and a substantial increase in CO_2 concentration in the atmosphere at the end of Pleistocene (Ehleringer et al., 2005). Environmental heterogeneity resulted in more and more species diversity and therefore, the people-plant relationship in relation to domestication changed and became more strengthened. However, the focus remained the transformation of fragile form into stable forms of crop cultivars, e.g., wheat, and barley.

Domesticated plants can be broadly classified into six categories: i) cereals and grasses, ii) pulses and other legumes, iii) root and tuber crops (onions, potatoes, and arums), iv) oil plants (palms, barssicas, and sunflower), v) fruits and nuts (apples, peaches, plums, citrus fruits, and many tropical and temperate nuts), and vi) vegetables and spices (brassicas, gourds, and many others). However, in most cases, propagation, and survival of domesticated plants are totally human dependent.

Consequences of domestication were enormous, not only in increasing the production of cultivated crops but also had a strong impact on ecological and health-related issues. However, dependence on a few crop plants only significantly increased the risk of famine, i.e., due to complete crop failure. Habitat destruction and degradation due to agricultural field preparation changed the community structure and increased the soil erosion. Dependence on a few crops with high starch content increased the health risks like dental health, high cholesterol, weak bones, etc. (Diamond, 2002; Price, 2009).

Domestication of crop plants occurred independently all over the world. Major centers of diversity include South and East Asia, Near East, the Mediterranean region and Africa in the Old World and South America, eastern, Mexico, and North America in the New World. Wheat and barley were the most important of the cereals of the Old World, which were domesticated in the Near East (Zohary et al., Abbo et al., 2009). The considerable botanical investigation has provided a great deal of background on these and other crops. One of the ancestral species of bread wheat, the einkorn wheat, was domesticated at about 9000 B.C., whereas another wild relative of wheat, the emmer wheat, in 8000 B.C. However, domestication of bread wheat comes by about 6000 B.C., considerably later than its ancestral species. Several cereals like sorghum, African rice and millets were domesticated datings back to 6000–1200 B.C. in Africa, mainly near the equator, Sahara desert, and Nile Valley. These represent at least four regional complexes (Abbo et al., 2010). Some species of millets (foxtail and broomcorn) were domesticated in Asia at about 6500 B.C., hemp at 3500 B.C. and Asian rice at 5000 B.C. (Crawford, 2006)

Domestication began independently in South America, Mexico, and eastern North America in the New World. The earliest domesticated crops in South America were maize, beans, tomato, gourds, chili pepper, grain amaranths, avocado, and cacao (Erickson et al., 2005; Smith, 2006). Squash and gourds were domesticated at about 8000–7000 B.C., and beans and corn at 5000 B.C. (Anderson, 2012). Sunflower, many chenopods, and cucurbits were domesticated in North America at about 1500 B.C. (Heiser, 1979; Smith, 1989).

Most of the crop plants were domesticated once, i.e., with a single origin of domestication. The strongest and characteristic examples are of maize and einkorn wheat. However, multiple origins of domestication have been evidenced in several crop species. Asian rice (*Oryza sativa*) has at least two independent centers of origins, which is confirmed from gene sequence of two subspecies, indica, and japonica (Vitte et al., 2004; Londo et al., 2006). Common bean (*Phaseolus vulgaris*) was domesticated twice in South and Central Americas (Koinange et al., 1996; Chacon et al., 2005), whereas squash (*Cucurbita pepo*) in Mexico and southeastern North America (Sanjur et al., 2002).

Domestication syndrome, i.e., selection of a distinct array of traits, played a critical role in crop improvement. These traits were associated with reduced shattering, determinate growth habit, large inflorescence size, large fruit and seed size, and reduced seed dormancy (Doebley et al., 2006; Weeden, 2007). The possible changes in plant species due to domestication can be listed as: i) wider geographical range and adaptability to diverse environmental conditions, ii) simultaneous crop maturity, iii) lack of shattering or scattering of seeds, iv) increased size of fruits and seeds, v) annual habit of the crops, vi) loss of seed dormancy and photoperiodic response, vii) self pollination, viii) increased self-compatibility, ix) loss of defensive adaptations like toxins, spines, thorns, etc., x) improved palatability, xi) increased susceptibility for diseases and pests, xii) development of seedless fruits, and xiii) vegetative propagation (Ross-Ibarra et al., 2007; Vaughan et al., 2008).

Domestication brought about significant changes in crop natural ancestors and their wild relatives, which are generally according to the human need. Such differences include improved germination rate, synchronous germination, increased production, and decreased yield losses due to breaking up or shattering of fruits/seeds, although, domesticated crops are more prone to diseases and other abiotic hazards.

1.4 CENTERS OF ORIGIN AND DIVERSITY

The concept of centers of origin and center of diversity was first given by Vaviliv (1951), who identified eight centers of cultivated plants (Tables 1.1 and 1.2). He considered the regions of maximum diversity (center of diversity as a type of formation (center of origin) of a particular crop species. Domestication of ancient crops started in four nuclear centers of agricultural crops originated to 9000–10,000 years ago. The other crops domesticated much later with the spread of older crops and agriculture (Hawkes, 1983).

TABLE 1.1   Centers of Diversity Discovered by N.I. Vavilov

I. China-Japan	*Actinidia* spp. (gooseberries: *A. kolomikta, A. arguta, A. chinensis, A. polygama*), *Aleurites* (wood-oil trees: *A. fordii, A. montana, A. cordata*), *Alliums* spp. (onions: *A. macrostemon, A. chinensis*), *Armeniaca vulgaris* (wild apricot), *Arundinaria* spp. (canes), *Avena sativa* var. *nuda* (naked oats), *Bambusa* spp. (bamboos), *Boehmeria nivea* (ramie), *Brassica rapa* (turnip), *Camellia sinensis* (tea), *Castonea mollissima* (Chinese chestnut), *Cerasus* spp. (cherries), *Chaenomeles sinensis* (Chinese quince), *Cinnamomum camphora* (camphor tree), *Citrus* spp. (oranges: *C. ichangensis, C. junos, C. sinensis, C. reticulata*), *Colocasia esculenta* (taro), *Corylus* (hazels: *C. sieboldiana, C. chinensis, C. heterophylla*), *Cucumis melo* (cantaloupe), *Dioscorea batatas* (Chinese yam), *Diospyros* spp. (persimmons: *D. kaki, D. lotus*), *Echinochloa frumentacea* (Indian barnyard millet), *Eucommia ulmoides* (hardy rubber tree), *Fortunella* spp. (kumquats: *F. margarita, F.japonica, F. hindsii*), *Glycine* spp. (soybeans: *G. max, G. soja*), *Gossypium arboreum* var. *nanking* (Bluntleaf cotton), *Hordeum vulgare* (common barley), *Juglans* spp. (walnuts: *J. mansharica, J. ailantifolia*), *Malus* spp. (crabapples: *M. baccata* var. *mansharica, M. rockii, M. hupehensis, M. halliana, M. pumila, M. asiatica, M. prunifolia, M. spectabilis, M. micromalus, M. sieboldii*), *Morus* spp. (mulberry), *Oryza sativa* ssp.*japonica* (Japanese rice), *Panicum miliaceum* (Proso millet), *Perilla frutescens* (beefsteak plant), *Persica davidiana* (Chinese wild peach), *Phaseolus* spp. (beans: *P. angularis, P. vulgaris* var. *chinensis*), *Phyllostachys* spp. (bamboos), *Sasa* spp. (bamboos), *Poncirus trifoliate* (trifoliate orange), *Prunus* spp. (plums: *P. simonii, P. salicina*), *Pyrus* spp. (pears: *P. pyrifolia, P. bretschneideri, P. betulifolia, P. calleryana, P. phaeocarpa, P. ussuriensis* var. *aromatica, P. ussuriensis* var. *ovoidea, P. ussuriensis*), *Raphanus sativus* (radish), *Saccharum sinense* (Japanese cane), *Setaria italica* (foxtail millet), *Solanum melongena* (eggplant), *Sorghum bicolor* ssp. *bicolor* (grain sorghum), *Triticum aestivum* (bread wheat), *Vitis davidii* (spiny vitis), *Ziziphus jujuba* (jujube)
II. Indinesia-Indichina	*Aleurites moluccana* (Indian walnut), *Arenga pinnata* (sugar palm), *Artocarpus* spp. (jackfruits: *A. integer, A. heterophyllus*), *Bambusa* spp. (bamboos), *Caryota urens* (Toddy palm), *Citrus* spp. (oranges: *C. maxima, C. aurantium, C. macroptera, C. hystrix*), *Cocos nucifera* (coconut palm), *Colocasia esculenta* (taro), *Dendrocalamus* spp. (clumping bamboos), *Dioscorea* spp. (yams: *D. alata, D. batatas, D. osculenta, Dioscorea bulbifera*), *Durio zibethinus* (durian), *Garcinia* spp. (mangosteens: *G. mangostana, G. dulcis*), *Gigantochloa* spp. (bamboos), *Gossypium arboretum* (tree cotton), *Mangifera caesia* (Malaysian mango), *Metroxylon sagu* (sago palm), *Musa* spp. (bananas: *M. balbisiana, M. acuminata, M. textilis*), *Oryza* spp. (rices: *O. sativa* ssp.*javanica, O. sativa* var. *fatua, O. officinalis, O. minuta, O. meyeriana, O. ridleyi*), *Saccharum* spp (canes: *S. robustum, S. spontaneum*)
III. Australia	*Acacia* spp. (wattles), *Eremocitrus* spp. (limes), *Eucalyptus* spp. (gums), *Gossypium* spp. (cottons: *G. sturtii, G. robinsonii, G. costulatum, G. populifolium, G. cunninghamii, G. australe, G. bickii, G. pulchellum, Macadamia ternifolia* (Gympie nut), *Microcitrus* spp. (limes), *Nicotiana* spp. (tobaccos), *Oryza* spp. (rices: *O. australiensis, O. schlechteri, O. ridleyi*), *Trifolium subterranneum* (Subterranean clover)

TABLE 1.1 (Continued)

IV. India	*Arenga saccharifera* (sugar palm), *Bambusa* spp. (bamboos), *Canavalia ensiformis* (jack bean), *Cannabis indica* (Indian hemp), *Cicer arietinum* (chickpea), *Citrus* spp. (oranges: *C. assamensis, C. medica, C. latipes*), *Cocos nucifera* (coconut), *Corchorus capsularis* (Tossa jute), *Cucumis sativus* (cucumber), *Dendrocalamus* spp. (bamboos), *Dolichos biflorus* (horse gram), *Eleucine coracana* (finger millet), *Gossypium arboreum* (tree cotton), *Luffa acutangula* (angled luffa), *Mangifera indica* (mango tree), *Oryza* spp. (rices: *O. sativa* spp. *indica, O. coarctata*), *Piper nigrum* (black pepper), *Raphanus sativus* (radish), *Saccharum* spp. (canes: *S. officinarum, S. barberi, S. spontaneum*), *Sesamum indicum* (sesame), *Solanum melongena* (eggplant), *Sorghum bicolor* (sorghum), *Triticum sphaerococcum* (shot wheat), *Vigna* spp. (beans: *V. radiata* var.*sublobata, V. mungo, V. umbellata*)
V. Middle Asia	*Allium* spp. (onions: *A. cepa, A. sativum*), *Amygdalus* spp. (almonds: *A. communis, A. bucharica, A. vavilovii, A. petunnikovii, A. spinosissima, A. ulmifolia*), *Armeniaca vulgaris* (wild apricot), *Brassica campestris* (white mustard), *Carthamus tinctorius* (safflower), *Cerasus* spp. (cherries: *C. vulgaris, C. amygdaliflora, C. verrucosa, C. alaica, C. turcomanica, C. chodshaatensis, C. erythrocarpa, C. pseudoprostrata, C. microcarpa*), *Cicer arietinum* (chickpea), *Crataegus* spp. (hawthorns: *C. pontica, C. turkestanica, C. altaica, C. hissarica, C. songarica*), *Cucumis melo* (muskmelon), *Daucus carota* (carrot), *Elaeagnus* spp. (oleasters: *E. angustifolia, E. orientalis, E. songorica*), *Fragaria bucharica* (strawberry), *Gossypium herbaceum* (Levant cotton), *Juglanas regia* (Persian walnut), *Lathyrus sativus* (grass pea), *Lens culinaris* (lentil), *Linum usitatissimum* (common flax), *Malus* spp. (apples: *M. kirghisorum, M. sieversii, M. niedzwetzkyana, M. hissarica, M. linczevskii*), *Medicago sativa* (alfalfa), *Pisum sativum* (garden pea), *Pistacia vera* (pistachio), *Prunus* spp. (plums: *P. cerasifera, P. ferganica, P. tadzhikistanica*), *Pyrus* spp. (*P. korshinskyi, P. bucharica, P. regelii, P. vavilovii, P. turcomanica, P. boissieriana*), *Saccharum spontaneum* (wilde sugarcane), *Secale* spp. (ryes: *S. cereale* var. *afghanicum, S. cereale* var. *eligulatum*), *Spinacia oleracea* (spinach), *Triticum aestivum* (bread wheat), *Vigna radiata* var. *radiata* (mungbean), *Vitis vinifera* (common grape vine)
VI. Near Asia	*Aegilops* spp. (goat grasses), *Allium porrum* (garden leek), *Amygdalus* spp. (almonds: *A. georgica, A. turcomanica, A. fenzliana, A. urartu, A. scoparia, A. communis*), *Armeniaca vulgaris* (wild apricot), *Avena* spp. (oats: *A. sativa, A. bizantina*). *Beta* spp. (beets: *B. lomatogona, B. macrorrhiza, B. corolliflora, B. vulgaris*), *Brassica oleracea* (wild cabbage), *Castanea sativa* (sweet chestnut), *Cerasus* spp. (cherries: *C. avium, C. vulgaris, Cicer arietinum* (chickpea), *Cornus mas* (European cornel), *Corylus* spp. (hazelnuts: *C. avellana, C. maxima, C. pontica, C. colchica, C. iberica, C. cervorum, C. colurna*), *Cucumis* spp. (melons: *C. melo, C. sativus*), *Cydonia oblonga* (quince), *Daucus carota* (carrot), *Ficus carica* (common fig), *Gossypium* spp. (cottons: *G. areysianum, G. stocksii, G. incanum*), *Hordeum* spp. (barleys: *H. spontaneum, H. distichum, H. vulgare*), *Laurocerasus officinalis* (cherry laurel), *Lens* spp. (lentils: *L. culinaris, L. nigricans, L. kotschyana, L. orientalis*), *Linum usitatissium* (common flax), *Malus* spp. (apples: *M. orientalis, M. sieversii* var. *turkmenorum*), *Mandragora turcomanica*

TABLE 1.1   *(Continued)*

	(Himalayan mandrake), Medicago spp. (medick: M. daghestanica, M. dzhawakhetica M. falcata ssp. romanica, M. cancellata), Mespilus germanica (common medlar), Morus spp. (mulberries), Onobrychis spp. (sainfoins: O. altissima, O. transcaucasica), Pisum elatius (tall wild pea), Prunus spp. (plums: P. cerasifera, P. spinosa, P. domestica), Punica spp. (pomegranates: P. granatum, P. protopunica), Pyrus spp. (pears), Rosa spp. (roses), Secale spp. (ryes: S. montanum, S. kuprijanovii, S. vavilovii, S. cereale, S. ancestrale), Trifolium apertum (open clover), Triticum spp. (wheats: T. thaoudar, T. urartu, T. monococcum, T. araraticum, T. palaeo-colchicum, T. timopheevii, T. militinae, T. dicoccum, T. persicum, T. zhukovskyi, T. aestivum, T. spelta, T. macha, T. vavilovii), Vavilovia formosa (beautiful vavilovia), Vicia spp. (vetches: V. ervilia, V. sativa, V. pannonica, V. narbonensis), Vitis spp. (grape vines: V. vinifera ssp. sylvestris, V. vinifera, V. labrusca)
VII. Mediterranean	*Allium cepa (onion), Anethum graveolens (dill), Avena spp. (A. strigosa, A. brevis, A. bizantina, A. sterilis), Beta spp. (beets: B. webbiana, B. procumbens, B. patellaris, B. vulgaris), Brassica (brassicas: B. oleracea, B. napus (rapeseed), Ceratonia siliqua (carob tree), Cicer arietinum (chickpea), Citrus spp (oranges: C. limon, C. sinensis, C. aurantium), Cynara cardunculus var. scolymus (globe artichoke), Cyperus esculentus (yellow nutsedge), Daucas carota (carrot), Drimia maritima (sea squill), Ervum monanthos (one-flowered lentil), Hedysarum coronarium (French honeysuckle), Hordeum spp. (barleys: H. vulgare, H. vulgare f. distichon), Lathyrus sativus (grass pea), Laurus nobilis (bay laurel), Lavandula officinalis (lavender), Lens culinaris (lentil), Linum spp. (flaxes: L. usitatissimum, L. angustifolium), Lupinus spp. (lentils: L. angustifolius, L. luteus, L. albus, L. termis, L. pillosus, L. pilosus), Olea spp. (O. europaea, O. chrysophylla), Ornothopus sativus (common bird's-foot), Petroselinum sativum (parsley), Phalaris canariensis (Canary grass), Pisum sativum (garden pea), Prunus spp. (plums), Quercus suber (cork oak), Scorzonera hispanica (black salsify), Secale spp. (ryes: S. cereale, S. rhodopaeum), Trifolium spp. (clovers: T. repens, T. alexandrinum, T. vavilovii, T. israeliticum), Triticum spp. (wheats: T. monococcum, T. dicoccoides, T. turgidum, T. durum, T. dicoccum, T. polonicum, T. spelta), Vicia spp. (vetches: sativa, V. ervilia), Vitis sylvestris (wild grape vine)*
VIII. Africa	*Aloe spp. (aloes), Avena abyssinica (Ethiopian oat), Bambusa spp. (bamboos), Brassica carinata (Ethiopian mustard), Cajanus cajan (pigeonpea), Citrullus spp. (melons: C. lanatus, C. naudianus), Cola spp. (cola nuts: C. nitiga, C. acuminata), Coffea spp. (coffees: C. arabica, C. canephora, C. liberica), Elaeis guineensis (African oil palm), Gladiolus spp. (glads), Gossypium spp. (cottons: G. herbaceum, G. triphyllum, G. anomalum, G. somalense), Guaduella spp. (herbaceous bamboos), Hordeum spp. (barleys), Lagenaria siceraria (bottle gourd), Linum usitatissimum (common flax), Lupinus albus var. albus, Musa ensete (Ethiopian banana), Oryza glaberrima (African rice), Oxytenanthera spp. (bamboos), Pennisetum spicatum (wild millets), Phoenix dactylifera (date palm), Ricinus communis (castor oil plant), Secale africanum (African wheat), Sesamum spp. (sesames), Sorghum spp. (sorghums), Triticum spp. (wheats: T. durum, T. turgidum, T. dicoccum, T. polonicum), Vigna sinensis (cowpea), Voandzeia subterranea (Bambara groundnut)*

TABLE 1.1 (Continued)

IX. Europe-Siberia	*Amygdalus nana* (Chinese wild almond), *Apocynum sibericum* (clasping-leaved dogbane), *Armoracia rusticana* (horseradish), *Beta vulgaris* (sugar beet), *Brassica oleracea* ssp. *sylvestris* (wild mustard), *Canrabis sativa* (marijuana), *Cerasus* spp. (cherries: *C. avium*, *C. fruticosa*, *C. vulgaris*), *Fragaria* spp. (strawberries: *F. moschata*, *F. viridis*; *F. vesca*), *Grossularia* spp. (gooseberries: *G. reclinata*, *G. acicularis*), *Hippophae rhamnoides* (common sea-buckthorn), *Humulus lupulus* (common hop), *Juglans regia* (Persian walnut), *Linum utitatissimum* var. *elongatum* (flax), *Malus* spp. (crabapples: *M. sylvestris*, *M. domestica*, *M. baccata* var. *baccata*, *M. prunifolia*), *Medicago* spp. (alfalfas: *M. borealis*, *M. falcata*, *M. sativa*), *Pyrus* spp. (pears: *P. communis*, *P. rossica*), *Ribes* spp. (gooseberries), *Rubus* spp. (blackberry), *Trifolium* (clovers: *T. pratense*, *T. hybridum*, *T. repens*), *Vitis* spp. (wild grapes: *V. amurensis*, *V. sylvestris*)
X. Central Asia	*Agave* spp. (century plants: *A. sisalana*, *A. fourcroydes*), *Capsicum* spp. (chili peppers: *C. annum*, *C. frutescens*), *Carya pecan* (pecan), *Cucurbita* spp. (pumpkins: *C. pepo*, *C. pepo* var. *texana*, *C. moschata*, *C. mixta*, *C. lundelliana*), *Gossypium* spp. (cottons: *G. hirsutum*, *G. aridum*, *G. armourianum*, *G. harknessii*, *G. klotzschianum*, *G. thurberi*, *G. lobatum*, *G. tomentosum*), *Ipomoea batatas* (sweet potato), *Juglans mollis* (Mexican walnut), *Nicotiana* spp. (tobaccos: *N. tabacum*, *N. rustica*), *Persea americana* (avocado), *Phaseolus* spp. (beans: *P. vulgaris* var. *aborigineus*, *P. vulgaris*, *P. acutifolius*, *P. coccineus*, *P. lunatus*, *Sechium edule* (chayote), *Solanum* spp. (wild potatoes), *Theobroma cacao* (cacao tree), *Zea mays* (maize)
XI. South America	*Acca sellowiana* (pineapple guava), *Amaranthus mantegazzianus* (love lies bleeding), *Ananas comosus* (pineapple), *Annona* (sops: *A. cherimola*, *A. muricata*, *A. reticulata*, *A. squamosa*), *Arachis* spp. (peanuts: *A. hypogaea*, *A. monticola*), *Bertholletia excelsa* (Brazil nut), *Caravalia ensiformis* (jackbean), *Canna edulis* (arrowroot), *Capsicum* spp. (peppers: *C. baccatum* var. *pendulum*, *C. pubescens*), *Carica* spp. (pawpaws: *C. candamarcensis*, *C. papcya*), *Cerasus capuli* (wild cherry), *Chenopodium quinoa* (quinoa), *Cinchona calisaya* (cinchona), *Cucurbita* spp. (squashes: *C. maxima*, *C. maxima* ssp. *andreana*, *C. ficifolia*), *Cyphomandra betacea* (tree tomato), *Erythroxylon coca* (coca), *Fragaria chiloensis* (Chilean strawberry), *Gossypium* spp. (cottons: *G. peruvianum*, *G. raimondii*, *G. klotschianum*), *Hevea brasiliensis* (rubber tree), *Helianthus* spp. (sunflowers), *Hordeum* spp. (wild barleys: *H. bacomosum*, *H. chilense*, *H. compressum*, *H. muticum*, *H. stenostachys*, *H. lechleri*, *H. procerum*, *H. pusillum* ssp. *euclaston*), *Ilex paraguariensis* (mate tea), *Juglans australis* (nogal criollo), *Lupinus* spp. (lupines), *Lycopersicon* spp. (tomatoes: *L. esculentum*, *L. pimpinellifolium*, *L. peruvianum*, *L. hirsutum*, *L. chilense*), *Manihot esculenta* (cassava), *Oxalis tuberosa* (oca), *Passiflora* spp. (passionfruits: *P. edulis*, *P. quadrangularis*), *Phaseolus lunatus* (lima bean), *Solanum muricatum* (sweet pepino), *Solanum* spp. (potatoes: *S. tuberosum* ssp. *tuberosum*, *S. tuberosum* spp. *andigena*, *S. ajanhuiri*, *S. goniocalyx*, *S. phureja*, *S. stenotomum*, *S. x chaucha*, *S. x juzepezukii*, *S. curtilobum*, *S. acaule*, *S. commersonii*, *S. maglia*, *S. chacoense*), *Theobroma cacao* (cacao tree), *Tropaeolum tuberosum* (perennial nasturtium), *Ullucus tuberosus* (ulluco), *Zea mays* (maize)

TABLE 1.1 *(Continued)*

XII. North America	*Amygdalus* spp. (almonds), *Carya* spp. (hickory), *Cerasus bessey* (western sand cherry), *Fragaria virginiana* (Virginia strawberry), *Gossypium thurberi* (desert cotton), *Helianthus* spp. (sunflowers: *H. lenticularis*, *H. tuberosus*), *Hordeum* spp. (barleys: *H. pusillum*, *H. depressum*, *H. jubatum*, *H. arizonicum*), *Lupinus* spp. (lupines), *Juglans hindsii* (Hinds' black walnut), *Malus* spp. (apples), *Nicotiana* spp. (tobaccos: *N. bigelovii*, *N. quadrivalvis*), *Prunus* spp. (plums: *P. americana*, *P. nigra*, *P. angustifolia*, *P. munsoniana*, *P. texana*), *Ribes* spp. (currants), *Rubus* spp. (blackberries), *Solanum* spp. (wild potatoes: *S. jamesii*, *S. fendleri*), *Vitis* spp. (grape vines), *Zizania aquatica* (annual wildrice)

TABLE 1.2 Centers of Origin Discovered by N.I. Vavilov in the Old World

Old World

I. Chinese Center (mountainous regions of central and western China, and adjacent lowlands)		*Allium* spp. (onions: *A. chinense, A. fistulosum, A. sativum*), *Brassica* spp. (cabbages: *B. chinensis, B. rapa* ssp. *pekinensis*), *Cannabis sativa* (hemp), *Cinnamomum camphora* (camphor), *Cucumis sativus* (cucumber), *Dioscorea batatas* (Chinese yam), *Echinochloa frumentacea* (sawa millets), *Fagopyrum esculentum* (common buckwheat), *Glycine max* (soybean), *Hordeum vulgare* ssp. *vulgare* (six-rowed barley), *Juglans regia* (walnut), *Litchi chinensis* (litchi), *Malus asiatica* (Chinese apple), *Mucuna pruriens* var. *utilis* (velvet bean), *Panax ginseng* (ginseng), *Panicum* spp. (millets: *P. miliaceum, P. italicum*), *Papaver somniferum* (opium poppy), *Phaseolus angularis* (Adzuki bean), *Prunus* spp. (plums: *P. persica, P. armeniaca, P. pseudocerasus*), *Pyrus* spp. (pears: *P. serotina, P. ussuriensis*), *Raphanus sativus* (radish), *Saccharum sinense* (Chinese cane), *Sorghum bicolor* (sorghum)
II. Indian Center	A. Main Center (Hindustan, Includes Assam and Burma)	*Acacia nilotica* (babul acacia), *Bambusa tulda* (bamboo), *Cajanus cajan* (pigeon pea), *Cannabis indica* (hemp), *Carthamus tinctorius* (safflower), *Cicer arietinum* (chickpea), *Cinnamomum zeylanicum* (cinnamon tree), *Citrus* spp. (oranges: *C. sinensis, C. nobilis, C. medica*), *Cocos nucifera* (coconut palm), *Colocasia esculenta* var *antiquorum* (imperial taro), *Corchorus capsularis* (white jute), *Crotalaria juncea* (sann-hemp), *Croton tiglium* (purging croton), *Cucumis sativus* (cucumber), *Dioscorea alata* (purple yam), *Gossypium arboreum* (tree cotton), *Hibiscus cannabinus* (kenaf), *Indigofera tinctoria* (indigo), *Mangifera indica* (mango), *Oryza sativa* (rice), *Piper nigrum* (black pepper), *Raphanus sativus* (radish), *Saccharum officinarum* (sugar cane), *Santalum album* (sandalwood), *Sesamum indicum* (sesame), *Solanum melongena* (eggplant), *Tamarindus indica* (tamarind), *Vigna* spp. (beans: *V. mungo, V. radiata, V. umbellate, V. unguiculata*)
	B. Indo-Malayan Center (Indo-China and the Malay Archipelago)	*Aleurites moluccana* (candlenut), *Artocarpus communis* (breadfruit), *Citrus grandis* (pummelo), *Cocos nucifera* (coconut palm), *Coix lacryma-jobi* (job's tears), *Garcinia mangostana* (mangosteen), *Mucuna pruriens* var. *utilis* (velvet bean), *Musa* spp. (bananas: *M. acuminata, M. paradisiaca, M. sapientum*), *Myristica fragrans* (nutmeg), *Musa textilis* (Manila hemp), *Piper nigrum* (black pepper), *Saccharum officinarum* (sugarcane), *Syzygium aromaticum* (clove)

TABLE 1.2 *(Continued)*

Old World	
III. Central Asiatic Center (Northwest India (Punjab, Northwest Frontier Provinces and Kashmir), Afghanistan, Tadjikistan, Uzbekistan, and western Tian-Shan)	*Allium* spp. (onions: *A. cepa*, *A. sativum*), *Amygdalus communis* (almond), *Brassica juncea* (mustard), *Cannabis indica* (hemp), *Cicer arientinum* (chickpea), *Daucus carota* (carrot), *Gossypium herbaceum* (levant cotton), *Linum usitatissimum* (flax), *Malus pumila* (paradise apple), *Pistacia vera* (pistacia), *Pisum sativum* (pea), *Pyrus communis* (pear), *Sesamum indicum* (sesame), *Spinacia oleracea* (spinach), *Triticum* spp. (wheats: *T. aestivum*, *T. compactum*, *T. sphaerococcum*), *Vicia faba* (broad bean), *Vigna radiate* (mungbean), *Vitis vinifera* (grape)
IV. Near-Eastern Center (interior of Asia Minor, all of Transcaucasia, Iran, and the highlands of Turkmenistan)	*Avena* spp. (oats: *A. byzantina*, *A. sativa*), *Crataegus azarolus* (hawthorn), *Cydonia oblonga* (quince), *Ficus carica* (fig), *Hordeum* spp. (barleys: *H. vulgare* f. *distichon.*, *H. distichon* var. *nutans*), *Lens culinaris* (lentil), *Lupinus* spp. (lupines: *L. pilosus*, *L. albus*), *Malus pumila* (paradise apple), *Medicago sativa* (alfalfa), *Prunus cerasus* (cherry), *Punica granatum* (pomegranate), *Pyrus communis* (pear), *Secale cereale* (rye), *Trifolium resupinatum* (Persian clover), *Trigonella foenum graecum* (fenugreek), *Triticum* spp. (wheats: *T. monococcum*, *T. durum*, *T. turgidum*, *T. turgidum* ssp. *turanicum*, *T. aestivum*, *T. persicum*, *T. timopheevi*, *T. aestivum* ssp. *macha*, *T. aestivum* var. *vavilovianum*), *Vicia* spp. (vetches: *V. sativa*, *V. villosa*)
V. Mediterranean Center (borders of the Mediterranean Sea)	*Apium graveolens* (celery), *Asparagus officinalis* (asparagus), *Avena* spp. (oats: *A. byzantina*, *A. brevis*), *Beta vulgaris* (sugar beet), *Brassica* spp. (brassicas: *B. oleracea*, *B. campestris*, *B. napus*, *B. napus*, *B. nigra*), *Carum carvi* (caraway), *Cichorium intybus* (chicory), *Humulus lupulus* (hop), *Lactuca sativa* (lettuce), *Lathyrus sativus* (grass pea), *Linum* spp. (flaxes: *L. usitatissimum*, *L. bienne*), *Lupinus albus* (lupine), *Mentha piperita* (peppermint), *Phalaris canariensis* (Canary grass), *Olea europaea* (olive), *Ornithopus sativus* (common bird's-foot), *Pastinaca sativa* (parsnip), *Pimpinella anisum* (anise), *Pisum sativum* (pea), *Rheum officinale* (rhubarb), *Salvia officinalis* (sage), *Thymus vulgaris* (thyme), *Trifolium* spp. (clovers: *T. alexandrinum T. repens, T. incarnatum*), *Triticum* spp. (wheats: *T. durum, T. dicoccum, T. polonicum, T. spelta*)

TABLE 1.2 *(Continued)*

Old World

VI. Abyssinian Center (Abyssinia, Eritrea, and part of Somaliland)	*Abelmoschus esculentus* (okra), *Coffea arabica* (coffee), *Commiphora abyssinicia* (myrrh), *Hordeum sativum* (barley), *Eleusine coracana* (African millet), *Indigofera argente* (indigo), *Lepidium sativum* (garden cress), *Linum usitatissimum* (flax), *Pennisetum spicatum* (pearl millet), *Ricinus communis* (castor bean), *Sesamum indicum* (sesame), *Triticum* spp. (wheats: *T. durum abyssinicum, T. turgidum abyssinicum, T. dicoccum abyssinicum, T. polonicum abyssinicum*), *Vigna sinensis* (cowpea)
VII. South Mexican and Central American Center (southern sections of Mexico, Guatemala, Honduras, and Costa Rica)	*Amaranthus cruentus* (grain amaranth), *Anacardium occidentale* (cashew), *Canavalia ensiformis* (jack bean), *Capsicum* spp. (peppers: *C. annuum, C. frutescens*), *Carica papaya* (papaya), *Cucurbita* spp. (gourds: *C. moshata, C. ficifolia*), *Gossypium hirsutum* (upland cotton), *Ipomea batatas* (sweetpotato), *Maranta arundinacea* (arrowroot), *Nicotiana rustica* (Aztec tobacco), *Opuntia cochenillifera* (cochenial cactus), *Phaseolus* spp. (beans: *P. vulgaris, P. lunatus, P. acutifolius*), *Prunus seroina* (wild black cherry), *Psidium guayava* (guava), *Sechium edule* (Chayote), *Solanum lycopersicum* var. *cerasiforme* (cherry tomato), *Zea mays* (maize), *Theobroma cacao* (Cacao),
VIII. South American Center: A. Peruvian, Ecuadorean, Bolivian Center	*Canna edulis* (edible canna), *Capsicum frutescens* (pepper), *Cinchona calisaya* (quinine tree), *Cucurbita maxima* (pumpkin), *Gossypium barbadense* (Egyptian cotton), *Nicotiana tabacum* (tobacco), *Passiflora ligularis* (granadilla), *Phaseolus* spp. (beans: *P. lunatus, P. vulgaris*), *Physalis peruviana* (ground cherry), *Psidium guajava* (guava), *Solanum* spp. (nightshades: *S. muricatum, S. lycopersicum* var. *lycopersicum, S. tuberosum* ssp. *andigenum, Tropaeolum tuberosum* (edible nasturtium), *Vasconcellea pubescens* (mountain papeya), *Zea mays* var. *amylacea* (starchy maize)
B. Chiloe Center	*Fragaria chiloensis* (wild strawberry), *Solanum tubersum* (common potato)
C. Brazilian-Paraguayan Center	*Anacardium occidentale* (cashew), *Ananas comosus* (pineapple), *Arachis hypogaea* (peanut), *Bertholletia excelsa* (Brazil nut), *Hevea brasiliensis* (rubber tree), *Manihot esculenta* ssp. *esculenta* (manioc), *Passiflora edulis* (purple granadilla)

Vavilov listed 136 species in the Chinese center, which is the largest independent center that mainly includes central and western China. Some species of *Panicum*, buckwheat, soybean, radish, some species of *Pyrus* and *Prunus*, litchi, opium poppy, camphor, and hemp are the major cultivated species of China center. The Indian center has two sub-centers, one is the main center (includes India, Assam, Burma, and the Punjab and Sindh in Pakistan. In this center, 117 species were listed, which include rice, chickpea, few species of pulses, eggplant, cucumber, few species of oranges, tamarind, jute, sesame, safflower, black pepper, indigo, sandalwood, and cinnamon (Vavilov, 1951). The second sub-center, the Indo-Malayan center, includes Indo-China and the Malay Archipelago, where 55 crop species were listed. Important cultivated species were banana, coconut palm, sugarcane, and clove (Vavilov, 1951).

The Central Asiatic center includes a sub-Himalayan region of India and Pakistan, Afghanistan, Tajikistan, and Uzbekistan, where 43 species were listed. Important cultivated species are bread wheat, pea, lentil, chickpea, mungbean, mustard, sesame, hemp, cotton, garlic, onion, spinach, carrot, almond, grapes, pear, and apple (Vavilov, 1992; Abbo et al., 2010). The near Asian center includes Asia Minor, Transcaucasia, Iran, and Turkmenistan, where 83 species were listed. Important species are few wild relatives of bread wheat (einkorn, durum wheat and *Triticum timopheevi*), oat, rye, lentil, lupin, alfalfa, clover, fenugreek, fig, pomegranate, pear, quince, and cherry (Zohary et al., 2012).

The Mediterranean center includes borders of the Mediterranean Sea, where 84 plant species were listed (Barthlott et al., 2005). Important species are durum wheat, pea, clover, olive, flax, rapeseed, sugar beet, cabbage, turnip, lettuce, celery, chicory, caraway, and peppermint. The Abyssinian center includes Abyssinia and Somaliland, where 38 species were listed by Vavilov (Vavilov, 1951). Important cultivated species are emmer wheat, barley, sorghum, pearl millet, cowpea, flax, sesame, castor bean, coffee, okra, and indigo.

The Mexican and Central American center includes Mexico, Guatemala, Honduras, and Costa Rica. Important species of this center are maize, common bean, grain amaranth, pumpkin, upland cotton, sweet potato, chili pepper, cashew, papaya, guava, and cocoa (Ladizinsky, 1998; Gepts et al., 2012). The South American center includes 62 plant species and has three sub-centers (Vavilov, 1992). First is the Peru, Ecuador, and Bolivia center, which includes important plant species like lima bean, common bean, edible canna, tomato, pumpkin, chili pepper, quinine tree, and tobacco. The second Chile center and third Brazil and Paraguay center include major species like potato, peanut, pineapple, Brazil nut, cashew, and passion fruit (Barthlott et al., 2005) (Figure 1.1).

I. Ethiopia	VII. Thailand, Malaysia, Java
II. Mediterranean region	VIII. Mexico, Guatemala
III. Asia Minor, Caucasia, Iran	IX. Bolivia, Peru, Ecuador, Colombia
IV. Afghanistan, Turkistan, Pakistan	X. Southern Chile
V. Indo-Burma	XI. Brazil, Paraguay
VI. China	XII. North America

FIGURE 1.1 **(See color insert.)** Centers of origin of the world (I–VII discovered by N.I. Vavilov, IX–XII discovered after N.I. Vavilov).

1.5 PLANT GENETIC RESOURCES

Genetic resources of cultivated plants can broadly be categorized into: i) modern varieties, ii) farmers' varieties, and iii) landraces and wild/weedy relatives of crop plants. Modern varieties (or high-yielding cultivars) are generally the products of plant breeding that have a high degree of genetic uniformity. Farmers' varieties (or traditional varieties) are the product of breeding or selection that spread over many generations and have a high degree of genetic diversity. Landraces and wild relatives are generally well adapted to environmental hazards and resistant to a variety of biotic and abiotic stresses (Harlan, 1975; Ceccarelli et al., 1992a; Ceccarelli et al., 1992b). Adaptive features of this group are of keen interest for plant breeders and geneticists that have contributed significantly to improving productivity and tolerance level to environmental hazards and pests/diseases (Harlan, 1981; Fritsche-Neto and Borém, 2012).

Genetic resources are depleting at an alarming rate, mainly due to habitat loss and preference of modern generation to few cultivated species. Conservation of all kinds of genetic resources will certainly make available the genetic resources for the use of present and future generations. Exploitation of genetic diversity will ensure the future food security (Reid and Miller, 1989; Blixt, 1994; Naeem et al., 2012). Direct harvesting of genetic diversity has been practiced in many parts of the world, which are used for food and fodder, timber, dyes, and lattices, and herbal medicines (Jarvis, 1999; Mellas, 2000). Local landraces and farmers, varieties are preferred for their reduced pesticides and fertilizer requirement, and therefore, they are environment-friendly (Vandermeer, 1995; Pimentel et al., 1997).

Genetic diversity in crop plants is primarily due to three main reasons: i) long period of domestication and cultivation in specific areas, ii) a wide range of habitats and cultivation practices on them, and iii) natural selection due to biotic and abiotic factors (Glaszmann et al., 2010a). These factors, e.g., pests, and other diseases, and diverse climates and other edaphic factors) have induced genetic heterogeneity, adaptation, and resistance; and these induced characteristics are immensely important today for crop improvement. Wild ancestors of crop plants and other wild and weedy relatives are still available in many parts of the world with a significant amount of genetic variation and specific adaptive features. However, the key factors in this regard are to recognize the primitive form for specific adaptive features, which are extremely helpful for: i) increasing resistance against pests, diseases, and other abiotic stresses, ii) improving yields and nutrient quality, and iii) enhancing adaptation to extreme environments (Glaszmann et al., 2010b).

Plant genetic resources are the sum total of the whole genetic diversity of a crop species, and hence, can prove to be a basic material for all kinds of crop improvement programmes. It includes landraces, modern or obsolete cultivars, ancestral, and weedy species, and wild relatives of crop plants, whether the source may be indigenous to a particular region or exotic.

1.5.1   WILD AND WEEDY RELATIVES

Wild and weedy species are extremely useful in crop improvement as well as in increasing tolerance against a variety of biotic and abiotic stresses (Meilleur and Hodgkin, 2004; McCouch et al., 2007; Damania, 2008). These also contain tremendous genetic diversity, as domestication and cultivation practices have limited the genetic diversity in a widely cultivated crop plant by imposing genetic uniformity (Tanksley and McCouch, 1997; Vollbrecht and Sigmon, 2005).

In the recent past, wild, and weedy species have benefited modern and elite cultivars of most of the major crops by donating useful genes to them, through either conventional breeding or genetic engineering. It is therefore, necessary to conserve all the remaining genetic diversity as a high priority to ensure future food security (Heywood, 2008; Maxted and Kell, 2009). In addition, increased tolerance/resistance is also the area of focus of the modern agricultural practices. These also contributed significantly to breeding techniques like ploidy level, male sterility, efficient hybrid varieties, etc.

Today, well over 75% of the human dietary requirements are fulfilled by only nine crops, including cereals like wheat, rice, maize, barley, and millets, pulses, root, and tuber crops like potato and sweet potato, and sugarcane. However, major crops are also threatened due to the high degree of genetic uniformity within the species. All major crop plants have been originated in the tropical region, in particular Asia, Africa, and South America (Levine and Chan, 2012).

It is estimated that 75% of the total genetic diversity of cultivated crop plants has been lost until the beginning of this century. The key factor for this is the dependence on only a few crops, which resulted in a rapid shrinkage of genepool of agricultural crops. Commercial high yielding uniform varieties are rapidly replacing the traditional and primitive landraces as well as wild and weedy relatives, and this phenomenon is even more threatening in the tropical centers of diversity (Hopkins and Maxted, 2011).

The entire spectrum of germplasm of the existing crop plants includes: i) elite/high-yielding varieties/cultivars, which are the product of hybridization with shorter stature and high-yielding potential, but relatively sensitive to environmental stresses, ii) principal commercial varieties, well-adapted to local environments and have many desired traits according to recent commercial trends, generally produced through elite cultivars and under-trial lines, iii) minor varieties, unimproved varieties with high genetic variability, and tolerance to adverse environmental conditions, iv) special types, with unusual features (e.g., tolerant to edaphic factors) that enable them to persist in specific areas for long duration, v) obsolete types, which were major or special cultivars at one time but cultivation has been discontinued by the farmers, vi) breeding stocks, improved hybrid cultivars/strains with many desired traits, but occasionally cultivated due to some reasons, vii) mutants, which have been developed from induced mutational changes in known cultivars, viii) primitive cultivars, which possess primitive morphological features, and generally found in undisturbed places and less accessible areas, ix) weedy races, wild forms found as weeds in cultivated fields and adjacent areas, and x) wild species, i.e., wild ancestors of the cultivated crops (Harlan, 1975; Harlan, 1976; Harlan, 1981; Smartt and Simmonds, 1995).

Wild and weedy relatives are the ancestral species of all crop species, and therefore, some modern, particularly genetically modified crops can easily be hybridized with ancestral species. As a result, there is a great risk of genetic contamination, and this may harm the human health as well as the environment. Another potential danger with the hybridized material is that they can become more vigorous weeds, particularly in the centers of origin of their respective crops.

1.5.2   PRIMITIVE VARIETIES AND LANDRACES

Life of primitive varieties could be around 100–120 million years or even more and their role has been well characterized in the uptake and maintenance of their genetic diversity. They have been a source of attraction to the farmers in spite of their small grains and low production and nutritive value. Perhaps their long storage capability was their major quality of attraction when utilized (Zeven, 1998; Villa et al., 2005).

Primitive varieties have played a very broad role in overcoming the shortcomings, relating to environmental hazards. Primitive varieties have a common background of evolution under cultivation, but all of them have been differing in the range of environments, particularly physical, biological, and social environments. In this way, evolution has been very specific and characteristic. These environments have been affecting the primitive varieties and at some time molding them through innumerable adaptations, which ultimately form the basis of human life and civilization (Cleveland et al., 1994; Zeven, 1998).

Environment played a pivotal role in the development of landraces and primitive varieties, whether there are good or bad, but certainly different from place to place. This genepool has been exposed to a range of environments, which gave rise to a wide range of populations, thus enabling the plant populations to withstand environmental and biological stresses, giving rise to a range of new adaptations that had never existed before. Primitive varieties are distinct in relation to their genetic organization, adaptation potential, the degree of tolerance and productivity and at some time may open up the doors for recombination, breeding, and genetic engineering. Whenever, widely different germplasm was recombined, the end-product had been extremely useful and better footed against a number of hazards (Villa et al., 2005; Malik and Singh, 2006).

Landraces are the populations with huge genetic variability, and therefore, they can have potential variability to adapt to a variety of environments, and at the same time, can tolerate environmental hazards more successfully than the modern breeds. On the whole, systematic conservation of primitive varieties and landraces can ensure future food security and improvement of local agriculture.

1.5.3 MINOR CROPS

During the evolutionary history of crop plants, traditional farming practices have produced a priceless genepool of thousands of genotypes of major and minor crops, which is well adapted to local environments (Myers, 1994). Both natural and artificial selection force has been operating on them since long. However, mainly the genetic erosion has threatened the genetic resources of primitive varieties and landraces, and wild and weedy relatives. This ultimately threatened the future food security in the long seen (Oldfield and Alcorn, 1991; Brush and Meng, 1998).

The Mediterranean region is the center of diversity of many cereals like wheat, barley, and rye, but also the center of diversity of many minor cereal crops (Heywood, 1999). Under modern cultivation practices, a few high-yielding genetically improved cultivars are rapidly replacing traditional varieties and landraces, as well as many important minor crops (Keneni et al., 2012). These crops were extensively used in the past, but currently, these are underutilized or much neglected. Some of them were the staple food of many geographical regions, for example, hulled wheats like einkorn (*Triticum monococcum*), emmer (*T. dicoccon*), spelt (*T. spelta*), buckwheat, millets, and sorghum, where the cultivation of these species were very popular at one time (Wahl et al., 1984; Heywood, 1999; Michalová, 2003).

Quite a few minor wheat species are currently under cultivation on small scale in different parts of the world. *Triticum polonicum*, the tetraploid Polish wheat, is a high-protein wheat, and is cultivated in some parts of the Mediterranean region and is mainly used for bread-making. *Triticum turgidum*, the durum wheat, cultivated throughout Europe, is one of the most productive wheats. *Triticum compactum*, the hexaploid club wheat of temperate Europe and Australia, is excellent for baking and bread-making. *Triticum macha*, the hexaploid macha wheat is endemic to Caucasian region, and *T. carthlicum,* the carthlicum wheat, is cultivated in Georgia (Akhalkatsi et al., 2010; Akhalkatsi et al., 2012).

Other minor cereals include *Secale cereale* var. *multicaule,* the semi-perennial rye. It is cultivated occasionally in central European countries and used as a fodder and also for baking purposes. *Hordeum vulgare* ssp. *distichon* var. *nudum* and *H vulgare* ssp. *vulgare* var. *coeleste,* the naked barley, are cultivated throughout Europe, particularly in alpine regions (Weiss et al., 2004; De Wet, 2009). *Avena nuda,* the naked oat, is mainly cultivated in central Europe. It has multiple uses like food, coffee substitute, and fiber and papermaking. *Sorghum bicolor,* (sorghum) and *Panicum mili-aceum* (common millet) and *Setaria italica* (foxtail millet) are still the staple foods of desert inhabitants of Asia and Africa, and generally referred to as poor peoples' crops. In addition, many other minor crops are currently under cultivation, but on small scales (Rachie, 1975; Hammer et al., 2003; Weber and Fuller, 2008).

1.5.4 MAJOR CROPS

Centers of origin of all the major crop plants are restricted to tropical and sub-tropical regions of the world, in particular Asia, Africa, and South America. Among the major crop plants, wheat, and barley were originated in Near East, Asian rice and soybean in China, coffee, sorghum, and yam in Africa, potatoes, and tomatoes in South America, and maize in Central and South America (Doebley, 2004; Abbo et al., 2009).

Agriculture was started by 8,000 years ago in the Middle East and soon spread to Eastern Europe. In America, people used to eat kidney bean, lima bean, corn, cocoa, avocado, chili pepper, tomatoes, pumpkins, and squashes. Farmer's domesticated wheat, barley, legumes, and many other plant species in early days. About 7,000 years ago, cultivation practices of major crop plants were more common in the western Mediterranean region and central Europe. In Africa by 5,000 years ago, people consumed millets, sorghum, coffee, yam, and cassava. By 4,000 years ago, these practices spread to the British Isles. In the Andes, the potato was the staple food of the area. At the same time, cultivation of fruits like dates, figs, grapes, olives, pomegranate, along with several cereal crops was very common, and meanwhile, agriculture also spread in the Americas (Gray, 1973; Randhawa, 1980; Mazoyer and Roudart, 2006).

During the agricultural history, thousands of genetically different varieties and cultivars of major crop plants have been developed by careful selections, breeding, and promoting farmers' varieties, wild relatives and local landraces. A considerable amount of genetic diversity is essential in protecting crops from complete failure, in addition, to provide a suitable genepool to meet the human needs and overcome the adverse effects of environmental cues.

1.6   GENE BANKS

In gene banks, the genetic material of plants in the form of seeds and other plant parts can be propagated (generally by freezing cuts) and stored for long-terms. Seed banks play a vital role in conserving genetic resources of all major and minor crop plants as well as wild and weedy races and their wild relatives. Therefore, gene banks are critically important for conserving biodiversity. There are many gene banks all over the world, some are for specific crops, while others are general (Table 1.5). Moisture requirements for seeds must be less than 5% and temperature –18°C for the optimal storage conditions. Seed DNA slowly degrades with time, and therefore, must be regenerated regularly after a proper time interval to get fresh seeds for another round of long-term storage. The viability of seeds may remain for hundreds or even thousands of years (Nagel et al., 2009; Probert et al., 2009; Rodríguez et al., 2011).

There are some challenges that are being faced by the gene banks: i) re-plantation is regularly required to avoid the loss of viability, ii) biodiversity conservation in the gene banks is restricted to very limited areas of the world, iii) recalcitrant (or unorthodox) seeds are very difficult to store properly, iv) at present, cataloging, and data managing is not up to the mark (this must include proper identity of the sample, location from where it was collected, the size of the sample and viability status), v) storing facilities are too expensive for many diversity rich countries particularly of the third world (Li and Pritchard, 2009; Rodríguez et al., 2011; Walters et al., 2011).

The primary functions of gene banks are to: i) reconstruct breeding lines, ii) support populations conserved *in vivo*, iii) increase population size of small populations, iv) back-up in case of genetic disasters (loss of genetic diversity), v) develop new lines and cultivars, vi) modify or redirect evolution or selection of populations, and vii) the availability of genetic stock for research purposes. The stored material in gene banks can be categorized in line of FAO as: i) core collection, the material required for reconstruction of breeds or cultivars, and to be used in critical situations, ii) historic collection, the material dated from the core collection, iii) working collection, material used for research and development of new genotypes, and iv) evaluation collection, material used to evaluate the status of genetic diversity (Blixt, 1994).

The Consultative Group on International Agricultural Research (CGIAR) is a strategic alliance, which is involved in the union of organizations all over the world that are involved in agricultural research for sustainable development with the help of financial donors. Donors for CIGAR include governments (mainly of developing and industrialized countries), foundations, and international and regional organizations. Food and Agriculture Organization (FAO) of the United Nations, World Bank, United Nations Development Programme (UNDP), and International Fund sponsor the CGIAR for Agricultural Development (IFAD). The CGIAR is currently involved in the collection of over 600,000 accessions, mainly of major and minor food crops, forages, and agro-forestry species. Summary of different organizations working under the CGIAR umbrella is presented in Table 1.3.

1.7   THREATS TO PLANT GENETIC RESOURCES

Genetic erosion is the major threat to plant genetic resources; Main reasons for the high degree of genetic loss are the modern agricultural technologies and high-yielding new introductions of new crops varieties. Europe and Asia are relatively more affected by the drastic loss of genepool, where genetic uniformity is already responsible for a complete loss of some major crops. The situation will be much better if the farmers and breeders concentrate on conserving genetic resources, particularly from rich diversity areas (e.g., centers of origin) and thereafter incorporating desired traits in the modern cultivars (Ceccarelli et al., 1992a; Akhalkatsi et al., 2010; Akhalkatsi et al., 2012).

The key factor for rapid loss of genetic diversity is the illogical too much success with the high-yielding modern varieties, which have replaced the farmers' cultivars and landraces in many parts of the world (Brush, 1995a; Mercer and Perales, 2010). More so, better management practices have eliminated the wild and weedy races, hence, minimizing the chance of gene flow through hybridization, and ultimately restricting the genetic variability. The authorities, who enforced the cultivation of specific cultivars and the usage of selected seeds provided by them, have discouraged cultivation of highly diverse mixtures and landraces. Habitat destruction and degradation have also eliminated the wild relatives of crop plants, and hence resulted in high genetic erosion (Gilbert et al., 1999).

TABLE 1.3 Some Important Gene Banks of the World Working Under the CGIAR

Gene bank	Location	Major collections
The International Crops Research Institute for the Semi-Arid Tropics (ICRISAT)	Hyderabad, India	Chickpea, pigeonpea, groundnut, pearl millet, sorghum, and their wild relatives, along with six small millets (finger millet, foxtail millet, barnyard millet, kodo millet, little millet, and proso millet) 80,000 samples
International Rice Research Institute (IRRI) West African Rice Development Association (WARDA)	Monrovia, Liberia	African and Asian rice varieties (a regional cooperative rice research in collaboration with IITA and IRRI)
International Plant Genetic Research Institute (IPGRI)	Rome Italy	Genetic conservation.
The Asian Vegetable Research and Development Center (AVRDC)	Taiwan	Tomato, Onion, Peppers Chinese cabbage.
International Rice Research Institute (IRRI)	Los Banos, Philippines	Rice verities, cultivars, landraces, and wild relatives (over 42,000 collections)
Center International de-Mejoramients de maize Trigo (CIMMYT)	El Baton, Mexico	Maize and wheat (triticale, barely, sorghum) with over 8,000 maize collections
Center International de-agricultural Tropical (CIAT)	Cali, Colombia	Major and minor crops of tropical regions with the main focus on cassava, beans, maize, and rice (in collaboration with CIMMYT and IRRI)
International Institute of Tropical Agriculture, (IITA)	Ibadan, Nigeria	Grain legumes, roots, and tubers
Center International de-papa (CIP; also called International Potato Center)	Lima. Peru	Potato varieties and wild relatives
Bioversity International (BI)	Rome, Italy	Major and minor crops
Center for International Forestry Research (CIFOR)	Bogor, Indonesia	Forest species
International Center for Agricultural Research in the Dry Areas (ICARDA)	Aleppo, Syria	Major and minor crops
International Food Policy Research Institute (IFPRI)	Washington, D.C., United States	Major and minor crops
International Water Management Institute (IWMI)	Battaramulla, Sri Lanka	Agricultural crops
International Center for Research in Agroforestry (ICRAF; also called World Agroforestry Centre)	Nairobi, Kenya	Agro-forestry crops

The prime factor affecting the genetic erosion is the excessive genetic uniformity in crop plants. Improved cultivars are rapidly replacing local varieties and landraces all over the world. These cultivars are of high market values, and hence farmers are more attracted towards their cultivation (van de Wouw et al., 2010). Moreover, changes in the cropping pattern, habitat destruction, and increased intensity of abiotic stresses all over the world drastically affect the wild genepool. Weedy and wild races generally have gene complexes, which are missing in modern cultivars, and therefore, their loss may result in a severe genetic erosion (Akhalkatsi et al., 2012).

Conventional agriculture with a high degree of crop uniformity (i.e., the absence of genetic variability) has exposed the crop plants to an infestation of pests and diseases. This resulted in high management costs, as the crop requirement for pesticides, antimicrobial agents, fertilizers, and herbicides has increased, which may cause health relating problems and also the habitat degradation. Traditional varieties and landraces, on the other hand, received no such chemicals for perfect growth and survival in the past (Schmidt and Wei, 2006, van de Wouw et al., 2010).

Gene flow is a regular continuous and natural phenomenon, and when a gene is released through genetically modified (GM) varieties, it has a fair chance of escape in farmers' varieties or landraces, or wild and weedy races of the same crop. Hence, this can contaminate the genetic structure of other crops. This again is a great threat to genetic variability, as this can cause rapid erosion of existing genetic diversity (Conway, 2000; Pautasso, 2012).

Global climate change is a serious threat to plant genetic resources in the modern era. Species that are unable to adapt the changing environments are facing extinction. Climate change not only threatens the major food crops, but also locally adapted landraces and wild and weedy relatives, which are the valuable source of genetic diversity (Bawa and Dayanandan, 1998; Fuhrer, 2003; Mercer and Perales, 2010).

Since the beginning of domestication, genetic variability has increased within the crops and among the crops. Migration of crops plants to a variety of habitats and geographical regions, natural mutations and hybridization, and selections (whether conscious or unconscious) have gradually expanded the genetic variability during several thousand years (Scarascia-Mugnozza and Perrino, 2002). Introduction of new scientific principles and agricultural techniques influenced the agriculture by increasing food production and survival success of crop plants, and in this manner, high-yielding uniform crops have replaced local landraces, and wild and weedy races of crop plants, which has proven the major cause of genetic erosion. Genetic erosion can be categorized into two groups: i) extinction of a population, i.e., a complete loss of genetic resources, mainly due to habitat destruction, and ii) drastic change in the genetic structure of a population, that is due to habitat degradation (Akhalkatsi et al., 2010, van de Wouw et al., 2010).

Lack of conservational practices all over the world, particularly in the developing world with high genetic diversity, may be another cause of genetic erosion. Limited support for gene banks, lack of institutional interest towards conservational strategies, and more interest towards the adaptation of new modern agricultural policies (e.g., support uniform crops, discourage landraces and wild races, GM plants, etc) have increased the risk of genetic erosion enormously over the years (van de Wouw et al., 2010). Other major factors causing genetic erosion include: preference and marketing of uniform crops, more risk of pest and disease outbreaks, land clearing for urbanization, high population pressure (particularly in the developing countries), lack of conservational strategies for crop genetic resources; poor management, and lack of illogical breeding programs (Myers, 1994).

In conclusion, genetic vulnerability or erosion is a major threat to modern and GM crops, which includes genetic uniformity, conventional agricultural practices, introgressive hybridization of GM crops and climate change on the global scale. The end result will be increased food insecurity, disease spread, and ecosystem destruction. Sustainable utilization of biological resources must be given prime importance to avoid further loss of available genetic resources, and ultimately the future food security.

1.8 CONSERVATION OF PLANT GENETIC RESOURCES

The main objectives for genetic resources conservations are: i) to conserve threatened and endangered species, ii) conserve adaptive gene complexes, iii) conserve the total genetic diversity of a particular texon, and iv) conserve primitive and wild genepool.

Conservation of plant genetic resources can be done by either *in situ* (on site) or *ex situ* (off-site). The *in situ* conservation method is generally used for the conservation of wild and weedy relatives of crop plants. Setting-up natural reserves (or gene parks) can be helpful in conserving genetic variability of wild crop species, and this is accomplished by restricting disturbances in natural habitats. This practice is ideal for conserving the wild relatives and ancestral species of crop plants that have restricted distribution. However, when the distribution of wild species spreads over large areas, such practice is possible only in certain specific areas (Altieri and Merrick, 1987; Heywood and Dulloo, 2006; Heywood, 2008).

Although *in situ* conservation is a preferable method for conserving wild and weedy races, and for the forest species, *in situ* conservation of major and minor crop plants is also a point of interest for many plant scientists (Brush, 1995b). *In situ* conservation generally increases the amount of genetic diversity as it allows evolution to continue, and hence the gene flow.

In situ reserves can also be established for farmers' varieties and highly diversified landraces. It has been suggested by international organizations like IBPGR, IUCN, and WWF that certain areas, with high genetic diversity, should be restricted for the cultivation traditional agricultural crops, local primitive landraces, and other wild and weedy relatives of crop plants. However, this practice may face a lot of difficulties relating to financial and marketing problems, but this may surely be helpful in the future of food security (Heywood and Dulloo, 2006).

Ex situ conservation is generally used to protect genetic resources that are facing the danger of destruction, replacement or deterioration. *Ex situ* conservation methods include seed storage, DNA, and pollen storage, field genebanks and botanical gardens (Khoury et al., 2010). Seed storage is the most convenient among the methods of *ex situ* conservation that can preserve genetic variability for hundreds and thousands of years. Desiccation of seeds to low moisture content and storage at extremely low-temperature is the requirement of orthodox seeds. However, conventional seed storage strategy is not possible for recalcitrant seeds of many tropical and subtropical species that cannot tolerate desiccation and quickly lose their viability (Nellist, 1981; Cohen et al., 1991). Clonal propagation is preferred to conserve genetic variability of prominent genotypes that are sterile or cannot produce seeds easily (e.g., potato, sweet potato, banana, cassava, and yam) (Engelmann and Engels, 2002). *In vitro* techniques are either for short- or medium-term storage using nutrient gels for preservation, or long-term storage using liquid nitrogen, i.e., cryopreservation (Engelmann and Engels, 2002). Ultra-low-temperature ($-196°C$) is required for cryopreservation storage.

Periodic regeneration of the collected germplasm in gene banks is necessary, not only to maintain seed viability and multiply seed stocks for distribution. This practice, however, is expensive and laborious, and requires material resources and proper planning and management (Koo et al., 2004). Regeneration is also complimentary for a variety of gene bank functions, in particular, multiplication of existing gene pool, characterization of data, and eliminating diseased accessions (Brown et al., 1997; Khoury et al., 2010).

Conservation of genetic resources is a modern-day need and is of major importance for food security of the future generations. Such materials can be used in research and breeding programmes in the field of agriculture and forestry. The material suitable for conservation may be seeds, cells, tissues or organs, or vegetative propagules, which can be preserved environmentally controlled conditions like seed banks, DNA banks, gene banks, bio-banks, etc.

1.8.1   NEGLECTED AND UNDERUTILIZED CROPS

Promotion of cultivation, utilization, and commercialization of neglected and underutilized crops is of the prime importance of the modern era to secure food for the future. Underutilized crops are also known as minor, neglected, underexploited, underdeveloped, lost, novel, promising, and traditional crops. These crops are now subject of interest worldwide, as they have a great potential for raising the income of the farmers (Frison et al., 2000).

Geographical, social, and economic value of the underutilized crops has not been realized since the beginning of agriculture and domestication (Padulosi et al., 2002). It is quite often that a species could be underutilized in some specific regions of the world, but maybe not in the others. Cowpea (*Vigna unguiculata*) is a staple crop in sub-Sahara Africa, but considered as an underutilized crop in Mediterranean countries where it was cultivated widely in the past (Padulosi et al., 1987; Laghetti et al., 1990). Chickpea (*Cicer arietinum*) is an underutilized species in Italy, but the main pulse crop in Syria and many countries of Asia.

Many socio-economic species are the important component of the daily diet of millions of people (e.g., leafy vegetables in sub-Sahara Africa), but poor marketing makes them economically underutilized (Guarino, 1997). Rocket (*Eruca sativa*) is a highly-priced vegetable in Europe due to modern cultivation and commercial practices (Pimpini and Enzo, 1997), while it is among the cheapest vegetables in Egypt. It is a rich source of micronutrients (Mohamedien, 1995). Hulled wheats (*Triticum monococcum, T. dicoccum,* and *T. spelta*) are important crops in Italy and other European countries (Padulosi et al., 1996), but largely underutilized in remote areas of Turkey, where it is referred to as life-saving crops (Karagöz, 1996).

Underutilized crops are often referred to as 'new crops' or 'novel crops' (Vietmeyer, 1988), but in fact, these crops are usually used by native farmers over the generations. However, their cultivation has been ignored due to a change in preferences and leftover the traditional agricultural practices. The crop can be new in a particular area, as it might be introduced from distant geographical regions. One of the most common examples is of kiwi fruit, which is native to China, but is unknown outside its native range until recently when introduced in New Zealand (Ferguson, 1999). Similarly, custard apple has been introduced into Lebanon in the recent past.

Underutilized crops were once more widely cultivated in many parts of the world, but due to many factors relating to genetic, economic, and cultural reasons, these are not being used in the recent era. Global climate change is another important factor, because changed environments alter their adaptations, and therefore, they cannot compete with major crop species in the same environments. Neglected and underutilized crops were originally grown in their centers of origin or diversity, but many of them are still cultivated by local farmers due to their high genetic variation and adaptability potential. These are of prime importance for many crop improvement programmes (Varshney et al., 2011).

The contribution of neglected and underutilized crops cannot be ignored in relation to future food security and overall human health. Most prominent use of these crops is food consumption and other economic and medicinal uses. These crops are critical as sources of food nutrients and income generation, particularly of rural areas. However, all such crops received little attention in the past, and the major factors behind this are poor marketing/ commercialization, and lack of effective policy to exploit these resources.

1.8.2 CONVENTIONAL PLANT BREEDING

In conventional breeding, the genetic makeup of a crop plant is changed to develop new and better genetic recombination's in a new variety. These new varieties are generally bred for some specific traits like adaptations to different climatic conditions, improved yield, taste, and nutritional values, better resistance to pests and diseases, better water use efficiency and photosynthetic capacity, etc. In conventional breeding, two closely related varieties, or even species are sexually crossed to achieve new and better genetic combinations with favorable traits of both parental species by excluding the unwanted traits (Ashraf, 1994; Varshney et al., 2005).

Plant breeders' intentions in a parental cross are usually to achieve breeding success with maximum positive traits and minimum negative traits from both parents, but this might not be successful in the first cross. The most conventional method, in this case, is the 'back-crossing,' which involves the crosses between selected progeny with one of the original parents. This process may continue for generations to get desirable traits. Another method that is generally used in conventional breeding is 'wide crossing,' which involves crossing between totally unrelated species or even genera. However, this crossing can be achieved only by using some modern breeding techniques.

Selection is the most primitive and fundamental procedure in plant breeding, which has been continuing since the domestication started. Three major steps are involved in a selection procedure: i) to hunt maximum diversity from the original population, ii) testing of progeny of individual plant selections for several years under different conditions and continuously eliminating the unwanted selection, and ii) comparison of selected material from new genetic recombinations to highly developed commercial varieties for yield performance and other desired traits.

Hybridization is the most common breeding technique, which involves the combination of desired traits from different parents into the new single plant through cross-pollination. Steps involved in hybridization are: i) to generate homozygous inbred lines, generally by using self-pollinating plants, ii) the pure line so produced is out-crossed by combining with another inbred line, and iii) the resulting progeny are selected for the desired traits.

Polyploidy frequently occurs in most of our major crops under domestication, i.e., the plants with three or more complete sets of chromosomes. Polyploidy can be artificially induced in plants by the chemical treatment with colchicines, and this leads to a doubling of the chromosome number. Polyploid plants are generally larger and with high genetic variability, but the fertility level and growth rate are relatively low.

1.8.3 MOLECULAR TECHNIQUES AND GENETIC ENGINEERING

In the past, genetic diversity was assessed on the basis of phenotypic traits like plant height, branching pattern, seed size and weight, and agronomic traits like yield potential, disease resistance, stress tolerance, etc., and these characteristics were of direct interest to the scientists and farmers. This approach provided limited information about the adaptive features, and at the same time, it was strongly influenced by the environmental conditions.

The basic step in molecular genetics is the DNA-based recognition of genetic variation. This technique involves the use of cellular enzymes that act differently on DNA molecules. Discovery of restriction enzymes (or restriction endonucleases) has played a key role in molecular genetics, which can cut both strands of DNA independently. Restriction enzyme can recognize a unique DNA sequence (generally 4–6 base pairs in length) and cut the DNA strand at the specific restriction site, which results in millions of fragments of the DNA molecule. These DNA fragments vary considerably in length, but have the same sequences at the ends, and these can be separated by electrophoresis (Moose and Mumm, 2008).

Variation (or polymorphism) in DNA can be detected by several techniques; some are based upon the initial digestion of the DNA by restriction enzymes, whereas the others upon the use of different enzymatic reactions (the polymerase chain reactions). The development of the polymerase chain reaction (PCR) was a technological breakthrough in genome analysis, because it enabled the amplification of specific fragments from a total genomic DNA (Jain and Brar, 2010).

Molecular genetics has a revolutionary impact on plant breeding, which can induce a character of interest in a plant species and the genes controlling this specific trait can be isolated. Discovery of PCR-based markers (AFLP: amplified fragment length polymorphisms; microsatellites) has opened new vistas for the recognition of genes for specific traits. A number of these traits have been further characterized by different techniques (e.g., map-based cloning) and manipulated (marker-assisted selection and genetic engineering) in modern breeding programs. Techniques like DNA arrays and automation further strengthened the molecular techniques in plant breeding (Breyne et al., 2003; Fernie and Schauer, 2009).

Characterization of plant genetic resources using DNA markers (fingerprinting) is extremely useful in evaluating the genetic relationships among different accessions. Such information is now extensively used in the fields of evolution, ecology, and population genetics. Molecular genetics enables the plant scientists to more accurate verification of similarities and differences in the genetic makeup of genetic resources and in tracing particular genomic segments via pedigrees. Bioinformation, a computer-based hardware and software systems, also referred to as the next revolution in the world is of major importance for the precise utilization of plant genetic resources.

Biochemical methods based on protein and enzyme electrophoresis has been introduced in the early 1960s. This technique was useful in the analysis of genetic diversity, and this minimized influence of environmental conditions. DNA-based techniques have been introduced in the early 1990s, because of being very beneficial for identifying polymorphisms in DNA sequences. This technique can analyze the variation at DNA level excluding totally the environmental influences.

Two basic methods are used for molecular methods being used for recognizing DNA sequence variation: i) restriction enzymes that detect and cut specific short sequences of DNA (Restriction Fragment Length Polymorphism, RFLP), and ii) amplification of target DNA sequences by short oligonucleotide primers (Polymerase Chain Reaction, PCR). In particular, for the study of diversity, PCR-based techniques were very important among molecular techniques, e.g., Amplified Fragment Length Polymorphisms (AFLPs), Random Amplified Polymorphic DNAs (RAPDs), and Simple Sequence Repeats (SSRs, microsatellites). However, breeders and geneticists are now concentrated on newer and more powerful techniques for the detection of variation at specific gene loci, which perform more precise analysis of genetic variation.

Molecular genetic techniques are now being very frequently used for conservation, management, and utilization of plant genetic resources. The most common applications of these techniques are: i) to study genetic diversity in landraces, wild, and weedy relatives of crop plants, as efficient sampling strategies to capture maximum variation for conservation of genetic resources, ii) improve the conservation and management of genetic resources via filling gaps and redundancies in genepool, minimizing loss of genetic variation, and increasing efficiency of splitting uniform populations into distinct lines, and iii) identify groups from which core collection accessions can be selected or observe the effectiveness of different strategies for mining genetic diversity.

One of the major strategies used for screening germplasm resources is the use of molecular linkage maps and a breeding technique, also known as advanced backcross QTL method, which involves the examination and separation of alleles in wild or weedy relatives in the genetic background of a specific elite cultivar. Molecular linkage maps are used for: i) identification of chromosomal position of specific alleles, ii) transmission of these alleles into the progeny, iii) determines whether the wild species introgressions are associated with high-quality performance of the lines, and iv) purification of lines that contain only a specific transmitted QTL in an elite genetic background.

Molecular techniques or genetic engineering are now extensively applied in a number of research fields including biotechnology and medicines. As an outcome, many insects/disease-resistant and herbicide/stress tolerant cultivars have been introduced in the recent era. However, many controversies still existed relating to a safe release and utilization of genetically modified crops.

1.8.4   SUCCESS STORIES OF GM PLANTS

Genetic modification (GM) involves the direct insertion of a specific gene into a new and better plant variety, which is linked to a desirable trait. In contrary, the conventional breeding involves the direct recombination of genes because of the sexual crossing. The main advantages of GM plants are: i) to reduce the time for the development of new variety, ii) reduce chances of undesirable traits in a new variety, iii) utilization of genes from unrelated plants, or even other organisms, and iv) accessing and usage of wider choice genetic diversity. At present, GM plants can be effectively used as a precise technology to tackle health and environmental challenges, and hence, to ensure future food security.

The first success in the development of transgenic plants was made in 1980, and since then transgenic biotechnology has successfully and rapidly benefited the crop improvement programmes in agriculture (Engelmann, 2011). A great number of genetically GM crops have been released on commercial basis until now, the more common examples being of maize, soybean, cotton, papaya, tomatoes, canola, and many others that carry additional genes relating to desirable traits (i.e., yield, and nutrient quality, herbicide tolerance, insect resistance, or virus tolerance). Total area for the cultivation of GM crops has been increased more than 50-fold all over the world in the recent years, and in more than 20 countries (Bhattacharjee, 2009).

One of the major problems in relation to GM crops is the gene flow, i.e., transformed genes can move from the crop plants into their closely related wild and weedy species (Haspolat, 2012). Transgenes can create ecological problems, as genes for specific traits like resistance to biotic and abiotic stresses can enhance the ecological fitness of wild and weedy races (Snow, 2003; Rao, 2004). Such genes can easily backcross into the wild populations, as was recorded in the wild populations of sunflower (Whitton et al., 1997). Another example is of transgenic rice, which has caused weedy strains of *Oryza sativa* (e.g., red rice) and the wild relative, *O. rufipogon* contaminated with the GM transgenes due to introgressive hybridization (Gealy et al., 2003; Chen et al., 2004). Similarly, the spread of transgenic herbicide resistance can pose challenges for controlling closely related weeds of the crop plants (Snow et al., 1999). Transgenic species can also affect the ecosystem integrity in the following ways: i) crops with altered adaptations could change their abundance, ii) can influence unwanted effects on other related species, or iii) the ecosystems are affected indirectly by the transgenic plants (Jorgensen et al., 1999; Conway, 2000; Shivrain et al., 2009).

One of the major drawbacks of GM crops is the loss of genetic diversity of not only in landraces and farmers' cultivars, but also in wild and weedy relatives (Berthaud and Gepts, 2004). Transgenes can easily transmit and spread within the wild or weedy populations, hence contaminating the original populations of wild relatives, and can even lead to extinction, particularly that of endangered populations (Ellstrand and Elam, 1993). Examples of gene incorporation or extinction of the native populations have been recorded in date palm, olive, and coconut (de Caraffa et al., 2002). Moreover, the Taiwanese endemic wild population of wild rice, *Oryza rufipogon* ssp. *formosana* is near to extinction by the cultivation of GM rice crop (Small, 1984). Transgenic contamination has also been reported in Mexican landraces of maize (Quist and Chapela, 2001).

GM crops have so far proved to impose revolutionary development for improving nutrition and yield. Transgenic techniques have created new gene combinations with novel traits like resistance to pests, diseases, herbicides, as well as to a number of abiotic stresses. The beneficial impacts of the introduction of GM crops are: i) reduction in crop production cost and increased yield, ii) reducing toxic chemicals in the environment, i.e., pesticides, herbicides, fertilizers, etc, iii) environmental monitoring and phytoremediation, and iv) plant-based biopharmaceuticals.

1.9 CONCLUSION

Conventional breeding and genetic engineering largely depend upon the genetic variability of genetic resources for the improvement of yield and nutritional quality, and increased resistance against different biotic and abiotic stresses. Genetic diversity in major crop plants and their wild and weedy relatives are rapidly depleting, and many of them are at the verge of extinction, mainly due to anthropogenic activities like habitat destruction and degradation, reliability on only a few crops while ignoring the others, replacing of landraces and farmers' varieties by modern high-yielding elite varieties. Speciation started since domestication, and heterogeneity of environments resulted in a great magnitude of genetic variability in local landraces and farmers' varieties, as well as in their wild and weedy relatives and ancestral species. Since the genetic resources of crop plants are decreasing at a high rate, it is an urgent need of the modern era to conserve as much as possible the remaining genetic variability from all the sources. Additionally, promotion of the cultivation of underutilized and neglected crops and local landraces with a high amount of genetic diversity along with the modern elite cultivars will certainly enhance the food security. It will also be helpful in avoiding the crop species from complete disasters like crop failure due to environmental hazards.

KEYWORDS

- **conservation**
- **crop improvement**
- **plant genetic resources**
- **threats**
- **utilization**

REFERENCES

Abbo, S., Lev-Yadun, S., & Gopher, A., (2010). Agricultural origins: Centers and non-centers – A near eastern reappraisal. *Crit. Rev. Plant Sci., 29,* 317–328.

Abbo, S., Lev-Yadun, S., & Gopher, A., (2012). Plant domestication and crop evolution in the Near East: On events and processes. *Crit. Rev. Plant Sci., 31,* 241–257.

Abbo, S., Saranga, Y., Peleg, Z., Kerem, Z., Lev-Yadun, S., & Gopher, A., (2009). Reconsidering domestication of legumes versus cereals in the ancient Near East. *Quarter. Rev. Biol., 84,* 29–50.

Ahmad, P., Ashraf, M., Younis, M., Hu, X. Y., Kumar, A., Akram, N. A., & Al-Qurainy, F., (2012). Role of transgenic plants in agriculture and biopharming. *Biotechnol. Advan., 30,* 524–540.

Akhalkatsi, M., Ekhvaia, J., & Asanidze, Z., (2012). Diversity and genetic erosion of ancient crops and wild relatives of agricultural cultivars for food: Implications for nature conservation in Georgia (Caucasus). In: Tiefenbacher, J., (ed.), *Perspectives on Nature Conservation – Patterns, Pressures, and Prospects.* InTech Publishers.

Akhalkatsi, M., Ekhvaia, J., Mosulishvili, M., Nakhutsrishvili, G., Abdaladze, O., & Batsatsashvili, K., (2010). Reasons and processes leading to the erosion of crop genetic diversity in mountainous regions of Georgia. *Mountain Res. Develop., 30,* 304–310.

Altieri, M. A., & Merrick, L., (1987). In situ conservation of crop genetic resources through maintenance of traditional farming systems. *Econ. Bot., 41,* 86–96.

Anderson, M., (2012). *Early Civilizations of the Americas.* Rosen Education Service, New York, NY.

Anjum, N. A., Ahmad, I., Pereira, M. E., Duarte, A. C., Umar, S., & Khan, N. A., (2012). The plant family Brassicaceae. Springer, Germany.

Ashraf, M., & Akram, N. A., (2009). Improving salinity tolerance of plants through conventional breeding and genetic engineering: An analytical comparison. *Biotechnol. Advan., 27,* 744–752.

Ashraf, M., (1994). Breeding for salinity tolerance in plants. *Crit. Rev. Plant Sci., 13,* 17–42.

Barthlott, W., Mutke, J., Rafiqpoor, M. D., Kier, G., & Kreft, H., (2005). Global centers of vascular plant diversity. *Nova Acta Leopoldina, 92,* 61–83.

Bawa, K. S., & Dayanandan, S., (1998). Global climate change and tropical forest genetic resources. *Climatic Change, 39,* 473–485.

Berthaud, J., & Gepts, P., (2004). *Assessment of Effects on Genetic Diversity*. Maize and biodiversity: The effects of transgenic maize in Mexico. Commission for Environmental Cooperation, Montreal, Canada.

Bhattacharjee, R., (2009). Harnessing biotechnology for conservation and increased utilization of orphan crops. *ATDF J., 6*, 24–32.

Blixt, S., (1994). Plant genetic resources – Problems and prespective. *Biolog. Zent., 113*, 25–34.

Breto, M. P., Asins, M. J., & Carbonell, E. A., (1993). Genetic-variability in *Lycopersicon* species and their genetic-relationships. *Theor. Appl. Genet., 86*, 113–120.

Breyne, P., Dreesen, R., Cannoot, B., Rombaut, D., Vandepoele, K., Rombauts, S., et al. (2003). Quantitative cDNA-AFLP analysis for genome-wide expression studies. *Mol. Genet. Genom., 269*, 173–179.

Brown, A. H. D., Brubaker, C. L., & Grace, J. P., (1997). Regeneration of germplasm samples: Wild versus cultivated plant species. *Crop Sci., 37*, 7–13.

Brown, C. R., (2011). The contribution of traditional potato breeding to scientific potato improvement. *Potato Res., 54*, 287–300.

Brush, S. B., & Meng, E., (1998). Farmers' valuation and conservation of crop genetic resources. *Genet. Resour. Crop Evol., 45*, 139–150.

Brush, S. B., (1995a). *In situ* conservation of landraces in centers of crop diversity. *Crop Sci., 35*, 346–354.

Brush, S. B., (1995b). *In situ* conservation of landraces in centers of crop diversity: Implications of germplasm conservation and utilization. *Crop Sci., 35*, 346–354.

Brussaard, L., Caron, P., Campbell, B., Lipper, L., Mainka, S., Rabbinge, R., et al. (2010). Reconciling biodiversity conservation and food security: Scientific challenges for a new agriculture. *Curr. Opin. Environ. Sustain., 2*, 34–42.

Caruso, I., Lepore, L., De Tommasi, N., Dal Piaz, F., Frusciante, L., Aversano, R., et al. (2011). Secondary metabolite profile in induced tetraploids of wild *Solanum commersonii* Dun. *Chem. Biodiver., 8*, 2226–2237.

Ceccarelli, S., Valkoun, J., Erskine, W., Weigand, S., Miller, R., & Van Leur, J. A. G., (1992). Plant genetic resources and plant improvement as tools to develop sustainable agriculture. *Experimental Agriculture, 28*, 89–98.

Chacon, M. I., Pickersgill, B., & Debouck, D. G., (2005). Domestication patterns in common bean (*Phaseolus vulgaris* I..) and the origin of the Mesoamerican and Andean cultivated races. *Theor. Appl. Genet., 110*, 432–444.

Chen, L. J., Lee, D. S., Song, Z. P., Suh, H. S., & Lu, B. R., (2004). Gene flow from cultivated rice (*Oryza sativa*) to its weedy and wild relatives. *Ann. Bot., 93*, 67–73.

Cherry, J. H., Locy, R. D., & Rychter, A., (2000). Plant tolerance to abiotic stresses in agriculture: Role of genetic engineering. In: *Proceedings of the NATO Advanced Research Workshop Mragowa*, Poland (13–18 June, 1999).

Cleveland, D. A., Soleri, D., & Smith, S. E., (1994). Do folk crop varieties have a role in sustainable agriculture? *Bioscience, 44*, 740–751.

Cohen, J. I., Williams, J. T., Plucknett, D. L., & Shands, H., (1991). *Ex situ* conservation of plant genetic resources: Global development and environmental concerns. *Science, 253*, 866–872.

Conway, G., (2000). Genetically modified crops: Risks and promise. *Conserv. Ecol., 4*(1), 2.

Cooper, H. D., Hodgkin, T., & Spillane, C., (2001). *Broadening the Genetic Base of Crop Production*. CABI Publishing, Oxford.

Crawford, G. W., (2006). East Asian plant domestication. *Archaeol. Asia*, 77–95.

Cupic, T., Tucak, M., Popovic, S., Bolaric, S., Grljusic, S., & Kozumplik, V., (2009). Genetic diversity of pea (*Pisum sativum* L.) genotypes assessed by pedigree, morphological, and molecular data. *J. Food Agric. Environ., 7,* 343–348.

Damania, A. B., (2008). History, achievements, and current status of genetic resources conservation. *Agron. J., 100,* 9–21.

De Caraffa, V. B., Maury, J., Gambotti, C., Breton, C., Berville, A., & Giannettini, J., (2002). Mitochondrial DNA variation and RAPD mark oleasters, olive, and feral olive from Western and Eastern Mediterranean. *Theor. Appl. Genet., 104,* 1209–1216.

De Wet, J. M. J., Timothy, D. H., Hilu, K. W., & Fletch, G. B., (1981). Systematics of South American *Tripsicum* (Gramineae). *Amer. J. Bot., 68,* 269–276.

De Wet, J., (2009). The three phases of cereal domestication. In: Chapman, G. P., (ed.), *Grass Evolution and Domestication* (pp. 176–198). Cambridge University Press.

Diamond, J., (2002). Evolution, consequences, and future of plant and animal domestication. *Nature, 418,* 700–707.

Doebley, J. F., Gaut, B. S., & Smith, B. D., (2006). The molecular genetics of crop domestication. *Cell, 127,* 1309–1321.

Doebley, J., (2004). The genetics of maize evolution. *Ann. Rev. Genet., 38,* 37–59.

Eckardt, N. A., (2010). Evolution of domesticated bread wheat. *Plant Cell, 22,* 993–993.

Ehleringer, J. R., Cerling, T. E., & Dearing, M. D., (2005). *A History of Atmospheric CO$_2$ and its Effects on Plants, Animals, and Ecosystems.* Springer Verlag.

Ellstrand, N. C., & Elam, D. R., (1993). Population genetic consequences of small population-size – Implications for plant conservation. *Ann. Rev. Ecol. System., 24,* 217–242.

Engelmann, F., & Engels, J., (2002). *Technologies and Strategies for Ex Situ Conservation.* CAB International, Wallingford, UK, and IPGRI, Rome, Italy.

Engelmann, F., (2011). Use of biotechnologies for the conservation of plant biodiversity. *In Vitro Cell. Dev. Biol. -Plant, 47,* 5–16.

Erickson, D. L., Smith, B. D., Clarke, A. C., Sandweiss, D. H., & Tuross, N., (2005). An Asian origin for a 10, 000-year-old domesticated plant in the Americas. *Proceed. Nat. Acad. Sci., 102,* 18315.

Evenson, R. E., Gollin, D., & Santaniello, V., (1998). *Agricultural Values of Plant Genetic Resources.* CABI, Oxford, UK.

Ferguson, A. R., (1999). *New Temperate Fruits: Actinidia Chinensis and Actinidia Deliciosa.* ASHS Press Alexandria, VA.

Fernie, A. R., & Schauer, N., (2009). Metabolomics-assisted breeding: A viable option for crop improvement? *Trends Genet., 25,* 39–48.

Frankel, O. H., & Hawkes, J. G., (2011). *Crop Genetic Resources for Today and Tomorrow.* Cambridge Univ Press.

Fridman, E., & Zamir, D., (2012). Next-generation education in crop genetics. *Curr. Opin. Plant Biol., 15,* 218–223.

Frison, E., Omont, H., & Padulosi, S., (2000). GFAR and International Cooperation on Commodity Chains. In: *GFAR–2000 Conference.* GFAR, Dresden, Germany.

Fritsche-Neto, R., & Borém, A., (2012). *Plant Breeding for Abiotic Stress Tolerance.* Springer, The Netherlands.

Fuhrer, J., (2003). Agroecosystem responses to combinations of elevated CO$_2$, ozone, and global climate change. *Agriculture Ecosystems & Environment, 97,* 1–20.

Ge, S., Sang, T., Lu, B. R., & Hong, D. Y., (1999). Phylogeny of rice genomes with emphasis on origins of allotetraploid species. *Proceed. Nat. Acad. Sci., 96,* 14400–14405.

Gealy, D. R., Mitten, D. H., & Rutger, J. N., (2003). Gene flow between red rice (*Oryza sativa*) and herbicide-resistant rice (*O-sativa*): Implications for weed management. *Weed Technol., 17,* 627–645.

Gepts, P., (2000). *A Phylogenetic and Genomic Analysis of Crop Germplasm: Necessary Condition for its Rational Conservation and Use.* Kluwer Academic/Plenum Publishers New York, USA.

Gepts, P., Famula, T. R., Bettinger, R. L., Brush, S. B., Damania, A. B., & McGuire, P. E., (2012). Biodiversity in agriculture: domestication, evolution, and sustainability. In: *2nd Harlan Symposium.* Cambridge University Press, NY, University of California, Davis.

Gilbert, J., Lewis, R., Wilkinson, M., & Caligari, P., (1999). Developing an appropriate strategy to assess genetic variability in plant germplasm collections. *Theor. Appl. Genet., 98,* 1125–1131.

Gladis, T., & Hammer, K., (2002). The relevance of plant genetic resources in plant breeding. In: *Animal Breeding and Animal Genetic Resources.* Institute of Animal Science and Animal Behavior, FAL Mariensee.

Glaszmann, J. C., Kilian, B., Upadhyaya, H. D., & Varshney, R. K., (2010b). Accessing genetic diversity for crop improvement. *Curr. Opin. Plant Biol., 13,* 167–173.

Gray, L. C., (1973). *History of Agriculture in the Southern United States to 1860.* P. Smith Publisher.

Gross, B. L., & Olsen, K. M., (2010). Genetic perspectives on crop domestication. *Trends Plant Sci., 15,* 529–537.

Guarino, L., (1997). Traditional African vegetables. In: *Proceedings of the IPGRI International Workshop on Genetic Resources of Traditional Vegetables in Africa: Conservation and Use.* International Plant Genetic Resources Institute, Rome, Italy, ICRAF-HQ, Nairobi, Kenya.

Hammer, K., Arrowsmith, N., & Gladis, T., (2003). Agrobiodiversity with emphasis on plant genetic resources. *Naturwissenschaften, 90,* 241–250.

Hammer, K., Diederichsen, A., & Spahillari, M., (1999). Basic studies toward strategies for conservation of plant genetic resources. In: *Proceedings of the Technical Meeting on the Methodology of the FAO World Information and Early Warning System on Plant Genetic Resources* (pp. 29–33).

Harlan, J. R., (1975). Our vanishing genetic resources. *Science, 188,* 618–621.

Harlan, J. R., (1981). *Evaluation of Wild Relatives of Crop Plants FAO,* Rome.

Harlan, J., (1976). Genetic resources in wild relatives of crops. *Crop Sci., 16,* 329–333.

Haspolat, I., (2012) Genetically modified organisms and biosecurity. *Ankara Universitesi Veteriner Fakultesi Dergisi, 59,* 75–80.

Hawkes, J. G., (1983). The diversity of crop plants. Harvard University Press., Massachusetts.

Hawkes, J. G., (1990). The potato: Evolution, biodiversity, and genetic resources. Smithsonian Institution Press, Washington, D. C.

Hawkes, J., Maxted, N., & Ford-Lloyd, B., (2000) The *ex situ* conservation of plant genetic resources. Kluwer Academic Publishers.

Heiser, C. B., (1979). Origins of some cultivated new world plants. *Ann. Rev. Ecol. System., 10,* 309–326.

Heywood, V. H., (2008). Challenges of *in situ* conservation of crop wild relatives *Turk. J. Bot., 32,* 421–432.

Heywood, V., & Dulloo, E., (2006). *In situ Conservation of Wild Plant Species, a Critical Global Review of Good Practices.* IPGRI, Rome.

Heywood, V., (1999). The Mediterranean region – A major center of plant diversity. *Opti. Mediterr, 38,* 1–15.

Hopkins, J., & Maxted, N., (2011). *Crop Wild Relatives: Plant Conservation for Food Security.* Natural England Research Report NERR037. University of Birmingham, Peterborough.

Hyten, D. L., Song, Q., Zhu, Y., Choi, I. Y., Nelson, R. L., Costa, J. M., Specht, J. E., Shoemaker, R. C., & Cregan, P. B., (2006). Impacts of genetic bottlenecks on soybean genome diversity. *Proceed. Nat. Acad. Sci., 103,* 16666–16671.

Jain, S. M., & Brar, D. S., (2010). *Molecular Techniques in Crop Improvement.* Springer Verlag, Germany.

Jarvis, D. I., (1999). Strengthening the scientific basis of *in situ* conservation of agricultural biodiversity on-farm. *Bot. Lithuan., 2,* 79–90.

Jenkins, J., (1948). The origin of the cultivated tomato. *Econ. Bot., 2,* 379–392.

Jing, R. C., Vershinin, A., Grzebyta, J., Shaw, P., Smykal, P., Marshall, D., et al. (2010). The genetic diversity and evolution of field pea (*Pisum*) studied by high throughput retrotransposon based insertion polymorphism (RBIP) marker analysis. *BMC Evolution. Biol., 10,* 44.

Jing, R., Ambrose, M. A., Knox, M. R., Smykal, P., Hybl, M., Ramos, A., et al. (2012). Genetic diversity in European *Pisum* germplasm collections. *Theor. Appl. Genet., 125,* 367–380.

Jorgensen, R. B., Andersen, B., Snow, A., & Hauser, T. P., (1999). Ecological risks of growing genetically modified crops. *Plant Biotechnol., 16,* 69–71.

Karagöz, A., (1996). Agronomic practices and socioeconomic aspects of emmer and einkorn cultivation in Turkey. In: *Hulled Wheats: Promotion of Conservation and use of Valuable Underutilized Species* (pp. 172–177). IPGRI, Castelvecchio Pascoli, Tuscany, Italy.

Keneni, G., Bekele, E., Imtiaz, M., & Dagne, K., (2012). Genetic vulnerability of modern crop cultivars: Causes, mechanism, and remedies. *Int. J. Plant Res., 2,* 69–79.

Khoury, C., Laliberte, B., & Guarino, L., (2010). Trends in *ex situ* conservation of plant genetic resources: A review of global crop and regional conservation strategies. *Genet. Resour. Crop Evol., 57,* 625–639.

Koinange, E. M. K., Singh, S. P., & Gepts, P., (1996). Genetic control of the domestication syndrome in common bean. *Crop Sci., 36,* 1037–1045.

Konig, A., Cockburn, A., Crevel, R. W. R., Debruyne, E., Grafstroem, R., Hammerling, U., et al. (2004). Assessment of the safety of foods derived from genetically modified (GM) crops. *Food Chem. Toxicol., 42,* 1047–1088.

Koo, B., Pardey, P. G., Wright, B. D., Bramel, P., Debouck, D., Van Dusen, M. E., et al. (2004). *Saving Seeds: The Economics of Conserving Crop Genetic Resources Ex Situ in the Future Harvest Centers of the CGIAR.* IFPRI – SGRP and CABI Publishing.

Kovach, M. J., & McCouch, S. R., (2008). Leveraging natural diversity: back through the bottleneck. *Curr. Opin. Plant Biol., 11,* 193–200.

Ladizinsky, G., (1998). *Plant Evolution Under Domestication.* Springer, The Netherlands.

Laghetti, G., Padulosi, S., Hammer, K., Cifarelli, S., Perrino, P., Ng, N., & Monti, L., (1990). Cowpea (*Vigna unguiculata* L. Walp.) germplasm collection in southern Italy and preliminary evaluation. In: *Cowpea Genetic Resources* (pp. 46–57). IITA.

Levine, J., & Chan, K. M. A., (2012). Global human dependence on ecosystem services. In: Koellner, T., (ed.), *Ecosystem Services and Global Trade of Natural Resources: Ecology, Economics, and Policies* (p. 11). Routledge, Taylor & Francis Group, London.

Li, D. Z., & Pritchard, H. W., (2009). The science and economics of ex situ plant conservation. *Trends Plant Sci., 14,* 614–621.

Londo, J. P., Chiang, Y. C., Hung, K. H., Chiang, T. Y., & Schaal, B. A., (2006). Phylogeography of Asian wild rice, *Oryza rufipogon,* reveals multiple independent domestications of cultivated rice, *Oryza sativa. Proceed. Nat. Acad. Sci., 103,* 9578–9583.

Malik, S., & Singh, S., (2006). Role of plant genetic resources in sustainable agriculture. *Ind. J. Crop Sci., 1,* 21–28.

Matsuoka, Y., (2005). Origin matters: Lessons from the search for the wild ancestor of maize. *Breed. Sci., 55,* 383–390.

Maxted, N., & Kell, S. P., (2009). *Establishment of a Global Network for the In Situ Conservation of Crop Wild Relatives: Status and Needs.* FAO, Rome.

Maxted, N., Ford-Lloyd, B. V., & Hawkes, J. G., (1997). *Plant Genetic Conservation: The In Situ Approach.* Chapman and Hall, London.

Maxted, N., Kell, S., & Magos, B. J., (2011). *Options to Promote Food Security: On-Farm Management and In Situ Conservation of Plant Genetic Resources for Food and Agriculture.* Food and Agriculture Organization of the United Nations Rome, Italy.

Maxted, N., White, K., Valkoun, J., Konopka, J., & Hargreaves, S., (2008). Towards a conservation strategy for *Aegilops* species. *Plant Genet. Res. Character. Utiliz., 6,* 126–141.

Mazoyer, M., & Roudart, L., (2006). *A History of World Agriculture: From the Neolithic Age to the Current Crisis.* Monthly Review Press, New York.

McCouch, S. R., Sweeney, M., Li, J. M., Jiang, H., Thomson, M., Septiningsih, E., et al. (2007). Through the genetic bottleneck: *O-rufipogon* as a source of trait-enhancing alleles for *O-sativa. Euphytica, 154,* 317–339.

Meilleur, B. A., & Hodgkin, T., (2004). In situ conservation of crop wild relatives: Status and trends. *Biodiver. Conserv., 13,* 663–684.

Mellas, H., (2000). Morocco. *Seed-Supply Systems: Data Collection and Analysis.* IPGRI, Rome, Italy.

Mercer, K. L., & Perales, H. R., (2010). Evolutionary response of landraces to climate change in centers of crop diversity. *Evolution Appl., 3,* 480–493.

Michalová, A., (2003). Minor cereals and pseudocereals in Europe Report of a Network Coordinating Group on Minor Crops, *Maccarese,* Rome, Italy.

Mohamedien, S., (1995). *Rocket Cultivation in Egypt.* Lisbon, Portugal.

Moose, S. P., & Mumm, R. H., (2008). Molecular plant breeding as the foundation for 21st-century crop improvement. *Plant Physiol., 147,* 969–977.

Myers, N., (1994). Protected areas – protected from a greater 'what'? *Biodiver. Conserv., 3,* 411–418.

Naeem, S., Duffy, J. E., & Zavaleta, E., (2012). The functions of biological diversity in an age of extinction. *Science, 336,* 1401–1406.

Nagel, M., Vogel, H., Landjeva, S., Buck-Sorlin, G., Lohwasser, U., Scholz, U., & Börner, A., (2009). Seed conservation in ex situ genebanks – genetic studies on longevity in barley. *Euphytica, 170,* 5–14.

Nellist, M., (1981). Predicting the viability of seeds dried with heated air. *Seed Sci. Technol., 9,* 439–455.

Oldfield, M. L., & Alcorn, J. B., (1991). *Conservation of Traditional Agroecosystems.* Boulder, San Francisco, and London, Westview.

Padulosi, S., Cifarelli, S., Monti, L. M., P., P., (1987). *Cowpea Germplasm in Southern Italy.* FAO, Rome.

Padulosi, S., Hammer, K., & Heller, J., (1996). Hulled wheats: Promotion of conservation and use of valuable underutilized species. In: *Proceedings of the First International Workshop on Hulled Wheats* (p. 262). International Plant Genetic Resources Institute, Rome, Italy, Castelvecchio Pascoli, Tuscany, Italy.

Padulosi, S., Hodgkin, T., Williams, J., & Haq, N., (2002). Underutilized crops: Trends, challenges, and opportunities in the 21st century. In: Engels, J. M. M., (ed.), *Managing Plant Genetic Diversity*. CABI-IPGRI.

Pautasso, M., (2012). Challenges in the conservation and sustainable use of genetic resources. *Biol. Lett., 8*, 321–323.

Peng, J. H. H., Sun, D. F., & Nevo, E., (2011). Domestication evolution, genetics, and genomics in wheat. *Mol. Breed., 28*, 281–301.

Perez-de-Castro, A. M., Vilanova, S., Canizares, J., Pascual, L., Blanca, J. M., Diez, M. J., et al. (2012). Application of genomic tools in plant breeding. *Curr. Genom., 13*, 179–195.

Pimentel, D., Wilson, C., McCullum, C., Huang, R., Dwen, P., Flack, J., et al. (1997). Economic and environmental benefits of biodiversity. *Bioscience, 47*, 747–757.

Pimpini, F., & Enzo, M., (1997). *Present and Future Prospects for Rocket Cultivation in the Veneto Region*. International Plant Genetic Resources Institute, Rome, Italy.

Prakash, S., Bhat, S., Quiros, C., Kirti, P., & Chopra, V., (2010). Brassica and its close allies: Cytogenetics and evolution. *Plant Breed. Rev., 31*, 21–187.

Price, T. D., (2009). Ancient farming in eastern North America. *Proceed. Nat. Acad. Sci., 106*, 6427–6428.

Probert, R. J., Daws, M. I., & Hay, F. R., (2009). Ecological correlates of *ex-situ* seed longevity: A comparative study on 195 species. *Ann. Bot., 104*, 57–69.

Quist, D., & Chapela, I. H., (2001). Transgenic DNA introgressed into traditional maize landraces in Oaxaca, Mexico. *Nature, 414*, 541–543.

Rachie, K. O., (1975). *The Millets: Importance, Utilization, and Outlook*. International Crops Research Institute for the Semi-Arid Tropics, Hyderabad, India.

Randhawa, M. S., (1980). A history of agriculture in India. *Indian Council of Agricultural Research*. New Delhi, India.

Rao, N. K., (2004). Plant genetic resources: Advancing conservation and use through biotechnology. *Afr. J. Biotechnol., 3*, 136–145.

Reid, W. V., & Miller, K. R., (1989). *Where is the World's Biodiversity Located?*. World Resources Institute, Washington, D. C.

Rodríguez, M. V., Toorop, P. E., & Benech-Arnold, R. L., (2011). Challenges facing seed banks and agriculture in relation to seed quality. *Method. Mol. Biol., 773*, 17.

Ross-Ibarra, J., Morrell, P. L., & Gaut, B. S., (2007). Plant domestication, a unique opportunity to identify the genetic basis of adaptation. *Proceed. Nat. Acad. Sci., 104*, 8641–8648.

Sang, T., & Ge, S., (2007). The puzzle of rice domestication. *J. Integ. Plant Biol., 49*, 760–768.

Sanjur, O. I., Piperno, D. R., Andrés, T. C., & Wessel-Beaver, L., (2002). Phylogenetic relationships among domesticated and wild species of *Cucurbita* (Cucurbitaceae) inferred from a mitochondrial gene: Implications for crop plant evolution and areas of origin. *Proceed. Nat. Acad. Sci., 99*, 535–540.

Scarascia-Mugnozza, G. T., & Perrino, P., (2002). *The History of Ex Situ Conservation and Use of Plant Genetic Resources* (1st edn.). Cabi Publishing, Oxford.

Schmidt, M. R., & Wei, W., (2006). Loss of agro-biodiversity, uncertainty, and perceived control: A comparative risk perception study in Austria and China. *Risk Anal., 26*, 455–470.

Shivrain, V. K., Burgos, N. R., Gealy, D. R., Sales, M. A., & Smith, K. L., (2009). Gene flow from weedy red rice (*Oryza sativa* L.) to cultivated rice and fitness of hybrids. *Pest Manag. Sci., 65*, 1124–1129.

Small, E., (1984). Hybridization in the domesticated-weed-wild complex. In: Grant, W. F., (ed.), *Plant Biosystematics*. Academic Press Toronto.

Smartt, J., & Simmonds, N. W., (1995). *Evolution of Crop Plants*. Longman Scientific & Technical Press, London, UK.

Smith, B. D., (1989). Origins of agriculture in eastern North America. *Science, 246,* 1566.

Smith, B. D., (2006). Eastern North America as an independent center of plant domestication. *Proceed. Nat. Acad. Sci., 103,* 12223–12228.

Snow, A. A., Andersen, B., & Jørgensen, R. B., (1999). Costs of transgenic herbicide resistance introgressed from *Brassica napus* into weedy *B. rapa. Mol. Ecol., 8,* 605–615.

Snow, A., (2003). Genetic engineering: Unnatural selection. *Nature, 424,* 619.

Tanksley, S. D., & McCouch, S. R., (1997). Seed banks and molecular maps: unlocking genetic potential from the wild. *Science, 277,* 1063–1066.

van de Wouw, M., Kik, C., van Hintum, T., van Treuren, R., & Visser, B., (2010). Genetic erosion in crops: Concept, research results, and challenges. *Plant Genet. Res. Character. Utiliz., 8,* 1–15.

Vandermeer, J., (1995). The ecological basis of alternative agriculture. *Ann. Rev. Ecol. System., 26,* 201–224.

Varshney, R. K., Bansal, K. C., Aggarwal, P. K., Datta, S. K., & Craufurd, P. Q., (2011). Agricultural biotechnology for crop improvement in a variable climate: hope or hype? *Trends Plant Sci., 16,* 363–371.

Varshney, R. K., Graner, A., & Sorrells, M. E., (2005). Genomics-assisted breeding for crop improvement. *Trends Plant Sci., 10,* 621–630.

Vaughan, D. A., Lu, B. R., & Tomooka, N., (2008). The evolving story of rice evolution. *Plant Science, 174,* 394–408.

Vaughan, D., Balazs, E., & Heslop-Harrison, J., (2007). From crop domestication to super-domestication. *Ann. Bot., 100,* 893–901.

Vavilov, N. I., (1951). The origin, variation, immunity, and breeding of cultivated plants. *Chron. Bot., 13,* 1–336.

Vavilov, N. I., (1992). *Origin and Geography of Cultivated Plants*. Cambridge University Press, Cambridge, UK.

Vietmeyer, N., (1988). The new crops era. In: Janick, J., & Simon, J. E., (eds.), *Advances in New Crops*. Timber Press Portland, OR.

Villa, T. C. C., Maxted, N., Scholten, M., & Ford-Lloyd, B., (2005). Defining and identifying crop landraces. *Plant Genet. Res. Character. Utiliz., 3,* 373–384.

Vitte, C., Ishii, T., Lamy, F., Brar, D., & Panaud, O., (2004). Genomic paleontology provides evidence for two distinct origins of Asian rice (*Oryza sativa* L.). *Mol. Genet. Genom., 272,* 504–511.

Vollbrecht, E., & Sigmon, B., (2005). Amazing grass: Developmental genetics of maize domestication. *Biochem. Soc. Transact., 33,* 1502–1506.

Wahl, I., Anikster, Y., Manisterski, J., & Segal, A., (1984). Evolution at the center of origin. In: Bushnell, W. R., & Roelfs, A. P., (eds.), *The Cereal Rusts* (pp. 39–77). Academic Press, Orlando, Florida.

Walters, C., Hill, L., Koster, K., & Bender, J., (2011). Comparisons of seed longevity under simulated aging and genebank storage conditions using Brassicaceae seeds. In: *Meeting Abstract* (p. 279). International Society for Seed Science, Bahia, Brazil.

Weber, S. A., & Fuller, D. Q., (2008). Millets and their role in early agriculture. *Pragdhara, 18,* 69–90.

Weeden, N. F., (2007). Genetic changes accompanying the domestication of *Pisum sativum*: Is there a common genetic basis to the 'domestication syndrome' for legumes? *Ann. Bot., 100,* 1017–1025.

Weiss, E., Kislev, M. E., Simchoni, O., & Nadel, D., (2004). Small-grained wild grasses as staple food at the 23000-year-old site of Ohalo II, Israel. *Econ. Bot., 58,* 125–134.

Whitton, J., Wolf, D. E., Arias, D. M., Snow, A. A., & Rieseberg, L. H., (1997). The persistence of cultivar alleles in wild populations of sunflowers five generations after hybridization. *Theor. Appl. Genet., 95,* 33–40.

Wilson, E. O., (1992). *The Diversity of Life.* Penguin Press, London.

Wllkes, H., (1977). Hybridization of maize and teosinte, in Mexico and Guatemala and the improvement of maize. *Econ. Bot., 31,* 254–293.

Wolfenbarger, L. L., & Phifer, P. R., (2000). The ecological risks and benefits of genetically engineered plants. *Science, 290,* 2088–2093.

Zeven, A. C., (1998). Landraces: a review of definitions and classifications. *Euphytica, 104,* 127–139.

Zohary, D., Hopf, M., & Weiss, E., (2012). *Domestication of Plants in the Old World: The Origin and Spread of Domesticated Plants in Southwest Asia, Europe, and the Mediterranean Basin* (4th edn.). Oxford University Press, Oxford, UK.

CHAPTER 2

WHEAT PRODUCTION TECHNOLOGY: A PREAMBLE TO FOOD SECURITY IN PAKISTAN

IJAZ RASOOL NOORKA[1] and J. S. (PAT) HESLOP-HARRISON[2]

[1]Department of Plant Breeding and Genetics, University College of Agriculture, University of Sargodha, Pakistan, Tel.: 0092300–6600301, E-mail: ijazphd@yahoo.com

[2]Department of Biology, Molecular, and Cytogenetic Lab, University of Leicester, United Kingdom

ABSTRACT

Today's agriculture is primarily probed to please plentiful and sustainable production to fulfill the escalating population needs. The world is on the brink of depleting renewable resources. It revealed that the outlook for developing societies is at risk in the provision of food and fiber production. However, after a quantum jump in wheat production after the green revolution paved the path towards food security but the aftermath since the last 50 years, the wheat production is at mere slow than targets. Pakistani wheat is a political crop having deep effects in society and governments. This study reviews the advent of wheat production, constraints, policy, and recommendations so that to maintain the cropping intensity, pattern, and yields to feed the ever-growing population through the adoption of biotechnology practices that holds enormous promise to break the barriers in increasing food production and harness the growing body of science and innovations in the area to ensure food security as well as law and order situation in most populated area of the world. Agricultural biotechnology, development of drought-resistant varieties and water saving techniques are key to crop maximization.

2.1 BACKGROUND, HISTORY, AND ECONOMIC IMPORTANCE

Where to start....? "The matter of food is a heck of a lot more thirsty and thorny than the charts and speeches show!"

The growing population of human, animals, and associated needs are demanding a firm concomitant influx of increase in staple foods, fiber, and fuel by generating a series of positive efforts to combat degradation of ecosystem, genetic diversity, soil fertility, yield stability under the prevailing set of climatic changes (Noorka and Afzal, 2009). Since historical times, agriculture remained a prime hope to satisfy the needs of humankind and the basic right to get plentiful and nutritive food.

2.2 HOW CEREALS SAVED HUMANITY

A historical case is the time of the Holy Prophet Hazrat Yousuf (A.S. Joseph) when he dreamed that a famine was approaching: He recommended storage of cereals for seven years consumption. Archaeological evidence from the middle-ages has shown occasional regional population collapses (Shennan et al., 2013), most likely due to famine and disease. Similarly the Great Hunger of 1044, Great Irish Famine, 1846–1850, and the Bengal Famine, 1943 all diverted human attentions towards research for food, feed, fiber, and fuel security (United Kingdom, 1997). In present times, cereals are among the most significant economic commodities in the world (Daly, 1996). World cereal production – wheat, rice, and maize, along with barley, oats, rye, millet, and sorghum– is about 2billion tons per year. The species are members of the grass family Gramineae and the major source of food calories and protein in most of the world, not least because of the concentration of energy and nutrients in the dry grain (typically <15% moisture, achieved without artificial heat sources), which is easily stored and transported, advantages exploited since the start of agriculture. Cereal foods are important constituents of the regular diet, providing carbohydrates, proteins, dietary fibers and vitamins (Ahmed et al., 2011). Pakistan's population has a great reliance on cereals to encounter the everyday requisite of food energy. Cereals account for 47 present of whole calorie source per capita, and the role of cereals to protein stock per capita is 46 present (FAO, 2013). Typical production statistics for cereals describe to crops reaped for dry grain: Other uses include forage, hay or harvesting green for food, feed, or silage (Fischer and Maurer, 1978; Khush, 1995), as well as fiber for animal bedding, roofing, construction or fuel. No other food could have helped man to the extent of cereals: they have the potential and required nutrition as required, along with a few pseudo-cereals such as buckwheat (*Fagopyrum*, Polygonaceae), quinoa (*Chenopodium*, Amaranthaceae) and the legumes, where the seeds also have a high energy density and easy storage. Humans have shown considerable ingenuity in growing cereals even in adverse conditions (even growing them on the roof of huts in floods; Chahal and Gossal, 2002; Noorka and Haidery, 2011).

2.2.1 MIRACLE SEEDS

After World War II, it was hoped that the battle against hunger was won: harvests in many nations, mechanization, and improved transport, notably the United States, created apparent food surpluses in the West. The number and proportion of people working in agriculture dropped rapidly. Agronomy (precision planting, fertilization, rotations, crop protection) played a substantial part in yield improvement, and post-harvest losses were reduced by timely harvest, combine harvesters (combining reaping, threshing, and winnowing) and dry, pest-proof storage and transport. However, by the 1960s, large parts of the world were suffering from major food shortages. Subsequently, improved yield (in combination with continuing agronomic improvements) from "Miracle seeds," with high genetic potential, pioneered by Dr. Norman Borlaug, provided a basis for green revolution. It brought the reality that densely populated countries of Asia and America scold attain self-sufficiency.

2.3 WHEAT A POLITICAL CROP IN PAKISTAN

Wheat (*Triticum sp.*) is a global cultivated grass originating from the Fertile Crescent area of the Near East. Wheat grain is a staple food used to make flour for leavened, flat, and steamed breads, biscuits, cookies, cakes, breakfast cereal, pasta, noodles, or biofuel (GOP, 2013). Wheat is cultivated to a limited degree as a fodder crop for cattle, and the hay can be utilized as food for livestock or as a building material for roofing thatch or a component of walls (Bates et al., 2008; Batool et al., 2013). Wheat is the main staple food element of Pakistan's residents and the principal grain harvest of the country. It contributes 13.1% to the value added in agriculture and 2.8% to GDP: the 2012 wheat crop is 23.4 million tons, with year-on-year increases of more than 2% in yield in the last 20 years driving the production increase (Figure 2.1) (Economic Survey of Pakistan, 2013; FAO, 2014).

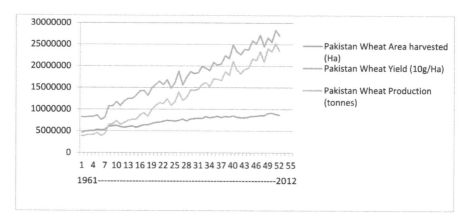

FIGURE 2.1   (See color insert.) Pakistan wheat Production trends since 1961 to 2012.
Source: (Economic Survey of Pakistan, 2013; FAOSTAT, 2014).

Wheat is the most important cereal in the political horizon of Pakistan as for all politicians had to emphasize on the availability of food in accordance with the need of its people. Former rulers Muhammad Ayub Khan, Yahya Khan, Zulfiqar Ali Bhutto, Zia-ul-Haq, Muhammad Nawaz Sharif, Benazir Bhatto, Pervaiz Musharaf and again Muhammad Nawaz Sharif, as well as all others, had a main concern with the cereals sufficiency and their government revolved around the self-sufficiency in wheat.

2.4   GREEN REVOLUTION FOR PAKISTAN

Pakistan has a place of pride since independence period by achieving a marvelous milestone of green revolution of 1960. Pakistan enjoyed the full benefit of the green revolution of 1960, due to which country has taken a right direction and this effort not only made Pakistan self-sufficient in wheat production but also an exporter. This has paved a path towards positive direction of the economy. The collective wisdom and integrated efforts made by farmers, scientists, researchers, planners, and farmers in developing a world of technical, policy-based that refocused on agricultural research to ensure food security (Mugabe, 2003; Afzal et al., 2011; Chassy, 2003) revealed that majority of world is at nutritionally risk. The United Nations Food and Agriculture Organization (FAO) estimate that two out of five children in the developing societies are stunted, one in three is under-weight, and one in ten is 'wasted' due to under-nourishment (Dupont and William, 2000).

2.5　WHY WHEAT FOR STAPLE FOOD?

A staple food generally considered to a food that is liked and consumed in much quantity. It constitutes a dominant diet portion of the society. It ensures quantitative and qualitative and generally a significant proportion of the intake of other nutrients as well (Byerlee, 1992). The Staple foods used in the world are Corn/Maize, Rice, Wheat, Potatoes, Cassava, Soybeans, Sweet Potatoes, Banana, Sorghum, etc. Wheat is considered the world's most favored staple food particularly in South Asia, in terms of total production and consumption. Wheat alone provides more nourishment for humans than any other food source (Morris and Byerlee, 1998).

2.6　ORIGIN AND FORMS OF WHEAT

Wheat considered to be originated in the Levant and the Ethiopian Highlands is now cultivated in a versatile environment. Other commercially minor but nutritionally-promising species of naturally evolved wheat include black, yellow, and blue wheat, however, the hexaploid form *Triticum aestivum* was introduced at the dawn of agriculture in 10,000 BC. Its quality accounts for poor people chapatti making properties of wheat (Mujeeb-Kazi and Hettel, 1995; Rajaram and Borlaug, 1997; Anderson, 2000). The list of top wheat producing countries is presented in Table 2.1.

TABLE 2.1　The World's Top Producers of 2012 by Order in Million-Metric-Ton

Country	Production in Million-metric Tons
EU (European Union)	134.5
China	125.6
India	94.9
United States	61.8
France	40.3
Russia	37.7
Australia	29.9
Canada	27.0
Pakistan	23.5

2.7   WHEAT PRODUCING ZONES OF PAKISTAN

In Pakistan, extensive work has been done on the germplasm collection, exploration, and research activities pre-dominantly on wheat by National Agricultural Research Council (NARC) Islamabad and Ayub Agricultural Research Institute (AARI), Faisalabad, Pakistan.

Pakistan has been distributed into ten production zones, reflecting the wide adaptation of wheat and the diverse climate of the country from Himalayan mountain ranges to sub-tropical coastal regions. The zoning is chiefly centered on cropping pattern, disease occurrence, and climatic aspects. Nevertheless, production zones must be re-entered.

2.8   CROPPING PATTERNS AND YIELD GAP

In Pakistan, wheat is grown in diverse cropping systems through crop rotations and fallow lands. However, the development of semi-dwarf wheat varieties has revolutionized in yield potential and substantial research facilities as well as farmer outreach demonstrations, by which farmers have reaped maximum yield potential of the cultivars (Safdar et al., 2013; Noorka and Heslop-Harrison, 2014). Wheat production in the country, however, has been a challenge to defeat low productivity variables and harvesting losses. Furthermore, farmers are not alert of up-to-date technologies because of fragile extension services system. However, by the cluster shown below the yield gap may be minimized (Figure 2.2).

FIGURE 2.2   (See color insert.) Wheat crop production technology major concerns for successful crop production.

2.9   CROP MANAGEMENT

Across Pakistan, farmers generally do late wheat cultivation due to their packed rice-wheat, sugarcane-wheat, and cotton-wheat cropping pattern. Presently, only 20% of wheat is being cultivated at what was regarded as the optimal planting time (15th October to 15th November). Farmers are probably choosing their cropping pattern to optimize their return, and it is clear that the genes controlling the development of varieties selected a decade ago may no longer be optimal for current rotations and requirements. The avoidance of fallow periods, with disadvantages of erosion and loss of soil structure or moisture, mechanization allowing planting and harvest at any time, and lack of opportunity for expansion of agricultural lands mean that demand and indeed the necessity of multiple or continuous cropping is likely to increase.

2.9.1   SEED TREATMENT AND SEED RATE

The government institutes, adaptive research station advised the farmers to use healthy and clean seed of having good genetic potential and approved varieties up to 50–60 kg/ac. It is a precautionary measure to treat the seed with an appropriate systemic fungicide. Commonly tractor driven automatic rabidrill, zero tillage drill or pore/Kera and even broadcast/ Chatta method are used. The weeds like *Phalaris minor* (Dumbisitti), *Avenafatua* (Wild oats), *Chenopodium* (Bathu) & *Convolvulus* (Lehli), *Medicago* spp. (Maina) *Lathyrus* spp. (Matri), *Asphodelus tenuifolius* (Piazzi), *Carthamus oxycantha* (Pohli) etc.are controlled by cultural methods like Daab and Bar Harrow as well as chemical methods with selective weedicides for broad and narrow-leaved weeds (Mehmood et al., 2014)

2.9.2   FERTILIZER APPLICATION PER ACRE

Fertilizers have given a good response to increasing the wheat yield, so it may be decided according to the previous grown crop, cropping intensity, water availability, and crop health. A general dose of chemical fertilizer is given in Table 2.2.

TABLE 2.2   General Dose of Chemical Fertilizer With Respect to the Type of Soil in Pakistan

Type of soil	(kg)			Bags		
	N	P	K	DAP	Urea	Potash
Poor	52	46	25	2.0	1.75	1.0
Medium	42	34	25	1.50	1.50	1.0
Fertile	32	23	25	1.0	1.25	1.0
Progressive farmer's dose	64	46	25	2.0	2.0	1.0

2.9.3 IRRIGATION

The decision about the irrigation, number of irrigations, when to apply, how to apply, how much to apply would be decided on the availability of irrigation water, crop health, soil condition, and ex-cultivated crop. Normally four to five irrigations can be made keeping in view the critical stages of wheat plant growth (Table 2.3).

TABLE 2.3   Phenology of Wheat and Irrigation Scheme

Crop Stage	Days after Sowing (DAS)
Crown Root Initiation	22–28
Jointing Stage	52–60
Boot Stage	92–105
Pollination Stage	112–122
Dough Stage	130–140

2.9.4 MAIN INSECT, PEST, ATTACK ON WHEAT CROP

Insect/pest attack severely damaged the wheat crop, so it is essential to save the crop by proper identification and remedies (Tables 2.4 and 2.5, Figure 2.3).

TABLE 2.4 Insect-Pest, Mode of Attack and its Biological Control in Wheat Crop

Casual organism	Identity	Way of attack	Control
Wheat Weevil	*Sitophilus granaries* Having dark brown color and sharpen mouth	It attacks the germinated crops at initial stages.	Appropriate pesticides may be used
Aphid	A small insect having a greenish color and slow movement	It sucks juice from leaves, and spikes, sever attack in the month of March	Use ladybird beetle, Chrysopaand cold water. Do not use pesticides
Jassid	Greenish colored active insect	Suck juice from leaves	-same-
Cutworm	It attacks mainly at night and hibernates in stubbles	It cuts the germinating plants	By irrigating the field the attack may be minimized
Shoot Fly	Small in size than a normal house fly	Its maggot enter and cut the soft twigs	The attack may be minimized by cultivating wheat before December

TABLE 2.5   Common Diseases That Causes the Sever Harm to Wheat Crop

Disease	Casual organism	Symptoms	Control
Loose Smut	*Ustilago tritici*	The blackish colored powder appears and destroys the grains	Sow healthy seeds and use a fungicide on seed
Flag Smut	*Urocystis tritici*	Symptom appears after forty days of sowing. Reduce grain formation	Sow healthy seeds and use a fungicide on seed
Foot Rot	*Helminthosporium-sativum*	It attacks twice and effects the grain formation	Use resistant varieties and healthy seeds
Rusts	*Puccinia striiforms Puccinia recondite Puccinia graminis*	Winter rains and moist condition enhance the disease, spots appear on the leaves and stems	Use resistant varieties against rust disease

FIGURE 2.3   Wheat crop grown with canal water and tube well irrigation.

2.10   FUTURE VIEW OF WHEAT PRODUCTION

The area under wheat has increased to 8.3 million hectares and centered on past 18 years wheat data, the area growth rate was 0.04. While a marginal increase in area is possible, it is unlikely and largely undesirable for social, economic, and environmental reasons. Urban, transport, and industrial uses continue to require more former agricultural land. Expansion of farming into wild or minimally cultivated areas – with larger machinery for example – removes many biodiversity reserves. Also, expansion of the area requires water resources: some will check the discharge of surface fresh water to the sea, but water resource limitations and conflicting political opinions will limit more water availability, and dams and lakes require more land. So, the primary recourse is to exploit current agricultural land more, and expand the yield of current and new wheat varieties, and raising the "crop per drop" of water. The gap between the genuine yield and the potential yield is about 82 present for wheat. This gap can be lessened through improved agronomic practices, weed control, appropriate sowing and balanced use of fertilizer. The subsequent year, yield gap can be lessened by elevating yields between 2 to 12% as is predicted in the wheat yield expansion thrust of the government. The yield gap can further be bridged in the year to come by elevating yields by 8 to 27% and proliferation by 20 to 42% by the succeeding year (Jat and Gerad, 2014). It is need of the hour that we have to take the advantage of biotechnological approaches in wheat improvement for end-use quality attributes (Jeffreys, 1985; Gianessi, 2002; Schmidhuber and Tubiello, 2007; Anhalt et al., 2009).

2.11   SWOT (STRENGTHS WEAKNESSES OPPORTUNITIES THREATS) ANALYSIS

i)   Strengths:

The strength of any breeding or crop improvement will surely depend upon the following components:

- Infrastructure;
- Germplasm;
- Seed Centre;
- Genomics and knowledge of genes;
- Improved trialing and statistical approaches;
- Modeling of crop and gene-combination responses;
- Better trained farmers and extension workers (MSc level);
- Better information about varieties, agronomy, and markets for farmers.

ii)   Weaknesses:

We should be careful about the weakness to tackle the crop improvement constraints:

- Lack of land resources;
- Lack of skilled/unskilled manpower— a shortage of breeders;
- Farmer-friendly policies/ unintended farmers issues/ crop support prices;
- Lack of access to capital for improved cultivation and harvest equipment.

iii)   Opportunities:

- Staple food crop;
- Highly receptive end-user;
- Leading edge to the organization.

iv)   Threats:

- Diseases – fungal, bacterial, and viral; rust is a particular challenge in the 2010s;
- Shrinking funds;
- Climate change;
- Lack of coordination among various research groups;
- Urbanization removing land;
- Lack of appropriate political management of water resources.

2.12   CONCLUSION AND FUTURE PROSPECTS

Wheat domestic production had been varying and mostly lowers than the need for import. Thus, justifiable wheat production appears authoritative for food security. In order to meet the country's wheat needs, there is a requirement to exploit the edge of information and expertise, and progress farm organization practices and mechanization plus to raise yield to feed the big community if the population keeps growing at its established trend rate, by 2030 Pakistan will need another 8 million tonnes of wheat per annum. The history envisages that low production and food insecurity, devise, and implement the most unwanted tension and course of action. However, by the employment of novel and innovative approaches to increase the crop productivity involving traditional plant breeding and modern biotechnological approaches to develop superior genetic stocks with high-quality grain, high fertile tillering, more spikelets/spike and the bumper yield to ensure food for all and to meet Global needs (IPCC, 2007).

2.13   RECOMMENDATIONS AND FUTURE POLICIES

The wheat crop may be planted well in time to minimize insect/pest and disease attack and to increase crop yield. Zero tillage sowing technology and budget saving water expertise may enhance wheat farming. Certified and pure seed and stable use of fertilizer is need of the hour to increase the wheat yield. Development of drought and salt tolerant will surely add rain-fed areas in the mainstream for crop maximization.

ACKNOWLEDGMENT

The authors are thankful to Dr. Abdur Rahman (Late), Professor Emeritus, Ex-Vice Chancellor, University of Agriculture, Faisalabad, and Dr. Trude Schwarzacher, Molecular Cytogenetics and Plant Cell Biology, University of Leicester, the United Kingdom for their kind guidance to prepare this chapter.

KEYWORDS

- **agricultural biotechnology**
- **developing societies**
- **food situations and roles**

REFERENCES

Afzal, M., Noorka, I. R., & Tabasum, S., (2011). Triumphant and sustainable wheat production in the wake of water stress to mitigate climate change and to ensure food security in farming system of Pakistan. OIDA. *Int. J. Sus. Dev., 2* (7), 119–120.

Ahmed, N., Chowdhry, M. A., Khaliq, I., & Maekaw, M., (2011). The inheritance of yield and yield components of five wheat hybrid populations under drought conditions. *Indo. J. Agri. Res., 8* (2), 53–59.

Anderson, M. B., (2000). Vulnerability to disaster and sustainable development: A general framework for assessing vulnerability. In: Pielke, Jr. R., & Pielke, Sr. R. (eds.), *Storms* (Vol. 1, pp. 11–25). London, U.K., Routledge.

Anhalt, U. C. M., Heslop-Harrison, J. S., Piepho, H. P., Byrne, S., & Barth, S., (2009). Quantitative trait loci mapping for biomass yield traits in a *Lolium* inbred line derived F2 population. *Euphytica*, *170*, 99–107. doi: 10. 1007/s10681–009–9957–9.

Bates, B. C., Kundzewicz, Z. W., Wu, S., & Palutikof, J. P., (2008). *Climate Change and Water*. Technical Paper of the *Intergovernmental Panel on Climate Change*. Geneva, Switzerland: IPCC Secretariat.

Batool, A., Noorka, I. R., Afzal, M., & Syed, A. H., (2013). Estimation of heterosis, heterobeltiosis, and potency ratio over environments among pre and post green revolution spring wheat. *J. Basic Appl. Sci., 9*, 36–43.

Byerlee, D., (1992). Technical change, productivity, and sustainability in irrigated cropping systems of South Asia: Emerging issues in the post-Green Revolution era. *J. Int. Develop., 4* (5), 477–496.

Chahal, G. S., & Gossal, S. S., (2002). *Principles and Procedures of Plant Breeding*. Oxford, UK: Alpha Science International.

Chassy, B., (2003). The role of agricultural biotechnology in world food aid. Electron. *J. U. S. Depart. State, 8*, 3.

Daly, D. C., (1996). The leaf that launched a thousand ships. *Natural History, 1/96*, 24–25. New York, NY.

Dupont, K., & William, F., (2000). *The 21st Century – An Agribusiness Odyssey, D. W.* Brooks Lecture, University of Georgia, College of Agricultural and Environmental Sciences. Available at: www.dupont.com/whats-new/speeches/kirk/dwbrooks.htm.

Economic Survey of Pakistan, (2013–2014). *Ministry of Finance*, Government of Pakistan, Islamabad. www.growpk.com

Fischer, R. A., & Maurer, R., (1978). Drought resistance in spring wheat cultivars. I. Grain yield responses. *Aust. J. Agric. Res., 29*, 897–912.

Food and Agriculture Organizations (FAO), (2013). *The State of Food and Agriculture 2012*, Rome: FAO of the United Nations.

Food and Agriculture Organizations (FAO), (2014). *FAO Statistic. faostat.fao*.org, Rome: FAO of the United Nations.

Gianessi, L., (2002). *Plant Biotechnology: Current and Potential Impact for Improving Pest Management in U. S. Agriculture*. Washington, DC: National Center for Food and Agricultural Policy.

Government of Punjab, (2013). *Agriculture Marketing Information Service*, Lahore.

Intergovernmental Panel on Climate Change, (2007). *Impacts, Adaptation, and Vulnerability. Contribution of Working Group II to the Fourth Assessment Report of the IPCC*. Cambridge, U. K.: Cambridge University Press.

Jat, M. L., & Gerard, B., (2014). Nutrient Management and Use Efficiency in Wheat Systems of South Asia. *Adv. Agron., 125*, 171–259.

Jeffreys, A. J., Wilson, V., & Thein, S. L., (1985). Hypervariable minisatellite regions in human DNA. *Nature, 314*, 67–73.

Khush, G. S., (1995). Modern varieties – their real contribution to food supply and equity. *Geo. Journal, 35* (3), 275–284.

Mehmood, Z., Ashiq, M., Noorka, I. R., Ali, A., TabasumS., & Iqbal, M. S., (2014). Chemical control of monocot weeds in wheat (*Triticum aestivum* L.) *Amer. J. Plant Sci., 5* (9), 1272–1276. http://dx.doi.org/10. 4236/ajps. 2014. 59140.

Morris, M., & Byerlee, D., (1998). "Maintaining Productivity Gains in Post-Green Revolution Asian Agriculture." In: Eicher, C. K., & Staatz, V. J., (eds.), *International Agricultural Development* (3rd edn.). Hopkins University Press, Baltimore, and London.

Mugabe, J. O., (2003). *Agricultural Biotechnology in Africa: Building Public Confidence and Scientific Capacity for Food Production.* FAO, Rome.

Mujeeb-Kazi, A., & Hettel, G. P., (1995). Utilizing Wild Grass Biodiversity in Wheat Improvement – 15 Years of Research in Mexico for Global Wheat Improvement. *Wheat Special Report No. 29.* Mexico, DF: CIMMYT.

Noorka, I. R., & Afzal, M., (2009). Global climatic and environmental change impact on agricultural research challenges and wheat productivity in Pakistan. *Earth Sci. Fron., 16* (SI), 99–100.

Noorka, I. R., & Haidery, J. R., (2011). Conservation of genetic resources and enhancing resilience in water stress areas of Pakistan to cope with vagaries of climate change. In: Rang, A., et al. (eds.), *Proc. Int. Conf. Preparing Agriculture for Climate Change.* Ludhiana, India.

Noorka, I. R., & Heslop-Harrison, J. S., (2014). Water and crops: Molecular biologists, physiologists, and plant breeders approach in the context of evergreen revolution. *Handbook of Plant and Crop Physiology*, pp. 967–978. CRC Press, Taylor, and Francis, USA.

Rajaram, S., & Borlaug, N. E., (1997). *Approaches to Breeding for Wide Adaptation, Yield Potential, Rust Resistance and Drought Tolerance,"* Paper presented at Primer Symposia International de Trigo, Cd. Obregon, Mexico, April 7–9.

Safdar, E., Noorka, I. R., Tanveer, A., Tariq, S. A., & Rauf, S., (2013). Growth and yield of advanced breeding lines of medium grain rice as influenced by different transplanting dates. *J. Animal Plant Sci., 23* (1), 227–231

Schmidhuber, J., & Tubiello, F. N., (2007). Global food security under climate change. *Proc. Natl. Acad. Sci. USA, 104* (50), 19703–19708.

Shennan, S., Downey, S. S., Timpson, A., Edinborough, K., Colledge, S., Kerig, T., et al. (2013). Regional population collapse followed initial agriculture booms in Mid-Holocene Europe. *Nature Communications, 4.* URL http://dx.doi.org/10. 1038/ncomms3486.

United Kingdom, (1997). *Eliminating World Poverty: A Challenge for the 21st Century.* White Paper presented to Parliament by the Secretary of State for International Development.

CHAPTER 3

CAPSAICIN PRODUCTION IN CELL SUSPENSION CULTURES OF *CAPSICUM ANNUUM* L.

BENGU TURKYILMAZ UNAL and CEMIL ISLEK*

Nigde Omer Halisdemir University, Arts and Sciences Faculty, Biotechnology Department, 51240 Nigde-Turkey, E-mail: bturkyilmaz@ohu.edu.tr

Corresponding author: cislek@ohu.edu.tr

3.1 INTRODUCTION

Capsicum is known to humanity since the occurrence of civilization in the Western Hemisphere. It has been used as food since 7500 BC (MacNeish, 1964). The native people had collected the wild chili peppers and selected them for various types known today. The history of chilli's coincides with the voyage of Columbus, who first introduced this plant to Europe, it then spread in Africa and Asia (Heiser, 1976; Bosland, 1996).

Genus *Capsicum* includes almost 22 wild and 5 domesticated species including *C. annuum, C. baccatum, C. chinense, C. frutescens,* and *C. pubescens* (Bosland, 1994). *Capsicum annuum* L. is the most common of the 20 species of pepper and is cultivated as a vegetable as well (Ozturk et al., 2009)

The pungency in *Capsicum* varies to a great extent in species/varieties. This is regulated by several factors including geographical, environmental, and genetic (Rahman and Inden, 2012; Rodriguez-Burruezo et al., 2010). Based on these variations *Capsicum* varieties are popularized regionally, each having its unique aromatic and culinary features as well as variable color and pungency; all are highly consumed vegetable/spice globally (Guzman et al., 2011; Tewksbury et al., 2006; Wahyuni et al., 2011).

Depending on the amount of capsaicinoids contained, the pepper species can be differentiated as: hot, medium-hot, and sweet (Legin, 1996). In many countries, types with small fruits and pungent varieties are named "chilies", while those with larger fruits and less pungent are named as "capsicums" or "paprika." Fruits of *Capsicum* spp. have different vernacular names such as aji, bell pepper, cayenne pepper, chili pepper, (hot) red pepper, pimento, Tabasco (Basu and De, 2003; Thampi, 2003). It is not always certain that these popular names will be given to certain taxa (Eich, 2008).

Turkey ranks third in the world pepper production (FAO, 2014). Peppers have been mostly cultivated in the Mediterranean, Aegean, Marmara, and Southeastern Anatolian regions of Turkey (Beyaz et al., 2009). They are widely cultivated as a field crop in Kahramanmaras (especially Pazarcik and Turkoglu), Urfa, Adiyaman, Diyarbakir, and Hatay and as greenhouse crop in Antalya, Icel, Mugla, and Izmir in Turkey (Ustun, 1990). The best product is grown in Kahramanmaras (Akinci and Akinci, 2004), providing 10% of red pepper production of Turkey with 1372 ha area (TUIK, 2010).

3.1.1 GENERAL USE OF CAPSAICINOIDS

Capsaicin is used as a food additive, a counter-irritant in medicine, cosmetic as well as self-defense products (e.g., pepper spray) and oral herbal supplement (Bley et al., 2012). Antioxidant properties have also been determined (Sudha and Ravishankar, 2002; Ochi et al., 2003; Kogure et al., 2002).

Capsaicin or capsaicinoids in hot pepper are used internally in flatulence and atonic dyspepsia. On the other hand, their use for external purposes is more significant, e.g., for the relief of pains against diabetic polyneuropathy, lumbago, post-herpetic neuralgia or rheumatic diseases in the form of creams, liniments, ointments or plasters (Eich, 2008; Evans and Tears, 2002; Ravishankar et al., 2003).

Suzuki and Iwai (1984) have summarized the pharmacological features and effects on the cardiovascular system, gastrointestinal tract, respiratory system, and thermoregulation, as well as anesthesia.

The anticancer activity of capsaicin and its saturated analog dihydrocapsaicin has been reported for the first time in 2002 (Luo et al., 2011). Recently, capsaicin has been determined to have a large anti-proliferative effect on various prostate cancer cell lines due to new mechanisms of action. Capsaicin markedly reduced the growth of prostate cancer tissue as weight and size when administered orally. It has caused apoptosis or cell death in 80% of the growing prostate cancer cells in mice (Mori et al., 2006). The results of this study have revealed that capsaicinoids can be used in the treatment of prostate cancer. Furthermore, based on experiments with male rats, it was stated that the capsaicin would be useful in avoiding cancer (e.g., human colon cancer) (Yoshitani et al., 2001).

Capsaicinoids increase low-density lipoprotein cholesterol resistance and/or fat oxidation when incubated together (Ahuja et al., 2006). Studies on fat oxidation and weight loss are still underway.

There are also many patents on insecticides, animal-repellents (e.g., cats, dogs, etc.), and insect repellents containing capsaicinoids. Moreover, this applies to the formulations of sterilizing and disinfecting surfaces or killing microorganisms on contact with regard to food and meat stuff as well as for their processing equipment and facilities (Neumann, 2001).

3.1.2 CHEMICAL CONSTITUENTS OF CAPSICUM

The most prominent component of *Capsicum* was first isolated in 1876 by Thresh (1876) and was called capsaicin. Nelson (1919) isolated crude capsaicin from commercial Cayenne pepper. Four years later, the final structure was detected to be 8-methylnon-6-enoyl-vanillylamide (Nelson and Dawson, 1923). Kosuge et al. (1958, 1960) discovered 6,7-dihydrocapsaicin as a natural congener. The mixture of capsaicin and dihydro derivative was called capsaicinoid by these authors (Eich, 2008). However, in later years this term was used as the general term for capsaicin and its congeners. Since the chemical properties and solubility of capsaicin are similar to capsaicinoids, "natural capsaicin" has long been considered as a unique substance (Legin, 1996).

Capsaicinoids are unique natural compounds; their formation is limited to the genus *Capsicum*. The characteristic pungent taste of the fruits of this genus is caused by 25 metabolites named capsaicinoids. They are secondary metabolites with very common use, about 1/4 of the world population consumes red peppers every day (Tewksbury et al., 2006).

All the capsaicinoids are acid amides of C_{9-11} branched-chain fatty acids and vanillylamine. The main difference between the various capsaicinoids is in the length of the aliphatic side chain, the branching point, the absence or presence of a double bond and their relative pungency (Figure 3.1) (Díaz et al., 2004). The majority is pungent, but there are also non-pungent forms, named as capsinoids (Ochi et al., 2003).

The main irritant component is the alkaloid capsaicin, an amide derivative of vanillylamine and 8-methylnon-trans-6-enoic acid (Bennett and

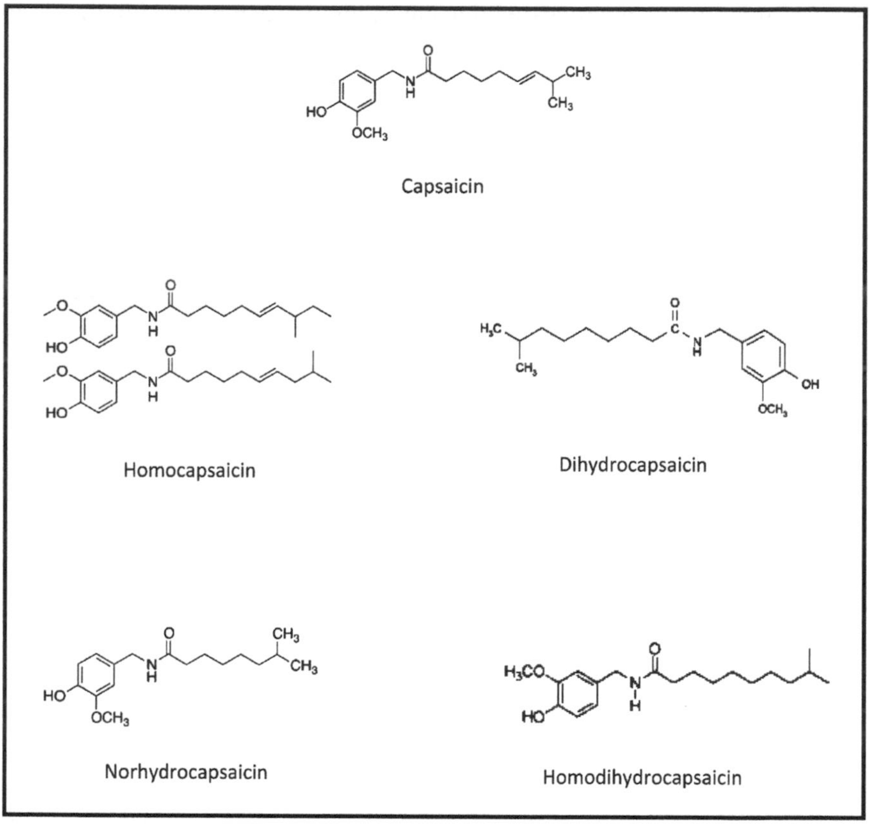

FIGURE 3.1   Names and chemical structures of major capsaicinoids.

Kirby, 1968; Bernal et al., 1993). The capsaicin content of *Capsicum* ranges from 0.1 to 1% w/w (Islek, 2009).

Zamski et al. (1987) explored the synthesis and the translocation of the capsaicinoids in secreting cells. They found that they were synthesized in the inner compartment, migrated along the cytoplasm with sacs and merged with plasmalemma.

Capsaicinoids are localized in the vacuole of the epidermal cells and septa of *Capsicum* spp. It was found that capsaicin accumulations reached a maximum when fruits reached maximum size in *C. annuum*. Bennett and Kirby (1968) found Tritium-labeled l-phenylalanine and vanillylamine to be precursors in feeding experiments with ripening seeds of *Capsicum*. Eich (2008) has expressed a high pungent pepper variety which has 147 times more vanillylamine than a low one.

3.1.3   BIOSYNTHESIS OF CAPSAICINOIDS

There are two pathways in the biosynthesis of capsaicinoids: fatty acid metabolism and phenylpropanoid pathway (Figure 3.2). The phenolic structure comes from the phenylpropanoid pathway, in which phenylalanine is the precursor (Ochoa-Alejo and Gomez-Peralta, 1993). Valine pathway is more crucial than the phenylpropanoid pathway in regulating capsaicin biosynthesis with elicitors (Gururaj et al., 2012).

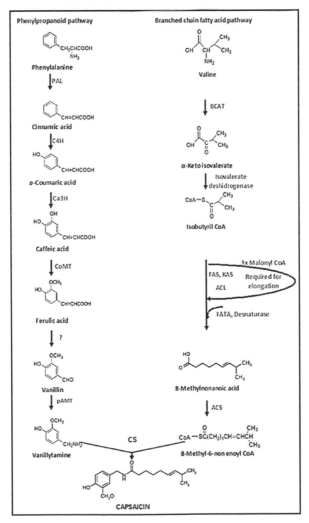

FIGURE 3.2   Pathway of capsaicin biosynthesis in *Capsicum* showing the principal enzymes and intermediates.

Capsaicin is produced from C. annuum by cell culture method (Johnson et al., 1991). The amount of capsaicin increases several times on immobilization of cells (Lindsey and Yeoman, 1983). Johnson et al. (1991) and Sudha and Ravishankar (2003a, b) investigated the positive effects of microbial factors on capsaicin production. Some scientists state that exogenous cellulase treatment may lead to the formation of secondary metabolites (Ellialtioglu et al. 1998; Ma, 2008).

Jasmonic acid (JA) and methyl jasmonate (MeJa) treatments result in the elicitation of capsaicin production (Sudha and Ravishankar 2003a,b). It is well known they are signal molecules in plants exposed to abiotic and biotic stresses (Enyedi et al., 1992; Creelman and Mullet, 1995), having either inhibitory or stimulant effects, in the form of morphological, biochemical and physiological changes (Sembdner and Parthier, 1993) and also in defence metabolism (Gundlach et al., 1992).

It is also known that heavy metal treatments as an abiotic elicitor have positive effects on increasing capsaicin production (Islek and Turkyilmaz Unal, 2015; Islek et al., 2017).

3.2   IN VITRO CULTURE STUDIES

3.2.1   ESTABLISHMENT OF IN VITRO CULTURES AND MEDIA

Capsicum annuum L. seeds were obtained from the Eastern Mediterranean Transitional Zone Agricultural Research Station. The seeds were sterilized in 70% ethanol for three minutes, in sodium hypochlorite for twenty minutes and then washed with sterile dH2O. Approximately 40 mL of sterile agar medium was placed in magenta containers. Pepper seeds were placed in each of the outer surface sterilized container (Islek, 2009). Covers of magenta planted were closed and isolated from outside by wrapping with a stretch film (Figure 3.3a).

To germinate the pepper seeds, hormone-free Murashige and Skoog (1962) (MS) basic nutrient medium was used. Pepper seedlings were obtained by germinating the seeds. These were used as the source of explant under photoperiodic conditions of 16 hours of light and 8 hours dark in a climate room and placed in the magenta boxes under sterile conditions after completing a four-weeks of incubation time (Figure 3.3b). 1.0 mg/L 2,4-D with 0.1 mg/L kinetin and 3% saccharose with 0.7% agar were added to obtain callus from hypocotyl explants; MS medium was used. After adjusting the medium pH to 5.7, hypocotyl explants were placed horizontally on the MS medium, sterilized in the autoclave. Developing callus tissues (Figure 3.3c) were transferred to the MS nutrient medium with 1.0 mg/L 2,4-D, 0.1 mg/L kinetin, 3% saccharose, and 0.7% agar to the sub-culture. Callus tissues in the magenta boxes were left to develop (Figure 3.3d) in the climate room under same photoperiodic conditions (Islek et al., 2014).

FIGURE 3.3   Callus production from the pepper seed (a) The cultivation of pepper seeds, (b) Four weeks of pepper seedlings, (c) Developing callus from hypocotyl explant, (d) Developing callus from the subculture.

MS nutrient medium was added as 1.0 mg/L 2,4- D with 0.1 mg/L kinetin and 3% saccharose used for callus occurrence. These were also used in the suspension culture; only agar addition to the medium was not done. Callus tissues developed in the magenta boxes was added to 2 g of each of the flasks which are 100 mL; 40 mL liquid nutrient medium was added to each under aseptic conditions. Erlenmeyer flasks with cell suspensions were placed on a shaking incubator (Figure 3.4), adjusted at 25°C. The velocity of the shaker was set to the 100 rev/min. (Islek, 2009).

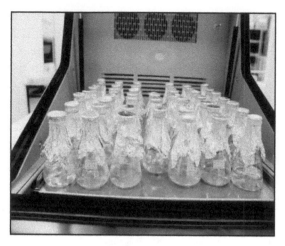

FIGURE 3.4   Cell suspension cultures on shaking incubator.

3.2.2 *STIMULATION OF CAPSAICIN PRODUCTION*

3.2.2.1 *IMMOBILIZATION OF CELLS*

About 1 g of free cells from each of 1-month-old suspension cultures from control and treatment groups were separated from the medium by filtration and suspended in a sodium alginate solution. These cells were then extruded into calcium chloride dihydrate through sterilized glass pipette glued to a peristaltic pump (Figure 3.5) (Islek et al., 2014).

FIGURE 3.5   Immobilized cells in MS medium.

3.2.2.2   ELICITOR PREPARATION AND INOCULATION

The study was performed in Erlenmeyer containing 40 mL of fresh medium. Each Erlenmeyer was inoculated with the 1 g fresh cell. After 14 days growth, sterile elicitor (cellulase, xanthan, curdlan, salicylic acid, and methyl jasmonate) was added. dH_2O was used instead of elicitor in control groups. The cultures were kept on a shaker incubator at 25°C for different days. Cell and liquid phase samples were collected, stored at −70°C and then capsaicin analysis was performed. The experiments were performed in triplicate (Islek, 2009).

3.2.3   CAPSAICIN EXTRACTION AND DETERMINATION

Cell and medium were separately extracted three times with ethyl acetate for extraction of capsaicin. After that, it was centrifuged, and supernatants were collected at the base. Supernatants were collected to the evaporator filled balloons and with the help of a vacuum ethyl acetate part under low pressure was evaporated near to the dryness. Residues were solved at the acetate and labeled on the bottles. It has been stored under −70°C until examining in these examples.

Before injecting into HPLC, the samples were filtered through a 0.45-μm Millipore filter. The amount of capsaicin obtained from callus and media was determined by HPLC at the C18 nucleosil column and 280 nm. The isocratic mobile phase was used as methanol: water (1% acetic acid) (39:60 v/v). The flow rate was 1 mL/min. (Islek, 2009).

3.3   RESULTS

In our studies, cellulase, xanthan, curdlan, salicylic acid and methyl jasmonate increased capsaicin production depending on application time and concentration (Islek, 2009). For example, in the study of Islek et al. (2014), the highest capsaicin concentration for the immobilized cells was 362.91 μg/g fresh weight at the 30 μg/mL cellulase treatments at the 24th hour.

3.4   SUMMARY AND CONCLUSIONS

There are two pathways of capsaicin biosynthesis: the phenylpropanoid pathway and the metabolic pathway of valine biosynthesis. The first stage of aromatic biosynthesis occurs by phenylpropanoid metabolism in all plants.

This suggests that the externally applied stimulator promotes the metabolic pathway in this direction.

The pepper to be used for capsaicin production is generally grown in 4–5 months. Therefore, the production process of capsaicin takes a long time and should be performed within a limited time. Free and immobilized cell cultures of pepper under in vitro conditions are a continuous way of obtaining capsaicin. It is supported by studies that immobilized cells in the culture medium are more advantageous than conventional methods. *Capsicum* cells immobilized to the culture medium can lead to much higher amounts of capsaicin synthesis.

KEYWORDS

- **capsaicin biosynthesis**
- ***Capsicum annuum* L.**
- **cellulase**
- **curdlan**
- **methyl jasmonate**
- **salicylic acid**
- **xanthan**

REFERENCES

Ahuja, K. D., Kunde, D. A., Ball, M. J., & Geraghty, D. P., (2006). Effects of capsaicin, dihydrocapsaicin, and curcumin on copper-induced oxidation of human serum lipids. *J. Agr. Food Chem., 54* (17), 6436–6439.

Akinci, S., & Akinci, I. E., (2004). Evaluation of red pepper for spice *(Capsicum annuum* L.) Germplasm resource of Kahramanmaras Region (Turkey). *Pak. J. Biol. Sci., 7* (5), 703–710.

Aydın, S., Islek, C., & Turkyilmaz Unal, B. (2017). Effect of Silver Nitrate Solvent on Total Protein, Total Phenolic and Some Antioxidant Enzyme Activities in Cell Suspension Culture of *Capsicum annuum* L. *Turkish Journal of Agriculture-Food Science and Technology,* 5(12), 1544–1549.

Aziz, A., Poinssot, B., Daire, X., Adrian, M., Bezier, A., Lambert, B., Joubert, J. M., & Pugin, A., (2003). *Laminarin elicits* defense responses in grapevine and induces protection against *Botrytis cinerea* and *Plasmoparaviticola. Mol. Plant-Microbe Interact., 16,* 1118–1128.

Basu, S. K., & De, A. K., (2003). *Capsicum:* Historical and botanical perspectives. *Capsicum: The Genus Capsicum. 33,* 1–15.

Bennett, D. J., & Kirby, G. W., (1968). Constitution and biosynthesis of capsaicin. *J. Chem. Society C: Organic,* 442−446.

Bernal, M. A., Calderon, A. A., Pedreno, M. A., Munoz, R., Ros Barcelo, A., & Merino de Caceres, F., (1993). Capsaicin oxidation by peroxidase from *Capsicum annuum* (variety *annuum)* fruits. *J. Agr. Food Chem., 41* (7), 1041−1044.

Beyaz, A., Ozguven, M. M., Ozturk, R., & Acar, A. I., (2009). Volume determination of Kahramanmarasred pepper (*Capsicum annuum* L.) by using an image analysis technique. *J. Agr. Mach. Sci., 5* (1).

Bley, K., Boorman, G., Mohammad, B., McKenzie, D., & Babbar, S., (2012). A comprehensive review of the carcinogenic and anticarcinogenic potential of capsaicin. *Toxicol. Pathol., 40* (6), 847−73.

Bosland, P. W., (1994). Chillis: History, cultivation, and uses. In: Charalambous, G., (ed.), *Spices, Herbs, and Edible Fungi* (pp. 347−366). Elsevier Publ., New York.

Bosland, P. W., (1996). *Capsicums: Innovative Uses of an Ancient Crop.* Progress in new crops. ASHS Press, Arlington, VA, 479−487.

Creelman, R. A., & Mullet, J. E., (1995). Jasmonic acid distribution and action in plants: regulation during development and response to biotic and abiotic stress. *Proceed. Nat. Acad. Sci., 92* (10), 4114−119.

Díaz, J., Pomar, F., Bernal, A., & Merino, F. (2004). Peroxidases and the metabolism of capsaicin in *Capsicum annuum* L. *Phytochemistry Reviews,* 3(1-2), 141−157.

Eich, E. (2008). Phenylalanine-derived metabolites/phenylpropanoids. *Solanaceae and Convolvulaceae: Secondary Metabolites: Biosynthesis, Chemotaxonomy, Biological and Economic Significance (A Handbook),* 271−342.

Ellialtioglu, S., Ustun, A. S., & Mehmetoglu, U., (1998). Determining the most suitable medium composition to obtain in vitro callus formation of some pepper varieties. *II: Kizilirmak International Science Congress,* 5158, Kinkkale, Turkey.

Enycdi, A. J., Yalpani, N., Silverman, P., & Raskin, I., (1992). Localization, conjugation, and function of salicylic acid in tobacco during the hypersensitive reaction to tobacco mosaic virus. *Proceed. Nat. Acad. Sci., 89* (6), 2480−2484.

Evans, W. C., & Trease, E. T., (2002). *Pharmacognosy* (15th edn.). W. B. Saunders Company Limited pp. 135−150.

FAO, (2014). Food and Agriculture Organization of the United Nations, Rome, Italy. http://www.fao.org (2014), Accessed 12th Aug 2015.

Gundlach, H., Muller, M. J., Kutchan, T. M., & Zenk, M. H., (1992). Jasmonic acid is a signal transducer in elicitor-induced plant cell cultures. *Proceed. Nat. Acad. Sci., 89* (6), 2389−2393.

Gururaj, H. B., Giridhar, P., & Ravishankar, G. A., (2012). Laminarin as a potential non-conventional elicitor for enhancement of capsaicinoid metabolites. *Asian J. Plant Sci. Res., 2* (4), 490−495.

Guzman, I., Bosland, P. W., & O'Connell, M. A., (2011). Heat, color, and flavor compounds in *Capsicum* fruit. In: Gang, D. R., (ed.), *The Biological Activity of Phytochemicals, Recent Advances in Phytochemistry* (Vol. 41, pp. 109−126). Springer Science + Business Media, Springer: New York, Dordrecht, Heidelberg, London.

Heiser, C. B., (1976). Peppers *Capsicum* (Solanaceae). In: Simmonds, N. W., (ed.), *The Evolution of Crops Plants* (pp. 265−268). Longman Press, London.

Islek, C., (2009). The effects of some elicitors on capsaicin production in freely suspended cells and immobilized cell cultures of *Capsicum annuum* L. *PhD Thesis,* Ankara University, Ankara, Turkey (In Turkish).

Islek, C., & Turkyilmaz Unal B., (2015). Copper Toxicity in Capsicum annuum: Superoxide Dismutase and Catalase Activities, Phenolic and Protein Amounts of in-vitro-Grown Plants. *Polish Journal of Environmental Studies, 24*(6).

Islek, C., Ustun, A. S., & Koc, E., (2014). The effects of cellulase on capsaicin production in freely suspended cells and immobilized cell cultures of *Capsicum annuum* L. *Pak. J. Bot., 46*(5), 1883–1887.

Johnson, T. S., Ravishankar, G. A., & Venkataraman, L. V., (1991). Elicitation of capsaicin production in freely suspended cells and immobilized cell cultures of *Capsicum frutescens. Food Biotechnol., 5,* 197–205.

Kosuge, S., Inagaki, Y, & Niwa, M., (1958). Pungent principles of *Capsicum.* 2. The pungent principles (in Japanese), *Nippon Nogei Kagaku Kaishi, 32* (9), 720–722.

Kosuge, S., Inagaki, Y., & Takatoshi, G., (1960). Pungent principles of red pepper. (in Japanese), *Nippon Nogei Kagaku Kaishi, 34* (10), 811–814.

Legin, G. Y, (1996). Capsaicin and it's analogous: properties, preparation, and applications (a review). *Pharm. Chem. J., 30,* 60–68.

Lindsey, K., & Yeoman, M. M., (1983). Novel experimental systems for studying the production of secondary metabolites by plant tissue cultures. In: Mantell, S. H., & Smith, H., (eds.), *Plant Biotechnology.* Cambridge University Press, London, pp. 39–66.

Luo, X. J., Peng, J., & Li, Y. J., (2011). Recent advances in the study on capsaicinoids and capsinoids. *Eur. J. Pharmacol., 650,* 1–7.

Ma, C. J., (2008). Cellulase elicitor induced accumulation of capsidiol in *Capsicum annuum* L., suspension cultures. *Biotechnol. Lett., 30,* 961—965.

MacNeish, R. S., (1964). Ancient Mesoamerican civilization. *Science, 143,* 531–537.

Mori, A., Lehmann, S., O'Kelly, J., Kumagai, T., Desmond, J. C., Pervan, M., et al. (2006). Capsaicin, a component of red peppers, inhibits the growth of androgen-independent, p. 53 mutant prostate cancer cells. *Cancer Res., 66* (6), 3222–3229.

Murashige, T., & Skoog, F., (1962). A revised medium for rapid growth and bioassays with tobacco tissue cultures. *Physiol. Plant., 15,* 473–497.

Nelson, E. K., & Dawson, L. E., (1923). The constitution of capsaicin, the pungent principle of *Capsicum.* III. *Jo. Amer. Chem. Soc., 45* (9), 2179–2181.

Nelson, E. K., (1919). The constitution of Capsaicin, the pungent principle of *Capsicum. J. Amer. Chem. Soc., 41* (7), 1115–1121.

Neumann, R., (2001). Disciplining peasants in Tanzania: From state violence to self-surveillance in wildlife conservation. *Violent Environ.,* pp. 305–327.

Ochi, T., Takaishi, Y, Kogure, K., & Yamauti, I., (2003). Antioxidant activity of a new capsaicin derivative from *Capsicum annuum. J. Nat. Prod., 66,* 1094–1096

Ochoa-Alejo, N., & Gomez-Peralta, J. E., (1993). Activity of enzymes involved in capsaicin biosynthesis in callus tissue and fruits of chili pepper *(Capsicum annuum* L.). *J. Plant Physiol., 141* (2), 147–152.

Ozturk, M., Gucel, S., Sakcali, S., Dogan, Y., & Baslar, S., (2009). Effects of temperature and salinity on germination and seedling growth of *Daucus carota* cv. Nantes and *Capsicum annuum* cv. Sivri and flooding on *Capsicum annuum* cv. Sivri. Chapter 6, In: Salinity and Water Stress: Improving Crop Efficiency (Eds.M.Ashraf et al.), 51–64 Springer Verlag-Tasks for Vegetation Science: 44.

Rahman, M. J., & Inden, H., (2012). Effect of nutrient solution and temperature on capsaicin content and yield contributing characteristics in six sweet pepper (*Capsicum annuum* L.) cultivars. *J. FoodAgri. Environ.,* 10 (1), 524–529.

Ravishankar, G. A., Suresh, B., Giridhar, P., Rao, S. R., & Johnson, T. S., (2003). Biotechnological studies on *Capsicum* for metabolite production and plant improvement. *Capsicum: The Genus Capsicum*, pp. 96–128.

Rodriguez-Burruezo, A., Kollmannsberger, H., Gonzalez-Mas, M. C., Nitz, S., & Nuez, F., (2010). HS-SPME comparative analysis of genotypic diversity in volatile fraction and aroma contributing compounds of *Capsicum* fruits from the annuum-chinense-frutescens complex. *J. Agric. Food Chem., 58,* 4388–4400.

Sembdner, G. A. P. B., & Parthier, B., (1993). The biochemistry and the physiological and molecular actions of jasmonates. *Ann. Rev. Plant Biol., 44* (1), 569–589.

Sudha, G., & Ravishankar, G. A., (2002). Involvement and interaction of various signaling compounds on the plant metabolic events during defense response, resistance to stress factors, formation of secondary metabolites and their molecular aspects. *Plant Cell Tissue Organ Culture, 71,* 181–212

Sudha, G., & Ravishankar, G. A., (2003a). Putrescine facilitated enhancement of capsaicin production in cell suspension cultures of *Capsicum frutescens. J. Plant Physiol., 160,* 339–346.

Sudha, G., & Ravishankar, G. A., (2003b). Influence of methyl jasmonate and salicylic acid in the enhancement of capsaicin production in cell suspension cultures of *Capsicum frutescens* Mill. *Curr. Sci., 85,* 1212–1217.

Suzuki, T., & Iwai, K., (1984). Constituents of red pepper spices: chemistry, biochemistry, pharmacology, and food science of the pungent principle of Capsicum species. In: Brossi, A., (ed.), *The Alkaloids: Chemistry and Pharmacology* (Vol. 23, pp. 227–299). Academic Press, Orlando, FL.

Tewksbury, J. J., Manchego, C., Haak, D. C., & Levey, D. J., (2006). Where did the chili get its spice? Biogeography of capsaicinoid production in ancestral wild chili species. *J. Chem. Ecol., 32* (3), 547–564.

Thampi, P. S. S., (2003). A glimpse of the world trade in Capsicum. *Capsicum: The genus Capsicum,* 16.

Thresh, J. C., (1876). Capsaicin, the active principle of *Capsicum* fruits. *Pharm. J., 7,* 21.

TUIK, (2010). *The Summary of Agricultural Statistics.* Turkish Statistical Institute, Turkey. http://tuik.gov.tr.

Ustun, A. S., (1990). The physiological and biochemical investigation of the causes of the resistance to the root rot (Phytophthoracapsici Leon.) disease in peppers. PhD Thesis, University of Ankara, 121 p. Turkey.

Vera, J., Castro, J., Gonzalez, A., & Moenne, A., (2011). Seaweed polysaccharides and derived oligosaccharides stimulate defense responses and protection against pathogens in plants. *Marine Drugs, 9* (12), 2514–2525.

Wahyuni, Y., Ballester, A. R., Sudarmonowati, E., Bino, R. J., & Bovy, A. G., (2011). Metabolite biodiversity in pepper *(Capsicum)* fruits of thirty-two diverse accessions: variation in health-related compounds and implications for breeding. *Phytochemistry, 72* (11), 1358–1370.

Yoshitani, S. I., Tanaka, T., Kohno, H., & Takashima, S., (2001). Chemoprevention of azoxymethane-induced rat colon carcinogenesis by dietary capsaicin and rotenone. *Int. J. Oncol., 19* (5), 929–939.

Zamski, E., Soham, O., Palevitch, D., & Levy, A., (1987). Ultrastructure of capsinoid-secreting cells in pungent and non-pungent red pepper (*Capsicum annuum* L.) cultivars. *Bot. Gaz., 148,* 1–6.

CHAPTER 4

DROUGHT-INDUCED CHANGES IN GROWTH, PHOTOSYNTHESIS, AND YIELD TRAITS IN MUNGBEAN: ROLE OF POTASSIUM AND SULFUR NUTRITION

SHAHID UMAR[1], NASER A. ANJUM[2], PARVAIZ AHMAD[3], and MUHAMMAD IQBAL[1,4]

[1]Department of Botany, Jamia Hamdard (Hamdard University), New Delhi, 110062, India, E-mail: s_umar9@hotmail.com

[2]Department of Botany, Aligarh Muslim University, Aligarh, 202002, India

[3]Department of Botany, S.P. College, Srinagar, Jammu and Kashmir, India

[4]Formerly Department of Botany, Jamia Hamdard (Hamdard University), New Delhi, 110062, India; Department of Plant Production, College of Food & Agricultural Sciences, King Saud University, PO Box 2460, Riyadh 11451, Saudi Arabia

ABSTRACT

Mungbean [*Vigna radiata* (L.) Wilczek] is the major pulse crop, stands next in importance to soybean, and is mostly grown as a summer crop in India. This study tested the role of two major plant nutrients namely potassium (K) and sulfur (S) in the control of the drought stress-induced impairments in *V. radiata* (cultivar Pusa, 9531) grown under varying soil moisture conditions. Drought stress was imposed by skipping one irrigation each at pre-flowering (soil moisture content, SMC, 15.3%), flowering (SMC, 13.2%) and post-flowering (pod development, SMC, 10.09%) stages in a pot culture experiment. K and S were applied to drought-exposed plants as basal dose @ 60 and 40 mg kg^{-1} soil, respectively. Moisture stress imposed at each of the three growth stages decreased chlorophyll content, soluble protein, photosynthetic rate, yield, and its attributes. K improved the tested parameters more when applied at pre-flowering and post-flowering (pod development) stages; whereas, S mitigated the drought stress effects when applied at flowering stage of *V. radiata*. Of the phenological stages exposed to drought stress, the flowering stages were found most moisture stress-sensitive that was followed by vegetative and post-flowering (pod development) stages.

4.1 INTRODUCTION

Mungbean [*Vigna radiata* (L.) Wilczek] is a potential crop in the Indian sub-continent due to its ready market, N$_2$-fixation capability, early maturity and the ability to fit well in crop rotation programme (Anjum, 2006). Among naturally occurring abiotic stresses, drought is probably the principal cause of crop failure worldwide, adversely affecting average yields for major crops. A large percentage of the world's crops are exposed to chronic or sporadic periods of drought and that affect virtually every aspect of growth, physiology, and metabolism in plants (Boyer, 1982; Shinozaki et al., 2002; Zhu, 2002). However, the stress response varies depending on the intensity, rate, and duration of the water deficit-exposure and the crop growth stage (Nam et al., 2001; Anjum, 2006; Thalooth et al., 2006). It has been amply proved that providing adequate moisture at some of the moisture-sensitive phenological stages of various pulse crops improves their N$_2$-fixation capability and productivity (Kasturikrishna and Ahlawat, 1999).

There is increasing evidence that plants suffering from environmental stresses like drought have a larger internal requirement for potassium (K) (Cakmak and Engels, 1999; Khan, 1991; Umar, 2006). K nutrition in plants stimulates root growth, regulates the closure and opening of stomata and maintains the photosynthetic CO$_2$ fixation (Saxena, 1985; Elstner and Osswald, 1994). Another plant nutrient sulfur (S) is an essential macronutrient after N, P, and K and plays a vital role in the regulation of plant growth and development (Ernst, 1998). S fertilization is a feasible technique to enhance the uptake and fertilizer-use efficiency of N, P, and K (Umar, et al., 1997) and suppress the plant uptake of undesirable, toxic elements (Aulakh, 2002).

An understanding of sensitive growth stage for drought stress and water requirements of crops is necessary to achieve maximum yield (Thalooth et al., 2006). Although research reports the effect of drought stress at different plant growth stage in leguminous crops including *V. radiata*, little is known about the mitigation of drought-induced alterations at different phenological stages with the use of mineral nutrients. Given above, the present investigation explores sensitive growth stage of *V. radiata* for drought stress and water requirements and examines the drought stress-induced alterations in the growth characteristics, contents of chlorophyll (Chl) and protein, the rate of photosynthesis and yield traits and incorporates the two important plant nutrients in the conventional agronomic package for *V. radiata* cultivation.

4.2 MATERIALS AND METHODS

4.2.1 EXPERIMENTAL MATERIALS, PROCEDURE, AND SOIL CHARACTERISTICS

Seeds of *V. radiata* cultivar Pusa 9531 were sown in 30-cm-diameter earthen pots filled with 8 kg of soil. The soil was sandy loam in texture, with pH 7.8, E.C. 0.38 dsm, 0.43% organic carbon, 70 ppm available K and 5 ppm available S. A sufficient amount of nitrogen (N) and phosphorus (P) at the rate of 120 and 30 mg kg^{-1} soil, respectively was applied at the time of sowing. S and K (@ 40 and 60 mg kg^{-1} soil, respectively) were applied as a treatment in the form of solution to mungbean plants five days before drought stress imposition at various growth stages. The sources of N, P, K, and S were urea, single super phosphate, gypsum, and muriate of potash (KCl_2), respectively. After germination, three plants per pot were maintained up to maturity. Pots were kept in the greenhouse of the Department of Botany, Hamdard University, New Delhi (India) under semi-controlled condition. A polythene plastic film was used to thwart the effects of rainfall, which allowed the transmittance of 90% of the visible wavelength (400–700 nm) under natural day and night conditions with a day/night temperature 25/20 ± 4°C and relative humidity of 70 ± 5%.

4.2.2 APPLICATION OF WATER DEFICIT STRESS AND K OR S APPLICATION

Drought stress was given to the plant as per the following schedule:

a) No stress (C): crop was irrigated on alternate days.

b) Drought at pre-flowering (*DPre-fl*): stress was imposed at flowering (15 d after sowing) by withholding water for 5 days, and then watered normally (without K/S).

c) Drought at pre-flowering (*DPre-fl*): stress was imposed at flowering (15 d after sowing) by withholding water for 5 days, and then watered normally (with K/S).

d) Drought at flowering (*DFl*): stress was imposed at flowering (30 d after sowing) by withholding water for 5 days, and then watered normally (without K/S).

e) Drought at flowering (*DFl*): stress was imposed at flowering (30 d after sowing) by withholding water for 5 days, and then watered normally (with K/S).

f) Drought at pod development (*DPod*): stress was imposed by withholding water for 5 days at pod development (50 d after sowing), and then watered normally (without K/S).

g) Drought at pod development (*DPod*): stress was imposed by withholding water for 5 days at pod development (50 d after sowing), and then watered normally (with K/S).

Soil moisture content was measured gravimetrically on a dry weight basis at the time of pre-flowering, flowering, and post-flowering (pod development) stages (Table 4.1). The treatments were arranged in a randomized complete block design, and each treatment was replicated five times.

TABLE 4.1   Soil Moisture Content (SMC, %) at no Drought (C), Drought at Pre-Flowering (*DPre-FL*), Flowering (*DFl*) and Pod Development (*Dpod*) Stages of Mungbean (*Vigna radiate* L. Wilczek). Values are Means of Five Replications ± SE (N = 3)

Irrigation treatments	Soil moisture content (SMC, %)
No drought (C)	20.78 ± 1.04
Dpre-Fl	15.31 ± 0.76
DFl	13.22 ± 0.66
Dpod	10.09 ± 0.51

4.2.3   MEASUREMENTS

Leaf area was measured with a leaf area meter (LI–3000A, LI-COR, Lincon, NE). Plant dry mass was determined after drying the plant at 80°C to a constant weight with the help of an electronic balance (SD–300). Net photosynthetic rate (P_N) was recorded in fully expanded leaves of second youngest nodes using infra-red gas analyzer (IRGA, LI-COR, 6400, Lincon, NE) on a sunny day between 10:00–11:00h. Chl content was estimated in fully expanded young leaves at each stage using the method given by Hiscox and Israelstam (1979). Estimation of soluble protein content was done according to Bradford (1976) using bovine serum albumin (BSA, Sigma) as standard. The leaf area, plant dry mass, net photosynthetic rate, Chl content and soluble protein content were estimated at 20, 35, and 55 days after sowing; while, 100-seeds weight and seed yield plan r^{-1} were determined at harvest.

4.2.4   STATISTICS

Data were subjected to ANOVA test using SPSS (Statistical Procedure for Social Sciences, version 10.0 Inc., Chicago, USA) and the results are presented as means ± standard error (SE).

4.3   RESULTS AND DISCUSSION

4.3.1   EFFECT OF DROUGHT STRESS

Imposing drought stress at various phenological stages caused perceptible variations in growth (leaf area, dry mass accumulation), contents of Chl and protein, and the rate of photosynthesis and yield attributes (100-seeds weight and seed yield). *V. radiata* with no drought stress had significantly higher leaf area, dry mass accumulation, the contents of Chl and protein, the rate of photosynthesis and yield attributes (100-seeds weight and seed yield per plant) as compared with drought stress at any of the three stages (Table 4.2). Drought stress at pre-and post-flowering stages had pronounced adverse effects on the growth attributes. Yield attributes decreased significantly at pod development (post-flowering) stage. Moreover, skipping irrigation at pre-flowering was found drastic for the accumulation of dry mass (84%), leaf area (69%) and protein content (44%) over control. However, skipping irrigation at flowering caused significant reductions in photosynthetic rate (83%), 100-seeds weight (165% and seed yield per plant 96%) over respective controls.

TABLE 4.2 Dry Mass Accumulation, Leaf Area, Chlorophyll Content, Soluble Protein Content, Net Photosynthetic Rate, 100-Seeds Weight and Seed Yield as Influenced by Drought Stress at Pre-Flowering (*Dpre-Fl*) or Flowering (*DFl*) or Pod Development (*Dpod*) Imposed for 5 Days and Potassium (K) and Sulfur (S) Application in Mungbean (*Vignaradiata* L. Wilczek). Values are Means ± SE (N=3). *Indicates Significant Differences at P<0.05 With a Column

Irrigation Treatments	K/S application (mg kg⁻¹ soil)	Dry mass (g)	Leaf area (cm$_2$)	Chlorophyll content (mg g⁻¹ fresh weight, f.w.)	Soluble protein content (mg g⁻¹ f.w.)	Net photosynthetic rate (μ mol CO$_2$ m⁻¹ s⁻¹)	100-seeds weight (g)	Seed yield pl⁻¹ (g)
DPre-Fl	Control	1.82 ± 0.33*	147.11±10.11*	1.17 ± 0.15*	3.73 ± 0.41*	13.21 ± 1.03*	5.89 ± 0.07*	8.26 ± 0.15*
	DPre-Fl	0.99 ± 0.22	87.02 ± 9.12	0.76 ± 0.11	2.59 ± 0.21	9.62 ± 1.33	4.12 ± 0.05	4.69 ± 0.07
	DPre-Fl+K$_{60}$	1.53 ± 0.88	151.07 ± 12.06	1.04 ± 0.14	2.88 ± 0.32	10.61 ± 2.11	5.65 ± 0.10	8.24 ± 0.11
	DPre-Fl+S$_{40}$	1.32 ± 0.43	140.66 ± 9.32	0.96 ± 0.09	2.49 ± 0.18	10.47 ± 2.16	4.47 ± 0.12	7.98 ± 0.13
DFl	Control	9.26 ± 0.65*	362.55±14.54*	1.37±0.16*	4.06 ± 0.39*	19.99 ± 3.43*	5.89 ± 0.07*	8.26 ± 0.15*
	DFl	6.24 ± 0.34	291.8 ± 13.21	0.88 ± 0.07	3.21 ± 0.55	10.94 ± 3.21	2.22 ± 0.04	4.2 ± 0.07
	DFl+ K$_{60}$	7.46 ± 0.29	321.11 ± 16.32	0.97 ± 0.11	3.45 ± 0.31	11.16 ± 4.10	5.04 ± 0.09	9.76 ± 0.11
	DFl+ S$_{40}$	8.02 ± 0.65	360.64 ± 14.44	1.22 ± 0.13	3.69 ± 0.65	13.32 ± 2.87	5.73 ± 05	10.18 ± 0.13
DPod	Control	16.65 ± 1.02*	303.12±10.65*	1.30 ± 0.16*	2.36 ± 0.16*	14.86 ± 4.38*	5.89 ± 0.07*	8.26 ± 0.15*
	DPod	10.6 ± 0.89	192.18 ± 8.54	0.76 ± 0.06	2.24 ± 0.32	10.14 ± 3.55	2.3 ± 0.09	4.23 ± 0.08
	DPod+ K$_{60}$	16.34 ± 0.56	201.13 ± 8.99	1.16 ± 0.13	2.29 ± 0.26	12.76 ± 2.18	4.11 ± 0.04	8.21 ± 0.11
	DPod+ S$_{40}$	16.14 ± 0.87	185.21 ± 7.92	1.03 ± 0.08	2.26 ± 0.15	11.61 ± 4.24	3.98 ± 0.06	7.97 ± 0.10

4.3.2   ROLE OF K AND S APPLICATION

The application of K and S ameliorated the drought-induced alterations when applied at all the growth stages of the plant. In particular, K might be utilized maximally when applied at pre- and post-flowering (pod development) stages. Notably, maximum increase in dry mass accumulation (35%), leaf area (43%) and the content of protein (10%) was found when K was applied at pre-flowering over drought affected *V. radiata*; while 100-seeds weight and seed yield significantly increased by 56% and 57%, respectively when K was applied at flowering stage of *V. radiata*. The application of S was supposed to be utilized maximally when applied to the drought-affected *V. radiata* at flowering stage. The most of the parameters studied significantly improved with S application at flowering stage. The 100-seeds weight, seed yield per plant, Chl content, the rate of photosynthesis and protein content ignorantly improved by 62%, 58%, 28%, 18%, and 13%, respectively over drought-affected *V. radiata* (Table 4.2).

The depressing effect of drought on plant growth was previously shown to be associated with an increase in osmotic pressure (OP) in the root zone which tends to decrease the synthesis of metabolites, reduce the translocation of mineral nutrients from soil to the plant as well as decrease division and elongation of the cells. These results are in harmony with those obtained in the studies of Thalooth et al. (2006) and Anjum (2006). The reduction in the contents of photosynthetic pigments in plants, subjected to water stress has widely been demonstrated previously (Pastori and Trippi, 1993a,b; Baisak et al., 1994; Krause et al., 1995). These authors concluded that the loss in the content of Chl was mainly due to a reduction in the lamellar content of the light harvesting Chl a/b protein. Makhmudov (1983) and Prakash and Ramachandran (2000) postulated that moisture stress would have either inhibited biosynthesis of the precursors of Chl or degraded Chl at a high rate, which reduced the total Chl content (Khanna-Chopra et al., 1980; Yang et al., 2001). The rate of photosynthesis was decreased/stopped under severe water stress condition in wheat (Secenji et al., 2005) and several other plant species (Yordanov et al., 2003). The content of protein decreased in the present study. Drought stress enhances the protein degradation because of protease activity (Palma et al., 2002) and polysome activity and/or due to the generation of reactive oxygen species (Baisak et al., 1994; Kramer and Boyer, 1995; Deleu et al., 1999).

Regardless of irrigation treatments, results in the table reveal also that K and S application significantly increased the growth and yield attributes the contents of Chl and protein and the rate of photosynthesis. Differences among the effect of K/S application (s) were recorded. Both K and S were superior at their own places. K was superior in all the features when applied at pre-flowering and post-flowering (pod development) stages. However, S was found superior in action when applied at flowering stage of the crop plant (Table 4.2). The adequate and balanced supply of mineral nutrients has been shown to play a vital role in sustaining food security. Enhancement effect on the parameters studied might be attributed to the favorable influence of K and S nutrients on metabolism and biological activity and its stimulating effect on photosynthetic pigments and activities of several enzymes which in turn encourage mitigation of drought-induced alterations (Anjum, 2006; Marschner, 1995; Singh et al., 1994; Yang et al., 2001). K has been found promising to promote root growth to explore more soil water (Umar, 2006). It increases the stomatal resistance and thereby helps plants to reduce water loss through transpiration (Sinha, 1978). Potassium increases leaf area, Chl content, net assimilation rate (NAR), improves the water content of the plant and stabilizes the nitrate reductase activity (Khan, 1991; Umar, 2006). The cumulative effect of these nutrients was reflected in the present study by improved plant growth, photosynthetic traits and higher yield of *V. radiata* under drought stress condition. S is involved in the light reaction of photosynthesis as an integral part of ferredoxin, a non-haem iron S protein (Marschner, 1995). Maximum utilization of S has previously been reported when applied at flowering stage in rapeseed crop (Ahmad et al., 2005; Anjum, 2006). Application of S increased the seed yield and attributing characters in other crops also (Singh, et al., 1994; Kar and Babulkas, 1999; Patel, et al., 2002).

It can be concluded that *V. radiata* flowering stage is drought-sensitive, and is followed by pre-flowering and post-flowering stages. However, optimum moisture should be maintained at pod-development (post-flowering) stage to achieve maximum yield of the produce. In addition, the present study also indicates that application of K and S had positive effects on growth parameters, yield, the contents of Chl and protein and the rate of photosynthesis. K application surpassed S when applied at pre- and post-flowering stages; whereas, the later was found beneficial when applied at flowering stage of *V. radiata*.

ACKNOWLEDGMENTS

NAA is deeply indebted to Hamdard National Foundation (HNF), New Delhi, India for financial assistance in the form of Research Fellowship (DSW/HNF–18/2006).

KEYWORDS

- drought stress
- mungbean
- phenological stages
- potassium
- soil moisture content
- sulfur

REFERENCES

Ahmad, A., Khan, I., Anjum, N. A., Diva, I., Abdin, M. Z., & Iqbal, M., (2005). Effect of timing of sulfur fertilizer application on growth and yield of rapeseed. *J. Plant Nutr.*, *28*, 1049–1059.

Anjum, N. A., (2006). Effect of abiotic stresses on growth and yield of *Brassica campestris* L. and [*Vigna radiata* (L.) Wilczek] under different sulfur regimes. *Ph.D. Thesis*, Jamia Hamdard, New Delhi, India.

Aulakh, M. S., (2002). Crop responses to sulfur nutrition. In: Abrol, Y. P., & Ahmad, A., (eds.), *Sulfur in Plants*. Kluwer Academic Publishers, The Netherlands, pp. 341–358.

Baisak, R., Rana, D., Acharya, P., & Kar, M., (1994). Alterations in the activities of active oxygen scavenging enzymes of wheat leave subjected to water stress. *Plant Cell Physiol.*, *35*, 489–495.

Boyer, J. S., (1982). Plant productivity and environment. *Science*, *218*, 443–448.

Bradford, M. M., (1976). A rapid and sensitive method for the quantization of microgram quantities of protein utilizing the principle of protein-dye binding. *Anal Biochem.*, *72*, 248–254.

Cakmak, I., & Engels, C., (1999). Role of mineral nutrients in photosynthesis and yield formation. In: Rengel, A., (ed.), *Mineral Nutrition of Crops: Mechanisms and Implications* (pp. 141–168). The Haworth Press, New York, USA.

Deleu, C., Coustaut, M., Niogret, M. F., & Larher, F., (1999). Three new osmotic stress-regulated cDNAs identified by differential display polymerase chain reaction in rapeseed leaf discs. *Plant Cell Environ.*, *25*, 979–988.

Elstner, E. F., & Osswald, W., (1994). Mechanism of oxygen activation during plant stress. *Proc Royal Soc Edinburg, Sec B*, *102*, 131–154.

Ernst, W. H. O., (1998). Sulfur metabolism in higher plants: Potential for phytoremediation. *Biodegradation*, *9*, 311–318.

Hiscox, J. H., & Israelstam, G. F., (1979). A method for extraction of chlorophyll from leaf tissues without maceration. *Can. J. Bot.*, *57*, 1332–1334.

Kar, D., & Babulkas, P. S., (1999). The response of safflower (*Carthamus tinctorius* L) to sulfur and zinc in Versitols. *Indian J. Dryland Agric. Res. Dev.*, *14*, 17–18.

Kasturikrishna, S., & Ahlawat, P. S., (1999). Growth and yield response of pea (*Pisum sativum* L.) to moisture stress, phosphorus, sulfur, and zinc fertilizers. *Indian J. Agron.*, *44*, 588–596.

Khan, N. A., (1991). Ameliorating water stress by potassium in mustard. *Plant Physiol. Biochem.*, *18*, 80–83.

Khanna-Chopra, R., Chaturverdi, G., Aggarwal, P., & Sinha, S., (1980). Effect of potassium on growth and nitrate reductase during water stress and recovery in maize. *Physiol. Plant.*, *49*, 495–500.

Kramer, P. J., & Boyer, J. S., (1995). *Water Relations of Plants and Soils*. Academic Press, San Diego, USA.

Krause, G., Virgo, H. A., & Winter, K., (1995). *High Susceptibility to Photoinhibition of Young Leaves of Tropical Forest Trees*, *197*, 583–591.

Makhmudov, S. H. A., (1983). A study on chlorophyll formation in wheat leaves under moisture stress. *Field Crops Abst.*, *39*, 1353.

Marschner, H., (1995). *Mineral Nutrition of Higher Plants*. Academic Press, New York.

Nam, N. H., Chauhan, Y. S., & Johansen, C., (2001). Effect of timing of drought stress on growth and grain yield of extra-short -duration peagonpea lines. *J. Agric. Sci.* (Cambridge), *136*, 179–189.

Palma, J. M., Sandalio, L. M., Corpas, F. J., Romero-Puertas, M. C., McCarthy, I., & Del Rio, L. A., (2002). Plant proteases, protein degradation, and oxidative stress: role of peroxisomes. *Plant Physiol. Biochem.*, *40*, 521–530.

Pastori, G. M., & Trippi, V. S., (1993a). Cross-resistance between water and oxidative stresses in wheat leaves. *J. Agric. Sci.* (Cambridge), *120*, 289–294.

Pastori, G. M., & Trippi, V. S., (1993b). Antioxidative protection in a drought-resistant maize strain during leaf senescence. *Physiol. Plant.*, *87*, 227–231.

Patel, P. T., Patel, G. A., Sonani, V. V., & Patel, H. B., (1999). Effect of sources and levels of sulfur on seed and oil yield of safflower (*Catharanthus tinctorius* L). *J. Oilseed Res.*, *19*, 76–78.

Prakash, M., & Ramachandran, K., (2000). Effect of moisture stress and antitranspirants on leaf chlorophyll, soluble protein and photosynthetic rate in brinjal plants. *J. Agron. Crop. Sci.*, *184*, 153–156.

Saxena, N. P., (1985). *The Role of Potassium in Drought Tolerance*. Potash review, No. 5 International Potash Institute, Bern, *16*, 1–15.

Secenji, M., Lendvai, A., Hajosne, Z., Dudits, D., & Gyorgyey, J., (2005). Experimental system for studying long-term drought stress adaptation of wheat cultivars. *Acta Biol. Szeged.*, *49*, 51–52.

Shinozaki, K. Y., Kasuga, M., Liu, Q., Nakashima, K., Sakuma, Y., Abe, H., et al. (2002). *Biological Mechanism of Drought Stress Response*. JIRCAS Working Report 1–8.

Singh, S. K., Singh, R., & Singh, H. P., (1994). Effect of sulfur on growth and yield of summer mooing [*Vigna radiata* (L.) Wilczek]. *Legume Res.*, *17*, 53–56.

Sinha, S. K., (1978). Influence of potassium on tolerance to stress. In: Sekhon, G. S., (ed.), *Potassium in Soil and Crops* (pp. 223–240). Potash Research Institute of India, New Delhi.

Thalooth, A. T., Tawfik, M. M., & Magda-Mohamed, H. A., (2006). Comparative study of the effect of foliar application of zinc, potassium, and magnesium on growth, yield, and some chemical constituents of mungbean plants grown underwater stress conditions. *World J. Agric. Sci.*, *2*, 37–46.

Umar, S., (2006). Alleviating adverse effects of water stress on the yield of sorghum, mustard, and groughd by potassium application. *Pak. J. Bot.*, *38*, 1373–1380.

Umar, S., Devnath, G., & Bansal, S. K., (1997). Groundnut pod yield and leaf spot disease as affected by potassium and sulfur nutrition. *Indian J. Plant Physiol., 2*, 59–64.

Yang, J., Zhang, J., Wang, Z., Zhu, Q., & Liu, L., (2001). Water deficit-induced senescence and its relationship to the remobilization of pre-stored carbon in wheat during grain filling. *Agron. J., 93*, 196–206.

Yordanov, I., Velikova, V., & Tsonev, T., (2003). Plant responses to drought and stress tolerance. *Bulgarian J. Plant Physiol. SI*, 187–206.

Zhu, J. K., (2002). Salt and drought signal transduction in plants. *Annu. Rev. Plant Biol., 53*, 247–273.

CHAPTER 5

COTTON, WHITE GOLD OF PAKISTAN: AN EFFICIENT TECHNIQUE FOR BUMPER CROP PRODUCTION

IJAZ RASOOL NOORKA[1], MUHAMMAD SHAHID IQBAL[2], MÜNIR ÖZTÜRK[3], MUHAMMAD RAFIQ SHAHID[2], and IHSAN KHALIQ[4]

[1]Department of Plant Breeding & Genetics, University College of Agriculture, University of Sargodha, Pakistan, E-mail: ijazphd@yahoo.com

[2]Cotton Research Institute, Faisalabad, Punjab, Pakistan

[3]Department of Botany Ege University, Izmir, Turkey

[4]Department of Plant Breeding & Genetics, University of Agriculture, Faisalabad, Pakistan

ABSTRACT

Pakistani agriculture is at the crossroad to fulfill burgeoning population needs. Cotton itself and its by-products revealed provision of food and fiber production. However, it's per acre yield demands employment of advanced production technology like seed selection, delinting, best-adapted method of sowing, weeds infestation, timely application of organic and inorganic fertilizer, water scouting, integrated pest management, safe picking, and proper storage. This study reviews the advent of cotton production, constrictions, and recommendations to ensure the best crop stand.

5.1 BACKGROUND, HISTORY, AND ECONOMIC IMPORTANCE

Cotton (*Gossypium hirsutum* L.) is an important trade crop of the world, grown on an area of more than 33.1 million hectares (Johnson et al., 2014).

Cotton is grown in various countries over five continents (Afzal and Ali, 1983; Kannan et. al., 2004; Rahman et al., 2012). Agricultural crops are considered as the backbone of the economy for Asian countries (Ahmad, 1999; Noorka and Heslop-Harrison, 2014). It is also a service generated industry due to the highest number (nearly 250 million) people are directly or indirectly involved from sowing to marketing and by-products of cotton throughout the world (Abro et. al., 2004; Ethridge et al., 2006).

Pakistan is ranked fifth in mass production country of cotton. Cotton is Pakistan's major cash and fiber crop. It has fundamental importance in our national economy (Economic Survey of Pakistan, 2013–14). Cotton not only provides the raw material for cloth and oil industry but also the share of cotton or its items is very high in Pakistan's total exports (Pan et al., 2007; Iqbal et al., 2013). This cash crop is not only profitable for the farmers but also provides livelihood to a large population. Although Pakistan ranks good in cotton production, but our average acreages yields are less than cotton producing many countries of the world in spite of our soil texture, promising varieties, and favorable climatic conditions. Progressive farmers in Punjab are producing 60–70 months yield per acre while in Sindh, several farmers produce more than Punjab. Our farmers can increase their average per acre yield if they keep in mind suitable land and its good seedbed preparation, selections of pure good quality seeds, seed rate, time of sowing, hoeing, eradication of weeds, balanced use of fertilizers, irrigation, Cotton Leaf Curl Disease (CLCuD) management, insect pest control and timely picking.

5.2   SUITABLE LAND SELECTION AND GREEN MANURING

For the cultivation of cotton fertile soil with good water holding capacity considered best. Sowing should be done on beds in salty (kalrathi) soil. Prepare land in such a way that there should be no stubbles of previous crop and stones. Land should be leveled for balanced irrigation. The roots of cotton should penetrate deep into the soil and for this purpose use chisel plow after every 2–3 year so that thick layer of soil breaks because healthy roots produce healthy plants. Green manuring crops should be buried into the soil almost 30 days before sowing and then irrigate the field 10 days after burial, so that green manuring crop rotted efficiently. Apply ½ bags of urea at the time of burial to enhance the rotting process (Noorka and Shahid, 2013).

5.3 HISTORICAL OUTLOOK AND ADAPTATION REGARDING SELECTION OF SEED, ITS PREPARATION AND SEED RATE

Since the start of Agriculture, the best practice is the selection of seed. It is the seed who gave a good crop stand (Noorka and Khaliq, 2007). For obtaining the best germination and good production registered, healthy, pure, and delinted seed should be used (Noorka et al., 2013). For delinting, 1-litre HCL per 10 kg of seed is used. After delinting wash it 3–4 times with water. It is very necessary to test the germination before sowing and if there is less germination percentage then definitely increase seed rate. In case of drill sowing use, 10 kg/acre delinted seed to get maximum plant population while in case of bed sowing use 8 kg/acre delinted seed. Apply Imidacloprid @ 10 g per kg of seed before sowing. It saves the crop from sucking insects at its initial stages. Seed treatment with fungicides is required in case of early march sowing.

5.4 THE CRITICAL FACTOR: SELECTION OF SUITABLE VARIETY AND TIME OF SOWING

As some farmers are growing cotton at early to get rid of cotton leaf curl virus. For early sowing recommended sowing time is 1st March and do not sow before it, because the temperature of the soil should be 20°C for good germination. The process of germination remained slow due to low-temperature and attack of fungus may be increased which affects the process of germination. Remember that at 15°C, the process of photosynthesis slows down and seedlings start dying (Figure 5.1 and Table 5.1).

TABLE 5.1 Varieties and Their Time of Sowing

Variety	Sowing Time	Variety	Sowing Time
IR–3701	April 15 to May 15	GN-HYBRID–2085	April 15 to May 15
AA–703	March 1 to April 15	TARZAN–1	March 1 to May 15
MG–6	April 1 to May 15	MNH–886	March 1 to May 15
SITARA–008	March 1 to April 15	NS–141	March 1 to May 15
NEELUM–121	March 1 to April 30	FH–114	March 1 to May 15
A-ONE	March 1 to May 15	IR–3	March 1 to May 15
AA–802	April 15 to May 15	CIM–598	March 1 to May 15
IR–1524	April 15 to May 15	SITARA–009	March 1 to May 15

FIGURE 5.1   Selection of suitable variety and time of sowing is important.

5.4.1   PLANTING DENSITY AND METHODS OF SOWING

For general awareness of the Plant to plant &bed to bed distance and plant population for the cultivation of cotton is given in Table 5.2.

TABLE 5.2   Plant Population for the Cultivation of Cotton

Sowing Time	Inter-Plant Distance	Bed to Bed Distance	Plant population/acre
Early (March)	12–15 inches	2 feet and 6 inches	14,000–17,500
Mid (April)	9–12 inches	2 feet and 6 inches	17,500–23,000
Late (May)	6–9 inches	2 feet and 6 inches	23,000–35,000

1) **Flat sowing**: Sowing of cotton should be done by using the tractor driven drill in lines by maintaining the depth of seeds at 2–3-inch depth. However, for students/researchers limited space experiments, Men driving hand drill can be used easily.

Benefits of flat sowing:

The benefits of flat sowing are as follows:

- Easy cultural practices like weeding, hoeing, etc.;
- Easy weed control;
- Balance crop height;
- Balance utilization of fertilizers;
- Easy spraying;
- Lessen boll and bud shedding;
- More yield.

2) **Bed sowing:** The two methods of bed sowing are:

   a) **Mechanical sowing:** Land should be leveled in case of mechanical sowing. Mechanical sowing should be operated on dry beds after putting seed in the machine. The level of water should be 2–3 inches below the seed for better germination of seed. 2nd irrigation should be applied 3–4 days after 1st irrigation.

   b) **Sowing by hand:** In the case of hand sowing, apply 6–7 inches water in channels. After watering sow seed 1 inch above water level by hand. Gaps should be filled after 2nd irrigation of water.

Benefits of bed sowing

- Good germination percentage and plants population;
- More area of sowing in less time;
- Water saving method;
- Less loss due to heavy rain;
- Less seed rate;
- Easy spray;
- Better plant growth;
- Easy weed control.

Measures to reduce rain loss in bed sowing: it should be kept in mind that bed sowing in Kalrathi (Salt effected) and affected area, start again sowing if there is raining 2–3 days after sowing. Avoid sowing in low lying areas. Perform hoeing and weeding after rain.

5.5   WEEDS OF COTTON, CONTROL BY INTEGRATED APPROACH

The process should be done 20–25 days after sowing. Perform roughing after 1ˢᵗ irrigation and remove virus-infected plants. It has been seen that attack of virus or insect pests starts from weeds aroundwater channels, ridges, and roads. So clean water channel, ridges, and roads before sowing.

Major weeds of cotton: Nut Grass (*Cyperus rotundas*),Horse purslanes, (*Trianthemamonogyna*), *Amaranthus polygamous Viridis*, Cotton Weed (*Digeraarvensis*), Field bindweed (*Convolvulus arvensis*), Common Purslane (*Portulaca oleracea*), *Echimochloacolonum*, Halfa grass *(Desmostachyabipinnata)*, *Corchorus depressus*, goosegrass (*Eleusineindica*), False Daisy (*Eclipta alba*), *Corchorus trilocularis, Xanthium strumarium, Sorghum halepense*, Bindii (*Tribulusterrestris*) and Bermuda grass (*Cynodondactylon*).

Demerits of weeds:

- Reduce production;
- Shelter for insect pests;
- Absorb water, nutrients, and air, light, etc.;
- Cause CLCuD;
- Secrete chemical compounds that harm plants;

Weed Control by integrated approach:

1) **By hoeing:** There are several benefits of weed control by hoeing. It retains moisture in the soil. Air passage through the soil.

   a) **Dry method:** In the dry method, hoeing should be done after sowing and before first irrigation. 1–2 dry hoeing is enough.

   b) **Wet method:** In the wet method, perform hoeing after each irrigation at water conditions. Hoeing should be done in exact water condition (Table 5.3).

2) **Herbicides:** Two types of herbicides may be used. Pre-emergence herbicides should be applied equally followed by planking

   a) Pre-emergence
   b) Post-emergence

Instructions for the use of herbicides: Prepare the land efficiently and there should be no stubbles of the previous crop. Use T-jet or foam nozzle and calibrate spray machine. Be aware, there should be no part without spray. Spray equally. The pressure of the sprayer should be equal. The nozzle of the spray machine should be accurate. Use a standard dose of herbicides. Spray in morning or evening. Don't spray in the fast air.

TABLE 5.3 A Comparison of Sowing Methods

Method of sowing	Irrigation
Line sowing	1st irrigation should be done 30–35 days after sowing and remaining after each 12–15 days interval
Bed sowing	1st irrigation 3–4 days after sowing, 2nd 6–9 days after sowing and remaining after each 15 days interval

5.6 IRRIGATION SCHEDULING

Keeping in mind the points of soil fertility, methods of sowing, variety, environmental conditions and conditions of the crop, irrigation should be applied. Symbols of water shortage appear first on the upper part of the plants that cause the leaves blue color, flowering at the tip, reduction in growth (Chapagain et al., 2006; Noorka et al., 2013).

5.7 FERTILIZERS, NUTRIENT AVAILABILITY IN DIVERSE FERTILIZERS

Soil analysis is very important for measuring the quantity of fertilizers application. Soil analysis report can be seen in Tables 5.4 and 5.5.

TABLE 5.4 A Comparison of N, P, and K Availability in Different Soils

Analysis	Nutrient availability
Organic matter	**Nitrogen**
Less than 0.86%	Less
0.86% to 1.29%	Average
More than 1.29%	Satisfactory
Phosphorus availability (ppm)	**Phosphorus**
Less than 7 ppm	Less
7 to 14 ppm	Average
15 to 21 ppm	Satisfactory
More than 21 ppm	Sufficient
Potash availability	**Potash**
Less than 80 ppm	Less
80 to 180 ppm	Average
More than 180 ppm	Sufficient

94

Crop Production Technologies for Sustainable Use and Conservation

TABLE 5.5 Nutrients Availability in Different Fertilizers

Fertilizer	Bag wt (kg)	Nutrients (%)			Nutrients per bag (Kg)		
		N	P_2O_5	K_2O	N	P_2O_5	K_2O
Urea	50	46	-	-	23	-	-
Ammonium nitrate	50	26	-	-	13	-	-
Triple superphosphate	50	-	46	-	-	23	-
Single superphosphate	50	-	18	-	-	9	-
Single superphosphate	50	-	14	-	-	7	-
Di-ammonium phosphate	50	18	46		9	23	
Potassium sulfate	50	-	-	50	-	-	25
Potassium chloride	50	-	-	60		-	30
Calcium amonium nitrate	50	26	-	-	13	-	-

N = Nitrogen, P_2O_5 = Phosphorus, K_2O = Potash.

5.8   INSECT PEST AND DISEASES OF COTTON CROP

1) **Pests:** Whitefly, Jassid, Aphid, Spotted bollworm, Pink bollworm, Thrips, *Helicoverpaarmigera*, Mite, Armyworm, Mealy Bug, Dusky Cotton Bug, Bollworms.
2) **Diseases of cotton:** Cotton Leaf Curl Virus, Angular Leaf Spot or Bacterial Leaf Blight, Root Rot, Boll Rot, Plant Wilt.

5.9   PICKING AND STORAGE OF COTTON

a) Start picking when 40–50% boll opening occur.
b) Start picking at 10 am when there is less dew.
c) Picking should be done downward to upward.
d) Only pick good opened bolls.
e) Do not place lint in moist soil.
f) Separate last picking.
g) Pick upland cotton at 15–20 days interval.
h) Put different varieties separately.
i) Pick pink bollworm affected cotton separately.
j) Later on picking sun drying cotton, place cotton in dry, ventilated, and airy store.
k) Store cotton in cotton made bags or either jute bags in such a way that air can easily pass through them.

5.10   RECOMMENDATIONS AND FUTURE POLICIES

The cotton crop is very sensitive to environmental hazards. For successful crop production the farmers and extension staff training regarding seed selection, the best method of sowing, integrated pest management, water management, the environmental friendly technique may be employed to get maximum yield per acre basis. Now we should ensure certified seed and balanced use of fertilizers to maximize cotton yield.

ACKNOWLEDGMENT

The authors are thankful to Dr. Abdur Rahman (Late) Professor Emeritus, Ex-Vice Chancellor, University of Agriculture, Faisalabad, and Dr. Trude Schwarzacher Molecular Cytogenetics and Plant Cell Biology, University of Leicester, the United Kingdom for their kind guidance to prepare this chapter.

KEYWORDS

- agriculture
- economy
- fertilizers
- food situations
- pest management

REFERENCES

Abro, G. H., Syed, T. S., Tunio, G. M., & Khuhro, M. A., (2004). Performance of transgenic Bt cotton against insect pest infestation. *Biotechnology, 3,* 75–81.

Afzal, M., & Ali, M., (1983). *Cotton Plant in Pakistan.* Ismail Aiwan-i-Science, Shahrah-i-Roomi, Lahore Pakistan.

Ahmad, Z., (1999). Pest problems of cotton, a regional perspective," *Proceedings of ICAC-CCRI, Regional Consultation Insecticide Resistance Management in Cotton,*" June 28-July 1st, Multan (Pakistan), 5–20.

Chapagain, A. K., Hoekstra, A. Y., Savenije, H. H. G., & Gautam, R., (2006). The water footprint of cotton consumption: An assessment of the impact of the worldwide consumption of cotton products on the water resources in the cotton producing countries. *Ecol. Econ., 60* (1), 186–203.

Economic Survey of Pakistan, (2013–2014). *Ministry of Finance*, Government of Pakistan, Islamabad. www.gov.pk.com.

Ethridge, D., Welch, M., Pan, S., Fadiga, M., & Mohanty, S., (2006). World cotton outlook: Projections to 2015/16. In: *Beltwide Cotton Conferences, San Antonio, Texas-January* (pp. 3–6).

Iqbal, M., Hussain, Z., Noorka, I. R., Akhtar, M., Mahmood, T., Suleman, S., & Sana, I., (2013). Revealed comparative and competitive advantages of white gold of Pakistan (cotton) by using balsa and white index. *Int. J. Agric. Appl. Sci., 5* (1), 64–68

Johnson, J., MacDonald, S., Meyer, L., Norrington, B., & Skelly, C., (2014). *Agricultural Outlook Forum 2014*. Presented Friday, February 21, (2014). The World and United States cotton outlook U. S. Department of Agriculture.

Kannan, M., Uthamasamy, S., & Mohan, S., (2004). Impact of insecticides on sucking pests and natural enemy complex of transgenic cotton. *Curr. Sci., 86,* 726–729.

Noorka, I. R., & Heslop-Harrison, J. S., (2014). Water and crops: Molecular biologists, physiologists, and plant breeders approach in the context of evergreen revolution. *Hand Book of Plant and Crop Physiology* (pp. 967–978). CRC Press, Taylor, and Francis, USA.

Noorka, I. R., & Khaliq, I., (2007). An efficient technique for screening wheat (*Triticum aestivum* L.)Germplasm for drought tolerance. *Pak. J. Bot., 39* (5), 1539–1546

Noorka, I. R., & Shahid, S. A., (2013). Use of conservation tillage system in the semiarid region to ensure wheat food security in Pakistan. Development in Soil Salinity Assessment and Reclamation. ISBN 978–94–007–5683–0. Springer.

Noorka, I. R., Tabassum, S., & Afzal, M., (2013). Detection of genotypic variation in response to water stress at seedling stage in escalating selection intensity for rapid evaluation of drought tolerance in wheat breeding. *Pak. J. Bot., 45* (1), 99–104.

Pan, S., Mohanty, S., Welch, M., Ethridge, D., & Fadiga, M., (2007). Effects of Chinese currency revaluation on world fiber markets. *Contem. Econ. Policy, 25* (2), 185–205.

Rahman, M., Shaheen, T., Tabbasam, N., Iqbal, M. A., Ashraf M., Zafar, Y., & Paterson, A. H., (2012). Cotton genetic resources. A review. *Agron. Sustain. Dev., 32,* 419–432.

CHAPTER 6

CORCHORUS SPECIES: JUTE: AN OVERUTILIZED CROP

M. MAHBUBUL ISLAM

Chief Scientific Officer & Head, Agronomy Division, Bangladesh Jute Research Institute, Dhaka–1207, Bangladesh, Tel.: +8801552416537, E-mail: mahbub_agronomy@yahoo.com; mahbubulislam63@gmail.com

ABSTRACT

Jute is grown in Bangladesh almost solely as a rainfed crop without any irrigation or drainage provisions. The status of jute crop of Bangladesh as regard to its production, cultivation areas, cultivation ratio, sowing, and harvesting time etc., the temperature, rainfall, day length, etc., was selected for its large concentration points of relation with jute production. The total jute acreage in 1972–73 was 2214.70 thousand acre and the production was 1181.00 thousand tonnes. It was 1908.00 thousand acres in 1986–87 and the area gone down to 500.00 thousand acres in 2006–07 from where now it is producing 1526 thousand tonnes from land areas of 708 thousand hectares in 2010–2011, respectively. In Bangladesh, there was pre-monsoon shower during the month of March and April thereby offering optimum condition for land preparation and sowing. During May and June, there were moderate and intermittent rain and shower providing enough moisture in the soil needed for growth of jute plants. Both species of jute were of photoperiodism-short day plant, the critical light period seems to be near about 12.50 hours. It was observed that high and medium high land where rain and floodwater do not stand are suitable for Tossa jute cultivation. Jute requires a warm and humid climate with temperature fluctuating between 24°C and 37°C. The permissible relative humidity favorable to growth ranges between 70% and 90%. Rainfall ranges from 250 mm to 270 mm are essential requisite for good growth and yield of jute. Proper seed rate is the main factor for plant population, growth, and for maximum yield. In terms of intercultural management, first weeding, mulching, and simultaneous thinning at 10–15 days after sowing, second weeding, mulching, and simultaneous thinning at 25–30 DAS, third weeding, mulching, and simultaneous final thinning and topdressing of urea fertilizer followed by hoeing 40–50 DAS were recommended. Besides, one tanabatch between 60–70 DAS and one katabatch around 90 DAS provide a good effect on fiber yield. The cropping patterns Jute-T. aman-Wheat, Potato-Jute-T. aman, Onion-Jute-T. aman, Vegetable-Jute-T. aman and Jute-T. aman-Mustard were sorted agronomically feasible and economically viable. The status of jute as a cash crop of Bangladesh is not at all satisfactory. Millions of people of Bangladesh depend on all affairs of the jute crop. Lack of proper government policy on jute, lack of production of jute, random closures of jute mills, failure to modernize the cultivation system and manufacturing units, mismanagement, and malpractice, fall of demand of jute in world market, use of alternative source to jute, etc. are found as problems in the development of jute fiber in Bangladesh. Proper Government policy can solve the problems in the jute sector of Bangladesh.

6.1   INTRODUCTION

Jute (*Corchorus* spp.) is now universally recognized that jute is the English version of the current Bengali word 'Pat,' a kind of fiber which is obtained from two species (annual and short day plants) of the genus *Corchorus* belonging to the family *Tiliaceae*. It is a common term used both for plant and the fiber obtained from the bark of the plants, *Corchorus capsularis* L. and *Corchorus olitorius* L. There are over 30 species, which belongs to the genus *Corchorus*. Jute *(Corchorus capsularis* and *Corchorus olitorius)*, Kenaf *(Hibiscus cannabinus)* and Roselle *(H. sabdariffavar (Altissima)* are vegetable bast fiber plants next to cotton in importance. In the trade, there are usually two names of jute, White, and Tossa. *Corchorus capsularis*is called White Jute and *Corchorus olitorius* is called Tossa Jute (Table 6.1). In India and Bangladesh Roselle is usually called Mesta. Jute fibers are finer and stronger than Mesta and are, therefore, better in quality. Depending on demand, price, and climate, the annual production of jute and allied fibers in the world remains around 3 million tonnes.

   The fiber finds its use in the producing as well as in consuming countries in the agricultural, industrial, commercial, and domestic fields. Sacking and Hessians (Burlap) constitute the bulk of the manufactured products. Sacking is commonly used as packaging material for various agricultural commodities viz., rice, wheat, vegetables, corn, coffee beans, etc. Sacking and Hessian Cloth are also used as packing materials in the cement and fertilizer manufacturing industries. Fine Hessian is used as carpet backing and often made into big bags for packaging other fibers viz. cotton and wool.

The long vegetable fiber jute plays a vital role in the economy of Bangladesh. Its yield, however, remained 1.5 to 2.07 tons ha^{-1} for the last two decades. During this period cultivated area under jute declined from 7% (Anonymous, 1969–70) total cultivable land to 4% (BBS, 1997; 2004) and as reported by FAO (2000) the total jute area was 355 thousand hectare with a total production of 630 thousand metric tons. The production of jute in India, China, Thailand, and Nepal has also showed a gradual decline during the period from 1990 to 2000 (FAO, 1993; 2000). Though in Bangladesh jute is grown by a large number of Jute farmers for their use and trade in order to earn cash, however, statistics of the last 30 years shows that the area has continuously fluctuated from year to year. Area ranged from 416 to 708 thousands hectare, whereas average area was 533.866 hectares during the years. In case of production of jute was lowest 711 and highest 1526 thousand metric tons in the year 1999–2000 and 2010–11 respectively which also fluctuated than the average production of jute 914.46 thousand metric ton during the years (Hussain, 2013; Islam, 2010).

Once, jute was one of the main exporting goods of Bangladesh. Bangladesh earned huge foreign currency by exporting jute. Before the 70s, until the ready-made garments appeared we earned a fabulous sum of foreign currency from jute. Owing to mismanagement and lack of foresight, we have already lost our golden age of jute. The present condition of jute as a cash crop in Bangladesh is very miserable. Why is it so? What are the reasons behind it? I have tried my level best to look into the above matter in my following study. Jute being the most important commercial crop plays a major role in our agriculture. The foreign exchange earnings from jute finance various development projects of Bangladesh. Jute also holds an important position in the industrial sector of the economy of Bangladesh. Jute is a versatile and environment-friendly biodegradable natural fiber widely grown in Asia, particularly in Bangladesh, India, and China. It is an important cash crop in Bangladesh and India, which together accounts for about 84% of the world production of jute fiber (JDPC, 2006).

Jute is a plant with many uses. All plants in the *Corchorus* genus are considered jute, although two have particular economic and culinary value, *C. olitorius* L. and *C. capsularis* L. The leaves of these plants are simple, and they may have slightly serrated edges. When harvested young, jute leaves are flavorful and tender; older leaves tend to be more woody and fibrous, making them less ideal for consumption.

Jute fiber is produced mainly from two commercially important species, namely White Jute (*Corchourscapsularis* L.), and TossaJute (*Corchorus olitorius* L.). The center of origin of white jute is said to be Indo-Burma including South China, and that of tossa Africa. The word jute is probably coined from the word *jhuta* or *jota*, an Orrisan word. However, the use of *juttapotta* cloth was mentioned in both the Bible and Monushanghita-Mahabharat.

Jute grows under wide variation of climatic conditions and stress of tropic and subtropics. Jute is as old as civilization and has been used in almost as many applications as one can imagine. This paper reviews history, chemical constituents, plant morphology and the most interesting studies on the various biological activities of jute (*Corchorus* spp).

TABLE 6.1   Taxonomy of Jute

Rank	Scientific Name and Common Name
Kingdom	Plantae – Plants
Subkingdom	Tracheobionta – Vascular plants
Superdivision	Spermatophyta – Seed plants
Division	Magnoliophyta – Flowering plants
Class	Magnoliopsida – Dicotyledons
Subclass	Dilleniidae
Order	Malvales
Family	Tiliaceae – Linden family
Genus	Corchorus L. – corchorus
Species	*Corchorus olitorius* L. – nalta jute
English name	*Corchorus capsularis* L. – white jute
Bengali name	Jute
	Pat, Paat, Naila, etc.

6.2   HISTORY OF JUTE

Jute, the Golden Fiber, carries a glorious history in the packaging sector as well as in the economy of Bangladesh. Over the years, jute has been stretching its field to wide and diverse areas and thus has acquired a multi-dimensional role on consumer market despite facing strong competition from the synthetic sector. Jute is one of the most versatile fibers known to man. Raw jute fiber is obtained from two varieties of plant: Corchorus Capsularis (White jute) and Corchorus Olitorius (Tossa jute), both native to Bangladesh.

Jute is a rain-fed crop with little need for fertilizer or pesticides. The production is concentrated in Bangladesh and some in India, mainly Bengal. The jute fiber comes from the stem and ribbon (outer skin) of the jute plant. The fibers are first extracted by retting. The retting process consists of bundling jute stems together and immersing them in low, running water. There are two types of retting: stem and ribbon. After the retting process, stripping begins. Women and children usually do this job. In the stripping process, non-fibrous matter is scraped off, then the workers dig in and grab the fibers from within the jute stem. India, Pakistan, China are the large buyers of local jute while Britain, Spain, Ivory Coast, Germany, and Brazil also import raw jute from Bangladesh.

During the 19th and early 20th centuries, jute was indispensable. Its uses included sacking bag, ropes, boot linings, aprons, carpets, tents, roofing felts, satchels, linoleum backing, tarpaulins, sandbags, meat wrappers, sailcloth, scrims, tapestries, oven cloths, horse covers, cattle bedding, electric cable, even parachutes. Jute's appeal lay in its strength, low-cost, durability, and versatility.

Today, jute can be defined as an eco-friendly natural fiber with versatile application prospects ranging from low-value geo-textiles, hand/shopping bags to high-value carpet, apparel, composites, decorative upholstery, furnishings, fancy non-woven for new products, decorative color boards, and many more such jute diversified products.

6.2.1   WHITE JUTE (CORCHORUS CAPSULARIS)

Several historical documents during the era of Mughal Emperor Akbar (1542–1605) state that the poor villagers of India used to wear clothes made of jute. It also states that Indians, especially Bengalis, used ropes and twines made of white jute from ancient times for various uses.

6.2.2   TOSSA JUTE (CORCHORUS OLITORIUS L.)

Tossa jute (*Corchorus olitorius* L.) is an Afro-Arabian variety. It is quite popular for its leaves that are used as an ingredient in a mucilaginous potherb called molokhiya, popular in certain Arab countries. The Book of Job in the Hebrew Bible mentions this vegetable potherb as Jew's mallow. Tossa jute fiber is softer, silkier, and stronger than white jute. This variety astonishingly showed good sustainability in the climate of the Ganges Delta. Along with white jute, Tossa jute has also been cultivated in the soil of Bengal from the start of the 19th century. Currently, Bangladesh is the largest global producer of the Tossa jute variety.

6.3 THE HISTORICAL BACKGROUND OF JUTE

Also known as, the Golden fiber, Jute is a natural fiber comprised of silky and golden shine. It is one of the most cheapest and economical vegetable fiber after cotton, obtained from the skin or bast of plant's stem. Recyclable, 100% biodegradable and eco-friendly jute has low extensibility and high tensile strength. Jute is the versatile natural fiber widely used as a raw material in many textile, non-textile, packaging, construction, and agricultural applications.

6.3.1 JUTE IN ANCIENT TIMES

Jute has been used since ancient times in Africa and Asia to provide cordage and weaving fiber from the stem and food from the leaves. In several historical documents (Ain-e-Akbari by Abul Fazal in, 1590) during the era of the great Mughal Emperor Akbar (1542 –1605) states that the poor villagers of India used to wear clothes made of jute. The weavers, who used to spin cotton yarns as well, used simple handlooms and hand spinning wheels. History also states that Indians, especially Bengalis, used ropes and twines made of white jute from ancient times for household and other uses.

Chinese papermakers from very ancient times had selected almost all the kinds of plants as hemp, silk, jute, cotton, etc. for papermaking. QiuShiyu, the researcher of the Harbin Academy of Sciences and expert on Jin history, concluded that Jews used to take part in the work of designing "jiaozi, "made of coarse jute paper. A small, piece of jute paper with Chinese characters written on it has been discovered in Dunhuang in Gansu Province, in north-west China. It is believed it was produced during the Western Han Dynasty (206 BC–220 AD).

6.3.2 PERIOD FROM 17TH CENTURY

The British East India Company was the British Empire Authority delegated in India from the 17th century to the middle of 20th century. The company was the first Jute trader. The company traded mainly in raw jute during the 19th century. During the start of the 20th century, the company started trading raw jute with Dundee's Jute Industry. This company had monopolistic access to this trade during that time. Margaret Donnelly, I was a jute mill landowner in Dundee in the 1800s. She set up the first jute mills in India. The Entrepreneurs of the Dundee Jute Industry in Scotland were called The Jute Barons.

In 1793, the East India Company exported the first consignment of jute. This first shipment, 100 tons, was followed by additional shipments at irregular intervals. Eventually, a consignment found its way to Dundee, Scotland where the flax spinners were anxious to learn whether jute could be processed mechanically.

Starting in the 1830's, the Dundee spinners learned how to spin jute yarn by modifying their power-driven flax machinery. The rise of the jute industry in Dundee saw a corresponding increase in the production and export of raw jute from the Indian sub-continent, which was the sole supplier of this primary commodity.

Period 1: The ancient period from the year 1855, during the era of great Mughal emperor Akbar, poor villagers of India were used to wear jute clothes. Since ancient times, ropes, and twines, used by Bengali Indians are made up of white jute for varied household applications. In addition, Chinese paper makers have used all forms of plants like jute, hemp, cotton to make paper.

6.3.3   PERIOD FROM 1855

Calcutta (now Kolkata) had the raw material close by as the jute growing areas were mainly in Bengal. There was an abundant supply of labor, ample coal for power, and the city was ideally situated for shipping to world markets. The first jute mill was established at Rishra, on the River Hooghly near Calcutta in 1855 when Mr. George Acland brought jute-spinning machinery from Dundee. Four years later, the first power-driven weaving factory was set up.

By 1869, five mills were operating with 950 looms. Growth was rapid and, by 1910, 38 companies operating 30,685 looms exported more than a billion yards of cloth and over 450 million bags. Until the middle 1880's, the jute industry was confined almost entirely to Dundee and Calcutta. France, America, and later Germany, Belgium, Italy, Austria, and Russia, among others, turned to jute manufacturing in the later part of the 19th century.

In the following three decades, the jute industry in India enjoyed even more remarkable expansion, rising to command leadership by 1939 with

68,377 looms, concentrated mainly on the River Hooghly near Calcutta. These mills alone have proved able to supply the world demand.

The earliest goods woven of jute in Dundee were coarse bagging materials. With longer experience, however, finer fabrics called burlap, or hessian as it is known in India were produced. This superior cloth met a ready sale and, eventually, the Indian Jute Mills began to turn out these fabrics. The natural advantage these mills enjoyed soon gave Calcutta world leadership in burlap and bagging materials and the mills in Dundee and other countries turned to specialties, a great variety of which were developed.

Period 2: The period from the 17th century to the middle of the 20th century, the British Empire authority was delegated by the British East India Company which was the first jute trader. This company traded the raw jute. During the start of 20th century, Margaret Donnelly I was a jute mill landowner in Dundee who had set up first jute mill in India. East India Company exported the first consignment of Jute in the year 1793. In the country, Scotland, flax spinners were trying to learn whether jute could be mechanically processed. At the beginning of the year 1830, Dundee spinners have determined to spin of Jute yarn by transfiguring their power-driven flax machinery. The major jute growing areas were mainly in Bengal at the Kolkata side. When Mr. George Acland had brought jute-spinning machinery from Dundee to India, the first power-driven weaving factory was established at Rishra, on the River Hooghly near Calcutta in the year 1855. By the year 1869, five mills were established with around 950 looms. The growth was so fast that, by the year 1910, 38 companies were operating around 30,685 looms, rendering more than a billion yards of cloth and over 450 million bags. Until the middle of the year 1880, the jute industry has acquired almost whole of Dundee and Calcutta. Later in the 19th century, manufacturing of jute has started in other countries also like in France, America, Italy, Austria, Russia, Belgium, and Germany.

6.3.4 JUTE INDUSTRY AFTER 1947

After the fall of British Empire in India during 1947, most of the Jute Barons started to evacuate India, leaving behind the industrial setup of the Jute Industry. Most of the jute mills in India were taken over by the Marwaris business persons.

Period 3: The period from the 19th century till 1947, outstanding expansion in jute industry has been noticed in the 19th century. Throughout the year 1939, around 68,377 looms were established on the River Hooghly near Calcutta. The prime commodities woven by jute are coarse bagging materials, produced by finer fabrics also known as hessian or burlap. The handlooms established in Calcutta, give this place world-class leadership in burlap and other bagging materials.

In East Pakistan after partition in 1947 lacked a Jute Industry but had the finest jute fiber stock. As the tension started to rise between Pakistan and India, the Pakistani felt the need to setup their own Jute Industry. Several groups of Pakistani families (mainly from West Pakistan) came into the jute business by setting up several jute mills in Narayanganj of then East Pakistan, the most significant ones are: Bawanis, Adamjees, Ispahanis, and Dauds. After the liberation of Bangladesh from Pakistan in 1971, most of the Pakistani owned Jute Mills were taken over by the government of Bangladesh. Later, to control these Jute mils in Bangladesh, the government built up Bangladesh Jute Mills Corporation (BJMC).

Period 4: The period after the year 1947, after getting Independence, most of the Jute barons had started to quit India, leaving the setup of jute mills. Marwaris businesspersons took most of them. During the year 1947, after the partitioning, East Pakistan had the finest stock of jute. The tension had already begun between India and Pakistan, now Pakistani people felt the need of jute industry. From then onwards, different groups of Pakistani families have joined the jute business by establishing many mills in Narayanganj. The Pakistani were in general, Bawanis, Adamjees, Ispahanis, and Dauds. In the year 1971, the liberation of Bangladesh took place from Pakistan, thus most of the jute mills were taken over by the Bangladesh government. Later, the government had built BJMC (Bangladesh Jute Mills Corporation) to control and handle jute mills of Bangladesh.

6.4   DESCRIPTION OF JUTE CROP

Annual, much-branched herb 90–120 cm tall; stems glabrous. Leaves 6–10 cm long, 3.5–5 cm broad, elliptic-lanceolate, apically acute or acuminate, glabrous, serrate, the lower serratures on each side prolonged into a filiform appendage over 6 mm long, rounded at the base, 3–5 nerved; petioles 2–2.5 cm long, slightly pubescent, especially towards the apex; atipulessubulate, 6–10 mm long. Flowers pale yellow; bracts lanceolate; peduncle shorter than the petiole; pedicles 1–3, very short. Sepalsca 3 mm long, oblong, apiculate. Petals 5 mm long, oblong spathulate. Style short; stigma microscopically papillose. Capsules 3–6.5 cm long, linear, cylindric erect, beaked, glabrous, 10-ribbed, 5-valved; valves with transverse partitions between the seeds. Seeds trigonous, black (Kirtikar and Basu, 1975).

6.4.1 GEOGRAPHICAL DISTRIBUTION

Jute is grown in Bangladesh, India, Myanmar, Nepal, China, Taiwan, Thailand, Vietnam, Cambodia, Brazil, and some other countries. Bangladesh used to enjoy almost a monopoly of this fiber commercially. Although jute is grown in almost all the districts of Bangladesh, Faridpur, Tangail, Jessore, Dhaka, Sirajganj, Bogra, and Jamalpur are considered the better growing areas. Total area under the crop cultivation of Bangladesh in the year 2010 was 7.08 lac ha and the total production was 84 lac bales.

6.4.2 FEATURES OF JUTE

Jute fiber is 100% biodegradable and recyclable and thus environment-friendly. They are possibly the world's largest source of lingo-cellulosic bast fibers, which is extracted from plants by a natural microbial process known as retting (Pan et al., 2000; Roy et al., 2002; Mohiuddin et al., 1987).

- It is a natural fiber with golden and silky shine and hence called *The Golden Fiber.*
- It is the cheapest vegetable fiber procured from the bast or skin of the plant's stem.
- It is the second most important vegetable fiber after cotton, in terms of usage, global consumption, production, and availability.
- It has high tensile strength, low extensibility and ensures better breathability of fabrics. Therefore, jute is very suitable for agricultural commodity bulk packaging.
- It helps to make the best quality industrial yarn, fabric, net, and sacks. It is one of the most versatile natural fibers that have been used in raw materials for packaging, textiles, non-textile, construction, and agricultural sectors. Bulking of yarn results in a reduced breaking tenacity and an increased breaking extensibility when blended as a ternary blend (Basu et al., 2005).
- The jute plant is derived from a relative of the hemp (Cannabis) plant. However, jute is very free from narcotic elements or odor.
- The best source of jute in the world is the Bengal Delta Plain in the Ganges Delta, most of which is occupied by Bangladesh.
- Advantages of jute include good insulating and antistatic properties, as well as having low thermal conductivity and moderate moisture

regain. Other advantages of jute include acoustic insulating properties and manufacture with no skin irritations (Pan et al., 2000).

- Jute has the ability to be blended with other fibers, both synthetic
  and natural, and accepts cellulosic dye classes such as natural, basic,
  vat, sulfur, reactive, and pigment dyes. As the demand for natural
  comfort fibers increases, the demand for jute and other natural
  fibers that can be blended with cotton will increase (Sreenath et
  al., 1996; Basu et al., 2005). The resulting jute/cotton yarns will
  produce fabrics with a reduced cost of wet processing treatments.
  Jute can also be blended with wool. By treating jute with caustic
  soda, crimp, softness, pliability, and appearance is improved, aiding
  in its ability to be spun with wool. Liquid ammonia has a similar
  effect on jute, as well as the added characteristic of improving flame
  resistance when treated with flame-proofing agents (Basu et al.,
  2005; Pan et al., 2000).
- Some noted disadvantages include poor drapability and crease
  resistance, brittleness, fiber shedding, and yellowing in sunlight.
  However, preparation of fabrics with castor oil lubricants result
  in less yellowing and less fabric weight loss, as well as increased
  dyeing brilliance. Jute has a decreased strength when wet, and
  also becomes subject to microbial attack in humid climates. Jute
  can be processed with an enzyme in order to reduce some of its
  brittleness and stiffness. Once treated with an enzyme, jute shows
  an affinity to readily accept natural dyes, which can be made from
  marigold flower extract. In one attempt to dye jute fabric with this
  extract, the bleached fabric was mordanted with ferrous sulfate,
  increasing the fabric's dye uptake value. Jute also responds well
  to reactive dyeing (Chattopadhyay et al., 2004). This process is
  used for bright and fast colored value-added diversified products
  made from jute.
- Dioxaneacidolysis lignin of jute stick was isolated. Jute seed cake
  was low in protein and high in lysine, isoleucine, and fiber content
  (Ahmed et al., 2001).

6.4.3 JUTE PLANT

Jute fiber is obtained from two closely related, annual herbaceous species,
Corchorus cpasularis L. and *Corchorus olitorius* L., belonging to the family
Tiliaceae. The genus corchorus includes about 40 species distributed mostly

in tropical regions. The largest number of species is found in Africa, while only eight occur in the Indo-Pakistan area (Kundu, 1951).

6.4.3.1   CORCHORUS CAPSULARIS L. DESHI JUTE

It is generally 1.5 to 3.7mtall, can withstand waterlogging in later stages, herbaceous annual, 3–5-month duration for fiber yield depending upon sowing time. It`s stem cylindrical, green to dull coppery red to pink, periderm in the basal portions in later stages.

Its plants may be branched or unbranched, axillary buds may or may not develop into branches.

Leaves are glabrous, 5–13 cm by 2.5–8.2 cm length and breath. Ovate oblong, acuminate, coarsely toothed, lowermost pair of serrations enlarged and ends in hairy appendages. Petiole 4–8 cm, various from green to pink in different varieties, stipules 0.5–2.0 cm or more, foliaceous in some varieties, tip colored or green.

Flowers are small generally in extra-axillary cymes in groups of 2–5 or more; 0.3–0.5 cm long and 0.5–0.6 cm wide; sepals 5, colored or green, petals 5, yellow or pale yellow, stamens 20–30; anther yellow to pale yellow, ovary rounded, 5-carpelled, syncarpous, ovals axile, usually 10 in each locales in 2 rows, giving about 50 ovules in each ovary; style 2–4 mm; stigma pubescent. Anthes is one or two harms after sunrise.

Capsules rounded, 1.0–1.5 cm in diameter, wrinkled, rarely smooth, muricate, 5-locular, seed- 7–10 in 2 rows in each ocular, without transverse partitions, 30–50 in each fruited seeds are small, chocolate brown, 4–5 faced about 300 per gram.

6.4.3.2   C. OLITORIUS L. (TOSSA JUTE)

It plants generally 1.5–4.5 m tall; cannot withstand waterlogging; herbaceous annual, 4–5 months duration depending on the time of sowing, flowers prematurely if sown very early.

Stems cylindrical, green, light or dark red, fewer shades of color than *capsularis;* no priderm but lenticules in later stages. Its plant branched, but branches normally develop less vigorously.

Leaves are glabrous, 7–18 cm by 4–89 cm, oblong, acuminate coarsely toothed, lowermost pair of serrations more enlarged than in *capsularis* and hairy appendages longer, Petiole 4–9 cm, various from green to dark red, stipules 0.5–1.5 cm or slightly more, tip colored or green, base colored except in the full green types. Flowers generally in extra-axillary cymes in groups of 2–5, about 1 cm in length, 2–2.5 times the size of *capsularis* flower, sepals 5–6, colored or green, tips prolonged in flower buds. Pleats 5–6, yellow, entire or split. Stamens 30–60, anthers yellow, ovary inengated, 5-varely 6-corpelled, syncarpous, ovules axile, usually 40 in each locale in 1 row giving about 200 ovules in each ovary, style 3–5 mm, stigma globular, entire, pubescent.

Anthesis is for an hour or less before sunrise. Capsule elongated, 6–10 cm long, 0.3–0.8 cm in diameter, ridged lengthwise, 5–6 locular, seed 25–40 in a single row in each loculus, with transverse partitions between each seed and 140–200 in each fruit. Seeds are smaller than those of *capsularis*, green to steel grey to even black, about 500 per g.

6.4.4 JUTE LEAVES: DESCRIPTIONS OF DIFFERENT SPECIES

i.   *Tossa jute (Corchorus olitorius L.):* It is an annual or biennial herb, erect, stout, branched, to 1.5 m high; rootstock woody. It's leaves lanceolate to ovate-lanceolate, subobtuse at the base, serrate at the margin with basal-most serrations extending into filiform processes, acute at apex, glabrous except sparsely hairy nerves, 3–5 nerved; petioles 2–3 cm long, pubescent; stipules subulate, 8–12 mm long, glabrous.

ii.   *White jute (Corchorus capsularis L.):* It is annual, much branched, spreading herbs; stems pilose, often reddish. It's leaves 2.5–7 x 1.5–3.5 cm, ovate to elliptic-lanceolate, base rounded, margins serrate, the basal pair of serrations ending in setae or not, apex acute, basally 3–5-nerved; petioles up to 3 cm long; stipules 4–8 mm long, setaceous (Figure 6.1).

There are so many jute varieties of both White and Tossa jute are developed by BJRI. Namely White Jute: D–154, CVL–1, CVE–3, CC–45, BJRIDeshi Pat–5, BJRIDeshi Pat–6, BJRIDeshi Pat–7 and BJRIDeshi Pat–8. Leaf of White jute is full green, stipule green, petiole green, leaf ovate-lanceolate, leaf length-breadth ratio 2:1 and Tossa Jute: O–4, O–9897, OM–1, BJRITossa Pat–4, BJRITossa Pat–5 and BJRITossa Pat–6. Leaves of Tossa jute are fully deep green, ovate. Leaf length-breadth ratio 2.7:1. Very recently, BJRI developed one vegetable jute variety named Vegetable Jute.

White Jute (CVL-1) leaf   **Tossa Jute (O-9897) leaf**

FIGURE 6.1   Tossa and white jute varieties of Bangladesh.

6.4.5   FUTURE PROSPECT OF JUTE LEAF AS VEGETABLE AND MEDICINE

There are so many jute and allied fiber crops varieties developed, released, and used at farmers' level for commercial cultivation. All those varieties leaves have both vegetable and medicinal values. Jute leaf has long been used as a remedy in many cultures. The jute leaf contains over 17 active nutrient compounds including many minerals, amino acids, and vitamins. Today, this multi-utility versatile plant part is considered to cure Mankind's different health problems. Jute leaf contains protein, calories, fibers, and as well as antitumor promoters; Phytol and Monogalactosyldiacylglycerol. At present, it is very important to an extension of the information and aware the people about this leaf's vegetable and medicinal use. In Bangladesh, both jute species are seasonally used as vegetables. Jute leaves have also prospect for export to some developed and developing countries across the globe. Bangladesh has an approximated prospect for exporting dry jute leaves to the tune of 0.18 to 0.25 million metric tons based on the present level of jute production. This figure may go up depending upon the research and development efforts and the efficiency of venturing market potentials of the world.

Jute is a fiber crop belongs to genus *Corchorus* of the Tiliaceae family with two cultivated species-*Corchorus capsularis* L. and *C. olitorius* L. Fiber is extracted from the bark of the plants. In 1970–80 decades about 15–16 lakh hectare of the total cultivable land was occupied by jute has now (2010–11) been reduced to about 6.00 to 6.76 lakh hectare. However, the national average yield is increased from 1.59 to 1.98 tons per hectare. It is happened due to use of high yielding jute varieties and production technologies, which together contributed toward higher yield. Improved technologies such as improved seed, line sowing, recommended fertilizer and plant protection measures increase the fiber yield of jute by 20, 23, 27 and 13%, respectively over conventional practices. Jute is still contributing about 4% GDP to the national economy and earns about 5% of foreign exchange as well.

6.5 CULTIVATION OF JUTE

6.5.1 SOIL AND CLIMATE

Bangladesh forms the largest delta in the world and is situated between 88.50° and 92°50′ East longitudes and between 20°50′ and 26°50′ north. Tropical monsoon rain, drench the land and the rivers. The topography of the lands has been divided into high land, medium land, medium low land, and low land, very low land and hilly land. The new grey alluvial soil, good depth receiving silt from the annual floods, is nature's best gift, but jute is widely grown in sandy loams, clay loams with varying soil management practices. Sandy soils and heavy clay soil are unsuitable for jute production at the same time soil with a low pH give a poor crop, the optimum pH being around 6.4. High and medium high land where rain and floodwater do not stand is suitable for Tossa jute cultivation. In the seedling stage, waterlogging is not tolerated by both species. To concern soil type, silt is the best for jute and allied crop production. However, it can be grown in any intermediate soil other than the extreme sand and extreme clay soil. The extreme sand soil is not suitable, because, its water holding capacity is very poor and jute plants suffer from moisture stress in this soil. In the extreme clay soil, proper root growth cannot take place due to the hardness of soil, which ultimately affects plant growth. Jute requires a warm and humid climate with temperature fluctuating between 24°C and 37°C. The permissible relative humidity favorable to growth ranges between 70 and

90%. Rainfall is one of the most important factors for growing jute and the ranges from 250–270 mm are essential requisite for good growth and yield of jute (Islam and Rahman, 2008).

6.5.2 LAND PREPARATION AND FERTILIZATION

Jute seed requires fine tillage for its proper germination, as its seeds are very small. A number of plows also varies depending on soil type. If it is of light soil, three plows may be enough. For heavy soil increased number of plows must be needed. The main point is that the soil is to bring to fine tillage by plow and harrowing. A number of plows may vary from 3–6 depending on the texture of the soil. During land preparation weeds and plant, debris is to be collated and removed from the land. Otherwise, the crop field will be heavily infested by weeds, which involve excess weeding cost. Jute field also needs proper leveling of soil. During land preparation, the land should be supplied organic fertilizer before 2–3 weeks of seed sowing. This also may be applied at the time of opening the land. It should always remember, the soil of Bangladesh is miserably deficient in organic matter, which plays a vital role for improving the physical health and nutrient elements of soil available to the plants. Most of the cultivable lands of Bangladesh are certainly deficient in organic matter. The jute field soil is to replenish with the supplement of organic matter during land preparation. Bangladesh soil is deficient in six elements in its present condition. Those are N, P, K, S, Zn, and Mg. Due to intensive cultivation, Bangladesh soil is getting deficient with more nutrient elements. The following chart indicates the emergence of nutrient deficiencies on a time scale in Bangladesh. At the present situation, jute crop for its potential yield requires N, P, K, Zn, S, and Mg nutrient elements. Time and amount of fertilizer application for jute fiber production in medium soil fertility and medium production target are given in Table 6.2.

6.5.3 PLANTING TIME OF FIBER

Jute is a photoperiod sensitive and short day plant. Any early planting provides premature flowering, reduces plant growth and yield of fiber. It accomplishes its vegetative growth in the long days and induces flowers in the short days. In Bangladesh condition, day length goes above 12 h from 22 March and 12.50 h from 14 April. Therefore, all the varieties, except

TABLE 6.2 Time and Amount of Fertilizer Application of Jute Fiber Production (Medium soil Fertility and Medium Production Target. kg/ha)

Variety	Dry cow dung 2–3 weeks before sowing	Chemical fertilizer (Amount and time of application)					
		Urea		TSP	MP	Gypsum	Zinc sulfate
		At sowing date	After 45 days of sowing	At sowing date			
White jute: CVL–1, CC–45, CVE–3, BJC–7370, BJC–83 Tossa jute: O–4	*	83	83	25	30	45	11
O–9897 (Tossa jute)	*	100	100	50	60	95	11
OM–1 (Tossa jute)	*	88	88	50	40	95	11
O–72 (Tossa jute)	*	83	83	50	40	95	11
White jute: CVL–1, CC–45, CVE–3, BJC–7370, BJC–83 Tossa jute: O–4	5000	27	88	-	-	-	-
O–9897 (Tossa jute)	5000	45	100	-	10	50	-
OM–1 (Tossa jute)	5000	33	88	-	-	50	-
O–72 (Tossa jute)	5000	61	61	-	10	50	-

O–4, can be safely planted after 22 March and O–4 to be planted after 14 April, because, O–4 is more sensitive to photoperiod. Sowing should be ensured within a month thereafter. However, some new varieties such as CC–45, BJC–7370, BJC–83, BJC–2142, O–9897, OM–1, O–72 and O–795 can be planted 15 days earlier without any fear from premature flowering, but that is not suggested, since growth of jute plant is also related to thermal sensitivity. It provides luxuriant growth under a temperature between 30°C to 40°C and relative humidity over 70%, and growth rate becomes stunted gradually when the temperature comes below 30°C. So planting too early obviously brings no beneficial effect rather it involves extra cost due to cultural management, nutrient, and soil moisture. To concern *C. capsularis* L. varieties planting around 30 March register optimum fiber yield and provides room for other crops to fit into 3-crop pattern. However, some recently developed high yielding varieties can be planted on one month ahead. To concern *C. olitorius* L. varieties, the conventional variety like O–4 (less photoperiod sensitive), planting should be done after 14 April. However, varieties are to be planted starting from 15 March, will help incorporate Tossa jute into three crop pattern. Released year, date of sowing and yield at farmers field of some important jute varieties developed by BJRI is given in Table 6.3.

TABLE 6.3 Released Year, Date of Sowing and Yield at Farmers Field of Some Important Jute Varieties Developed by BJRI

Name of variety	Released year	Date of sowing	Yield at farmers field (t/ha)
White jute (*Corchorlus capsularis* L.)			
1. CVL–1	1977	30 March–15April	5.16
2. CVE–3	1977	30 March–15April	4.52
3. BJRIDeshi pat–5 (BJC–7370)	1995	15 March–15April	3.25
4. BJRIDeshi pat–6 (BJC–83)	1995	30 March–15April	3.00
5. BJRIDeshi pat–7 (BJC–2142)	2007	March–June	2.75
6. BJRIDeshi pat–8 (BJC–2197)	2013	25 March- 25 April	3.15
Tossa jute (*Corchorus olitorius* L.)			
1. O–9897	1987	30 March–30 April	4.67
2. BJRITossa pat–3 (OM–1)	1995	20 March–30 April	4.50
3. BJRITossa pat–4 (O–72)	2002	15 March–30 April	4.81
4. BJRITossa pat–5 (O–795)	2008	15 March–30 April	3.40
5. BJRITossa pat–6 (O–3820)	2013	End March- April	3.25

Akter et al. (2009) reported that it was observed less photosensitive than the others having a wider sowing range from 14 March to April. Agronomic trials proved no early flowering occurred even seeding could be done one week before 14 March and crop could be harvested at 110 days of field duration to fit the crop in the three-cropped pattern. Potential yield of the variety O–795 was about 5.0 tha^{-1} by maintaining the plant population 3.5–4.0 lac. ha^{-1}, however in farmers field the average yield was observed 3.4 tha^{-1}, was calculated 10% higher than the check cv. O–72.

Khatun et al. (2009) reported that it was observed less photosensitive than the check variety CVE–3, having a wider sowing range from the third week of March to Mid-April. Agronomic trials proved the absence of premature flowering in appropriate sowing and harvested on 110 days of field duration to fit the crop in the three-cropped pattern. The variety BJC–2142 is suitable for Faridpur jute growing zone. Potential yield of the variety was observed 4.0 tha^{-1}, maintaining the plant population 3.50–4.50 lac.ha^{-1}, however in farmers field the average yield was observed 2.50 tha^{-1}, was calculated 3.24% higher than cv. CC–45 and 9.33% higher than cv. CVE–3.

If jute is sown untimely the growth is retarded and yield is reduced. When sown earlier, premature flowering occurs and yield is drastically reduced with deterioration in quality. Generally, the sowing of *capsularis* varieties starts from late February to April in low lying areas that retain moisture of the previous flood or monsoon. On the other hand, in the case of *olitorius* jute seeds are sown from 15 March to April.

6.5.4   SEEDING RATE

To get optimum plant population and desired yield, seeding rate was optimized and found that seeds having 80% viability, 5 kgha^{-1} of Tossa jute, 7 kgha^{-1} of deshi and 12–15 kgha^{-1} of Kenaf and Mesta seeds could offer desired population and optimum yield. For seed crop seed rate depends on the soil condition and planting methods followed. As a thumb rule, for Deshi jute 4.0–4.5 kg/ha in line sowing and 5.0–5.5 kg/ha in broadcast sowing should to be maintained. On the other hand for Tossa jute 3.0–3.5 kg/ha in line sowing and 4.0–4.5 kg/ha in broadcast sowing should be maintained. As jute seeds have no other alternative uses, it is to sow even if it is of 50% viability by adjustment the seed rate.

Proper seed rate is the main factor for plant population, the growth of jute plant and for obtaining maximum yield. Quality seeds of an improved variety itself provide 20% additional yield of the crop (Hossain et al., 1994).

Farmers commonly use 7–8 kg/ha for *C. capsularis* and 5–6 kg/ha for *C. olitorius* (tossa). For broadcast seeding, seed rates are the flowing, which has the germinability of seed being 80% or above. For *C. capsularis* (white), 7.0 to 9.0 Kg per hectare and for *C. olitorius* (tossa), 5.0 to 7.0 Kg per hectare. The seed rate in the case of line sowing depends on the spacing and a plant-to-plant distance. The seed rate is recommended for line sowing were for *C. capsularis* (white), 6.0 to 7.0 Kg per hectare and for C. *olitorius* (tossa), 4.0 to 5.0 Kg per hectare.

6.5.5   INTERCULTURAL OPERATIONS

Timely weeding and thinning are a most important operation to obtain a good establishment and healthy crop. First weeding, mulching, and simultaneous thinning at 20–25 days after sowing (DAS), second weeding, mulching, and topdressing of urea fertilizer followed by hoeing at 40–45 DAS, thinning by tanabach at 60–70 DAS, thinning by kata bach at 85–90 DAS and, entomological, and pathological care as and when necessary are most effective for higher yield. The highest plant height, base diameter, fiber, and stick yield were produced when improved seed, line sowing method, the recommended dose of fertilizer and plant protection measures were applied (Table 6.4).

TABLE 6.4   Intercultural Operation That are to be Attended in Jute Cultivation

Intercultural operation	Time of operation
a. First weeding, mulching and simultaneous thinning	20–25 DAS*
b. Second weeding, mulching, and topdressing of urea fertilizer followed by hoeing	40–45 DAS
c. Tanabach	60–70 DAS
d. Kata bach	85–90 DAS
e. Entomological care	When necessary
f. Pathological care	When necessary

*DAS = Days after sowing

Weed possess a major problem in jute cultivation and weeding operation constitute about one-third of the total cost of production. Some of the weeds are found major in jute field are listed here. *Cyperus rotundus, Cynodon dactylon, Echinocloa colonum, Digitarias angunalis,*

Cyperusiria, Eleusine indica, Panicum disticum. In Deshi jute Khudesama, Mutha, Durba, Fuskabegun, Katanotey, Anguli, Knotgrass/Gitlaghas, Gaicha, Nuniasak, Matichech, and Helenchawere infested in deshi jute growing areas. Among the growing weeds 6 major species at Manikganj, 5 at Kishoreganj and Chandina were observed. Manikganj, Kishoreganj, and Chandina jute fields were highly infested by Mutha which was followed by Durba. In Tossa jute the weed species viz., Khudesama, Mutha, Basketgrass, Durba, Fuskabegun, Knotgrass/Gitlaghass, Chechra, Chapra, Nuniasak, Shialleza, Gaicha, Shaknotey, Chanchi, and Hajardana were found. Among the infested weeds 8 major species were observed at Monirampur, 10 at Faridpur and 4 at Rangpur station. Jute field of Monirampur station was highly infested by Khudesama whereas Faridpur and Rangpur were infested by Mutha. It was calculated that twenty two type of weeds viz., Khudesama, Basketgrass, Mutha, Durba, Fuskabegun, Katanotey, Shaknotey, Chapra, Shialleza, Knotgrass/Gitlaghass, Borododhia, Sutododhia, Chechra, Hajardana, Gaicha, Chanchi, Nuniasak, Helencha, Anguli, Matichech, Borosama, Angtagrown which were identified at both deshi and tossa jute growing areas. Presently, Jute has identified as a major segment for herbicide use. Only after 2009 herbicide application was started mostly in Jute of Bangladesh; at present 20 products of only Fenoxaprop-ethyl, Metamifop, and Quizalofop-p-ethyl have been recommended for weed control for Jute crop, which was mostly using against lower infesting weeds. In jute cultivation weeding and labor, the requirement is very important in terms of cost. An approximate number of labor requirements for jute cultivation are given in Table 6.5.

TABLE 6.5   Labor Requirement in Different Operations of jute Cultivation

Operations	Labor requirement	
	Hectare	Country total (Million)
Land preparation	33.53	18.56
Seed sowing	2.56	1.41
Fertilizer and manuring	2.95	1.63
Weeding and thinning	81.30	45.00
Harvesting and carrying	41.64	22.69
Binding, retting, stripping, and washing	71.72	39.69
Drying and storing	19.78	10.95
Total	253.48	139.95

6.6   HARVESTING OF JUTE

6.6.1   HARVESTING TIME OF JUTE FIBER

Harvesting of jute usually begins from the middle of July and continues up to end of September depending on the time of sowing and the characteristics of the variety grown. Different results of the investigations reveal that fiber yield increases with the increases of field duration until 140 DAS, but fiber quality degrades. Delay the harvest increases the fiber yield but produced coarser fiber. On the other hand, harvest the crop too early result in a lower yield and weakened fiber. To make a compromise between yield and quality, field duration and harvest time of both jute species has been accorded and fixed on 120 days after sowing. On average tossa jute varieties could be harvest 10 days before than deshi jute varieties for good quality and as well as optimum yield.

6.6.2   IMPROVED METHOD OF JUTE RETTING

Jute retting is a process by which jute fibers are extracted from the jute plants, *Corchorus Capsularis*, and *Corchorus outorius*. It constitutes a very important part of the whole process of raw jute fiber production. The quality of jute fiber greatly depends on this process. Fiber strength, specky areas, runners, unretted or barky fibers, cuttings in the basal pans, entangled woody cores, dark or shamla color and presence of dirt or dust are some of the characters which are considered in assessing the quality of Jute fiber.

Bangladesh Jute Research Institute recommended the following ideal location specific jute retting methods for the production of quality fiber at the farm level. Field service personnel for mass adoption by the jute growers are disseminating the methods.

6.6.3   SELECTION OF LAND FOR JUTE CULTIVATION

Land for jute cultivation having facilities for retting water nearby should be preferably selected so that there may not be any problem of retting water for jute retting after the harvest.

6.6.4   HARVEST TIME AND JUTE QUALITY

Quality of jute fiber has a direct relationship with the stage of harvest. It was observed that if the plants were harvested just before the flowering quality of fiber improved but the- yield decreased a little bit. On the contrary, if the plants were harvested during the fruiting stage, yield increased but the quality deteriorated. However, little loss in yield because of the early harvest was compensated by the higher value of the improved quality fiber. The plants should be harvested at the early flowering stage in order to obtain good quality fiber is really in greater demand.

6.6.5   ASSORTING

While harvesting, the thick, thin, small, and big jute plants should be assorted and bundled separately. The bundle weight should normally be about 10 kg. The bundles of thick, thin, long or small plants should be retted separately in separate jaks (rets). The bundles should be tied loosely. Retting takes longer time if the bundles are tied tightly. The retting microbes and water cannot enter easily in tight bundles,

6.6.6   DEFOLIATION

The bundled jute plants should be kept stacked for 3–4 days in the jute field for defoliation. During this period the leaves defoliate and the plants become ready for next operation.

6.6.7   MALLETINGOR IMMERSION OF THE BOTTOM PARTS UNDERWATER

After defoliation, the bottom parts of the jute plants (45 cm from the bottom) should be immersed underwater for 4–5 days or they may be malleted by a wooden beater. After this operation, the bundles should be steeped underwater for retting. Either of the two methods reduces the percentage of the cuttings (hard basal parts) of the fiber. It should be noted that these two methods should be practiced separately.

6.6.8   SELECTION OF RETTING WATER

Transparent, clear, and mild streaming water is ideal for retting. Such water is normally found in some beels, haors or in vast water sources of canals, etc. Rettings should be conducted in such water if found nearby.

6.6.9   PREPARATION OF JAK (RET)

While preparing the jaks, the bundles of jute plants should be arranged in such a way that there remains ample, space for the easy movement of water and retting microbes. Jaks are to be made in different layers of jute bundles. In one layer, the bundles are to be arranged side by side lengthwise, then in the second layer they should be arranged crosswise perpendicular to, the first one and in the third, they should be placed as like as the first one and soon.

6.6.9.1   APPLICATION OF UREA

Application of urea (0.01% of the green weight of plants) accelerates the retting speed and improves the fiber color when retting is conducted in the stagnant water of small ponds/ditches. About one kg of urea may be applied in 100 bundles of green jute plants, each bundle weighing 10kg. The urea may be added either in a water solution or sprinkled directly in the layers of the jute bundles in the jaks.

6.6.9.2   IMMERSION OF JAKS (RET)

It is found that the retting microbes are plenty in the upper surface of retting water and the temperature is also favorable therein. So the jaks should be immersed in such a way that they do not go so deep as to come in contact with the clay of the bottom of retting pools. There should be about 10cm. water above the jaks. For immersing the jaks earth chunk, logs of Banana or Mango tree or any other green logs should be avoided as they convert the fiber color into shamla (dark). Stone boulder or, concrete slabs are ideal for immersing the jak underwater. Polythene-wrapped earth chunks can also be used safely as it protects the plants to come in contact with clay. Water hyacinth, soty plants, rice-straw, any kind of aquatic weeds, etc. should be used to cover the jaks properly. It is recently found that if the jaks could be covered very thoroughly with any water weeds no extra weights for steeping the jak is at all needed.

6.6.9.3 DETERMINATION OF END POINT OF RETTING

Determination of the proper endpoint of retting is very important. Fibers become weak if over- retted while under-retted fibers are barky and of poor in quality. So rettings are to be stopped all the lime when the fibers are well-separated and strong. To ensure it, retting plants should be tested for the endpoint of retting after 10/12 days from the day of steeping. It may be done by taking 2/3 plants out of the jaks, peeling fiber strands and examining them for proper separation of the fibers and strength. The test may also be done in a better way by cutting 3–4cm retted bark from the middle part of the retting plant and putting it in a small bottle filled with water. The bottle is then shaken well and the dirty water is drained off. The bottle is again filled with clear water and the fibers are examined for separation. If the fibers are found well separated, retting is to be considered complete. It is to be noted that slightly under retted fiber is better than over retted fiber. Over-retted fibers are of no use.

6.6.10 FIBER EXTRACTION

The fibers can be extracted from the retted plants in two ways: (1) by taking one or two plants at a time sitting on the dry land or (2) by standing in knee-deep water and using a bamboo frame, a bunch of plants is taken out of the jak and the fibers are extracted from the whole bunch at a lime by breaking the jute sticks at about 45cm from the bottom. In either of the two ways, if the bottom parts of the retted plants are scraped off by hand prior to extraction or beaten with a wooden beater the amount of cuttings are reduced significantly. Better fibers are obtained if extraction of fibers is done on the ground taking single or two plants at a time.

6.6.10.1 WASHING OF FIBERS

After extraction, the fibers are to be washed thoroughly in clean water so that broken jute sucks, a cuticular layer of barks, clay or any other dirt get free from fibers.

6.6.10.2   DRYING OF FIBERS

After washing, the fiber should be dried well in a bamboo frame or by hanging in any way, so that it encounter mud or dust Fibers should not be dried spreading on the ground. Mud or dust not only lowers the quality of the fiber but also creates health hazards for jute mill workers. Jute fibers are to be dried well before storing. Wet fibers are much susceptible to microbial degradation, which weakens the strength. So, wet fiber should never be stored.

6.7   RELATIONSHIP BETWEEN CLIMATIC FACTORS AND JUTE PRODUCTION

Result revealed that jute production is so nicely matched with the ecological attributes of Bangladesh that there is pre-monsoon shower during the month of March and April thereby offering optimum condition for land preparation and sowing. Following that, there were moderate and intermittent rain and shower during May and June providing enough moisture in the soil needed for growth of jute plants. Heavier rains follow during July and August, which fill the ditches and ponds with sufficient water needed for retting of jute plants after harvest.

Jute is mainly cultivated in Kharif–1 season (March-June). Tossa jute (*Corchorus olitorius* L.) fiber and seed are commercially superior to those of white jute (*C. capsularis* L.) in both national and international market. Tossa jute shows premature flowering resulting in yield with low-quality fiber when seeded in relatively low-temperature in early March. Tossa jute seed generally used to collect from the normal fiber-yielding crop seeded during April-May. This seed-yielding crop is severely affected by natural calamities like flood, water, cyclone, heavy rainfall, insect pest infestation, etc. during July-August almost every year and as a result, there is an acute shortage of seed for sowing in the next season. Hence, it becomes necessary to grow Tossa jute and under late sowing, a condition in the off-season during July-August for white and August-September for Tossa jute. Bangladesh Jute Research Institute released so many Tossa jute varieties O–9897, OM–1 and 0–72 that are more or less similarly tolerant to comparatively short day and low-temperature. Now the varieties do not initiate premature flower even when sown in early March and grown well under a temperature of October-November for seed production.

6.7.1   JUTE CULTIVATION AFFECTED BY AIR QUALITY

Jute is the fast-growing field crop. Its relative growth rate varies from 0.115 to 0.05 gg^{-1} day^{-1}. The net assimilation rate of the leaves varies from 0.0006 to 0.0001 cm^{-2} day^{-1}.during 30 to 120 days of its growing period (Khandakarr et al., 1990). Its carbon dioxide consumption is 0.23 mg CO_2 $m^{-2}S^{-1}$ on an average to a maximum of 0.44 mg CO_2 $m^{-2}S^{-1}$ (Palit and Bhattacharyya, 1988).

In a theoretical calculation, one ha of jute plants can consume to about 14.66 tons of CO_2 at the rate of 0.34 mg $m^{-2}S^{-1}$ in the lifespan of about 100 days. Therefore, it can be said that the jute crop is a good environmental cleaner that reduces the pressure of CO_2 in the atmosphere and thus lowers the rise of temperature of the atmosphere and reducing the possible green-house effect. On the contrary, it also refreshes the atmosphere with the release of fresh oxygen at the rate of 10.66 tons ha^{-1}. To a theoretical calculation, the total jute crop in Bangladesh removes/sweeps to about million tons of carbon dioxide in one jute season. In the same, process it freshness the air with the release of 10 tons oxygen ha^{-1} and in total 5.90 million tons of oxygen is released to Bangladesh atmosphere. (Calculation based on the basic equation of photosynthesis to produce the required dry weight in total. This calculation also agrees with the finding of Palit and Bhattacharyya, 1988). A picture (see Figure 6.2) drawn in relation with jute and environment.

6.8   CHALLENGES FOR JUTE

Jute growers, more specifically marginal jute growers, are entangled in a complex web of interrelated problems, which have two major effects: Unfavorable market price and Production inefficiency (both lower productivity and higher cost). Consequently, jute growers remain poor, as their income from jute is too low.

6.8.1   UNFAVORABLE MARKET PRICE

Jute growers in Bangladesh are getting an unfavorable market price for various reasons. Some of the reasons are:

Price volatility: India is world's largest producer of jute; at the same time, it is the largest consumer too. Indian govt. has implemented a mandatory packaging order, which stipulates that certain food crop must use jute packaging. As a result, India has a strong local demand, which also makes it feasible for the govt. to implement Minimum Support Price (MSP) for the farmers. Consequently, farmers in India are getting a better price. However, in Bangladesh, there is no policy of the govt. to increase domestic demand or set MPS for the jute growers. As a result, the price of jute in the local market largely depends on world demand situation, as this sector is dependent on export.

Jute Cultivation and The Environment

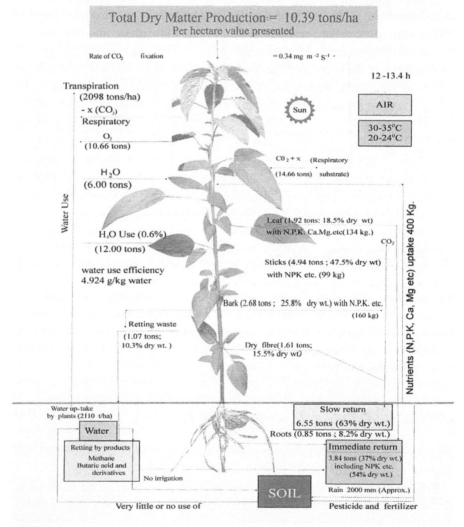

FIGURE 6.2 (See color insert.) Natural conservation through jute cultivation (*Source*: Khandakar et al., 1995).

Weak bargaining power: Most farmers are in serious liquidity crisis during jute harvesting season, which is between two rice-growing seasons. So, they are bound to sell their jute as quickly as possible at whatever price is offered. Moreover, most of the jute growers are marginal; they produce in very small quantity. Cash crisis and small quantity leave them with almost no bargaining power over price. Growers could gain bargaining power if they were organized in associations. However, no such organization of grower exists.

Multiple intermediary levels: Typically, jute trading involves multiple intermediary levels. A Faria (village level trader) who sells it to the local trader/purchase center who sell it to the jute mills or raw jute exporter collects often jute. These intermediaries are much richer than the growers are, so they can wait for the price to increase, but our jute growers cannot afford to wait. Therefore, ultimately the price that the jute growers get is too low.

Government procurement policy: State-owned jute mills procure the substantial amount of jute at a declared price. However, the procurement is usually done long after the jute harvesting time. By that time, jute is already in the store of the trader, who gets the benefit of the govt. declared price, not the jute growers.

Low-quality: Sometimes, the price gets even lower when the fiber is not of good quality. Farmers often fail to attain the desired quality mainly because of low-quality seed and inappropriate retting.

6.8.2   PRODUCTION INEFFICIENCY

The causes of production inefficiency are:

a) **Low-quality seed:** National Annual demand for jute seed is around 4000 MT, of which only about 20% is supplied by the Bangladesh Agricultural Development Corporation (BADC). The rest is imported from India; a considerable part of it is actually smuggled. Quality of the imported seed is questionable. Also, sometimes local traders counterfeit seed by mixing the old lot with the new. But the importers do not take any responsibility of quality as the seed is not sold under their brand name. Low-quality seed often results in low production, lower quality and sometimes even crop failure.

b) **Limited technical know-how:** Though average current yield has increased over the years, jute growers are not getting maximum yield because of limited technical know-how. They do not have the knowledge of soil management, appropriate doses of fertilizers, disease, and pest control. Govt. extension offices have limited manpower. It is not possible for them to reach out to wider grower communities to provide these services.

c) **The scarcity of water:** Jute is traditionally grown and harvested during the monsoon. So usually, growers would not need to irrigate jute fields or would not face scarcity of water for retting and washing jute fiber. But water is getting scarcer day by day. Now, growers need to irrigate their fields, increasing the cost of production. They are facing a water crisis during harvesting. As a result, the quality of the fiber is going down because of poor quality water for retting and washing. Often, growers need to carry jute to a different location where there is water, further increasing their cost. But they do not know the new technologies of retting that require less water without compromising quality. All these factors lead to a low income of jute growers and the continuation of poverty.

6.9   PROBLEMS TO JUTE INDUSTRY

There are so many problems regarding the jute industry. Very recently Mr. Abdul Latif Siddique, Minister, Ministry of Textiles & Jute, Government of Bangladesh, stated eight major problems to export Jute and Jute made products:

i.   Not having regular ships;
ii. No Mother Vessel can come to Bangladesh;
iii. Insufficient Feeder ships;
iv. Indiscipline in the schedule of feeder ships;
v.   The unstable rate of Ship-fare;
vi. Staffing problem in the port for goods;
vii. Tax and Tariff rule of a different country in the terms of import;
viii. Imposing Tariff of Indian Govt. while exporting Jute from Bangladesh to India.

6.10   THE GOVERNMENT ACTIONS TO OVERCOME PROBLEMS

The Jute Minister of the country recently told that the government has taken seven steps to overcome the ongoing problems of the jute industry. He stated that Bangladesh Jute Mills Corporation (BJMC) has taken initiatives to with-draw all the cases run by credit-court against many loan-loaded jute mills.

The Minister mentioned that in the current fiscal year till last April govt. has gained about 1,410 crores in foreign currency, by exporting 15 lakh BELL JUTE. As the accumulated account from July 2010 to April 2011, the profit was around BDT 89 lakh (Bangladesh Bank, 2007).

6.11   RECOMMENDATIONS

Jute is grown in Bangladesh almost solely as a rain-fed crop without any irrigation or drainage provisions. In this regard, jute is a better choice than its major competing crop Aus rice because jute is comparatively less affected by drought or stagnation of water. Moreover, 3–4 hundred thousand hectares of land in Bangladesh are suitable for growing no other crop but only jute in the Kharif season (April-September). Most rarely jute suffers total damage due to calamities like drought, excessive rainfall or flood. It seems that jute is somewhat naturally insured against common seasonal hazards. All the facts, as discussed above, lead to a logical appreciation that our agri-environment is ideal for jute cultivation. It may be concluded that jute and the environment in Bangladesh are mutually supplementary to each other. However, there is no room for complacency. It is yet to fetch the real benefit from the match of jute and the environment in Bangladesh.

Based on the findings of this study the following recommendations may be put forward for policy formulation with a view to improving the existing marketing system of raw jute.

- The jute agricultural research bodies could be given the additional function of establishing and administering a programme of education and advice to farmers on the best methods of cultivation, harvesting, and retting.
- The key factor of increasing raw jute export depends on the increase in total production of jute. Forecasting of jute price, target acreage, yield, and production before sowing will be helpful to the farmers in allowing them to adjust their jute acreage. Extend the availability and improve the distribution of good quality certified seed. To do this one necessary step may be the encouraging establishing a series of small cold stores by the private sector in the more remote jute growing areas.
- Jute production problems may be removed by ensuring supply of inputs, insecticides, and pesticides and credit in proper time. Jute

traders with the information center should connect the Internet, computer, and other modern facilities. With successful operation of buffer stock, price instability may be reduced. A comprehensive buffer stock scheme by the government with a view to stabilizing prices may, therefore, be attempted. Price support programme should be ensured through effective policy measures.

- Government should provide training facility on grading, retting, practices of balanced use of fertilizer and assorting through DAE and BJRI.
- Proper jute policy should be prepared and implemented.
- The mismanagement and malpractice in the jute sector must be removed.
- Supply of good seeds for the farmers, timely supply of fertilizer must be ensured.
- Modern scientific method of jute cultivation should be implemented.
- The government should provide market information for the farmers so that the farmers can get the profitable price form jute.
- Production cost of jute should be reduced into a minimum level so that we can compete with India in overseas market.
- The prescription given by different International agencies like IMF, ADB etc. should not be taken without proper examination.
- Honest and dedicated persons should be appointed at the jute factory. Labor politics should be controlled.
- Diversified and new items from jute should be manufactured to cope with the present demand of the world market.

KEYWORDS

- challenges
- harvesting
- jute industry
- production inefficiency

REFERENCES

Ahmed, Z., Banu, H., Akhter, F., & Ken Izumori, K., (2001). A simple method for d-xylose extraction from jute stick and rice husk. *OnLine J. Biol. Sci., 1,* 1001–1004.

Akter, N., Islam, M. M., Begum, H. A., Alamgir, A., & Mosaddeque, H. Q. M., (2009). BJRI Tossa–5 (O–795), An Improved Variety of *Corchorus olitorius* L. *Eco-friendly Agril. J., 2* (10), 864–869.

Anonymous, (1969–1970). Jute Board, Ministry of Commerce, Govt. of Pakistan. *The Jute Season (1969–70), An Annual Review*, p. 28.

Bangladesh Bank, (2007). Statistical Department, Bangladesh Bank, *Economic Trends, XXXII, 2,* 22–29.

Basu, G., Sinha, A. K., & Chattopadhyay, S. N., (2005). Properties of jute-based ternary blended bulked yarns. *Man-Made Textiles India, 48,* 350–353.

BBS, (1997). *Statistical Yearbook, Bangladesh Bureau of Statistics*. Ministry of planning, statistics division, Govt. of Peoples Republic of Bangladesh, Dhaka, Bangladesh.

BBS, (2004). *Statistical Yearbook of Bangladesh*. Ministry of Planning, Government of the People's Republic of Bangladesh, Dhaka, Bangladesh.

Chattopadhyay, S. N., Pan, N. C., & Day, A., (2004). A novel process of dying of jute fabric using reactive dye. *Textile Industry India, 42,* 15–22.

FAO, (1993). *Jute Kenaf and Allied Fibers. Statistics. Jute*, CCP (July/St/1993/1), Rome, Italy.

FAO, (2000). *Jute Kenaf and Allied Fibers, Statistics*, Jute, CCP (July/St/2000/1), Rome, Italy.

Hossain, M. A., Haque, S., Sultana, K. S., Islam, M. M., & Khandkar, A. L., (1994). Research on late jute seed production. *Pub. Seed Tech. Res. Team, Bangladesh Jute Res. Inst., Dhaka.,* pp. 176–178.

Hussain, M., (2013). *Jute in Bangladesh, Bangladesh Jute Research Institute*, Manik Mia Avenue, Dhaka- 1207.

Islam, M. M., & Rahman, M., (2008). In: *Hand Book on Agricultural Technologies of Jute, Kenaf, and Mesta Crops* (p. 92). Bangladesh Jute Research Institute, Manikmia Avenue, Dhaka–1207, Bangladesh.

Khandakar, A. L., Begum, S., & Haque, H., (1995). *Jute Cultivation and the Environment*. Agronomy Division, Bangladesh Jute Research Institute, Manikmia Avenue, Dhaka–1207, Bangladesh. pp. 23.

Khandakar, A. L., Begum, S., & Hossain, M. A., (1990). Comparative growth analysis of jute varieties (*Corchorus capsularis* L. and *C. olitorius* L.). *Bangladesh J. Bot., 19* (1), 33–39.

Khatun, R., Islam, M. M., Hussain, M. A., Parvin, N., & Sultana, K., (2009) Performance study of newly developed jute variety BJRI Deshi–7 (BJC–2142). *Int. J. Sustain. Agril. Tech., 5* (4), 12–18.

Kirtikar, K. R., & Basu, B. D., (1975). *Indian Medicinal Plants*. Published by Dehra Dun: Bishen Singh Mahendra Pal Singh.

Mohiuddin, G., Talukder, S. H., & Hasib, S. A., (1987). Chemical constituents of jute cuttings. *Bang. J. Jute Fiber Res., 6,* 75–81.

Palit, P., & Bhattacharyya, A. C., (1988). Photosynthesis and yield of jute in relation to foliar arrangements. *Proceedings of the International Congress of Plant Physiology*, Held on February 15–20. New Delhi, India.

Pan, N. C., Day, A., & Mahalanabis, K. K., (2000). Properties of Jute. *Indian Textile J., 110,* 16.

Roy, T. K. G., Chatterjee, S. K., & Gupta, B. D., (2002). Comparative studies on bleaching and dyeing of jute after processing with mineral oil in water emulsion vis-a-vis self-emulsifiable castor oil. *Colorage., 49,* 27–33.

Sreenath, H. K., Arun B. S., Vina, W. Y., Mahendra, M. G., & Thomas, W. J., (1996). Enzymatic polishing of jute/cotton blended fabrics. *J. Ferm. Bioeng., 81,* 18–20.

CHAPTER 7

RECENT DEVELOPMENTS IN THE BIOSYSTEMATICS AND MOLECULAR BIOLOGY OF SUGARCANE

KHUSHI MUHAMMAD[1,2], WAQAR AHMAD[1], HABIB AHMAD[1], EZED ALI[1], JAVED IQBAL[2], YONG-BAO PAN[3], and KHALID REHMAN HAKEEM[4]

[1]*Department of Genetics, Hazara University Mansehra–21300, Pakistan, E-mails: khushisbs@yahoo.com; wahmadhu@gmail.com; drhahmad@gmail.com; ali.ezed@gmail.com*

[2]*School of Biological Sciences, University of the Punjab, Lahore Pakistan, E-mail: javediqbal1942@yahoo.com*

[3]*USDA-ARS, Sugarcane Research Unit, 5883 USDA Road, Houma, Louisiana, 70360, USA, E-mail: yongbao.pan@ars.usda.gov*

[4]*Department of Biological Sciences, Faculty of Science, King Abdulaziz University, Jeddah, Saudi Arabia, E-mail: kur.hakeem@gmail.com*

ABSTRACT

Sugarcane is an important industrial crop cultivated on one million ha of land producing 63 million tons of sugar and providing job opportunities to millions of people both in urban and rural areas of Pakistan. The biosystematics of sugarcane shows that the crop evolved through inter-specific hybridization and continuous selection of the introgressed genes for the desirable traits. The modern sugarcane is a hybrid derived from *Saccharum officinarum* (Noble clone), *S. sinese* (Chinese clone), *S. barberi* (North Indian clone) and *S. spontaneum* (Wild clone). The sugarcane is polyploid with repeated aneuploid additions, has a complex genome. The chromosome number in the commercial cultivars ranges between 100–130 2n with the genome size ranging from 760 to 926 Mbp. Sugarcane productions are severely affected by a number of factors among which pathogenic diseases are very evident. Among the most significant sugarcane diseases are brown leaf rust, ratoon stunt, red rot, sugarcane mosaic, smut, stem canker, leaf spot, and pokkah boeng. The most economical, effective, and justified ways to enhance sugarcane yield is through breeding improved varieties resistant to insects and pest diseases. Review of the available literature shows that during the last three decades a lot of development has been made with respect to its genome constitution and genotyping of the improved cultivars and associated land acres. In recent years, the use of molecular markers in sugarcane crop has increased the precision of identifying desired genes for different agronomic traits and their introgression in improved genotypes. Recently, different attempts were made to find out the association of molecular markers, i.e., Random Amplified Polymorphic DNA (RAPD), Simple Sequence Repeats (SSR) and Amplified Fragment Length Polymorphism PCR (AFLP) with disease responses. This paper provides a synthesis of the most cited result available with respect to modern development in the sugarcane crops.

7.1   INTRODUCTION

Sugarcane (*Saccharum* spp.) is a multiuse cash crop. It provides brown sugar and white sugar to the growers. It provides fodder for livestock, bagasse for fuel, and the stubble root as organic manure and crop residues as compost (Khoso, 1992; Bhatti and Soomro, 1996). In recent years sugarcane is used to produce a bio-fuel ethanol as a renewable energy source. The demand for power production from plant materials, during the recent past, has increased the economic interest in sugarcane significantly. As an important industrial crop sugarcane (*Saccharum* spp.) is cultivated over 105 countries of the world within 30° of the equator (Bull and Glasziou, 1979). In 2016, sugarcane was cultivated in more than 23.7 million hectares in tropical and subtropical regions of the world, producing up to 1.683 billion metric tons (t) sugar (Anonymous, 2017). Among the 105 countries of the world growing sugarcane, Pakistan ranks 5th in the area, 12th in production, and 60th in yield. Although, Pakistan happens to be the world's 6th largest grower of sugarcane it ranks among lower yield countries. The area under sugarcane cultivation in Pakistan is estimated at 1.25 million hectares producing about 72 million metric tons of canes (Anonymous, 2017).

In recent year, sucrose is being fermented to produce ethanol (Schubert, 2006). In Brazil, a large amount of sugarcane crop about 47% was used for the production of ethanol, yielding 17.8 million liters (Goldemberg and

Guardabassi, 2009). In Brazil, approximately 40% of the fuel used in cars was ethanol in 2006 (Orellana and Neto, 2006). For this purpose, industries in the USA and Brazil have developed procedures to use cornstarch and sugarcane sucrose, respectively, to produce bioethanol (Pimentel and Patzek, 2005; Macedo et al., 2008). As a result, these two countries are now the chief producers of this biofuel in the world (Anonymous, 2012).

7.2 ORIGIN AND DOMESTICATION OF SUGARCANE

Review of the available literature on sugarcane shows that the origin and the main center of diversity of *Saccharum officinarum* L. are thought to be the Indo-Myanmar, China border with New Guinea (Daniels and Roach, 1987). Aitken et al. (2005) have supported this view by amplified fragment length polymorphism (AFLP) marker study. It is known that the commercial sugarcane cultivars have been evolved through recurrent selection and improved further through interspecific crosses within *Saccharum* genus, primarily involving crosses between *S. officinarum* and *S. spontaneum* (Cox et al., 2000; Lakshmanan et al., 2005). The noble cultivars are rich in sucrose but have poor disease resistance and it is purely cultivated or garden species since 8000 B.C. (Sreenivasan et al., 1987; Fauconnier, 1993). It has been noted that the noble cultivar *S. officinarum* was domesticated from *S. robustum* in New Guinea and dispersed in the Pacific and the mainland of Asia during human migration (Brandes, 1956). The scenario of evolution and domestication was discussed by Grivet et al. (2006) based on molecular data (Figure 7.1).

To attain noble cultivars of sugarcane, a series of backcrosses were made between the interspecific hybrids and the *S. officinarum* parent to minimize the negative effects of *S. spontaneum* and to retain the high sucrose producing ability of *S. officinarum*. Sreenivasan et al. (1987) described the term 'nobilization' (Figure 7.2). The nobilization of sugarcane is a major infringe in sugarcane improvement programs in the form of improved productivity, high disease resistance, and high ratooning ability but the successful use of nobilization because of narrow and complex genetic base of sugarcane cultivars (Deren, 1995; Arro, 2005). A few clones were used during the initial interspecific hybridizations (Arcenueaux, 1967) and most of the modern sugarcane cultivars are showing similarity to those few parents used during nobilization. Currently, the trend of making similar crosses is followed by breeders in the improvement of sugarcane but it is a long selection cycle of 12–15 years. Hence, the narrow genetic diversity among sugarcane genotypes is an important growing concern of the sugarcane breeders.

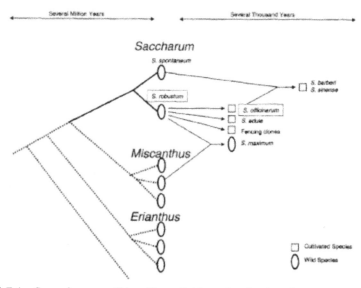

FIGURE 7.1 Scenario compatible with available molecular data for sugarcane evolution and domestication (adapted from Grivet et al., 2006). Ancestors that gave rise to current genera Saccharum, Erianthus. Miscanthus, and others diverged in the course of evolution, probably several million years ago. Only members of the Saccharurn clade contributed directly to sugarcane cultivars. Allopatric speciation gave rise to two species, *S. spontaneum* west of Sulewesi and *S. robustum* east of Sulawesa. Human-domesticated *S. robusturn* in the equatorial environment, probably in New Guinea, contributed *S. officinarum* cultivars for sugar. *S. edule* cultivars for vegetables, and possibly other cultivars for others use (fencing, constructions). *S. barberl* and *S. slnense* cultivars resulted from natural hybridization between *S. officinarum* cultivars transported by humans and local *S. spontaneum* populations in subtropical regions. *S. maximum* is at least partially the result of a Saccharum-Miscanthus inter gene-rich-hybridization event.

Nobilization

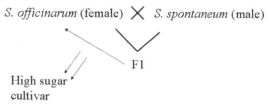

FIGURE 7.2 A detail about the nobilization of sugarcane cultivars (Adapted from Roach, 1972).

7.3   DIFFUSION OF SUGARCANE

It is thought that the noble cultivar has hybridized with wild sugar canes of India and China, to produce the 'thin' canes. The cultivated canes belong to two main groups: (a) thin, hardy north Indian types *S.barberi* and the Chinese *S.sinenses* and (b) thick, juicy noble canes *S.officinarum*. Arabs were responsible for much of its spread as they took it to Egypt around 640 Ad and they carried it with them as they advanced around the Mediterranean. Sugarcane spread by this means to Syria, Cyprus, and Crete, eventually reaching Spain around 715 AD. Around 1420 the Portuguese introduced sugar cane into Madeira, from where it soon reached the Canary Islands, the Azores, and West Africa. Columbus transported sugar cane from the Canary Islands to the Dominican Republic in 1493. The sugarcane cultivars were taken to Central and South America from the 1520s onwards, and later to the British and French West Indies (Figure 7.3).

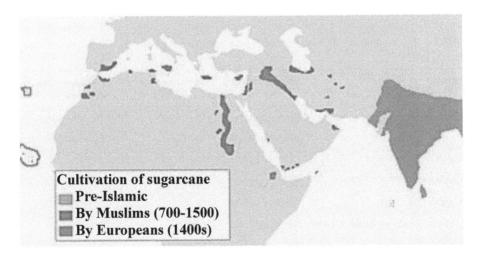

Cultivation of sugarcane
Pre-Islamic
By Muslims (700-1500)
By Europeans (1400s)

FIGURE 7.3   (See color insert.) Adapted from Wikipedia. The westward diffusion of sugarcane in pre-Islamic times (shown in red), in the medieval Muslim world (green) and by Europeans (violet).

7.4   TAXONOMY OF SUGARCANE

The taxonomy and lineage of sugarcane are complicated and known as '*Saccharum* complex' because it comprises of five genera, i.e., *Saccharum*, *Erianthus* section *Ripidium*, *Miscanthus* section *Diandra*, *Narenga*, and

Sclerostachya share similar characteristics. These five closely related genera share high levels of polyploidy and aneuploidy that is a challenge for both the taxonomist and molecular biologist (Daniels and Roach, 1987; Sreenivasan et al., 1987). Sugarcane refers to the genus *Saccharum* (family Poaceae, tribe Andropogoneae), first Linnaeus established by two species, i.e., *S. officinarum* and *S. spicalum* L. in 1753.

The original classification of Linnaeus' has been revised from two to six species (Table 7.1) i.e., *S. officinarum,* known as the noble cane; *S. spontaneum* L., *S. robustum* E.W. Brandes & Jeswiet ex Grassl, and *S. edule* Hassk., classified as wild species; and *S. sinense* Roxb. and *S. barberi* Jeswiet, classified as ancient hybrids (Buzacott, 1965; Daniels and Roach, 1987). The details about members of genus *Saccharum* are described in Table 7.1. Native to warm temperate to tropical regions of Asia, they have stout, jointed, fibrous stalks that are rich in sugar, and measure two to six meters (six to nineteen feet) tall. All sugar cane species interbreed, and the major commercial cultivars are complex hybrids.

TABLE 7.1   A Summary of Members of Genus *Saccharum* (Buzacott, 1965; Daniels and Roach, 1987)

Species	Classification	Sugar content	Chromosome number
S. spontaneum L.	Wild species	Very low	2n = 40–128
S., robustum Brandes and Jeswiet ex Grass	Wild species	Very low	2n = 60–200
S. officinarum L.	Noble canes	High	2n = 80
S. barberi Jeswiet	Ancient hybrid	Low	2n = 111–120
S. sinese Roxb. Ambend. Jeswiet	Ancient hybrid	Low	2n = 80–124
S. edule Hassk.	Cultivated species	Low-Compact Inflorescence eaten as a vegetable	2n = 60–80 with aneuploid forms

7.5   THE SUGARCANE GENOME

The complexity and size of the sugarcane genome is a major challenge in the genetic and molecular study. Much artificial interspecific hybridization between *S. officinarum,* as a female, and *S. spontaneum* and, to a lesser extent, *S. barberi* as the pollen donor are responsible to derive modern cultivars. The hybrids were backcrossed to *S. officinarum* to recover the thick sugar-containing stalks of this species. This process was accelerated through the selection of hybrids derived from 2n transmission of *S. officinarum* chromosomes (Bremer, 1961). Previous studies revealed that these modern cultivars are derived by the interbreeding of interspecific hybrids and it is estimated that 19 *S. officinarum* clones, a few *S. spontaneum* clones, and one *S. barberi* clone were involved in these interspecific crosses (Arcenueaux, 1967).

The genome of modern sugarcane cultivars is complex and highly polyploid ($\approx$12X $\approx$130) and interspecific origin (D'Hont, 2005). The high level of ploidy, the aneuploidy, the bispecific origin of the chromosomes, the existence of structural differences between chromosomes of the two origins, and the presence of interspecific chromosome recombinants are the distinctiveness of sugarcane genome. GISH (Genome *in situ* hybridization) studies of chromosomes preparations revealed that 15–25% of their chromosomes were inherited from *S. spontaneum*, that the recombination between homoeologous chromosomes is possible (D'Hont et al., 1996; Piperidis and D'Hont, 2001; Cuadrado et al., 2004; D'Hont, 2005). The detail about chromosomes complement has been described in Figure 7.4.

In diploids, the genome size are generally given for a basic set of the chromosome but in polyploidy, i.e., sugarcane, the genomic does not correspond to the basic set chromosome. Molecular cytogenetics (Piperidis and D'Hont, 2001; Cuadrado et al., 2004) and genetic mapping studies (Grivet et al., 1996; Hoarau et al., 2001) showed that modern cultivars typically display 70–80% more complex crops with a large genome such as bread wheat (34 Gb/2C) or sugarcane (10 Gb/2C; D'Hont, 2005). For *S. officinarum* (2n=8x=80), the genome size has been determined as 7440 mega base pairs (Mbp) for a somatic cell and to 926 Mbp for monoploid genome (X=10) (D'Hont and Glaszmann, 2001). The genome size *S. spontaneum* with 2n=8x=64 has been estimated as 6,080 Mbp for a somatic cell and to 760 Mbp for the monoploid genome (X=8) (D'Hont and Glaszmann, 2001). D'Hont (2005) estimated the genome size of somatic cells of typical modern cultivar R570 (2n=about 115) 10,000 Mb. The genome structure of typical modern cultivars is represented in Figure 7.5.

In addition, due to its genetic complexity, this species has received very little research investment despite its economic importance, and molecular resources have just recently begun to be developed (Grivet and Arruda, 2002). So, the selective breeding has been in practice to achieve desired goals and the gene pool exploitation is limited in traditional breeding programmes (Mariotti, 2002).

GENETICS AND ORIGIN OF CROPS

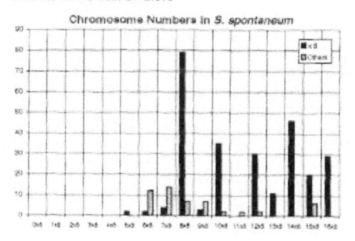

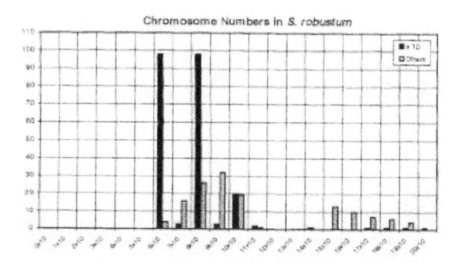

FIGURE 7.4 Chromosome complements in wild *Saccharum* species. The upper graph gives the frequency of chromosome numbers observed in *S. spotaneum* accession collected worldwide (data from Panjeand Babu, 1960). Multiple of 8 are figured in black: other numbers are in gray. The lower diagram gives the frequency of chromosome numbers in *S. robustum* accessions collected over the range distribution of the species. Data are from Price (1965). Multiples of 10 are figured in black and others are in grey.

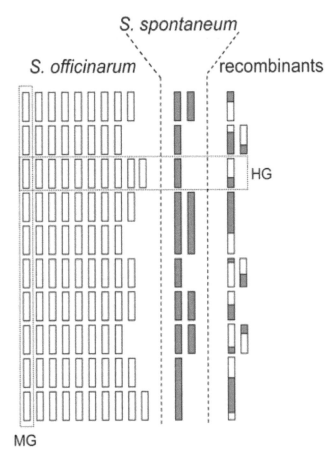

FIGURE 7.5 Adapted from Le Cunff et al. (2008): Schematic of a typical modern sugarcane cultivar genome. Each bar represents a chromosome; open boxes represent regions originating from *S. officinarum* and shaded boxes regions from *S. spontaneum*. Chromosomes aligned in the same row are hom (oe)ologous and represent a homology group (HG). Chromosomes assembled in the dotted vertical rectangle correspond to a monoploid genome (MG) of *S. officinarum*. The key characteristics of this genome are the high level of ploidy, the aneuploidy, the bispecific origin of the chromosomes, the existence of structural differences between chromosomes of the two origins, and the presence of interspecific chromosome recombinants.

7.6   SYNTENY WITH OTHER GRASSES

Comparative mapping has shown that the chromosome organization in grasses is highly conserved (Ming et al., 1998; Paterson et al., 2009), this is due to the species belonging to this family have evolved independently from a common ancestor. Comparative mapping and sequence analysis of nucleotides of alcohol dehydrogenase genes in sugarcane and sorghum have shown 95% nucleotide similarity (Dufour et al., 1997; Guimaraes et al., 1997; Jannoo et al., 2004, 2007). A comparative genetic in grasses is described in Figure 7.6. Many studies are focusing for comparative genetic mapping and strengthen the concept that sorghum has the simplest syntenic relationship with sugarcane (Dufour et al., 1997; Ming et al., 1998; Asnaghi et al., 2000). Consequently, access to the sorghum genome and to the largest dataset of the sugarcane transcriptome will be of considerable use in the genetic improvement of sugarcane.

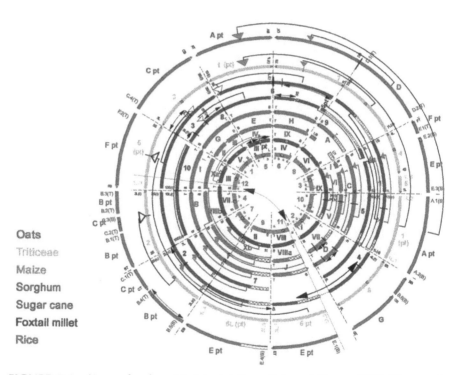

Oats

Triticeae

Maize

Sorghum

Sugar cane

Foxtail millet

Rice

FIGURE 7.6 (See color insert.) Adapted from Gale and Devos, 1998 (Comparative genetics in the grasses). A consensus grass comparative map. The comparative data have been drawn from many sources: Oats-wheat-maize-rice (Kurata et al., 1994); wheat-rice (Kurata et al., 1994; Van Deynze et al., 1995); maize-wheat (Ahn and Tanksley, 1993); maize-sorghum-sugarcane (Grivet et al., 1994; Dufour et al., 1997); and foxtail millet-rice (Devos et al., 1998). Arrows indicate inversions and transpositions necessary to describe present-day chromosomes. Locations of telomeres (.) and centromeres (h) are shown where known. Hatched areas indicate chromosome regions for which very little comparative data exist. Chromosome nomenclatures and arm (shortylong or topybottom) designations are as described by O'Donoughue et al. (1992) for oats, Pereira et al. (1994) for sorghum, Dufour (1997) for the sugarcane genomes, Wang et al. (1997) in foxtail millet, and Singh et al. (1996) in rice.

7.7 ANALYSIS OF MOLECULAR VARIATION IN SUGARCANE GENOME

In 1997, International Consortium for Sugarcane Biotechnology (ICSB) made an effort to develop and evaluate simple sequence repeats (SSRs) or microsatellite sequences as a marker system for sugarcane. Markers were developed from an enriched microsatellite library and were used to differentiate between sugarcane cultivars due to their ability to have large numbers of alleles (Cordeiro et al., 2000). Ever since, SSR marker system has been used for a number of applications both in sugarcane research and breeding. In previous studies, alleles generated by 72 SSR primer pairs were reported for construction of genetic map constructed on the Australian hybrid cultivar, Q165 (Aitken et al., 2004); for validation of genes introgression into F_1 hybrids produced by crosses between *S. spontaneum* and elite commercial clones (Pan et al., 2004); the fertility verification of F_1 hybrids produced by making crosses between *S. officinarum* and *E. arundinaceus* as well (Cai et al., 2005); and molecular markers were used to register and confirm sugarcane varieties by the United States Department of Agriculture (USDA) (Tew et al., 2003). SSR markers have also been used to depict useful information about Saccharum complex (Cordeiro et al., 2003; Cai et al., 2005) as well as relationships between clonal cultivars of hybrid canes (Pan et al., 2003). A fingerprint database based on SSR marker is now used for identification of cultivars and progeny validation for a breeding programme.

The studies on molecular marker have been used to unravel the complexity of sugarcane and for better understanding of its complex genetic make-up (Bonierbale et al., 1988; Wu et al., 1992; D'Hont et al., 1994; Sills et al., 1995; Ming et al., 2001; Rossi et al., 2003; Muhammad et al., 2013). The use of single dose (markers present on one chromosome only) and double dose (marker present on two chromosomes) markers for mapping and QTL analysis (Ming et al., 2001, 2002; Hoarau et al., 2002; Aitken et al., 2004), and large-scale EST sequencing projects by SUCEST-Sugar Cane EST Genome Project (Vettore et al., 2001), SASRI-South African Sugar Research Institute (Carson and Botha, 2000), UGA-University of Georgia, USA (Ma et al., 2004), and CSIRO- Australia's Commonwealth Scientific and Industrial Research Organization (Casu et al., 2004) are achievements for molecular development in this polyploidy crop. But still, the pace of sugarcane genomics studies has lagged behind that achieved with other agricultural crops (Ramsay et al., 2000; Delseny et al., 2001; Mullet et al., 2002; Muhammad et al., 2013).

7.8   DISEASES OF SUGARCANE

The reason for the decrease and low sugarcane yield in Pakistan involves many factors, one of the major ones being crop diseases. A number of diseases caused by different organisms (fungi, bacteria, viruses, and nematodes) and factors, i.e., environmental and physiological disorders and nutritional deficiencies are affecting sugarcane (Ryan and Egan, 1979). According to American phyto-pathological Society (APS), 64 different diseases of sugarcane, i.e., five bacterial, forty fungal, eight viral and mycoplasma-like organisms (MLO), three plant parasitic nematodes and eight are listed as disorders are reported in the world (http://www.apsnet.org/publications/commonnames/Pages/Sugarcane.aspx.). The diseases of sugarcane in Pakistan caused by fungi, bacteria, nematodes, and viruses have been summarized in Table 7.2.

TABLE 7.2   Disease of Sugarcane That Causes Yield Loss in Pakistan

Name	Causal agent	Control
Bacterial		
Leaf scalds	*Xanthomonas Albileneans*	Resistant cultivars
RSD	*Leifsoniaxyli subsp. xyli*	Disease free planting material
Red stripe	*Xanthomonasrubriieans*	Resistant cultivars
Fungal		
Rusts	*Pucciniamelanocephala*	Resistant cultivars
Yellow spot	*Helminthosporiumspp*	Resistant cultivars
Whip smut	*Ustilagoscitaminea*	Resistant cultivars, hot water treatment
Eyespot	*Helminthosporiumsacchari*	Resistant cultivars
Red rot	*Colletotrichumfalcatum*	Resistant cultivars
Stem canker	*Cytosporasacchari*	Resistant cultivars
PokkahBoeng	*Fusariummoniliforme*	Plants usually recover without the need for disease control
Viral		
Mosaic diseases	*Sugarcane mosaic virus*	Disease-free planting material and resistant cultivars

A wide range of fungal diseases of sugarcane has been recorded, including downy mildew, smut, rust, and red rot. Generally, fungal diseases are diagnosed by symptomatology, followed by isolation and examination of the organism. Rapid diagnostic techniques such as PCR, immunofluorescence, and ELISA are being developed to detect the fungal pathogens of many crops.

Recent development from sugarcane industry on disease detection (Grisham and Pan, 2005), DNA markers (Hameed et al., 2012; Muhammad et al., 2013; Que et al., 2014), molecular breeding technology (Tew and Pan, 2010) and genetic mapping of population will help to identify molecular markers that are closely linked to disease-resistant genes.

7.9   GENETIC MARKERS

The differences between individual organisms or species can be represented on the basis of genetic material, i.e., DNA packed into chromosomes. The chromosomes carry genes, which control the traits of plants, transferring from one parent to another. Scientists use a method to identify a gene called genetic markers. Generally, they do not represent the target genes themselves but act as 'signs' or 'flags' which may be referred to as gene 'tags.' Such markers are not distressing the trait of interest because they are closely linked or located near to genes of the trait. The genomic position on a chromosome occupied by genetic markers is called 'locus' (polural 'loci'). So, the locus is a specific piece of DNA with a known position on the genome.

The genetic markers are classified into three major types: (1) morphological markers based phenotypic traits or agronomical characters (assessed by visuals): (2) biochemical markers (isozymes) based on gene product; and (3) DNA (or molecular) markers based on polymorphism in DNA. The literature revealed the great understanding and the development of isozyme marker (Markert and Moller, 1959), and molecular markers (Botstein et al., 1980; Welsh and McClelland, 1990; Williams et al., 1990; Adams et al., 1991; Winter and Kahl, 1995; Jones et al., 1997).

DNA based markers are valuable tools for a large number of applications such as genotype identification, phylogenic analysis, population, and pedigree analysis, the screening of segregating population for linked markers, localization of a gene and improvement of plant varieties by marker-assisted selection (Fracaro et al., 2005; Ali et al., 2013). DNA marker technology has been used in commercial plant breeding programmes since the early 1990s and has proved helpful for rapid and efficient transfer of useful traits into ergonomically desirable varieties and hybrids (Tanksley et al., 1989). Markers linked to disease resistance loci can now be used for the marker-assisted selection (MAS) programmes, thus also allowing several resistance genes to be accumulated in the same genotype. In addition, markers linked to resistance gene may also be useful for cloning and sequencing the gene. Mapping and sequencing of plant genomes would help to elucidate gene function, gene regulation, and their expression. The schematic explanation was expressed by Huang and Roder (2004) and the detail about the marker, gene identification and molecular breeding of wheat is given in Figure 7.7. This is applicable to other crops, as well.

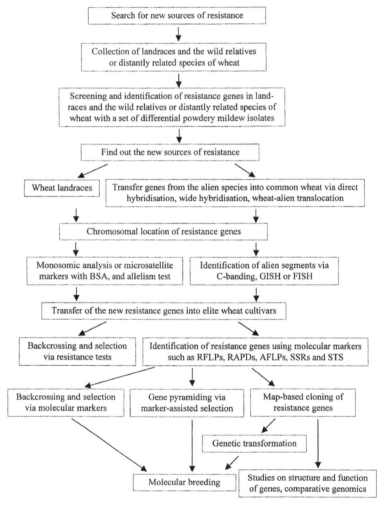

FIGURE 7.7   Adapted from Huang and Roder (2004): Scheme: from gene identification to molecular breeding.

The use of DNA markers has recently allowed genetic mapping in polyploids (DaSilva and Sobral, 1996). A novel genetic approach to direct mapping of polyploid plants was proposed by Wu et al. (1992). This approach is based on single-dose markers (SDMs). SDMs are present in one parent, absent in the other parent, and segregate 1:1 in the progeny. More recently, Ripol (1994) presented a methodology for mapping, multiple dose markers in polysomic polyploids, which greatly improved the accuracy of identification of homology groups. Restriction fragment length polymorphisms (RFLPs) were the first DNA markers used to construct genetic maps of higher organisms (Botstein et al., 1980).

7.9.1   TYPES OF MOLECULAR MARKERS

DNA markers may be broadly divided into three classes based on the method of their detection: (1) hybridization-based; (2) polymerase chain reaction (PCR) -based and (3) DNA sequence-based (Jones et al., 1997; Winter and Kahl, 1995). Molecular markers are based on naturally occurring polymorphism in DNA sequences, i.e., base pair deletion, substitution, and addition (Gupta et al., 1999).

In literature, many categories of molecular markers have been revealed, which are listed in alphabetical order as follows: amplified fragment length polymorphism (AFLP; Vos et al., 1995), diversity arrays technology (DArT; Jaccoud et al., 2001), DNA amplification fingerprinting (DAF; Caetano-Anolles et al., 1991), expressed sequence tags (EST; Adams et al., 1991), random amplified polymorphic DNA (RAPD; Williams et al., 1990), restriction fragment length polymorphism (RFLP; Botstein et al., 1980), sequence characterized amplified regions (SCAR; Paran and Michelmore, 1993), sequence specific amplification polymorphisms (S-SAP; Waugh et al., 1997), sequence tagged microsatelite site (STMS; Beckmann and Soller, 1990), sequence tagged site (STS; Olsen et al., 1989), short tandem repeats (STR; Hamada et al., 1982), simple sequence length polymorphism (SSLP; Dietrich et al., 1992), simple sequence repeats (SSR; Akkaya et al., 1992), single nucleotide polymorphism (SNP; Jordan and Humphries, 1994), variable number tandem repeat (VNTR; Nakamura et al., 1987) and start codon targeted markers (SCoT marker; Collard and Mackill, 2009; Que et al., 2014).

7.9.2   TAGGING GENES OF INTEREST

Molecular markers have provided a powerful tool to unravel the complex genome of sugarcane and enhance the determination of Mendelian bases for trait inheritance. In recent years, the use of molecular markers in this crop has increased rapidly. Polygenic characters which were previously very difficult to analyze using traditional plant breeding methods, would now be easily tagged using molecular markers. Only three genes have been mapped by now, two ruts resistance (Daugrois et al., 1996; Raboin et al., 2006) and one gene responsible for stalk color (Raboin et al., 2006).

The presence of a major resistance gene (*Bru*1) for brown rust in the sugarcane cultivar R570 was confirmed by analyzing segregation of rust resistance in a large population of 658 individuals, derived from selfing of clone R570. A subset of this population was analyzed using AFLP with bulked segregant analysis (BSA) approach to develop a detailed genetic map around the resistance gene and an AFLP marker flanking *Bru1* at 2 cm on both sides was identified (Asnaghi et al., 2004). Later, McIntyre et al. (2005) identified and validated molecular markers associated with Pachymetra root rot and brown rust resistance in a sugarcane cultivar Q.117 map and association-based approaches. Recently, Le Cunff et al. (2008) developed physical map using molecular markers and attempted cloning of map-based cloning of *Bru1* gene in sugarcane cultivar R570.

Work on molecular marker to find RGAs, develop SCAR, SCoT, and STS marker for rust, red rot and RSD (ratoon stunting disease)disease is being conducted in sugarcane resistant and susceptible cultivars grown in Pakistan (Hameed *at al.*, 2012; Muhammad et al., 2013; Ali and Muhammad, 2013). Moreover, we are elucidating chloroplast markers, i.e., rpoB, rbcL, and mtk, etc. to understand evolutionary tendency of modern sugarcane cultivars (Ali and Muhammad, 2014). The present work is the first elaborated study undertaken to differentiate between the rust-resistant/susceptible cultivars grown in Pakistan based on molecular marker studies.

KEYWORDS

- complex genome
- diseases
- molecular markers
- sugarcane

REFERENCES

Adams, M. D., Kelley, J. M., Gocayne, J. D., Dubrick, M., Polymeropoulos, M. H., Xiao, H., et al. (1991). Complementary DNA sequencing: expressed sequence tags and human genome project. *Science*, *252*, 1651–1656.

Ahn, S. N., & Tanksley, S. D., (1993). Comparative linkage maps of the rice and maize genomes. *Proc. Nat. Acad. Sci. U. S. A.*, *90*, 7980–7984.

Aitken, K., Jackson, P. A., & McIntyre, C. L., (2005). A combination of AFLP and SSR markers provides extensive map coverage and identification of homo (eo)logous linkage groups in a sugarcane cultivar. *Theor. Appl. Genet.*, *110*, 789–801.

Aitken, K., Jackson, P., Piperidis, G., & McIntyre, L., (2004). QTL identified for yield components in a cross between a sugarcane (*Saccharum* spp.) cultivar Q165[A] and a *S. officinarum* clone IJ76–514. In: Fischer, T., Turner, N., Angus, J., McIntyre, L., Robertson, M., Borrell, A., et al. (eds.), *New Directions for a Diverse Planet, Proceedings for the 4th International Crop Science Congress*. Brisbane, Australia.

Akkaya, M. S., Bhagwat, A. A., & Cregan, B. B., (1992). Length polymorphisms of simple sequence repeat DNA in soybean. *Genetics*, *132*, 1131–1139.

Ali, E., & Muhammad, K., (2013). *Estimation of Genetic Diversity Among Rust Resistance in Sugarcane*. BS (Hons) thesis, Department of Genetics, Hazara University, Mansehra Pakistan.

Ali, P., & Muhammad, K., (2014). Characterization of sugarcane commercially grown sugarcane genotype through rpob Marker. MSc Thesis, Department of Genetics, Hazara University, Mansehra Pakistan.

Ali, W., Muhammad, K., Nadeem, M. S. Inamullah, Ahmad, H., & Iqbal, J., (2013). Use of RAPD markers to characterize commercially grown rust resistant cultivars of sugarcane. *Int. J. Biosci.*, *3* (2), 115–121.

Anonymous, (Global Agriculture Information Network) report, (2017). Pakistan sugarcane annual report 2017. USDA foreign Agriculture Service. *GIAN Report Number: PK1708*.

Anonynmos, (2012). *Accelerating Industry Innovation – 2012 Ethanol Industry Outlook*. http://www.ethanolrfa.org/pages/annual-industry-outlook. Accessed 14 March 2012.

Arcenueaux, G., (1967). Cultivated Sugarcane of the world and their botanical derivation. *Proc. Int. Soc. Sugar Cane Technol.*, *12*, 844–854.

Arro, J. A., (2005). Genetic diversity among sugarcane clones using Target Region Amplification Polymorphism (TRAP) markers and pedigree relationships. *Master's Thesis*. Submitted to Louisiana State University, Baton Rouge, LA.

Asnaghi, C., Paulet, F., Kaye, C., Grivet, L., Glaszmann, J. C., & D'Hont A., (2000). Application of synteny across the Poaceae to determine the map location of a rust resistance gene of sugarcane. *Theor. Appl. Genet.*, *10*, 962–969.

Asnaghi, C., Roques, D., Ruffel, S., Kaye, C., Hoarau, J. Y., lismart, H. T., et al. (2004). Targeted mapping of a sugarcane rust resistance gene *Bru1* using bulked segregant analysis and AFLP markers. *Theor. Appl. Genet.*, *108*, 759–764.

Beckmann, J. S., & Soller, M., (1990). Toward a unified approach to genetic mapping of eukaryotes based on sequence tagged microsatellite sites. *Bio. Technol.*, *8*, 930–932.

Bhatti, I. M., & Soomro, A. H., (1996). Agriculture inputs and field crop in Sindh. *Agriculture Research Sindh*, Hydrabad.

Bonierbale, M. W., Plaisted, R. L., & Tanksely, S. D., (1988). RFLP maps based on a common set of clones reveal modes of chromosomal evolution in potato and tomato. *Genetics*, *120*, 1095–1103.

Botstein, D., White, R. L., Skolnick, M., & Davis, R. W., (1980). Construction of a genetic linkage map in man using restriction fragment length polymorphisms. *Am. J. Hum. Genet.*, *32*, 314–331.

Brandes, E. W., (1956). Origin, dispersal, and use in breeding of the Melanesian garden sugarcane and their derivatives, *Saccharum officinarum* L. *Proc. Int. Soc. Sugar Cane Technol.*, *9*, 709–750.

Bremer, G., (1961). Problems in breeding and cytology of sugarcane. I. A short history of sugarcane breeding: the original forms of *Saccharum*. *Euphytica*, *10*, 59–78.

Bull, T. A., & Glaszoius, K. T., (1979). Sugarcane. Chapter 4. In: Lovett, J. V., & Lazenby, A., (eds.), "*Australian Field Crops*" (pp. 95–113). Angus and Robertson Publishers.

Buzacott, J. H., (1965). Cane varieties and breeding. In: Kim, N. J., Mungomery, R. W., & Hughes, C. G., (eds.), "*Manual of cane growing*." (pp. 220–253). Sydney, Australia.

Caetano-Anolles, G., Bassam, B. J., & Gresshoff, P. M., (1991). DNA amplification fingerprinting using very short arbitrary oligonucleotide primers. *Biotechnology*, *9*, 553–557.

Cai, Q., Aitken, K., Deng, H. H., Chen, X. W., Fu, C., & Jackson, P. A., (2005). Verification of the introgression of *Erianthus arundinaceus* germplasm into sugarcane using molecular markers. *Plant Breed.*, *124*, 322–328.

Carson, D. L., & Botha, F. C., (2000). Preliminary analysis of expressed sequence tags for sugarcane. *Crop Sci.*, *40*, 1769–1779.

Casu, R. E., Dimmock, C. M., Chapman, S. C., Grof, C. P. L., McIntyre, C. L., & Bonnett, G. D., (2004). Identification of differentially expressed transcripts from maturing stem of sugarcane by *in silico* analysis of stem expressed sequence tags and gene expression profiling. *Plant Mol. Biol.*, *54*, 503–517.

Collard, B. C. Y., & Mackill, D. J., (2009). Start codon targeted (SCoT) polymorphism: A simple, novel DNA marker technique for generating gene-targeted markers in plants. *Plant Mol. Biol. Report.*, *27* (1), 86–93.

Cordeiro, G. M., Pan, Y. -B., & Henry, R. J., (2003). Sugarcane microsatellites for the assessment of genetic diversity in sugarcane germplasm. *Plant Sci.*, *165*, 181–189.

Cordeiro, G. M., Taylor, G. O., & Henry, R. J., (2000). Characterization of microsatellite markers from sugarcane (*Saccharum* sp.), a highly polyploid species. *Plant Sci.*, *155*, 161–168.

Cox, M., Hogarth, M., & Smith, G., (2000). Cane breeding and improvement. In: Hogarth, M., & Allsopp, P., (eds.), "*Manual of Cane Growing*" (pp. 91–108). Bureau of Sugar Experimental Stations, Indooroopilly, Australia.

Cuadrado, A., Acevedo, R., Diaz de la Espina, S. M., Jouve, N., & De la Torre, C., (2004). Genome remodelling in three modern *S. officinarum* 3 *S. spontaneum* sugarcane cultivars. *J. Exp. Bot.*, *55*, 847–854.

D'Hont, A., & Glaszmann, J. C., (2001). Sugarcane genome analysis with molecular markers: A first decade of research. *Proc. Int. Soc. Sugar Cane Technol.*, *24*, 556–559.

D'Hont, A., (2005). Unraveling the genome structure of polyploids using FISH and GISH: examples of sugarcane and banana. *Cytogenet. Genom. Res.*, *109*, 27–33.

D'Hont, A., Grivet, L., Feldmann, P., Rao, S., & Berding, N., (1996). Characterization of the double genome structure of modern sugarcane cultivars *Saccharum* spp by molecular cytogenetics. *Mol. Gen. Genet.*, *250*, 405–413.

D'Hont, A., Lu, Y., Gonzalez de Leon, D., Grivet, L., Feldmann, P., Lanaud, C., & Glaszmann, J. C., (1994). A molecular approach to unraveling the genetics of sugarcane, a complex polyploid of the Andropogoneae tribe. *Genome*, *37*, 222–230.

Daniels, J., & Roach, B. T., (1987). Taxonomy and evolution. In: Heinz, D. J., (ed.), *Sugarcane Improvement Through Breeding* (pp. 7–84). Elsevier press, Amsterdam.

DaSilva, J., & Sobral, B. W. S., (1996). Polyploid genetics. In: Sobral, B. W. S., (ed.), *The Impact of Plant Mol. Gen.*, (pp. 3–35). Birkhäuser, Boston.

Daugrois, J. H., Grivet, L., Roques, D., Hoarau, J. Y., Lombard, H., Glaszmann, J. C., & D'Hont. A., (1996). A putative major gene for rust resistance linked with a RFLP marker in sugarcane cultivar 'R570.' *Theo. Appl. Genet.*, *92*, 1059–1064.

Delseny, M., Salses, J., Cooke, R., Sallaud, C., Regad, F., & Lagoda, P., (2001). Rice genomics: Present and future. *Rice Genomics: Present and Future, 39*, 323–334.

Deren, C. W., (1995). Genetic base of U. S. mainland sugarcane. *Crop Sci.*, *35*, 1195–1199.

Devos, K. M., Wang, Z. M., Beales, J., Sasaki, T., & Gale, M. D., (1998). Comparative genetic maps of foxtail millet (*Setaria italica*) and rice (*Oryza sativa*). *Theor. Appl. Genet.*, *96*, 63–68.

Dietrich, W., Katz, H., Lincoln, S. E., Shin, H. S., Friedman, J., Dracopoli, N. C., & Lander, E. S., (1992). A genetic map of the mouse suitable for typing intraspecific crosses. *Genetics*, *131*, 423–447.

Dufour, P., (1996). Cartographice moleculaire du genome du sorgho (Sorghum bicolor L. Moench): Application en selection varietale: cartogrphie compare less andropogonees. *PhD. Thesis* (Universite' de Paris Sud, France).

Dufour, P., Deu, M., Grivet, L., D'Hont, A., Paulet, F., Glaszmann, C., & Hamon, P., (1997). Construction of a composite sorghum genome map and comparison with sugarcane, a related complex polyploidy. *Theo. Appl. Genet.*, *94*, 409–418.

Fauconnier, R., (1993). *Sugarcane*. Macmillan Press Ltd, London, UK.

Fracaro, F., Zacaria, J., & Echeverrigaray, S., (2005). RAPD based genetic relationships between populations of three chemotypes of *Cunila galioides* Benth. *Biochem. System. Ecol.*, *33*, 409–417.

Gale, M. D., & Devos, K. M., (1998). Comparative genetics in the grasses. *Proc. Nat. Acad. USA.*, *95*, 1971–1974.

Goldemberg, J., & Guardabassi, P., (2009). Are biofuels a feasible option? *Energy Policy*, *37*, 10–14.

Grisham, M. P., & Pan, Y. B., (2007). A genetic shift in the virus strains that cause mosaic in Louisiana sugarcane. *Plant Dis.*, *91*, 453–458.

Grivet, L., & Arruda, P., (2002). Sugarcane genomics: Depicting the complex genome of an important tropical crop. *Curr. Opin. Plant Biol.*, *5*, 122–127.

Grivet, L., D'Hont, A., Dufour, P., Hamon, P., Roques, D., et al. (1994). Comparative genome mapping of sugarcane with other species within the Andropogoneae tribe. *Heredity*, *73*, 500–508.

Grivet, L., D'Hont, A., Roques, D., Feldmann, P., Lanaud, C., & Glaszmann, J. C., (1996). RFLP mapping in cultivated sugarcane (*Saccharum* spp.): Genome organization in a highly polyploid and aneuploid interspecific hybrid. *Genetics*, *142*, 987–1000.

Grivet, L., Glaszmann, J. C., & D'Hont, A., (2006). Molecular evidences for sugarcane evolution and domestication. In: Montley, T., Zerega, N., & Cross, H., (eds.), *Darwin's Harvest*. New approaches to origin, evolution, and conservation of crops. Columbia University Press, USA.

Guimaraes, C. T., Sills, G. R., & Sobral, B. W. S., (1997). Comparative mapping of Andropogonae: *Saccharum* L. (sugarcane) and its relation to sorghum and maize. *Proc. Nat. Acad. Sci. U. S. A.*, *94*, 14261–14266.

Gupta, P., Varshney, R., Sharma, P., & Ramesh, B., (1999). Molecular markers and their applications in wheat breeding. *Plant Breed.*, *118*, 369–390.

Hamada, H., Petrino, M. G., & Kakunaga, T., (1982). A novel repeat element with Z-DNA-forming potential is widely found in evolutionarily diverse eukaryotic genomes. *Proc. Nat. Acad. Sci. USA*, *79*, 6465–6469.

Hameed, U., Muhammad, K., Pan, Y. -B., Afghan, S., & Iqbal. J., (2012). Use of simple sequence repeat (SSR) markers for DNA fingerprinting and diversity analysis of sugarcane (Saccharum spp.) cultivars resistant and susceptible to red rot. *Genet. Mol. Res.*, *11* (2), 1195–1204.

Hoarau, J. Y., Grivet, L., Offmann, B., Raboin, L. M., Diorflar, J. P., & Payet, J., (2002). Genetic dissection of a modern sugarcane cultivar (*Saccharum* spp.). II. Detection of QTLs for yield components. *Theor. Appl. Genet.*, *105*, 1027–1037.

Hoarau, J. Y., Offmann, B., D'Hont, A., Risterucci, A. M., Roques, D., Glaszmann, J. C., & Grivet, L., (2001). Genetic dissection of a modern sugarcane cultivar *Saccharum spp* I: Genome mapping with AFLP markers. *Theor. Appl. Genet.*, *103*, 84–97.

Huang, X. Q., & Roder, M. S., (2004). Molecular mapping of powdery mildew resistance genes in wheat: A review. *Euphytica*, *137*, 203–223.

Jaccoud, D., Peng, K., Feinstein, D., & Kilian, A., (2001). Diversity arrays: a solid state technology for sequence information independent genotyping. *Nucleic Acids Res.*, *29*, e25.

Jannoo, N., Grivet, L., Chantret, N., Garsmeur, O., Glaszmann, J. C., Arruda P., & D'Hont, A., (2007). Orthologous comparison in a gene-rich region among grasses reveals stability in the sugarcane polyploid genome. *Plant J.*, *50*, 574–585.

Jannoo, N., Grivet, L., David, J., D'Hont A., & Glaszmann, J. C., (2004). Differential chromosome pairing affinities at meiosis in polyploidy sugarcane revealed by molecular markers. *Heredity*, *93*, 460–467.

Jones, C. J., Edwards, K. J., Castaglione, S., Winfield, M. O., Sala, F., & Van de Wiel, C., (1997). Reproducibility testing of RAPD, AFLP, and SSR markers in plants by a network of European laboratories. *Mol. Breed.*, *3*, 381–390.

Jordan, S. A., & Humphries, P., (1994). Single nucleotide polymorphism in exon 2 of the BCP gene on 7q31–q35. *Human Mol. Genet.*, *3*, 1915.

Khoso, A. W., (1992). *Crops of Sindh* (5th edn.).

Kurata, N., Moore, G., Nagamura, Y., Foote, T., Yano, M., Minobe, Y., & Gale, M. D., (1994). Conservation of genome structure between rice and wheat. *Nat. Biotechnol.*, *12*, 276–278.

Lakshmanan, P., Geijkes, R. J., Aitken, K. S., Grof, C. L., Bonnet, G. D., & Smith, G. R., (2005). Sugarcane Biotechnology: the challenges and opportunities. *In Vitro Cellul. Develop. Biol. Plant*, *41*, 345–363.

Le Cunff, L., Garsmeur, O., Raboin, L. M., Pauquet, J., Telismart, H., Selvi, A., et al. (2008). Diploid/polyploid syntenic shuttle mapping and haplotype-specific chromosome walking toward a rust resistance gene (Bru1) in highly polyploid sugarcane (2n _ 12x _ 115). *Genetics*, *180*, 649–660.

Ma, H. M., Schulze, S., Lee, S., Yang, M., Mirkov, E., & Irvine, J., (2004). An EST survey of the sugarcane transcriptome. *Theor. Appl. Genet.*, *108*, 851–863.

Macedo, I. C., Seabra, J. E. A., & Silva, J. E. A. R., (2008). Greenhouse gases emissions in the production and use of ethanol from sugarcane in Brazil: The 2005/2006 averages and a prediction for 2020. *Biomass Bioenrg.*, *32*, 582–595.

Mariotti, J. A., (2002). Selection for sugarcane yield and quality components in subtropical climates. *Sugarcane Int.*, March/April: 22–26.

Markert, C. L., & Moller, F., (1959). Multiple forms of enzymes: Tissue, ontogenetic, and species specific patterns. *Proc. Natl. Acad. Sci. USA*, 45, 753–763.

McIntyre. C., Whan, V., Croft, B., Magarey, R., & Smith, G., (2005). Identification and validation of molecular markers associated with Pachymetra root rot and brown rust resistance in sugarcane using map- and Association-based approaches. *Mol. Breed.*, 11, 151–161.

Ming, R., Liu, S. C., Bowers, J. E., Moore, P. H., Irvine, J. E., & Paterson, A. H., (2002). Construction of a Saccharum consensus genetic map from 2 interspecific crosses. *Crop. Sci.*, 42, 570–583.

Ming, R., Liu, S. C., Lin, Y. R., Da Silva, J., Wilson, W., Braga, D., et al. (1998). Detailed alignment of *Saccharum* and *Sorghum* chromosomes: Comparative organization of closely related diploid and polyploid genomes. *Genetics*, 150, 1663–1682.

Ming, R., Liu, S. C., Moore, P. H., Irvine, J. E., & Paterson, A. H., (2001). QTL analysis in a complex autopolyploid: Genetic control of sugar content in sugarcane. *Genome Res.*, 11, 2075–2084.

Muhammad, K., Pan, Y. -B., Afghan, S., & Iqbal, J., (2013). Genetic variability among the brown rust resistant and susceptible genotypes of sugarcane by RAPD technique. *Pak. J. Bot.*, 45 (1), 163–168.

Mullet, J. E., Klein, R. R., & Klein, P. E., (2002). *Sorghum bicolor*-An important species for comparative grass genomics and a source of beneficial genes for agriculture. *Curr. Opin. Plant Biol.*, 5, 118–121.

Nakamura, Y., Leppert, M., O'Connell, P., Wolff, R., Holm, T., Culver, M., et al. (1987). Variable number tandem repeat (VNTR) markers for human gene mapping. *Science*, 235, 1616–1622.

O'Donoughue, L. S., Wang, Z., Roder, M. S., Kneen, B., Leggett, M., Sorrells, M. E., & Tanksley, S. D., (1992). An RFLP-based linkage map of oats based on a cross between two diploid taxa (*Avena atlantica* × *A. hirtula*). *Genome*, 35, 765–771.

Olsen, M., Hood, L., Cantor, C., & Botstein, D., (1989). A common language for physical mapping of the human genome. *Science*, 245, 1434–1435.

Orellana, C., & Neto, R. B., (2006). Brazil and Japan give fuel to ethanol market. *Nat. Biotechnol.*, 24, 232.

Pan, Y. B., Burner, D. M., Legendre, B. L., Grisham, M. P., & White, W. H., (2004). An assessment of the genetic diversity within a collection of *Saccharum spontaneum* L. with RAPD-PCR. *Genet. Res. Crop Evol.*, 51, 895–903.

Pan, Y. -B., Cordeiro, G. M., Richard, E. P., & Henry, R. J., (2003). Molecular genotyping of sugarcane clones with microsatellite DNA markers. *Maydica*, 48, 319–329.

Panje, R., & Babu, C. N., (1960). Studies on *Saccharum spontaneum*: Distrubution and geographical association of chromosome numbers. *Cytologia*, 25, 152–172.

Paran, I., & Michelmore, R. W., (1993). Development of reliable PCR-based markers linked to downy mildew resistance genes in lettuce. *Theo. Appl. Genet.*, 85, 985–993.

Paterson, A. H., Bowers, J. E., Feltus, F. A., Tang, H., Lin, L., & Wang, X., (2009). Comparative genomics of grasses promises a bountiful harvest. *Plant Physiol.*, 149, 125–131.

Pereira, M. G., Lee, M., Bramel-Cox, P., Woodman, W., Doebley, J., & Whitkus, R., (1994). Construction of an RFLP map in sorghum and comparative mapping in maize. *Genome*, 37, 236–243.

Pimentel, D., & Patzek, T. W., (2005). Ethanol production using corn, switchgrass, and wood, biodiesel production using soybean and sunflower. *Nat. Resour. Res.*, 14, 65–76.

Piperidis, G., & D'Hont A., (2001). Chromosome composition analysis of various *Saccharum* interspecific hybrids by genomic in situ hybridization (GISH). *Proc. Int. Soc. Sugar Cane Technol., 24,* 565–566.

Price, S., (1965). Interspecific hybridization in sugarcane breeding. *Proc. Int. Soc. Sugar Cane Technol., 12,* 1021–1026.

Que, Y., Pan, Y. -B., Lu, Y., Yang, C., Yang, Y., Huang, N., & Xu, L., (2014). Genetic analysis of diversity within a Chinese local sugarcane germplasm based on start codon targeted (SCoT) polymorphism. *Bio. Med. Res. Int.*, http://dx.doi.org/10. 1155/2014/468375.

Raboin, L. M., Oliveira, K. M., Le Cunff, L., Telismart, H., & Roques, D., (2006). Genetic mapping in sugarcane, a high polyploid, using bi-parental progeny: Identification of a gene controlling stalk color and a new rust resistance gene. *Theor. Appl. Genet., 112,* 1382–1391.

Ramsay, L., Macaulay, M., degli Ivanissevich, S., MacLean, K., Cardle, L., & Fuller, J., (2000). A simple sequence repeat-based linkage map of barley. *Genetics, 156,* 1997–2005.

Ripol, M. I., (1994). Statistical aspects of genetic mapping in autopolyploids. *MS. Thesis,* Cornell University, Ithaca, NY.

Roach, B. T., (1972). Nobilisation of sugarcane. *Proc. Int. Soc. Sugar Technol., 14,* 206–216.

Rossi, M., Araujo, P. G., Paulet, F., Garsmeur, O., Dias, V. M. Chen, H., Van Sluys, M. A., & D'Hont, A., (2003). Genomic distribution and characterization of EST-derived resistance gene analogs (RGAs) in sugarcane. *Mol. Genet. Genom., 269,* 406–419.

Ryan, C. C., & Egan, B. T., (1979). Rust. In: Ricaud, C., Egan, B. T., Gillaspie, A. G., & Hughes, C. G., (eds.), *Diseases of Sugarcane* (pp. 189–210). Academic Press, Amsterdam.

Schubert, C., (2006). Can biofuels take centrestage? *Nat. Biotechnolog., 24,* 777–784.

Sills, G., Bridges, W., Al-Janabi, S., & Sobral, B. W. S., (1995). Genetic analysis of agronomic traits in a cross between sugarcane (*Saccharum officinarum* L.) and its presumed progenitor (*S. robustum* Brandes, Jesw. Ex. Grassl). *Mol. Breed., 1,* 355–363.

Singh, K., Ishii, T., Parco, A., Huang, N., Brar, D. S., & Khush, G. S., (1996). Centromere mapping and orientation of the molecular linkage map of rice (*Oryza sativa* L.). *Proc. Nat. Acad. Sci. USA, 93,* 6163–6168.

Sreenivasan, T. V., Ahloohwalia, B. S., & Heinz, D. J., (1987). Cytogenyics. Chapter5. In: Heinz, D. J., (ed.), *Sugarcane Improvement Through Breeding* (pp. 211–153). Elsuvier Amsterdam.

Tanksley, S. D., Young, N. D., Paterson, A. H., & Bonierbale, M., (1989). RFLP mapping in plant breeding: New tools for an old science. *Biotechnology, 7,* 257–264.

Tew, T. L., & Pan, Y. B., (2010). Microsatellite (Simple Sequence Repeat) Marker-based Paternity Analysis of a Seven-Parent Sugarcane Polycross. *Crop Sci., 50,* 1401–1408.

Tew, T. L., White, W. H., Grisham, M. P., Dufrene Jr, E. O., Garrison, D. D., & Veremis, J. C., (2003). Registration of 'HoCP 13 96–540' Sugarcane. *Crop Sci., 45,* 785–786.

VanDeynze, A. E., Nelson, J. C., O'Donoughue, L. S., Ahn, S. N., Siripoonwiwat, W., Harrington, S. E., et al. (1995). Anchor probes for comparative mapping of grass genra. *Theo. Appl. Genet., 249,* 349–356.

Vettore, A. L., Da Silva, F. R., Kemper, E. L., & Arruda, P., (2001). The libraries that made SUCEST. *Genet. Mol. Biol., 24,* 1–7.

Vos, P., Hogers, R., Bleeker, M., Reijans, M., Van De Lee, T., Hornes, M., et al. (1995). AFLP: A new technique for DNA fingerprinting. *Nucleic Acids Res., 23,* 4407–4414.

Wang, Z. M., Devos, K. M., Liu, C. J., & Gale, M. D., (1997). A genetic map of melon (*Cucumis melo* L.) based on amplified length polymorphism (AFLP) markers. *Theo. Appl. Genet., 95,* 791–798.

Waugh, R., Bonar, N., Baird, E., Thomas, B., Graner, A., Hayes, P., & Powell, W., (1997). Homology of AFLP products in three mapping populations of barley. *Mol. Gen. Genet.*, *255*, 311–321.

Welsh, J., & McClelland, M., (1990). Fingerprinting genomes using PCR with arbitrary primers. *Nucleic Acids Res.*, *18*, 7213–7218.

Williams, J. G. K., Kubelik, A. R., Livak, K. J., Rafalski, J. A., & Tingey, S. V., (1990). DNA polymorphisms amplified by arbitrary primers are useful as genetic markers. *Nucleic Acids Res.*, *18*, 6531–6535.

Winter, P., & Kahl, G., (1995). Molecular marker technologies for plant improvement. *World J. Microbiol. Biotechnol.*, *11*, 438–448.

Wu, K., Burnquist, W., Sorrells, M. E., Tew, T. L., Moore, P. H., & Tanksley, S., (1992). The detection and estimation of linkage in polyploids using single-dose restriction fragments. *Theo. Appl. Genet.*, *83*, 294–300.

CHAPTER 8

COMPARATIVE STUDY OF THREE DIFFERENT FERTILIZERS ON YIELD AND QUALITY OF CAPSICUM

SAMEEN RUQIA IMADI[1], KANWAL SHAHZADI[2], and ALVINA GUL[1]

[1]*Atta-ur-Rahman School of Applied Biosciences, National University of Sciences and Technology, Islamabad, Pakistan, E-mail: alvina_gul@yahoo.com*

[2]*Department of Plant Sciences, Quaid-I-Azam University, Islamabad, Pakistan*

ABSTRACT

Capsicum is a widely used group of species of plants belonging to family Solanaceae. *Capsicum* fruits are used in food as for pharmaceutical purposes. There is a variety of medicinal properties of different species of *Capsicum*, which is the cause of its enhanced cultivation. The concentration of secondary metabolites and phenolic contents of *Capsicum* are considered the measure of the quality of *Capsicum*. Different fertilizers are used in different concentrations to enhance the production and quality of *Capsicum* species. Among these fertilizers most widely used are nitrogen fertilizers, potassium fertilizers, and phosphorus fertilizers. Besides these, micronutrients, and macronutrients are also used in certain concentrations to further enhance the yield and quality of crops. This chapter is intended to compare these three fertilizers and their effects on *Capsicum* crops. Finally, the chapter will look forward into the techniques, which can be used for fertilization of *Capsicum* crops without degrading the environment in a sustainable way.

8.1   INTRODUCTION

Capsicum is a commonly grown spice. It is usually referred to as bell pepper, red pepper, sweet pepper and sometimes simply as chili (Appireddy et al., 2008). *Capsicum* is basically a genus of spice plants which contains many species including *Capsicum annum, Capsicum, chinense, Capsicum bacatum, Capsicum mirabile, Capsicum frutescens* and many more (Medina-Lara et al., 2008). *Capsicum* can be used in different ways. It is used as a vegetable, for medicinal purposes and colorant. Medical purposes of *Capsicum* include its uses in treating gastrointestinal disorders, diarrhea, cramps, cardiovascular diseases, and excessive blood clotting and high cholesterol levels. A toothache, alcoholism, malaria, and fever can also be treated by *Capsicum*. This genus is also used for skin related problems. It is due to these purposes, that *Capsicum* genus is widely used. South Asian region is very famous for extensive growth of *Capsicum* and its uses botanically *Capsicum* is considered to be a fruit because it has seeds, but they are used for cooking and many culinary purposes. Hence, they are accepted as a vegetable in the world (Yamamoto et al., 2011). *Capsicum* annum plants are shown in Figure 8.1 and *Capsicum* flowers are shown in Figures 8.2 and 8.3.

FIGURE 8.1   *Capsicum annum* plant.

Bell pepper which is biologically known as *Capsicum* is a rich source of alkaloids especially capsaicin. Besides alkaloids; fatty acids, flavonoids, volatile oils, and carotene pigments are also present in bell pepper. It is also a rich source of vitamin C and zinc, which are vital minerals for the healthy immune system. Vitamin A, rutin, potassium, beta-carotene, calcium, and iron are also present in *Capsicum*. Bell pepper is also a source of magnesium, selenium, phosphorus, sodium, sulfur, and B complex vitamins (Agarwal et al., 2007). Use of chili fruit is determined by its capsaicinoid content. It is due to capsaicinoid content of *Capsicum* fruits that they possess beneficial effects in food and pharmaceutical applications (Gurung et al., 2011).

FIGURE 8.2   *Capsicum annum* plant at flowering stage.

FIGURE 8.3   *Capsicum annum* flower.

Fertilizers are a cost-effective way to increase yield and quality of culti-vated plants. Fertilizers act as insurance for low yields and promote micro-bial carbon utilization (Mulvaney et al., 2009). It is seen that soil moisture is highly enhanced by irrigation, which results in high yield and biomass of *Capsicum*. Usage of nitrogen fertilizer further enhances the yield of *Capsicum* (Yi-tao et al., 2007). Application of organic fertilizers on *Capsicum* plants results in overexpression of height of the plant, size of leave and dry weight of leaves (Karanatsidisand Berova, 2009). In an experiment, organic approaches and integrated nutrient management approaches were applied to *Capsicum* plants and it was observed that many folds in both cases increase quality as well as yield of Capsicum. Plots, which are treated with organic nutrients, are shown to possess high soil pH, high levels of organic carbon and increased microbial activities. Amount of urease, alkaline phosphatase, and dehydrogenase is also seen to be high in organic nutrient supply. It was also observed that levels of nitrogen, potassium, and phosphorus are high in integrated nutrient management as compared to organic nutrient supply (Appireddy et al., 2008).

Treatment of *Capsicum* plants with zinc micronutrients in the form of fertilizers result in an increase in yield of *Capsicum* fruit. Treatment with fertilizers fortified with iron results in increased level of seeds and fruits (Dongre et al., 2000). The yield of *Capsicum* significantly grows with the application of nitrogen fertilizers, potassium fertilizers and phosphorus fertilizers (Shi-min, 2005). Application of biofertilizers on *Capsicum* plants is also observed to increase their growth. Photosynthetic pigments are also enhanced and leaf gas exchange is improved (Berovaand Karanatsidis, 2008). Foliar fertilizers were applied to *Capsicum* plants. It was observed that these fertilizers were linked to high yield, enhanced chlorophyll content and high photosynthetic rate. They also increase the mass of leaves, vitamin C content, soluble sugars leading to the enhancement of quality of fruit (Rui-hai et al., 2009). Presence of nutrients in soil determines the amount of nutrients absorbed by crops. Fertilization is considered the most expensive cultural practice for increasing the yield and quality of crops (Gaskell and Smith, 2007).

It has been observed that many physiological diseases occur in *Capsicum* plants because of lack of fertilizer usage. Improper fertilization also leads to many diseases. Research on fertilizers is performed to produce high-quality and high yield sweet pepper (Zhuqing, 2007). Application of potassium phosphate and ammonium lignosulphonate fertilizers on *Capsicum* plants is associated with a decrease in the rate of diseases and infections. It not only helps the plant to prevent diseases but also is also involved in treating infections like those caused by bacterial spots (Abbasi et al., 2002). The deficiency of nitrogen fertilizer reduces growth as well as delay flowering of chili pepper. Potassium and phosphorus deficiencies are associated with a reduction in number of flowers and number of fruits. Calcium deficiency is seen to dry up the shoot tips and cause Chlorosis in young leaves. It has been observed that magnesium deficient plants become yellow and necrotic lesions (Roychoudhry et al., 1990).

8.2 NITROGEN FERTILIZERS

All the fertilizers, which use ammonia in them, are referred to as nitrogen fertilizers. Ammonia is considered an essential ingredient of these fertilizers. Sometimes ammonia is directly injected into soil as a fertilizer. These fertilizers are a cost-effective way of generation of high crop yield. Organic nitrogen is continuously decreased from the environment. Application of nitrogen fertilizers increases the productivity of soil by maintaining its level. Soil fertility is significantly enhanced with the use of nitrogen fertilizers. These are the most widely used fertilizers (Mulvaney et al., 2009). Slow release nitrogen materials can also be used to reduce leaching and losses of nitrogen from sandy soils (Guertal, 2000). Nitrogen fertilizers are available in different forms including nitrate fertilizer, ammonium fertilizer and urea fertilizer (Zhou et al., 2006). The rapid increase in anthropogenic production of nitrogen fertilizers is considered to be a major factor which accounts for the increase in growth of agricultural food production (Motavalli et al., 2008). Nitrogen is considered most important and limiting factor for production of crops. Agricultural world is incomplete without nitrogen fertilizers. Efficient use of nitrogen fertilizers can yield to the sustainability of the environment. Plants are able to bear with the challenging and ever-changing environment due to the dynamic nature of nitrogen. For evaluating the crop requirements of nitrogen, crop response to nitrogen needs to be measured. If nitrogen is continuously volatized, leached, ran off the surface or denitrified, then nitrogen recovery in crops is low. It is observed that recovery of nitrogen is less than 50% of worldwide crops (Fageriaand Baligar, 2005).

Slow release nitrogen fertilizers provide many potential benefits for the production of different vegetables. These fertilizers also tend to reduce environmental risks but also save many production costs (Guertal, 2009). Nitrogen applied largely in ammoniacal fertilizers proves as a cost-effective way for minimizing the low yield (Mulvaney et al., 2009).

Availability of nitrogen is considered to be a very important determinant of crop yield. Crop production is highly dependent on nitrogen availability (Ankumah et al., 2003). Conventional nitrogen fertilizers, as well as organic nitrogen fertilizers, are involved in enhancing the soil biota activity, mycorrhizal colonization, and leaf antioxidant content. Although organic nitrogen fertilizers are more effective as compared to conventional, nitrogen fertilizers but still they are involved in increasing tolerance of plants towards soil pathogens (Montalba et al., 2010). Nitrogen fertilizers are also involved in enhancing the amount of nitrates in soil that results in an increase in tolerance of plants toward plant pathogens (BaranauSkiené et al., 2003).

Nitrogen is known to be the most important and most costly nutrient to manage in the soil for high-quality and yield of crops. Wide varieties of nitrogen are available commercially but their use depends upon their cost, nitrogen content and nitrogen bioavailability (Gaskell and Smith, 2007). Nitrogen fertilizers have a significant effect on the height of *Capsicum* plants, lateral stem number as well as leaf chlorophyll content (Aminifard et al., 2013). It has been observed that among all nitrogen fertilizers, the absorption rates of nitrate fertilizers are significantly higher (Dai et al., 2001). Use of nitrogen fertilizers on sweet pepper increases the vigor of plants. The increase in vigouris confirmed by increased length of plants, number of leaves and shoots on plants, dry weight, and fresh weight of different organs of plants. The yield of healthy fruits with enhanced physical quality as well as chemical quality is also observed (Abdand Faten, 2009). It has been calculated after application of nitrogen fertilizers on *Capsicum* plants, that maximum growth rate and maximum accumulation of dry matter in fruit starts after 90 to 150 days of planting. This means that this period of plant life cycle is the time in which plants show maximum growth rate. Increasing the concentration of nitrogen in soil; increases the uptake of nitrogen by plants and stimulate the uptake of potassium as well as phosphorus fertilizers. It can be said that nitrogen plays a synergistic role in the uptake of potassium and phosphorus fertilizers (Qawasmi et al., 1999).

Nitrogen treatment provided to *Capsicum* plants grown underwater stress is shown to reduce the symptoms and effects of stress. The plant height, number of leaves, area of leaves, number of flowers and yield, which is significantly reduced due to the low moisture content in soil, is enhanced to a normal level by application of nitrogen fertilizers. There are huge benefits of applications of fertilizers on *Capsicum* (Abayomi et al., 2012).

8.2.1   EFFECTS OF NITROGEN FERTILIZERS ON YIELD AND QUALITY OF CAPSICUM

The height of *Capsicum* plants is significantly affected by usage of nitrogen fertilizers. Lateral stem number and chlorophyll content of leaves are also seen to be enhanced by nitrogen fertilizers. Nitrogen treatment is also seen to effect fruit volume and fruit weight (Aminifard et al., 2013). *Capsicum* plants which are fed with mixed nitrogen forms which also contain urea are observed to have higher shoot concentration of mineral ions like potassium, phosphorus, iron, and boron (Houdusse et al., 2007). *Capsicum* fruits which are grown under high levels of nitrogen fertilizers are shown to possess high firmness during cool storage (Martinez et al., 2002). Nitrogen fertilizers are observed to increase plant growth and fruit significantly while keeping high levels of capsaicin (Medina-Lara et al., 2008). Increase in quantity of nitrogen fertilizer in soil is associated with a decrease in the quantity of capsaicin content in the fruit of *Capsicum* (Changshan et al., 2005).

The function of nitrogen in plant growth and development is connected to carbon. Nitrogen bioavailability in plants controls the ratio of carbon to nitrogen (Aulakhand Malhi, 2005). With the application of nitrogen fertilizers, nitrate content of leaf of *Capsicum* plants is highly increased. Quality of *Capsicum* fruits is largely increased by use of nitrogen fertilizers (Wang et al., 2000). Application of nitrogen fertilizers on *Capsicum* plants enhances their yield. The plants which do not get enough nitrogen, produce low yield during harvest (Romic et al., 2003). Fruit quality and yield of *Capsicum* is increased by application of nitrogen fertilizers in the form of ammonia and nitrate. It was observed in an experiment conducted in hydroponics. It was observed that as soon as the concentration of ammonia increases and reaches above 2mmol per liter, the quality, as well as yield of bell pepper, decreases. It is due to the fact the excessive application of ammonia hinders with calcium uptake. It was also seen that occurrence of flat fruits increase with further increases in the concentration of ammonia (Bar-Tal et al., 2001).

It has been observed in an experiment that contents of vitamin C and soluble sugar decrease in sweet pepper, with the application of nitrogen fertilizers. The fruits which are treated with excessive nitrogen have less content of vitamin C as well as soluble sugars, as compared to control fruits. This also results in an increase in a dry matter of plant. It has been observed that certain nitrogen fertilizers including ammonia and urea are also involved an increase in yield of sweet pepper (LV et al., 2005). Addition of high amounts of nitrogen fertilizers in *Capsicum* plants leads to the development of high yield of fruit (Qawasmi et al., 1999). An experiment was conducted in which four different forms of nitrogen fertilizers including urea, ammonium sulfate, ammonium nitrate, and potassium nitrate were applied on *Capsicum* plants. It was explored that all of these four types of fertilizers increased plant growth parameters, fruit yield, and fruit quality. Ammonium nitrate was observed to be involved in the production of maximum yield (Owusu et al., 2000). Among all the nutrients, nitrogen is a most important nutrient for the production of pepper fruit (JinMyeon et al., 2009).

8.3 POTASSIUM FERTILIZERS

Among different mineral nutrients, potassium stands out, as it is a cation, which has the strongest influence on the quality and yield of plants. Fruit marketability, consumer preference and phytonutrients present in plants are highly dependent on the presence of potassium cations (Lester et al., 2010). Importance of potassium in ensuring the normal growth and development, production, quality, and yield of vegetables is well known (Bidariand Hebsur, 2011). Potassium is considered an essential nutrient and one of the most abundant cations in plant cells (Pacheco-Arjona et al., 2011).

Potassium sulfate fertilizers are shown to significantly promote the growth of pepper and reduce the lead content of *Capsicum* roots, shoots, and fruits. Low doses of potassium chloride fertilizers are observed to enhance the growth of *Capsicum* plants (Yanhong et al., 2009). Potassium fertilizers enhance the stress tolerance in *Capsicum* plants. Application of potassium fertilizers not only reduces the symptoms of stress but also helps the plant to combat it while increasing the quality as well as yield of crops (Veeranna et al., 2000).

8.3.1 EFFECTS OF POTASSIUM FERTILIZERS ON YIELD AND QUALITY OF CAPSICUM

Application of phosphorus fertilizers on *Capsicum* plants is seen to be associated with fruit length. Vitamin C content is also significantly enhanced by usage of phosphorus fertilizers (Gao et al., 2009). *Capsicum* fruits, which are grown under high levels of potassium fertilizers, are shown to possess high firmness during cool storage (Martinez et al., 2002). Usage of potassium fertilizers even in hydroponic conditions is seen to be significantly effecting plant growth, height, weight, stem diameter, dry weight and leaf area of *Capsicum* plants. Increasing levels of potassium also increase the yield of *Capsicum* fruits (Aldana, 2005). Potassium fertilizers not only affect growth and development of *Capsicum* plants but also play important role in resistance of plants towards many diseases. Improved potassium nutrition during the life cycle of *Capsicum* plants is associated with the strengthening of resistance towards diseases. It also regulates many physiological mechanisms including metabolic pathways like a phenolic pathway, carbon metabolism, nitrogen metabolism and active oxygen metabolism (Liu et al., 2006).

It has been observed that potassium fertilizers do not affect the quality of fruit except for the degree of shriveling of fruit. The degree of shriveling is directly proportional to the increasing amount of potassium fertilizers (Hochmuth et al., 1994). It has been observed that the application of potassium fertilizers on *Capsicum* plants is not associated with an enhancement in yield, the vigor of the plant, number of fruits per plant and single fruit weight. Potassium fertilizers are seen to enhance the growth period and fresh yield of red pepper (YuanYuan et al., 2007.). Potassium fertilizers significantly increase the quality of fruits by effecting total soluble solids, titratable acidity and ascorbic acid content (El-Masry, 2000). It is measured that application of potassium fertilizers increases the fruit parameters like fruit length, fruit weight, and vitamin C content. This is associated with enhanced quality of *Capsicum* fruits (El-Bassiony et al., 2010).

Potassium fertilizer was applied on hot pepper plants grown underwater stress conditions. It was observed that yield is significantly increased with the application of potassium fertilizers (Thakur et al., 2000). Application of potassium fertilizers at a concentration of 200 ppm on salt stressed *Capsicum* plants show that foliar application of these fertilizers leads to an increase in plant growth, plant biomass production as well as fruit yield. Total phenols in fruits and chlorophyll content in leaves also increased by treatment with potassium fertilizers. Application of potassium on *Capsicum* plants is somehow associated with mitigation of salt stress by increasing growth and yield of fruits (Hussein et al., 2012). Potassium fertilizer is widely used for increasing the quality of *Capsicum*, rather than increasing the yield. Potassium is observed to balance the acid to sugar ratio in *Capsicum* fruits. Deficiency of potassium may lead to irregular fruit color development in red pepper. Increase in the levels of potassium in soil results in an increase in the production of capsanthin in *Capsicum*; which increases its concentration. Carotenoid content in *Capsicum* is dependent on levels of potassium fertilizers in the soil (Bidari and Hebsur, 2011).

8.4   PHOSPHORUS FERTILIZERS

The growth of crops is continuously threatened by the limitation of phosphorus in most of the temperate and tropical soils. Limitation of phosphorus can be overcome by the use of phosphorus fertilizers, which are made up of salts of phosphate anions. These fertilizers have phosphates in them, which are absorbable forms of phosphorus. The fertilizers are efficient in providing excess phosphorus. Besides the management of phosphorus, soil type also determines the efficiency of phosphorus in it (Akinrinde and Adigun, 2005). Increasing the rates of phosphorus fertilizers in soil does not affect the growth of *Capsicum* plants but high levels of phosphorus results in the increased height of plants (Akinrinde and Adigun, 2005).

8.4.1   EFFECTS OF PHOSPHORUS FERTILIZERS ON YIELD AND QUALITY OF CAPSICUM

Usage of phosphorus fertilizers even in hydroponic conditions is seen to be significantly effecting plant growth, height, weight, stem diameter, dry weight and leaf area of *Capsicum* plants. Increasing levels of phosphorus also increase the yield of *Capsicum* fruits (Aldana, 2005). Increase in phosphorus levels in soil significantly enhance plant height, number of leaves per plant, number of branches per plant and leaf area in *Capsicum* plants. It also results in early flowering, maturity, and high yield (Alabi, 2006). It has been observed that treatment of *Capsicum* plants with 120 kg of phosphate fertilizer results in a remarkable increase of up to 10% in length of fruit. It also enhances the fruit shoulder, fruit wall and vitamin C content in *Capsicum* fruits (Gao et al., 2009). Phosphorus fertilizers significantly increase the quality of fruits by effecting total soluble solids, titratable acidity and ascorbic acid content (El-Masry, 2000). Crop yield and soil fertility are significantly increased because of phosphorus application. Crop production in different regions is largely dependent on phosphorus fertilizers (Liu et al., 2007a).

8.5   COMPARATIVE EFFECTS OF NITROGEN, POTASSIUM, AND PHOSPHORUS FERTILIZERS ON *CAPSICUM*

The interaction between nitrogen and phosphorus is one of the most important nutrient interactions for plant growth. Along with nitrogen, potassium, and phosphorus, sulfur interacts to increase the growth and yield of *Capsicum* plants (Aulakhand Malhi, 2005). The yield of *Capsicum* crops is significantly increased with the use of nitrogen, phosphorus, and potassium fertilizers. It is observed that the efficient combination of water and fertilizers result in enhancement of quality and yields of crops (Zhao et al., 2004). A study was conducted in which effects of nitrogen and potassium fertilizers were measured on nutrient content in red pepper. It was observed that increasing concentration of nitrogen fertilizers result in an increase of lycopene content in roots and beta-carotene levels in fruits. The lipophilic amino acid content was also seen to be enhanced by nitrate fertilizers. Nitrogen treatments were not observed to be associated with vitamin C, sugar, and total phenolic content in fruits. It was also observed that potassium treatments had no significant effects on the nutrient quality of red pepper. Nitrogen fertilizers significantly enhance nutrient quality of Capsicum fruits and not by potassium fertilizers; rather yield is enhanced by potassium fertilizers (Flores et al., 2004).

It is observed that organic nitrogen fertilizers, which are applied to Capsicum plants, have little effect as compared to the application of combined nutrient fertilizers (Ghonameand Shafeek, 2005). It is observed that application of nitrogen, phosphorus, and potassium fertilizers on crops result in enhancement of their nutrient content as well as vitamins. Phosphorus fertilizers are considered best suited for the yield of crops (Liu et al., 2007b). It is proved in experiments that application of nitrogen, phosphorus, and potassium fertilizers on *Capsicum* plants result in enhancement of soluble sugar content and vitamin C concentration in *Capsicum* fruits. Excessive use of these fertilizers may lead to a decrease in quality of fruits (Dong and Wei, 2009). The combined effects of nitrogen and phosphorus fertilizers on *Capsicum* plants were observed. It was seen that these fertilizers have significant effects on plants when combined with water (Juncang, 2000).

Combined treatment of different concentrations of nitrogen, potassium, and phosphorus fertilizers was applied on *Capsicum* plants. It has been observed that the highest concentration of combined nitrogen and potassium fertilizers is associated with the highest yield of *Capsicum* fruits. *Maximum* uptake of phosphorus is also seen to be highest in case of maximum concentration of nitrogen and potassium (Hari et al., 2007). When the *Capsicum* plants are treated with nitrogen, phosphorus, and potassium fertilizers in a ratio of 100:50:50 kg per hectare for a period of 105 days, it has been observed that plants show a high yield of green chili (Tumbare and Nikam, 2004). It has been explored that application of NPK fertilizers combined with zinc and boron, significantly decrease the yield of *Capsicum* plants. This is in contrast to the application of only nitrogen, phosphorus, and potassium fertilizers, which are involved in the enhancement of yield of the crop

(Liao et al., 2008). It has been observed that when integrated management approach for nitrogen, potassium, and phosphorus fertilizers was applied on *Capsicum* plants, it resulted in increased growth, yield, and quality of *Capsicum* fruits (Malik et al., 2011).

Rational irrigation by water and fertilizers can enhance the yield of *Capsicum* plants. Nitrogen fertilizer is observed to increase the yield whereas potassium fertilizers are seen to increase the production. Fruit growth and fruit weight are increased with the application of all three nitrogen, phosphorus, and potassium fertilizers. Vitamin C content is decreased by treatment of nitrogen fertilizers but is increased with potassium fertilizers. Increasing fertilizers concentrations are linked to an increase in water-soluble sugar and fat contents in *Capsicum* fruits. *Capsicum* quality and yield is explored to be best with increased irrigation in combination with high nitrogen, phosphorus, and potassium concentrations (Yang et al., 2009). Application of different fertilizers on *Capsicum* plants showed that nitrogen fertilizers are the one, which highly affects the quality and yield of *capsicum* fruits. Following the order of effects, it is observed that nitrogen is followed by potassium; which is further followed by phosphorus (Hay Yan et al., 2009).

It has been proved through research that *Capsicum* requires more nitrogen and potassium. It has been observed in a study that N, P, and K ratio of 1: 0.4: 0.85 is best suited for the production of highest quality and yield of *Capsicum* fruits (Fangyu and Rongjian, 2009). Application of nitrogen fertilizers has significant effects on plant growth and fruit yield whereas phosphorus application enhances yield and yield attributes (Chaudhary et al., 2007). The response of pepper plants towards nitrogen fertilizers is observed as an increase in plant height, fruit yield and quality of fruit with increased nutrient contents. Potassium fertilizers were however observed to be ineffective on plant yield. Rather they enhance the flowering of plants. Nitrogen treatment is also observed to enhance nitrogen, phosphorus, and potassium concentrations in plants (Ortas, 2013). Increasing concentration of potassium in soil; results in an increase in yield and growth of *Capsicum* fruit. But this increase is low as compared to the increase shown by the application of nitrogen fertilizers (Medhi et al., 1993).

Increase in nitrogen doses is associated with a significant increase in length and breadth of fruit and number of fruits per plant. The average weight of fruit is also increased at a nitrogen concentration of 150 kg nitrogen per hectare. However, the average weight of fruit and yield is significantly increased at phosphorus levels of 30 kg per hectare. 60 kg per hectare of phosphorus is observed to increase the length of fruits and number of fruits per plant (Roy et al., 2011). Effects of different fertilizers on yield and quality of *Capsicum* are given in Figure 8.4.

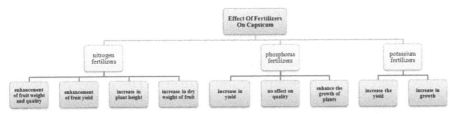

FIGURE 8.4 Effect of nitrogen, phosphorus, and potassium fertilizers on capsicum.

8.6 CONCLUSION

It has been seen that horticulture production largely focuses on increase in productivity of crops whether by intensification of fertilizers or by intensification of irrigation. This results in costs to the environment. It has been observed that high concentration of nitrogen and different fertilizer minerals in soil result in significant negative effects on quality of *Capsicum* as well as the content of secondary metabolites and vitamins. Sustainable production practices need to be implied for growing *Capsicum* crops. Nutrient content of *Capsicum* has been raised too many folds in recent years due to use of fertilizers. Integrated management approaches for fertilizers should be followed to enhance environmental stability as well as social sustainability (Stefanelli et al., 2010). Fertilizers, which use produce low emissions of greenhouse gases, should be applied for the cultivation of *Capsicum* crops. This will have a positive impact on global warming (Sistani et al., 2011).

Application of suitable nitrogen, potassium, and phosphorus fertilizers increases the yield as well as the quality of bell pepper (Shi-min et al., 2005). It has been confirmed through experiments that integrated nutrient management technique in terms of fertilizers prove to be the best technique for increasing the quality and yield of *Capsicum* plants. Application of integrated nutrient management technique results in highest yield and first class fruits. It is also seen that organic farming increases antioxidant activity but decreases chlorophyll content and beta-carotene (del Amor, 2007). Application of phosphate fertilizers may lead to accumulation of phosphate in the soil. This will result in the degradation of soil. Hence, it is suggested that the amount of phosphate fertilizers, which is recommended, should be reduced, as *Capsicum* can easily grow and produce a quality crop and high yield in low concentrations of phosphorus. Organic agriculture is considered the best way to produce high-quality crops with high yields without the use of any agrochemicals. Humic acid is the organic matter, which can serve as the best source of fertilization for plants. Foliar application of humic acid is observed to produce intense growth in plants with a high number of fruits per plant and high nutrient content fruits. Soil treatment of humic acid also produces a good yield of pepper (Karakurt et al., 2009). Hence, the humic acid approach can also be applied for sustainable agriculture as it cannot degrade the environment by phosphorus accumulation. Besides using phosphorus fertilizers, phosphate-solubilizing microorganisms can be used which are potent solubilizer of phosphate from soil. This avoids the accumulation of phosphorus in soil and does not make them toxic. This approach can be used for growing sustainable *Capsicum* crops (Khan et al., 2007).

KEYWORDS

- Capsicum
- fertilizers
- mineral nutrition
- yield

REFERENCES

Abayomi, Y. A., Aduloju, M. O., Egbewunmi M. A., & Suleiman, B. O., (2012). Effects of soil moisture contents and rates of NPK fertilizer application on growth and fruit yields of pepper (*Capsicum* spp.) genotypes. *Int. J. Agri. Sci.*, 2 (7), 651–663.

Abbasi, P. A., Soltani, N., Cuppels, D. A., & Lazarovitz, G., (2002). Reduction of bacterial spot disease severity on tomato and pepper plants with foliar applications of ammonium lignosulfonate and potassium phosphate. *Plant Dis.*, 86 (11), 1232–1236.

Abd El-al, & Faten, S., (2009). Effect of urea and some organic acids on plant growth, fruit yield and its quality on sweet pepper (*Capsicum annum*). *Res. J. Agric. Biol. Sci.*, 5 (4), 372–379.

Agarwal, A., Gupta, S., & Ahmed, Z., (2007). Influence of plant densities on productivity of bell pepper (*Capsicum annuum* L.) under greenhouse in high altitude cold desert of Ladakh. In: *Int. Symposium on Medicinal and Nutraceutical Plants*, 756, 309–314.

Akinrinde, E. A., & Adigun, I. O., (2005). Phosphorus-use efficiency by Pepper (*Capsicum frutescens*) and Okra (*Abelmoschusesculentum*) at different phosphorus fertilizer application levels on two tropical soils. *J. Appl. Sci.*, 5 (10), 1785–1791.

Alabi, D. A., (2006). Effects of fertilizer phosphorus and poultry droppings treatments on growth and nutrient components of pepper (*Capsicum annuum* L). *African J. Biotechnol.*, 5 (8), 671–677.

Aldana, M. E., (2005). Effect of phosphorus and potassium fertility on fruit quality and growth of Tabasco pepper (*Capsicum frutescens*) in hydroponic culture. *Doctoral Dissertation*, Louisiana State University.

Aminifard, M. H., Aroiee, H., Ameri, A., & Fatemi, H., (2013). Effect of plant density and nitrogen fertilizer on growth, yield, and fruit quality of sweet pepper (*Capsicum annum* L.). *Af. J. Agric. Res.*, 7 (6), 859–866.

Ankumah, R. O., Khan, V., Mwamba, K., & Kpomblekou, A. K., (2003). The influence of source and timing of nitrogen fertilizers on yield and nitrogen use efficiency of four sweet potato cultivars. *Agri, Ecosys Environ.*, 100 (2–3), 201–207.

Appireddy, G. K., Saha, S., Mina, B. L., Kundu, S., Selvakumar, G., & Gupta, H. S., (2008). Effect of organic manures and integrated nutrient management on yield potential of bell pepper (*Capsicum annuum*) varieties and on soil properties. *Arch. Agron. Soil Sci.*, 54 (2), 127–137.

Aulakh, M. S., & Malhi, S. S., (2005). Interactions of nitrogen with other nutrients and water: Effect on crop yield and quality, nutrient use efficiency, carbon sequestration, and environmental pollution. *Adv. Agron.*, 86, 341–409.

Baranau Skiené, R., Venskutonis, P. R., Viškelis, P., & Dambrauskiené, E., (2003). Influence of nitrogen fertilizers on the yield and composition of Thyme (*Thymus vulgaris*). *J. Agric. Food Chem.*, 51 (26), 7751–7758.

Bar-Tal, A., Aloni, B., Karni, L., Oserovitz, J., Hazan, A., Itach, M., et al. (2001). Nitrogen nutrition of greenhouse Pepper. I. Effects of nitrogen concentration and NO_3, NH_4 ratio on yield, fruit shape, and the incidence of blossom-end rot in relation to plant mineral composition. *Hort. Science*, 36 (7), 1244–1251.

Berova, M., & Karanatsidis, G., (2008). Influence of bio-fertilizer, produced by *Lumbricusrubellus* on growth, leaf gas-exchange and photosynthetic pigment content of pepper plants (*Capsicum annuum* L.). In: *IV Balkan Symposium on Vegetables and Potatoes 830* (pp. 447–452).

Bidari, B. I., & Hebsur, N. S., (2011). Potassium in relation to yield and quality of selected vegetable crops. *Karnataka J. Agric. Sci.*, 24 (1), 55–59.

Changshan, L., Jinling, W., & Guangjian, Y., (2005). Effects of nitrogen fertilizer in capsaicin content of pepper fruits. *J. Changjiang. Vegetables*, 7, 038.

Chaudhary, A. S., Sachan, S. K., & Singh, R. L., (2007). Effect of spacing, nitrogen, and phosphorus on growth and yield of *Capsicum* hybrid. *Int. J. Agri. Sci.*, 3 (1), 12–14.

Dai, T. B., Cao, W. X., & Jing, Q., (2001). Effects of nitrogen form on nitrogen absorption and photosynthesis of different wheat genotypes. *Chin. J. App. Ecol.*, 6, 011.

Dong, J. X., & Wei, C. X., (2009). Effects of N, P, K application on the yield and quality of hot pepper. *J. Mountain Agri. Biol.*, 5, 006.

Dongre, S. M., Mahorkar, V. K., Joshi, P. S., & Deo, D. D., (2000). Effect of micro-nutrients spray on yield and quality of chili (*Capsicum annuum* L.) varJayanti. *Agri. Sci. Digest.*, 20 (2), 106–107.

El Amor, F. M., (2007). Yield and fruit quality response of sweet pepper to organic and mineral fertilization. *Renew. Agri. Food Sys.*, 22 (3), 233–238.

El-Bassiony, A. M., Fawzy, Z. F., Abd El-Samad, E. H., & Riad, G. S., (2010). Growth, yield, and fruit quality of sweet pepper plants (*Capsicum annum* L.) as affected by potassium fertilization. *J. Am. Sci.*, 6 (12), 722–729.

El-Masry, T. A., (2000). Growth, yield, and fruit quality response in sweet pepper to varying rates of potassium fertilization and different concentrations of paclobutrazol foliar application. *Annals Agri. Sci.*, 38 (2), 1147–1157.

Fageria, N. K., & Baligar, V. C., (2005). Enhancing nitrogen use efficiency in crop plants. *Adv. Agron.*, *88*, 97–185.

Fangyu, X. X. P. S. Z., & Rongjian, L. Y. X. T. H., (2009). Effect models of nitrogen, phosphorus, and potassium fertilizer application in *Capsicum. J, Chin. Capsicum.*, *4*, 015.

Flores, P., Navarro, J. M., Garrido, C., Rubio, J. S., & Martínez, V., (2004). Influence of Ca^{2+}, K^+ and NO^{3-} fertilisation on nutritional quality of pepper. *J. Sci. Food Agri.*, *84* (6), 569–574.

Gao, S. T., Huang, L., Zhao, K., Liu, S. T., Li, W. X., & Zhou, X. L., (2009). Effects of different application rates of phosphate fertilizer on quality of pepper (*Capsicum annuum* L.). *Shandong Agri. Sci.*, *1*, 026.

Gaskell, M., & Smith, R., (2007). Nitrogen sources for organic vegetable crops. *Hort. Technology*, *17* (4), 431–441.

Ghoname, A., & Shafeek, M. R., (2005). Growth and productivity of sweet pepper (*Capsicum annum* L.) grown in plastic house as affected by organic, mineral, and Bio-N-fertilizers. *J. Agron.*, *4* (4), 369–372.

Guertal, E. A., (2000). Pre-plant slow-release nitrogen fertilizers produce similar bell pepper yields as split applications of soluble fertilizer. *Agron. J.*, *92* (2), 388–393.

Guertal, E. A., (2009). Slow-release nitrogen fertilizers in vegetable production: A review. *Hort. Technology*, *19* (1), 16–19.

Gurung, T., Techawongstien, S., Suriharn, B., & Techawongstien, S., (2011). Impact of environments on the accumulation of capsaicinoids in *Capsicum* spp. *Hort. Science.*, *46* (12), 1576–1581.

Hari, G. S., Rao, P. V., Reddy, Y. N., & Reddy, M. S., (2007). Effect of nitrogen and potassium levels on yield and nutrient uptake in paprika (*Capsicum annuum* L.) under irrigated conditions of Northern Telangana Zone of Andhra Pradesh. *Asian J. Horticulture*, *2* (1), 193–196.

HayYan, C., Xue, G., YuanBo, Z., ZhengHong, L., Hui, Y., & Jun, W., (2009). The effect of different combinations of nitrogen, phosphorus, and potassium fertilizer on yield of pepper. *Guizhou. Agric. Sci.*, *7*, 166–167.

Hochmuth, G., Shuler, K., Hanlon, E., & Roe, N., (1994). Pepper response to fertilization with soluble and controlled- release potassium fertilizers. *Proc. Fla. State Hort. Soc.*, *107*, 132–139.

Houdusse, F., Garnica, M., & Garćia-Mina, J. M., (2007). Nitrogen fertilizer source effects on the growth and mineral nutrition of pepper (*Capsicum annuum* L.) and wheat (*Triticum aestivum* L.). *J. Sci. Food Agric.*, *87* (11), 2099–2105.

Hussein, M. M., El-Faham, S. Y., & Alva, A. K., (2012). Pepper plants growth, yield, photosynthetic pigments, and total phenols as affected by foliar application of potassium under different salinity irrigation water. *Agric. Sci.*, *3* (2), 241–248.

JinMyeon, P., InBog, L., YunIm, K., & KiSung, H., (2009). Effects of mineral and organic fertilization on yield of hot pepper and changes in chemical properties of upland soil. *Korean J. Hort. Sci. Tech.*, *27* (1), 24–29.

Juncang, G. Y. L. J. T., (2000). Study on the effects of water-fertilizer coupling on hot pepper under drip irrigating in greenhouse. *J. Ningxia Agri. College*, *3*, 008.

Karakurt, Y., Unlu, H., Unlu, H., & Padem, H., (2009). The influence of foliar and soil fertilization of humic acid on yield and quality of pepper. *Acta Agr. Scand., Section B.*, *59* (3), 233–237.

Karanatsidis, G., & Berova, M., (2009). Effect of organic-N fertilizer on growth and some physiological parameters in pepper plants (*Capsicum Annum* L.). *Biotech. Biotechnol. Equip.*, *23* (1), 254–257.

Khan, M. S., Zaidi, A., & Wani, P. A., (2007). Role of phosphate-solubilizing microorganisms in sustainable agriculture – A review. *Agron. Sust. Develop.*, *27* (1), 29–43.

Lester, G. E., Jifon, J. L., & Makus, D. J., (2010). Impact of potassium nutrition on food quality of fruits and vegetables: a condensed and concise review of literature. *Better Crops.*, *94* (1), 18–21.

Liao, W. H., Liu, J. L., Jia, K., & Meng, N., (2008). Investigation of restrict yield nutrient and the balance of input and output N, P, K in rotation of Chinese cabbage and *Capsicum*. *Acta Agr. Boreali-Sinica*, *3,* 052.

Liu, E. K., Zhao, B. Q., Hu, C. H., Li, X. Y., & Li, Y. T., (2007a). Effects of long-term nitrogen, phosphorus, and potassium fertilizer applications on maize yield and soil fertility. *Plant Nutr. Fert. Sci.*, *5,* 006.

Liu, J-L, Liao, W-H., Zhang, Z-X., Zhang, H-T., Wang, X. J., & Meng, N., (2007b). Effect of phosphate fertilizer and manure on crop yield, soil P accumulation, and the environmental risk assessment. *Agric. Sci. Chin.*, *6* (9), 1107–1114.

Liu, X. Y., Ping, H. E., & Ji-yun, J., (2006). Advances in effect of potassium nutrition on plant disease resistance and its mechanism. *Plant Nutr. Fert. Sci.*, *3,* 025.

LV, C. S., Wang, J. L., Yu, G. J., & Chen, X. P., (2005). Effects of nitrogen fertilizer on quality and yield of pepper fruits. *J. Northeast Agric. Univ.*, *4,* 011.

Malik, A. A., Chattoo, M. A., Sheemar, G., & Rashid, R., (2011). Growth, yield, and fruit quality of sweet pepper hybrid SH-SP–5 (*Capsicum annum* L.) as affected by integration of inorganic fertilizers and organic manures (FYM). *J. Agric. Tech.*, *7* (4), 1037–1048.

Martinez, Y., Diaz, L., & Manzano, J., (2002). Influences of nitrogen and potassium fertilizer on the quality of'Jupiterpepper (*Capsicum annuum*) under storage. In*L XXVI International Horticultural Congress: Issues and Advances in Postharvest Horticulture 628*, pp. 135–140.

Medhi, R. P., Singh, B., & Parthasarathy, V. A., (1993). Effect of varying levels of nitrogen, phosphorus, and potassium on chilies. *Prog. Hort.*, *22* (1–4), 173–175.

Medina-Lara, F., Echevarría-Machado, I., Pacheco-Arjona, R., Ruiz-Lau, N., Guzmá-Antonio, A., & Martinez-Estevez, M., (2008). Influence of nitrogen and potassium fertilization on fruiting and capsaicin content in habanero pepper (*Capsicum chinense*Jacq.). *Hort. Science*, *43* (5), 1549–1554.

Montalba, R., Arriagada, C., Alvear, M., & Zúñiga, G. E., (2010). Effects of conventional and organic nitrogen fertilizers on soil microbial activity, mycorrhizal colonization, leaf antioxidant content, and Fusarium wilt in highbush blueberry (*Vacciniumcorymbosum* L.). *Sci. Hort.*, *125* (4), 775–778.

Motavalli, P. P., Goyne, K. W., & Udawatta, R. P., (2008). Environmental impacts of enhanced-efficiency nitrogen fertilizers. *Crop Manag.*, *7*(1).

Mulvaney, R. L., Khan, S. A., & Ellsworth, T. R., (2009). Synthetic nitrogen fertilizers deplete soil nitrogen: A global dilemma for sustainable cereal production. *J. Environ. Quality.*, *38* (6), 2295–2314.

Ortas, I., (2013). Influences of nitrogen and potassium fertilizer rates on pepper and tomato yield and nutrient uptake under field conditions. *Acad. J.*, *8* (23), 1048–1055.

Owusu, E. O., Nkansah, G. O., & Dennis, E. A., (2000). Effect of sources of nitrogen on growth and yield of hot pepper (*Capsicum frutescens*). *J. Tropic. Sci.*, *40* (2), 58–62.

Pacheco-Arijona, J. R., Ruiz-Lau, N., Medina-Lara, F., Minero-García, Y., Echevarría-Machado, I., Santos-Briones, C. D. L., & Martínez-Estévez, M., (2013). Effects of ammonium nitrate, cesium chloride and tetraethylammonium on high-affinity potassium uptake in habanero pepper plantlets (*Capsicum chinense*Jacq.). *Afr. J. Biotechnol.*, *10* (62), 13418–13429.

Qawasmi, W., Mohammad, M. J., Najim, H., & Qubursi, R., (1999). Response of bell pepper grown inside plastic houses to nitrogen fertigation. *Commn. Soil Sci. Plant Anal., 30* (17–18), 2499–2509.

Romic, D. Romic, M., Borosic, J., & Poljak, M., (2003). Mulching decreases nitrate leaching in bell pepper (*Capsicum annuum* L.) cultivation. *Agric. Water Manag., 60* (2), 87–97.

Roy, S. S., Khan, M. S. I., & Pall, K. K. K. K., (2011). Nitrogen and phosphorus efficiency on the fruit size and yield of *Capsicum. J. Exp Sci.,* 2 (1), 32–37.

Roychoudhry, A., Chatterjee, R., & Mitra, S. K., (1990). Effect of different doses of nitrogen, phosphorus, potassium, magnesium, calcium, and iron on growth and development in chili. *Indian Cocoa, Arecanut Spices J., 13* (3), 96–99.

Rui-hai, L., Qiwei, H., & Yangchun, Y., (2009). Effect of different foliar fertilizers on growth of *Capsicum annuum* L. *J. Nanjing Agr. Univ.,* 2, 017.

Shi-min, L., (2005). Effect of N, P, and K fertilizers application on the yield and economic efficiency of chili. *Soil Fert., 1,* 004.

Sistani, K. R., Jn-Baptiste, M., Lovanh, N., & Cook, K. L., (2011). Atmospheric emissions of nitrous oxide, methane, and carbon dioxide from different nitrogen fertilizers. *J. Environ. Qual., 40* (6), 1797–1805.

Stefanelli, D., Goodwin, I., & Jones, R., (2010). Minimal nitrogen and water use in horticulture: Effects on quality and content of selected nutrients. *Food Res. Int., 43* (7), 1833–1843.

Thakur, P. S., Thakur, A., & Kanaujia, S. P., (2000). Influence of bioregulators, bioextracts, and potassium on the performance of bell pepper (*Capsicum annuum*) varieties underwater-stress. *Ind. J Agric. Sci., 70* (8), 543–545.

Tumbare, A. D., & Nikam, D. R., (2004). Effect of planting and fertigation on growth and yield of green chili (*Capsicum annuum*). *Ind. J. Agric. Sci., 74* (5), 242–245.

Veeranna, H. K., Abdul Khalak, F. A. A., & Sujith, G. M., (2000). Effect of potassium fertigation on growth and yield of chili under moisture stress conditions. Spices and aromatic plants: challenges and opportunities in the new century. *Contributory Papers.* Centennial conference on spices and aromatic plants, Calicut, Kerala, India, pp. 117–121.

Wang, Q., Wang, L., He, C., Wang, H., Kui, X., & Jiang, Z., (2000). Study on accumulation effect and control measure of nitrate with application of excessive nitrogenous fertilizer for different vegetables. *Agro. Environ. Protect., 1,* 015.

Yamamoto, S., Matsumoto, T., & Nawata, E., (2011). *Capsicum*use in Cambodia: The continental region of Southeast Asia is not related to the dispersal route of *C. frutescens* in the Ryukyu Islands. *Econ. Bot., 65* (1), 27–43.

Yang, H., Jiang, H., Lai, W., Tu, X., & Zhan, Y., (2009). Study on effect of water coupling fertilizer on yield and quality in hot pepper. *J. Changjiang. Veg.,* 6, 018.

Yanhong, W., Mengjun, L., Shaoying, A., Shaohai, Y., & Jianwu, Y., (2009). Effect of potassium fertilizer on lead bioavailability of soil system and pepper. *Environ. Sci. Manag.,* 3, 041.

Yi-tao, Z., Yun-jiang, L., & Bo, G., (2007). Coupling effect of water and fertilizers on irrigation water use efficiency under the condition of *Capsicum*cultivation in protective field. *J. Jilin. Agric. Univ.,* 05.

YuanYuan, R., EnRang, Z., Huaqun, H., ShaoGang, Z., JinPei, M., & ZhongRong, W., (2007). The effect of potassium fertilizer on growth and yield of red pepper. *Southwest China J. Agric. Sci., 20* (5), 1044–1047.

Zhao, G. C., He, Z. H., Liu, L. H., Yang, Y. S., Zhang, Y., Li, Z. H., & Zhang, W. B., (2004). Study on the co-enhancing regulating effect of fertilization and watering on the main quality and yield in zongyou–9507 high gluten wheat. *Sci. Agric. Sin.*, *3,* 006.

Zhou, J-B., Xi, J. G., Chen, Z-J., & Li, S. X., (2006). Leaching and transformation of nitrogen fertilizers in soil after application of N with irrigation: A soil column method. *Pedosphere*, *16* (2), 245–252.

Zhuqing, Z., (2007). Studies on the fertilization technique for good-quality and high-yield hot pepper production. *J. Chin. Capsicum*, *3,* 020.

CHAPTER 9

COMPARATIVE STUDY OF THREE DIFFERENT FERTILIZERS ON YIELD AND QUALITY OF *SOLANUM*

SAMEEN RUQIA IMADI[1], KANWAL SHAHZADI[2], and ALVINA GUL[1]

[1]*Atta-ur-Rahman School of Applied Biosciences, National University of Sciences and Technology, Islamabad, Pakistan, E-mail: alvina_gul@yahoo.com*

[2]*Department of Plant Sciences, Quaid-I-Azam University, Islamabad, Pakistan*

ABSTRACT

Solanum is a widely used group of species of plants belonging to family Solanaceae. *Solanum* fruits are used for food purposes. The concentration of secondary metabolites and phenolic contents of *Solanum* are considered the measure of quality of *Solanum*. Moreover, number of tubers in case of potato and number of fruits in case of other plants are considered its yield. Different fertilizers are used in different concentrations to enhance the production and quality of *Solanum* species. Among these fertilizers most widely used are nitrogen fertilizers, potassium fertilizers and phosphorus fertilizers. Besides these, micronutrients and macronutrients are also used in certain concentrations to further enhance the yield and quality of crops. This chapter is intended to compare these three fertilizers and their effects on *Solanum* crops. Finally the chapter will look forward into the techniques which can be used for fertilization of *Solanum* crops without degrading the environment in a sustainable way.

9.1 INTRODUCTION

Solanum is known to be a genus of flowering plants. This genus includes food crops which have very high economic importance. Potato (*Solanum tuberosum*), tomato (*Solanum lycopersicum*) and brinjal (*Solanum melongena*) belong to this genus. Eggplant (*Solanum melongena*) is commonly known as aubergine, brinjal, and guinea squash. *Solanum melongena* is known to be one of the non-tuberous species of night shade family of Solanaceae (Moraditochaee et al., 2011). Production of potato has outstripped and even surpassed the production of many other staples foods in terms of both yield and cultivated areas. Fertilizers and organic manure have been used for increasing the production and quality of potatoes (Bertin et al., 2013). Balanced use of different nutrients is one of the most important factors in exploiting the yield and growth of *Solanum* plants (Pal et al., 2002). Low fertility of soil is considered to be a major constraint in production of *Solanum* species worldwide (Powon et al., 2006).

Fruits of *Solanum* species are a good source of many nutrients. *Solanum lycopersicum* is one of the most widely cultivated vegetable across the world. The fruit of this plant is commonly known as tomato. It is a good source of ascorbic acid and many anti-oxidants which include beta carotene and lycopene. These ingredients are involved in protecting against many cancers like prostate cancer (Khan et al., 2008). Different nutrients are applied to *Solanum* plants for enhancing their quality as well as yield. These nutrients are applied based on organic approaches, integrated nutrient management approaches and commercial approaches. Among many techniques of nutrient application as fertilizers, one of the techniques includes seaweed extract. This extract is an organic approach to provide necessary nutrients to plants. Seaweed extracts are not only used as nutrient supplements, but are also used for biostimulation and biofertilization. It has been observed that tomato plants treated with seaweed extract show enhanced germination which is characterized by low germination time and high germination index. Shoot length, root length and weight of plants are also observed to be increased by application of seaweed extract (Hernández-Herrera et al., 2014).

Use of soluble and chemical fertilizers for enhancing the supply of nitrogen, potassium, and phosphorus has been performed since decades to increase the quality and yield of *Solanum* species (Davenport et al., 2005). Application of fertilizers on *Solanum lycopersicum* results in enhancement of stem girth, fruit yield, plant height, number of leaves and number of branches (Ogbomo, 2011). It has been observed that application of selenium in young *Solanum tuberosum* plants results in accumulation of starch in upper leaves. Selenium is observed to possess positive effects on carbohydrate accumulation in potato as well as yield formation (Turakainen et al., 2004). The effects of calcium ammonium nitrate were observed on *Solanum* species. It was seen that calcium ammonium nitrate increase vitamin A content in 14 week old tissues. Vitamin C content is also observed to be enhanced by calcium ammonium nitrate in young as well as old tissues. It is proved that quality attributes of *Solanum* species increase significantly by fertilizers (Kipgoskei et al., 2003).

Fertilizers are seen to improve the height of plants, number of primary and secondary branches. It has been observed that the NPK ratio of 60: 20: 40 kg per hectare produce maximum number of fruits per plants, maximum individual fruit weight and enhanced yield of *Solanum* plants (Sundharaiya et al., 2000). It was explored in an experiment that application of integrated nutrient management approach for nitrogen, potassium, and phosphorus on *Solanum melongena*plants cast a positive impact on fruit size and yield hence enhancing the quality of crops (Srivastava et al., 2009). It has been observed that bioactive compounds of *Solanum* plants are not affected by application of organic fertilizers. There is no significant effect on total phenol and flavonoids content on tomato fruits, although antioxidant activity is halted (Riahi and Hdider, 2013).

Foliar fertilizers are observed to possess significant effects on plant height, tuber weight and total yield of potato plants. Minerals required by *Solanum tuberosum* plants include nitrogen, phosphorus, potassium, iron, copper, manganese, zinc, boron, and molybdenum (Jasim et al., 2013). Application of nitrogen, phosphorus, and potassium fertilizers in combination with vermicompost; result in enhancement of plant height, postharvest life of tomato fruits, leaf area, fruit density, leaf weight, fruit yield and fruit weight (Singh et al., 2010). Potato tuber yield, quality, and resistance against *Streptomyces scabies* is observed to be enhanced by sulfur fertilization. Application of magnesium micronutrient is also explored to be associated with decrease in rate of tuber infection and severity of infection. This also enhances the yield and quality of potato tubers (Klikocka, 2009).

Application of biofertilizers on potato plants was explored to possess low rates of enhancement of yield and quality of tubers. Rather when biofertilizers were used with nitrogen, phosphorus, and potassium fertilizers, the quality attributes and yield was significantly enhanced (Ghosh et al., 2000). It has been explored through experiments that application of nitrogen fertilizers to potato leads to an increase in magnesium uptake in plants. This phenomenon is also associated with delay in canopy senescence (Allison et al., 2001a). It has been observed that application of fertilizers on potato leads to increase in dry matter accumulation, tuber bulking rate and tuber yield (Sarkar and Mondal, 2005). Integrated nutrient management approaches including combination of both organic manure and commercial synthetic fertilizers assure improved quality potato production with high yield (Zaman et al., 2011).

The best nutrient management practices are developed in order to enhance yield and quality of tubers. The environmental losses of nutrients are also reduced by following these practices (Zebarth and Rosen, 2007). Fertilizers are involved in enhancement of chlorophyll content in potato plants. It has been observed in an experiment that different fertilizers increase the chlorophyll content in following order; nitrogen fertilizers, organic fertilizer, potassium fertilizer and phosphorus fertilizer (Juan-Guo, 2009).

9.2   NITROGEN FERTILIZERS

Adequate use of nitrogen is crucial to optimize quality as well as yield of potato (Cambouris et al., 2008). Management of nitrogen is important in maintaining the growth and development of potatoes. Best management practices of nitrogen enhance tuber yield and quality of potatoes (Zebarth and Rosen, 2007). Crop productivity is highly determined by nitrogen use efficiency. Accumulation as well as use of nitrogen is dependent on different elements which include species, cultivars, and irrigation (Sánchez-Rodríguez et al., 2011). Slow release nitrogen fertilizers provide maximum benefits for vegetable production around the world. There use not only enhances the quality and yield of *Solanum* species but they also reduce nitrogen leaching. Environmental risks are also reduced and cost in production of vegetables is also lowered by using slow release nitrogen fertilizers (Guertal, 2009).

It is a known fact that gene expression leads to synthesis of proteins and defining of metabolic systems. This also determines the capacity of plants for growth, development as well as yield production. The plants require resources for nutrients to manage these purposes. Crop production is highly enhanced by interaction between carbon dioxide and nitrate assimilation. Metabolism in potato is highly dependent on assimilation of nitrates to amino acids. Leaf growth, photosynthesis, cell division and all other related processes require nitrogen nutrient (Lawlor, 2002). Application of nitrogen fertilizers possess a significant effect on fruit yield, number of fruits per square meter, plant height, fruit length, fruit width and number of branches per plant in *Solanum melongena* plants. 80 kg per hectare of nitrogen spray produces the most yield in eggplants (Azarpour et al., 2012).

Base fertilizer application and top dressing nitrogen fertilizers are shown to possess most significant effects on yield of spring and autumn potatoes (Zheng et al., 2009). Water deficit stress is observed to cause a decline in uptake and accumulation of nitrogen in *Solanum* plants. Nitrogen use efficiency in plants is also decreased due to water stress (Sánchez-Rodríguez et al., 2011). Foliar application of nitrogen fertilizer in combination with nano iron chelate fertilizer show a significant effect on fruit yield, number of fruits per plant, height of plants and number of branches per plant. These effects are shown to be caused by interaction between nitrogen fertilizer and iron. High application of iron fertilizer results in highest yield in brinjal plants (Bozorgi, 2012).

It has been experimentally observed that application of nitrogen fertilizer increases tuber yield, tuber number per plant and leaf chlorophyll level of potato plants. *Maximum* yield of potato tubers is observed at 200 kg of nitrogen per hectare (Guler, 2009). Application of nitrogen fertilizers through banding technique is significantly important for increased production of potatoes with high-quality. Banding is the most efficient as well as economic technique for nitrogen fertilizer application (Khan et al., 2007). Since nitrogen is an essential element for potato growth, hence best management practice should be applied for its use. BMP actually limits the use of nitrogen fertilizer for potato production. This practice reduces the mobility of nitrogen and does not allow it to leave the field (Hutchinson et al., 2004).

Growth characteristics of *Solanum melongena* are enhanced by nitrogen fertilizers. These characteristics include plant height, leaf chlorophyll content and lateral stem number. Flowering factor including flower number and days to flowering of eggplant is also effected by nitrogen fertilizers. Nitrogen treatment is observed to decrease the time of 1st flowering as treated plants flowered earlier (Aminifard et al., 2010a). As the nitrogen uptake of potato roots increase, the dry matter of tubers is also increased (Vos, 1997).

9.2.1 EFFECTS OF NITROGEN FERTILIZERS ON YIELD AND QUALITY OF SOLANUM

Use of nitrogen fertilizers on potato plants by up to 160 kg per hectare; result in highest fresh tuber yield. With further increase in nitrogen concentration, nitrate content in fresh tuber and dry matter also increases (Shahbazi et al., 2009). Effects of nitrogen fertilizers were seen on the growth and yield of *Solanum* plants. It was explored that plant vegetative growth is significantly increased by nitrogen uptake. Plants which receive nitrogen concentration of 150 kg per hectare are shown to possess highest lateral stem number and leaf chlorophyll content. Good concentration of nitrogen was observed to enhance fruit weight as well as fruit yield. It can be concluded that nitrogen fertilization significantly enhance vegetative as well as reproductive growth of *Solanum* plants (Aminifard et al., 2010b).

Nitrogen rates are observed to positively enhance number of tubers in potato. Dry matter concentration of tuber is also increased by nitrogen uptake (Kavvadias et al., 2012). Application of nitrogen fertilizer at 160 kg per hectare on potato plants results in enhanced tuber yield, mean tuber weight, dry weight of tuber and nitrogen percentage of tubers (Saeidi et al., 2009). Increasing concentration of nitrogen fertilizers in soil is associated with an increase in yield of *Solanum melongema* plants. Number of fruits per plant is also increased (Aujla et al., 2007). Consecutive experiments were conducted to measure the growth parameters, quality, and yield of potato tubers by nitrogen application. It was observed that 50% of recommended dose of nitrogen, phosphorus, and potassium fertilizer and 50% of recommended dose of nitrogen fertilizer was significant for favorable enhancement of potato tuber yield, nutrient uptake and soil fertility (Kumar et al., 2012).

Optimum marketable yield of potatoes is obtained at nitrogen concentration of 600 kg per hectare. At this concentration of nitrogen fertilizer, tuber uptake of nitrogen is high, thus yield as well as quality of tubers is increased (Halitligil et al., 2002). It has been observed that primary and secondary metabolites in tomato fruits can be affected as a response to low concentration of nitrogen. These conditions may also halt vegetative growth of the plants which increase fruit irradiance. It leads to changes in composition of fruits and hence changes the quality (Bénard et al., 2009). Different cultivars of potato show different effects towards nitrogen fertilization. Bannock Russet potato generate high yield at high concentrations of nitrogen whereas Burbank Russet and Gem Russet require further high concentrations of nitrogen to generate commercially significant yield (Love et al., 2005).

Solanum villosum is a leafy vegetable. Application of nitrogen fertilizer results in increase in leaf area and shoots dry weight of this vegetable which is the indicator of enhancement of yield (Masinde et al., 2009). Yield optimization of potato tubers is one of the major prospects for which nitrogen fertilizers are used (Sparrow and Chapman, 2003). Tuber quality of potato is determined by specific gravity, ascorbic acid content, starch, protein content and nitrate. Increasing the rates of nitrogen fertilizer in soil; result in decrease in specific gravity and starch content of potatoes. Protein and nitrate content are enhanced by use of nitrogen fertilizers. Ascorbic acid content show negligible responses to excessive nitrogen fertilizers. Overall the quality of potato tuber is enhanced by use of nitrogen fertilizers (Lin et al., 2005).

Splitting of nitrogen fertilizer application is considered to be most adaptive approach for application of nitrogen in potato fields (Goffart et al., 2008). It has been explored through experiments that nitrate reductase concentration is highest in tubers in potato plants. As the uptake of nitrogen by tubers increase, levels of sucrose and starch increase by 40% and that of glucose and fructose are observed to decrease by as much as two folds (Mäck and Schjoerring, 2002). Effect of nitrogen fertilizers on yield and quality of *Solanum* species is described in Figure 9.1.

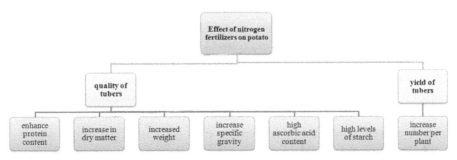

FIGURE 9.1   Effect of nitrogen fertilizers on yield and quality of potato (*Solanum tuberosum*).

9.3 POTASSIUM FERTILIZERS

High levels of potassium fertilizers are needed by potatoes for their optimum growth, enhanced tuber quality and yield production. It has been seen that tuber yield is significantly increased by periodic increases in potassium concentration in soil. Potassium effects height of plants, leaf area, chlorophyll content, carbohydrate content and specific gravity of tubers (Al-Moshileh and Errebi, 2004). Potassium is considered to be an essential fertilizer for growth of *Solanum lycopersicum* plants. Potassium increases the lycopene content and other quality characteristics of tomato. It has been observed that potassium treatment is also associated with increase in ascorbic acid content and beta carotene levels (Khan et al., 2008).

Responses of potato towards nitrogen are highly dependent on potassium uptake and thus potassium influence the yield of potato tubers. Low concentrations of exchangeable potassium in soil are involved in reduction of plant response towards nitrogen uptake (Milford and Johnston, 2009). Potassium nutrition is highly advantageous in enhancing potato plants tolerance towards salinity. *Solanum tuberosum* plants grown under highly saline conditions, when subjected to potassium fertilizer, are shown to enhance their tuber yield. Even low concentrations of potassium fertilizers significantly increase yield of tubers (Elkhatib et al., 2004). Crisp of potato is enhanced by potassium fertilizers. In an experiment, different forms of potato plants were subjected to different forms of potassium fertilizers. Objective of this experiment was to find out the potassium fertilizer which induces most crisp in tubers of potato. It was observed that potassium sulfate generates highest crisp which is followed by potassium chloride and further by potassium nitrate (Kumar et al., 2007).

Potassium sulfate is observed to possess positive effects on stomatal density, photosynthetic rate and stomatal conductivity in *Solanum lycopersicum* plants. Potassium chloride enhances the leaf thickness and rate of transpiration in spring tomato plants (Borowski et al., 2000). Excessive use of potassium sulfate on potato plants is associated with increase in height of plants, number of leaves per plant and leaf area per plant. Fresh and dry weight of whole plant is also increased. Increasing levels of potassium, increases the specific gravity of tubers, their starch content, protein content and dry matter, thus increase the quality of potatoes (Ahmed et al., 2009).

9.3.1   EFFECTS OF POTASSIUM FERTILIZERS ON YIELD AND QUALITY OF SOLANUM

Application of potassium fertilizers is associated with enhancement of growth, yield, and quality of *Solanum lycopersicum* fruits (Khan et al., 2008). Potassium fertilizer at concentration of 215 kg per hectare is considered to be significant for high yield and good quality of potatoes (AbdelGadir et al., 2003). Application of potassium fertilizers is of low concern in case of *Solanum tuberosum. However,* it is seen that increased concentration of potassium fertilizers may be associated with increased yield of potato tubers to some extent (Pervez et al., 2000). Application of potassium fertilizer at concentration of 150 kg per hectare results in an increase in yield of potato tubers in soils with low pH as compared to control plants (Haile and Boke, 2011).

Increase in levels of potassium fertilizers in plants result in an increase in tuber yield and tuber size of potatoes (Karam et al., 2009). Potassium fertilizer was applied to potato plants in combination with humic acid. It was explored that plant length, dry weight of shoots and leaves, number of leaves and number of branches per plants are significantly increased as a response to increase in concentration of potassium in soil. Increasing rates of potassium are also observed to enhance potato yield and tuber quality. Quality of tuber was determined by parameters including weight, gravity, size, diameter, and length. Nutritive value of potato tubers is also increased with increasing levels of potassium fertilizers (Mahmoud and Hafez, 2010).

Increasing concentrations of potassium in soil is associated with increase in growth parameters, tuber yield, economic parameters and yield components. Growth parameters which are enhanced by potassium fertilization include plant biomass and tuber bulking rate whereas yield components include tuber population and average tuber weight (Moinuddin et al., 2005a). Maximum tuber yield of potato is obtained by K_2O at a concentration of 353.4 kg per hectare (dos Anjos Reis and Monnerat, 2000). To measure the effects of potassium on potato plants, nitrogen, and phosphorus fertilizers were uniformly applied at concentration of 200 kg per hectare and 100 kg per hectare, respectively. Concentration of potassium fertilizer in the form of K_2O was increased after intervals. It was observed that many growth parameters including height of plant, biomass of shoot, biomass of tuber, tuber population and tuber weight are increased in the response (Moinuddin et al., 2005b).

Quality of tomato fruits is enhanced by use of potassium fertilizers. It was explored in an experiment that firmness of fruit, content of ascorbic acid, soluble solid content and lycopene content are enhanced in fruits treated with potassium fertilizer. It was also seen that the same fruits had low acidity, low weight loss and decreased decay percentage (Hewedy, 2000). Effects of potassium and boron fertilizers were determined on cinnamon loam soil. It was seen that production of potato was significantly enhanced by increasing concentration of potassium and boron or either one of the two (Zeyi and Fuling, 2005). Potassium application is associated with enhancement of tuber yield of drought-stressed Agria potato cultivar. Increase in yield by use of potassium fertilizer is observed to be about 29% (Khosravifar et al., 2008).

Specific gravity of potato tubers is increased by potassium fertilizer application. This is a factor in enhancement of quality of potatoes (Davenport, 2000). Four varieties of potatoes were subjected to potassium fertilizers and following results were found. Tuber yield is enhanced as a result of potassium application in Spunta and Derby cultivar. Shepody variety also shows an increase in tuber yield by increasing dry matter whereas Umatilla did not show any significant effects to potassium treatment

(Masaad et al., 2004). Potassium fertilizers at different concentrations were applied on potato plants grown in extremely acidic soils. It was observed that yield of potato tubers was highest at 100 kg per hectare of potassium fertilizer. Satisfactorily increase in yield of potato tubers was seen at 500 kg of potassium fertilizer per hectare (Adhikary and Karki, 2006).

Application of potassium fertilizer in proper amount on potato plants result in an increase in stomatal conductance, transpiration rate and photosynthetic rate in leaves (Guan-rong and Xiao-yan, 2009). Site specific application of potassium fertilizers leads to enhancement of quality of potatoes. Yield however, is not increased or affected by cite specific application of potassium fertilizers (Wijkmark et al., 2005). Fresh and dry weights of haulm of potato are increased by increasing concentrations of potassium fertilizers. 160 kg per hectare of potassium fertilizer is associated with highest tuber yield of potato. Vegetative growth, yield, and quality of potato is significantly increased by use of potassium fertilizers (Abou-Hussein, 2005). Effect of potassium fertilizers on yield and quality of *Solanum tuberosum* and *Solanum lycopersicum* are described in Figure 9.2 and Figure 9.3 respectively.

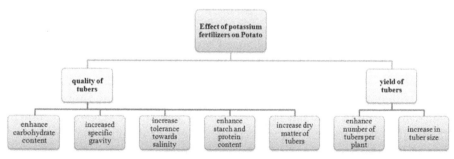

FIGURE 9.2   Effect of potassium fertilizer on yield and quality of potato (*Solanum tuberosum*).

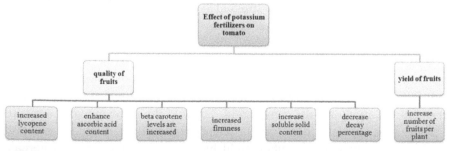

FIGURE 9.3   Effect of potassium fertilizer on yield and quality of tomato (*Solanum lycopersicum*).

9.4 PHOSPHORUS FERTILIZERS

Solanum tuberosum (potato) production is critically dependent on management of phosphorus fertilizers. It is due to the fact that potato needs relatively high volumes of phosphorus. Canopy development, root cell division, set of tubers and synthesis of starch are largely dependent on availability of phosphorus. Adequate concentrations of phosphorus are also required for optimizing the yields of potato tubers, solid content, nutritional quality as well as resistance to certain diseases (Rosen et al., 2014).

Phosphorus fertilization is seen to increase total leaf, stem, plant height and haulm yields. Tuber yield shows a significant increase with phosphorus fertilization. Tuber cadmium concentrations are also increased by excess phosphorus fertilizers (Maier et al., 2002). The active surface roots of potato help in phosphorus fertigation. Petiole phosphorus concentration is highly increased by application of phosphorus fertilizers. Significant increase in marketable yield is also obtained by phosphorus (Hopkins et al., 2010). Phosphorus fertilization was applied to *Solanum nigrum* plants, and its effects were determined. It was seen that phosphorus fertilization enhances most of the growth parameters of plants and leaves. Berry yield is also increased (Khan et al., 2000).

9.4.1 EFFECTS OF PHOSPHORUS FERTILIZERS ON YIELD AND QUALITY OF SOLANUM

It has been observed that phosphorus application significantly increases the yield and quality of potato tubers (Pal et al., 2002). Tuber dry weight is increased by 82% and fresh tuber yield is increased by up to 62% at high concentrations of phosphorus fertilizer. Application of phosphorus is significant for increase in yield of potatoes (Powon et al., 2006). The responses of tomato plants toward phosphorus fertilizers are significantly enhanced by irrigation. Interaction between irrigation and phosphorus fertilizers produces increased yield of tomatoes (Begum et al., 2000). Significant increases in tuber yield of potatoes have been observed as a result of application of phosphorus fertilizers. However, application of foliar phosphorus fertilizer was not observed to cast any significant changes in number of tubers or tuber yield of potatoes, thus this treatment is not recommended for *Solanum tuberosum* (Allison et al., 2001b).

Internal quality of potato is measured by tuber specific gravity. Nutrient management practices, highly determine the specific gravity of potato tubers. Tuber solids can be reduced by excessive application of nutrients and their excessive presence in soil. Phosphorus fertilizers are considered to be the matter of choice for improving specific gravity of potatoes (Laboski and Kelling, 2007). If phosphorus is used in excessive amount, it has been observed that it reduces yield as well as quality of potatoes. This reduction is not due to phosphorus toxicity. Rather excessive use of phosphorus leads to deficiency of zinc and other micronutrients which result in reduction of quality and yield. Hence over application of phosphorus fertilizers should be avoided (Hopkins and Ellsworth, 2003).

Yield of under sized tubers and total yield of tubers is enhanced by providing phosphorus nutrition in a controlled way using banding approach. Phosphorus application is associated with increase in number of small tubers and decrease in number of large tubers. Timing and source of phosphorus is not seen to be associated with tuber yield, number of tubers, size of tubers and petiole or tuber concentration of phosphorus (Rosen and Bierman, 2008). According to studies conducted in different eras, phosphorus shows a positive effect on yield of potatoes. However, this positive effect of phosphorus cannot be predicted because it may be associated with mineral deficiencies which can halt the advantages of phosphorus treatment (Mohr and Tomasiewicz, 2011). Dry matter of potato tubers is significantly increased by applying phosphorus fertilizer combined with zinc. This increase is due to interaction between zinc and phosphorus (Taheri et al., 2012). Effect of phosphorus fertilizers on yield and quality of *Solanum* species is described in Figure 9.4.

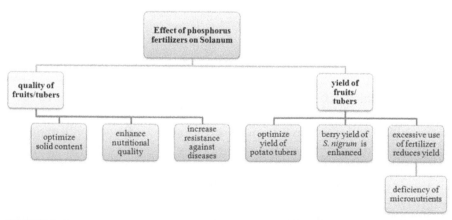

FIGURE 9.4   Effect of phosphorus fertilizers on yield and quality of *Solanum* species.

9.5 COMPARISON OF EFFECTS OF N, K, AND P FERTILIZERS

Effects of nitrogen, potassium, and phosphorus were observed on potato plants. It was seen that as the concentrations of nitrogen and potassium is increased, number of tubers and mass of tubers is also increased. This results in increased quality of potatoes (Kádár et al., 2000). Uptake of different fertilizers were compared in an experiment in which *Solanum nigrum* plants were spaced at 30*30 cm. it was observed that uptake of nitrogen was highest which was followed by phosphorus uptake. Potassium fertilizer uptake was observed to be lowest among all (Sivakumar and Ponnusami, 2011). Application of nitrogen, phosphorus, and potassium fertilizers in form of ammonium sulfate, calcium super phosphate and potassium sulfate, respectively at respective rates of 400, 300 and 200 kg per feddan showed that plant parameters are significantly enhanced. Fruit yield and vitamin C content in fruits is increased by this treatment. As compared to control, fruit yield, fruit quality and mineral content of tomato fruits was approximately doubled (Mahmoud and Amara, 2000).

Excessive use of nitrogen fertilizers lowers the quality of potato tubers whereas excessive use of potassium fertilizers enhances the quality of potato tubers. Yield however was increased by application of nitrogen fertilizers (Riley, 2000). It has been observed that application of 100 kg per hectare of combined nitrogen, phosphorus, and potassium fertilizer produces highest yield of potatoes (Ayeni, 2008). Shoot length, fruit weight, number of leaves, fruit number, dry matter weight of shoots and dry matter weight of roots in potato plants were significantly increased by nitrogen, phosphorus, and potassium fertilizers (Kashem et al., 2015). It has been observed that application of nitrogen fertilizers on potato plants enhance the yield of tubers but this enhancement is not as much as when this treatment is combined with PK fertilizers. This enhanced production of potato with the application of PK fertilizers in combination to nitrogen fertilizers is a sustainable approach (Singh et al., 2002).

Nitrogen and phosphorus fertilizers were applied to *Solanum tuberosum* plants in the form of potassium nitrate and phosphoric acid. It was observed that increasing the rates of nitrogen regardless of phosphorus is associated with increase in nitrate assimilation as well as total commercial yield of potatoes. Phosphorus however is involved in reducing foliar rates of nitrogen (López-Cantarero et al., 1997). It has been observed that application of nitrogen and potassium fertilizers at concentration of 160 kg per hectare and 120 kg per hectare, respectively generates interaction between these two ions. These interactions lead to the formation of highest yield in potato tubers (Singh and Raghav, 2000). Increase in yield of potatoes was observed while applying nitrogen, phosphorus, and potassium fertilizers. It was observed that nitrogen fertilizers lead to an increase of 38.7% in tuber yield whereas phosphorus fertilizers increase 10.7% yield. Potassium fertilizers are observed to enhance the yield by 23.6% (Bao-quan, 2008).

Interaction between micronutrients and phosphorus levels were observed in an experiment. It was seen that adding of zinc in Russet Burbank potato plants result in a decrease in root phosphorus levels whereas shoot phosphorus levels, copper, and iron levels are increased. Treating those plants with manganese showed a decrease in zinc concentration in all parts of plants except for roots (Barben et al., 2011). It has been observed that application of phosphorus in combination with sulfur is associated with enhancement of growth characters, total yield of tubers, dry matter and biomass and mineral content of tubers (nitrogen, potassium, and iron) (Mahmoud, 2000).

An experiment was conducted on tomato plants to measure the effects of different fertilizers. It was observed that application of zinc and phosphorus leads to early flowering whereas nitrogen application delays the flowering. Plots which have been applied with phosphorus and zinc showed minimum incidence of disease. However, disease rates were high in nitrogen treated plots. It has been observed that maximum number of fruits was obtained when nitrogen was applied at 150 kg per hectare, phosphorus at 100 kg per hectare and zinc at 10 parts per million. The plots which receive all these fertilizers together, showed a 100% increase in yield (Nawaz et al., 2012). It has been observed that application of nitrogen fertilizer enhances the protein content of potato tubers whereas application of phosphorus fertilizer enhances oil content. Potato crops should be fertilized with nitrogen fertilizer however phosphorus fertilizer should be used only if the soil is deficient in phosphorus (Oztürk et al., 2010).

9.6 CONCLUSION

It is a fact that fertile soil is base for food production and successful civilizations. The nutrients which are lost through harvest are added back to soil through fertilization. Enhancing the use and efficiency of fertilizers for *Solanum* species is very important because these have inefficient rooting system and are extremely sensitive to nutritional deficiencies. Many approaches can be applied to enhance the efficiency of fertilizers. These approaches include slow or controlled release fertilizers, addition of high charge density materials which does not allow chemicals to interfere with nutrient element, chelating of the fertilizer nutrients to increase their solubility and finally enhancement of solubility by modification of microsite pH (Hopkins et al., 2008). Inefficient usage of nitrogen, phosphorus, and potassium fertilizers has resulted into degradation of environment. For reducing the impact of fertilizer accumulation on environment, certain strategies can be applied. These strategies include precision agriculture, genetic manipulation, cover crops and slow release fertilizers. These approaches ensure sustainable use of fertilizers (Davenport et al., 2005).

NPK fertilizers applied to potato crops with crop residues of wheat and maize further enhance yield as well as quality of potatoes. This approach also leads to low-cost and hence can be considered as economically sustainable approach. The approach also implies integrated management of nutrients. Nutrient uptake and soil fertility is also increased with the treatment with this approach (Singh et al., 2010b).

KEYWORDS

- anti-oxidants
- fertilizers
- production enhancement
- *Solanum* species
- tubers
- yield

REFERENCES

AbdelGadir, A. H., Errebhi, M. A., Al-Sarhan, H. M., & Ibrahim, M., (2003). The effect of different levels of additional potassium on yield and industrial qualities of potato (*Solanum tuberosum* L.) in an irrigated arid region. *Amer. J. Potato Res.*, *80* (3), 219–222.

Abou-Hussein, S. D., (2005). Yield and quality of potato crop as affected by the application rate of potassium and compost in sandy soil. *Annals Agric Sci.*, *50* (2), 573–586.

Adhikary, B. H., & Karki, K. B., (2006). Effect of potassium on potato tuber production in acid soils of Malepatan, Pokhara. *Nepal Agric. Res. J.*, *7*, 42–48.

Ahmed, A. A., El-Baky, M. M. H. A., El-Aal, F. S. A., & Zaki, M. F., (2009). Comparative studies of application both mineral and bio-potassium fertilizers on the growth, yield, and quality of potato plant. *Res. J. Agric. Biol. Sci.*, *5* (6), 1061–1069.

Allison, M. F., Fowler, J. H., & Allen, E. J., (2001). Effects of soil- and foliar-applied phosphorus fertilizers on the potato (*Solanum tuberosum*) crop. *J. Agric. Sci.*, *137* (4), 379–395.

Allison, M. F., Fowler, J. H., & Allen, E. J., (2001). Factors affecting the magnesium nutrition of potatoes (*Solanum tuberosum*). *J. Agric. Sci.*, *137* (4), 397–409.

Al-Moshileh, A. M., & Errebi, M. A., (2004). Effect of various Potassium sulfate rates on growth, yield, and quality of potato grown under sandy soil and arid conditions. In: *IPI Regional Workshop on Potassium and Fertigation Development in West Asia and North Africa*, Rabat, Morocco: 24–28.

Aminifard, M. H., Aroiee, H., Fatemi, H., & Ameri, A., (2010a). Performance of eggplant (*Solanum melongena* L.) and sweet pepper (*Capsicum annuum* L.) in intercropping system under different rates of nitrogen. *Hort. Environ. Biotechnol.*, *51* (5), 367–372.

Aminifard, M. H., Aroiee, H., Fatemi, H., Ameri, A., & Karimpour, S., (2010b). Responses of eggplant (*Solanum melongena*L.) to different rates of nitrogen under field conditions. *J. Cent. Eur. Agric.*, *11* (4), 453–458.

Aujla, M. S., Thind, H. S., & Buttar, G. S., (2007). Fruit yield and water use efficiency of eggplant (*Solanum melongema* L.) as influenced by different quantities of nitrogen and water applied through drip and furrow irrigation. *Sci. Hort.*, *112* (2), 142–148.

Ayeni, L. S., (2008). Integrated application of cocoa pod ash and NPK fertilizer on soil chemical properties and yield of tomato. *Amer. -Eur. J. Sust. Agric.*, *2* (3), 333–337.

Azarpour, E., Motamed, M. K., Moraditochaee, M., & Bozorgi, H. R., (2012). Effects of bio, mineral nitrogen fertilizer management, under humic acid foliar spraying on fruit yield and several traits of eggplant (*Solanum melongena* L.). *Af. J. Agric. Res.*, *7* (7), 1104–1109.

Bao-quan, Y. A. O., (2008). NPK fertilizer response and optimum application rate for winter potatoes. *Fujian J. Agric. Sci.*, *2*, 017.

Barben, S. A., Hopkins, B. G., Jolley, V. D., Webb, B. L., Nichols, B. A., & Buxton, E. A., (2011). Zinc, manganese, and phosphorus interrelationships and their effects on iron and copper in chelator-buffered solution grown Russet Burbank potato. *J. Plant Nutr.*, *34* (8), 1144–1163.

Begum, M. M., Karim, A. J. M. S., Rahman, M. A., & Egashira, K., (2000). Effects of irrigation and application of phosphorus fertilizer on the yield and water use of tomato grown on a clay terrace soil of Bangladesh. *J. Faculty. Agric. Kyushu Uni.*, *45* (2), 611–619.

Bénard, C., Gautier, H., Bourgaud, F., Grasselly, D., Navez, B., Caris-Veyrat, C., Weiss, M., & Génard, M., (2009). Effects of low nitrogen supply on tomato (*Solanum lycopersicum*) fruit yield and quality with special emphasis on sugars, acids, ascorbate, carotenoids, and phenolic compounds. *J. Agric. Food Chem.*, *57* (10), 4112–4123.

Bertin, T., Ebenezar, A., Renata, Y., Zacharie, T., & Lazare, K., (2013). Impact of organic soil amendments on the physical characteristics and yield components of potato (*Solanum tuberosum* L.) in the highlands of Cameroon. *Middle-East J. Sci. Res.*, *17* (12), 1721–1729.

Borowski, E., Nurzyński, J., & Michalogić, Z., (2000). Reaction of glasshouse tomato to potassium chloride or sulfate fertilization on various substrates. *Annales Universitatis Mariae Curie-Skłodowska. Sectio EEE, Horticultura.*, *8*, 1–9.

Bozorgi, H. R., (2012). Study effects of nitrogen fertilizer management under nano iron chelate foliar spraying on yield and yield components of eggplant (*Solanum melongena* L.). *ARPN J. Agric. Biol. Sci.*, *7* (4), 233–237.

Cambouris, A. N., Zebarth, B. J., Nolin, M. C., & Laverdière, M. R., (2008). Apparent fertilizer nitrogen recovery and residual soil nitrate under continuous potato cropping: Effect of N fertilization rate and timing. *Canad. J. Soil. Sci.*, *88* (5), 813–825.

Davenport, J. R., (2000). Potassium and specific gravity of potato tubers. *Better Crops with Plant Food*, *84* (4), 14–15.

Davenport, J. R., Milburn, P. H., Rosen, C. J., & Thornton, R. E., (2005). Environmental impacts of potato nutrient management. *Am. J. Potato Res.*, *82* (4), 321–328.

Dos Anjos Reis, R., & Monnerat, P. H., (2000). Nutrient concentrations in potato stem, petiole, and leaflet in response to potassium fertilizer. *Sci. Agri.*, *57* (2), 251–255.

Elkhatib, H. A., Elkhatib, E. A., Khalaf Allah, A. M., & El-Sharkawy, A. M., (2004). Yield response of salt-stressed potato to potassium fertilization: A preliminary mathematical model. *J. Plant Nutr.*, *27* (1), 111–122.

Ghosh, D. C., Nandi, P., & Shivkumar, K., (2000). Effect of biofertilizer and growth regulator on growth and productivity of potato (*Solanum tuberosum*) at different fertility levels. *Ind. J. Agric. Sci.*, *70* (7), 466–468.

Goffart, J. P., Olivier, M., & Frankinet, M., (2008). Potato crop nitrogen status assessment to improve N fertilization management and efficiency: Past–present–future. *Potato Res.*, *51* (3–4), 355–383.

Guan-Rong, C. H. E. N., & Xiao-yan, G. S. M. Z., (2009). The effect of potassium application and water supplement period on photosynthetic characteristics and tuber yield of potato in semi-arid area. *J. Gansu Agric. Univ.*, *1*, 018.

Guertal, E. A., (2009). Slow-release nitrogen fertilizers in vegetable production: A review. *Hort. Technology*, *19* (1), 16–19.

Guler, S., (2009). Effects of nitrogen on yield and chlorophyll of potato (*Solanum tuberosum* L.) cultivars. *Bangladesh J. Bot.*, *38* (2), 163–169.

Haile, W., & Boke, S., (2011). Response of Irish potato (*Solanum tuberosum*) to the application of potassium at acidic soils of Chencha, Southern Ethopia. *Int. J. Agric. Biol.*, *13*, 595–598.

Halitligil, M., Akin, A., & Ylbeyi, A., (2002). Nitrogen balance of nitrogen–15 applied as ammonium sulfate to irrigated potatoes in sandy textured soils. *Biol. Fert. Soils.*, *35* (5), 369–378.

Hernández-Herrera, R. M., Santacruz-Ruvalcaba, F., Ruiz-López, M. A., Norrie, J., & Hernández-Carmona, G., (2014). Effect of liquid seaweed extracts on growth of tomato seedlings (*Solanum lycopersicum* L.). *J. Appl. Phycol.*, *26* (1), 619–628.

Hewedy, A. M., (2000). Effect of methods and sources of potassium application on the productivity and fruit quality of some new tomato hybrids. *Egypt. J. Agric. Res.*, *78* (1), 227–244.

Hopkins, B. G., Ellsworth, J. W., Shiffler, A. K., Bowen, T. R., & Cook, A. G., (2010). Pre-plant versus in-season application of phosphorus fertilizer for russet burbank potato grown in calcareous soil. *J Plant Nutr.*, *33* (7), 1026–1039.

Hopkins, B. G., Rosen, C. J., Shiffler, A. K., & Taysom, T. W., (2008). Enhanced Efficiency Fertilizers for Improved Nutrient Management: Potato (*Solanum tuberosum*). *Crop Manag.*, *7* (1), 0–0.

Hopkins, B., & Ellsworth, J., (2003). Phosphorus nutrition in potato production. In: *Proceedings of the Winter Commodity Schools*, *35*, 75–86.

Hutchinson, C. M., (2004). Influence of a controlled release nitrogen fertilizer program on potato (*Solanum tuberosum* L.) tuber yield and quality. In: *Meeting of the Physiology Section of the European Association for Potato Research*, *684*, 99–102.

Jasim, A. H., Hussein, M. J., & Nayef, M. K., (2013). Effect of foliar fertilizer (high in potash) on growth and yield of seven potato cultivars (*Solanum tuberosum* L.). *Euphrates J. Agric. Sci.*, *5* (1), 1–7.

Juan-guo, G. O. N. G., (2009). Effect of nitrogen, phosphorus, and potassium fertilizers and organic fertilizer on chlorophyll content in potato. *J. Anhui Agric Sci.*, *23*, 048.

Kádár, I., Márton, L., & Horváth, S., (2000). Mineral fertilisation of potato (*Solanum tuberosum* L.) on calcareous chernozem soil. *Növénytermelés*, *49* (3), 291–306.

Karam, F., Rouphael, Y., Lahoud, R., Breidi, J., & Colla, G., (2009). Influence of genotypes and potassium application rates on yield and potassium use efficiency of potato. *J. Agron.*, *8* (1), 27–32.

Kashem, M. A., Sarker, A., Hossain, I., & Islam, M. S., (2015). Comparison of the effect of vermicompost and inorganic fertilizers on vegetative growth and fruit production of tomato (*Solanum lycopersicum* L.). *Opn. J. Soil Sci.*, *5* (2), Article ID: 53820.

Kavvadias, V., Paschalidis, C., Akrivos, G., & Petropoulos, D., (2012). Nitrogen and potassium fertilization responses of potato (*Solanum tuberosum*) cv. Spunta. *Comm. Soil. Sci. Plant Anal.*, *43* (1–2), 176–189.

Khan, M. M. A., Bhardwaj, G., Naeem, M., Mohammad, F., Singh, M., Nasir, S., & Idrees, M., (2008). Response of tomato (*Solanum lycopersicum* L.) to application of potassium and triacontanol. In: *XI International Symposium on the Processing Tomato, 823*, 199–208.

Khan, M. M. A., Samiullah, A. S. H., & Afridi, R. M., (2000). Response of Black Nightshade (*Solanum nigrum* L.) to phosphorus application. *J. Agron. Crop Sci.*, *184* (3), 157–163.

Khan, S. M., Jan, N., Ullah, I., Younas, M., & Ullah, H., (2007). Evaluation of various methods of fertilizer application in potato (*Solanum tuberosum* L.). *Sarhad J. Agric.*, *23* (4), 889–894.

Khosravifar, S., Yarnia, M., Benam, M. B. K., & Moughbeli, A. H. H., (2008). Effect of potassium on drought tolerance in potato cv. Agria. *J. Food Agric. Environ.*, *6* (3/4), 236–241.

Kipgoskei, L. K., Akundabweni, L. S. M., & Hutchinson, M. J., (2003). The effect of farmyard manure and nitrogen fertilizer on vegetative growth, leaf yield and quality attributes of *Solanum villosum* (Black nightshade) in Keiyo district, rift valley. *Af. Crop Sci. Conference Proc.*, *6*, 514–518.

Klikocka, H., (2009). Influence of NPK fertilization enriched with S, Mg, and micronutrients contained in liquid fertilizer Insol 7 on potato tubers yield [*Solanum tuberosum* L.] and infestation of tubers with Streptomyces scabies and *Rhizoctoniasolani*. *J. Elementol.*, *14* (2), 271–288.

Kumar, M., Baishaya, L. K., Ghosh, D. C., Gupta, V. K., Dubey, S. K., Das, A., & Patel, D. P., (2012). Productivity and soil health of potato (*Solanum tuberosum* L.) field as influenced by organic manures, Inorganic fertilizers and biofertilizers under high altitudes of eastern Himalayas. *J. Agric. Sci.*, *4* (5), doi: 10. 5539/jas. v4n5p223.

Kumar, P., Pandey, S. K., Singh, B. P., Singh, S. V., & Kumar, D., (2007). Influence of source and time of potassium application on potato growth, yield, economics, and crisp quality. *Potato Res.*, *50* (1), 1–13.

Laboski, C. A. M., & Kelling, K. A., (2007). Influence of fertilizer management and soil fertility on tuber specific gravity: A review. *Am. J. Potato Res.*, *84* (4), 283–290.

Lawlor, D. W., (2002). Carbon and nitrogen assimilation in relation to yield: mechanisms are the key to understanding production systems. *J. Exp. Bot.*, *53* (370), 773–787.

Lin, S., Sattelmacher, B., Kutzmutz, E., Mühling, K., & Dittert, K., (2005). Influence of nitrogen nutrition on tuber quality of potato with special reference to the pathway of nitrate transport into tubers. *J. Plant Nutr.*, *27* (2), 341–350.

López-Cantarero, I., Ruiz, J. M., Hernandez, J., & Romero, L., (1997). Nitrogen metabolism and yield response to increases in nitrogen–phosphorus fertilization: Improvement in greenhouse cultivation of eggplant (*Solanum melongena* Cv. Bonica). *J. Agric. Food Chem.*, *45* (11), 4227–4231.

Love, S. L., Stark, J. C., & Salaiz, T., (2005). Response of four potato cultivars to rate and timing of nitrogen fertilizer. *Am. J. Potato. Res.*, *82* (1), 21–30.

Mäck, G., & Schjoerring, J. K., (2002). Effect of NO_3^- supply on N metabolism of potato plants (*Solanum tuberosum* L.) with special focus on the tubers. *Plant Cell Environ.*, *25*, 99–1009.

Mahmoud, A. R., & Hafez, M. M., (2010). Increasing productivity of potato plants (*Solanum tuberosum*L.) by using potassium fertilizer and humic acid application. *Int. J. Acad. Res.*, *2* (2), 83–88.

Mahmoud, H. A. F., & Amara, M. A. T., (2000). Response of tomato to biological and mineral fertilizers under calcareous soil conditions. *Bulletin of Faculty of Agriculture, University of Cairo.*, *51* (2), 151–174.

Mahmoud, H. A. F., (2000). Effect of sulfur and phosphorus on some eggplant cultivars under calcareous soil conditions. *Bulletin of Faculty of Agriculture, University of Cairo.*, *51* (2), 209–225.

Maier, N. A., McLaughlin, M. J, Heap, M., Butt, M., & Smart, M. K., (2002). Effect of current-season application of calcitic lime and phosphorus fertilization on soil pH, potato growth, yield, dry matter content, and cadmium concentration. *Commun. Soil Sci. Plant Anal.*, *33* (13–14), 2145–2165.

Masaad, R. C., Stephan, C., Rouphael, Y., Colla, G., Karam, F., & Lahoud, R., (2004). Yield and tuber quality of potassium treated potato under optimum irrigation conditions. In: *Meeting of the Physiology Section of the European Association for Potato Research, 684*, 103–108.

Masinde, P. W., Wesonga, J. M., Ojiewo, C. O., Agong, S. G., & Masuda, M., (2009). Plant growth and leaf N content of *Solanum villosum* genotypes in response to nitrogen supply. *Dynamic Soil, Dynamic Plant*, *3* (1), 36–47.

Milford, G. F. J., & Johnston, A. E., (2009). Potassium and nitrogen interactions in crop production. *Nawozy i Nawożenie.*, *34*, 143–162.

Mohr, R. M., & Tomasiewicz, D. J., (2011). Effect of phosphorus fertilizer rate on irrigated Russet Burbank Potato. *Commun. Soil Sci. Plant Anal.*, *42* (18), 2284–2298.

Moinuddin, S. K., & Bansal, S. K., (2005a). Growth, yield, and economics of potato in relation to progressive application of potassium fertilizer. *J. Plant Nutr.*, *28* (1), 183–200.

Moinuddin, S. K., Bansal, S. K., & Pasricha, N. S., (2005b). Influence of graded levels of potassium fertilizer on growth, yield, and economic parameters of potato. *J. Plant Nutr.*, *27* (2), 239–259.

Moraditochaee, M., Bozorgi, H. R., & Halajesani, N., (2011). Effects of vermicompost application and nitrogen fertilizer rates on fruit yield and several attributes of eggplant (*Solanum melongena* L.) in Iran. *World Appl. Sci. J.*, *15* (2), 174–178.

Nawaz, H., Zubair, M., & Derawadan, H., (2012). Interactive effects of nitrogen, phosphorus, and zinc on growth and yield of tomato (*Solanum lycopersicum*). *Afr. J. Agric. Res.*, *7* (26), 3792–3769.

Ogbomo, K. E. L., (2011). Comparison of growth, yield performance and profitability of tomato (*Solanum lycopersicon*) under different fertilizer types in humid forest ultisols. *Int. Res. J. Agric. Sci. Soil Sci.*, *1* (8), 332–338.

Oztürk, E., Kavurmaci, Z., Kara, K., & Polat, T., (2010). The effects of different nitrogen and phosphorus rates on some quality traits of potato. *Potato Res.*, *53* (4), 309–312.

Pal, S., Saimbhi, M. S., & Bal, S. S., (2002). Effect of nitrogen and phosphorus levels on growth and yield of brinjal hybrids (*Solanum melongena*). *Veg. Sci., 29* (1), 90–91.

Pervez, M. A., Muhammad, F., & Ullah, E., (2000). Effects of organic and inorganic manures on physical characteristics of potato (*Solanum tuberosum* L.). *Int. J. Agric. Biol., 2* (1–2), 34–36.

Powon, M. P., Aguyoh, J. N., & Mwaj, A. V., (2006). Growth and tuber yield of potato (*Solanum tuberosum* L.) under different levels of phosphorus and farm yard manure. *Agricultura. Tropica. Et. Subtropica., 39* (3), 189–194.

Riahi, A., & Hdider, C., (2013). Bioactive compounds and antioxidant activity of organically grown tomato (*Solanum lycopersicum* L.) cultivars as affected by fertilization. *Sci. Hort., 151* (28), 90–96.

Riley, H., (2000). Level and timing of nitrogen fertilizer application to early and semi-early potatoes (*Solanum tuberosum* L.) grown with irrigation on light soils in Norway. *Acta Agric. Scand., Section B – Soil and Plant Science, 50* (3), 122–134.

Rosen, C. J., & Bierman, P. M., (2008). Potato yield and tuber set as affected by phosphorus fertilization. *Am. J. Potato Res., 85* (2), 110–120.

Rosen, C. J., Kelling, K. A., Stark, J. C., & Porter, G. A., (2014). Optimizing phosphorus fertilizer management in potato production. *Am. J. Potato Res., 91* (2), 145–160.

Saeidi, M., Tobeh, A., Raei, Y., Roohi, A., Jamaati-e-Somarin, S., & Hassanzadeh, M., (2009). Evaluation of tuber size and nitrogen fertilizer on nitrogen uptake and nitrate accumulation in potato tuber. *Res. J. Environ. Sci., 3* (3), 278–284.

Sánchez-Rodríguez, E., Rubio-Wilhelmi, M. M., Blasco, B., Constán-Aguilar, C., Romero, L., & Ruiz, J. M., (2011). Variation in the use efficiency of N under moderate water deficit in tomato plants (*Solanum lycopersicum*) differing in their tolerance to drought. *Acta Physiol. Plant., 33* (5), 1861–1865.

Sarkar, B., & Mondal, S. S., (2005). Effect of integrated nutrient management on the growth and productivity of potato (*Solanum tuberosum*). *Environ. Ecol., 23S* (3), 387–391.

Shahbazi, K., Tobeh, A., Ebadi, A., Dehdar, B., Mahrooz, A., Jamaati-e-Somarin, S., & Shiri-e-Janagrad, M., (2009). Nitrogen use efficiency and nitrate accumulation in tubers as affected by four fertilization levels in three potatoes (*Solanum tuberosum* L.) cultivars. *Asian J. Biol. Sci., 2* (4), 95–104.

Singh, B., Pathak, K., Boopathi, T., & Deka, B., (2010a). Vermicompost and NPK fertilizer effects on morpho-physiological traits of plants, yield, and quality of tomato fruits: (*Solanum lycopersicum* L.). *Veg. Crop Res. Bullet., 73,* 77–86.

Singh, J. P., Lal, S. S., & Sharma, R. C., (2002). Productivity and sustainability of potato-based cropping systems with reference to fertilizer management in North-western plains of India. Potato, global research and development. *Proceedings of the Global Conference on Potato*, New Delhi, India, *2,* 929–934.

Singh, N. P., & Raghav, M., (2000). Response of potato to nitrogen and potassium fertilization under U. P. tarai conditions. *J. Ind. Potato Assoc., 27* (1/2), 47–48.

Singh, S. K., Kumar, D., & Lal, S. S., (2010). Integrated use of crop residues and fertilizers for sustainability of potato (*Solanum tuberosum*) based cropping systems in Bihar. *Ind. J. Agron., 55* (3), 203–208.

Sivakumar, V., & Ponnusami, V., (2011). Influence of spacing and organics on plant nutrient uptake of *Solanum nigrum. Plant Arch., 11* (1), 431–434.

Sparrow, L. A., & Chapman, K. S. R., (2003). Effects of nitrogen fertilizer on potato (*Solanum tuberosum* L., cv. Russet Burbank) in Tasmania. 1. Yield and quality. *Aust. J. Exp. Agric., 43* (6), 631–641.

Srivastava, B. K., Singh, M. P., Singh, S., Lata, S., Srivastava, P., & Shahi, U. P., (2009). Effect of integrated nutrient management on the performance of crops under brinjal (*Solanum melongena*) – pea (*Pisumsativum*) – okra (*Hibuscusesculentus*) cropping system. *Ind. J. Agric. Sci.*, *79* (2), 91–93.

Sundharaiya, K., Nainar, P., Ponnuswami, V., Jasmine, A. J., & Muthuswamy, M., (2000). Effect of inorganic nutrients and spacing on the yield of marunthukathiri (*Solanum khasianum* Clarke). *South Ind. Hort.*, *48* (1/6), 168–171.

Taheri, N., Sharif-Abad, H. H., Yousefi, K., & Roholla-Mousavi, S., (2012). Effect of compost and animal manure with phosphorus and zinc fertilizer on yield of seed potatoes. *J. Soil Sci. Plant Nutr.*, *12* (4), 705–714.

Turakainen, M., Hartikainen, H., & Seppänen, M. M., (2004). Effects of selenium treatments on potato (*Solanum tuberosum* L.) growth and concentrations of soluble sugars and starch. *J. Agric. Food Chem.*, *52* (17), 5378–5382.

Vos, J., (1997). The nitrogen response of potato (*Solanum tuberosum* L.) in the field: Nitrogen uptake and yield, harvest index and nitrogen concentration. *Potato Res.*, *40* (2), 237–248.

Wijkmark, L., Lindholm, R., & Nissen, K., (2005). *Uniform Potato Quality With Site-Specific Potassium Application. Precision Agriculture 05*. Papers presented at the 5th European Conference on Precision Agriculture, Uppsala, Sweden, 393–400.

Zaman, A., Sarkar, A., Sarkar, S., & Devi, W. P., (2011). Effect of organic and inorganic sources of nutrients on productivity, specific gravity and processing quality of potato (*Solanum tuberosum*). *Ind. J. Agric. Sci.*, *81* (12).

Zebarth, B. J., & Rosen, C. J., (2007). Research perspective on nitrogen bmp development for potato. *Amer. J. Potato Res.*, *84* (1), 3–18.

Zeyi, W., & Fuling, L., (2005). Effect of using potassium fertilizer and boron fertilizer on the increase of production and heightening of qualities of potatoes. *Chin. Agric. Sci. Bullet.*, *9*, 083.

Zheng, S. L., Yuan, J. C., Ma, J., Yang, C. Y., Wang, X. Q., & Deng, M. F., (2009). Comparative study on nitrogen application of spring and autumn potato (*Solanum tuberosum* L.). *Southwest Chin. J. Agric. Sci.*, *3*, 038.

I. Ethiopia
II. Mediterranean region
III. Asia Minor, Caucasia, Iran
IV. Afghanistan, Turkistan, Pakistan
V. Indo-Burma
VI. China

VII. Thailand, Malaysia, Java
VIII. Mexico, Guatemala
IX. Bolivia, Peru, Ecuador, Colombia
X. Southern Chile
XI. Brazil, Paraguay
XII. North America

FIGURE 1.1   Centers of origin of the world (I–VII discovered by N.I. Vavilov, IX–XII discovered after N.I. Vavilov).

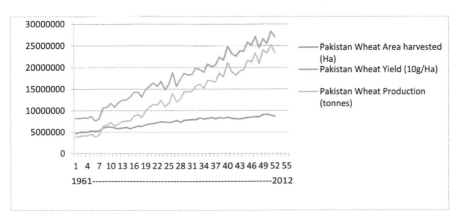

FIGURE 2.1   Pakistan wheat Production trends since 1961 to 2012.
Source: (Economic Survey of Pakistan, 2013; FAOSTAT, 2014).

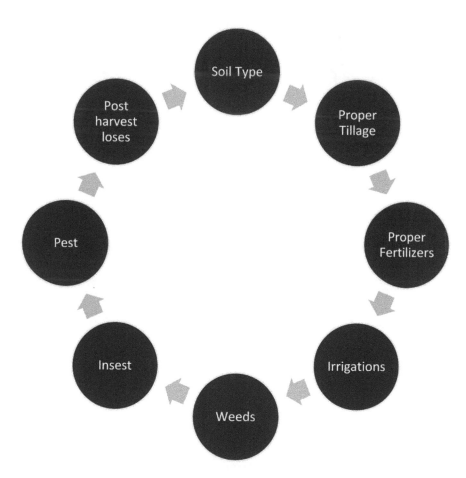

FIGURE 2.2   Wheat crop production technology major concerns for successful crop production.

Jute Cultivation and The Environment

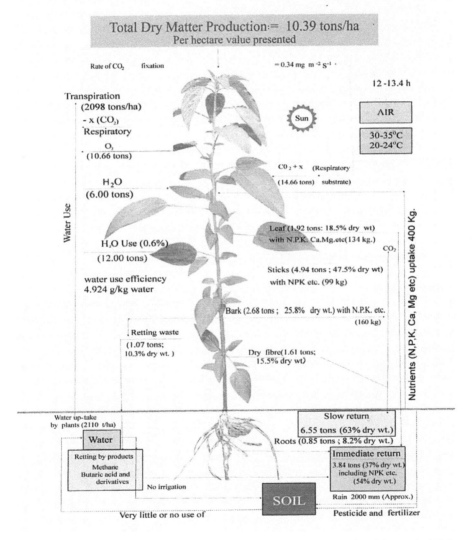

FIGURE 6.2 Natural conservation through jute cultivation (*Source*: Khandakar et al., 1995).

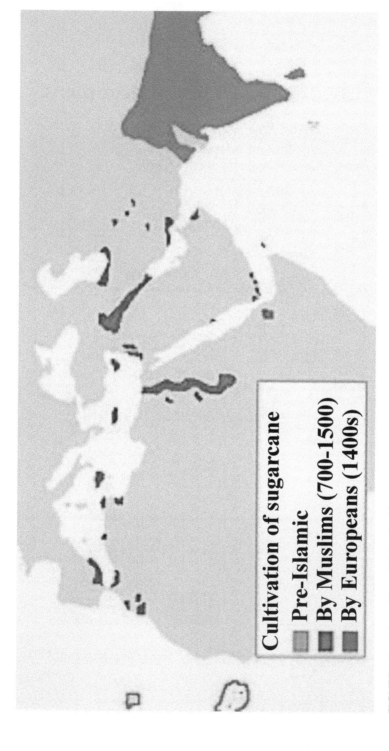

FIGURE 7.3 Adapted from Wikipedia. The westward diffusion of sugarcane in pre-Islamic times (shown in red), in the medieval Muslim world (green) and by Europeans (violet).

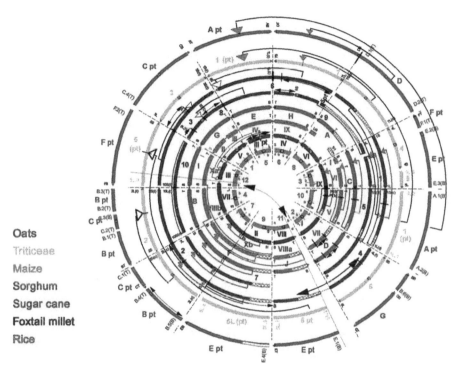

Oats

Triticeae

Maize

Sorghum

Sugar cane

Foxtail millet

Rice

FIGURE 7.6   Adapted from Gale and Devos, 1998 (Comparative genetics in the grasses). A consensus grass comparative map. The comparative data have been drawn from many sources: Oats-wheat-maize-rice (Kurata et al., 1994); wheat-rice (Kurata et al., 1994; Van Deynze et al., 1995); maize-wheat (Ahn and Tanksley, 1993); maize-sorghum-sugarcane (Grivet et al., 1994; Dufour et al., 1997); and foxtail millet-rice (Devos et al., 1998). Arrows indicate inversions and transpositions necessary to describe present-day chromosomes. Locations of telomeres (.) and centromeres (h) are shown where known. Hatched areas indicate chromosome regions for which very little comparative data exist. Chromosome nomenclatures and arm (shortylong or topybottom) designations are as described by O'Donoughue et al. (1992) for oats, Pereira et al. (1994) for sorghum, Dufour (1997) for the sugarcane genomes, Wang et al. (1997) in foxtail millet, and Singh et al. (1996) in rice.

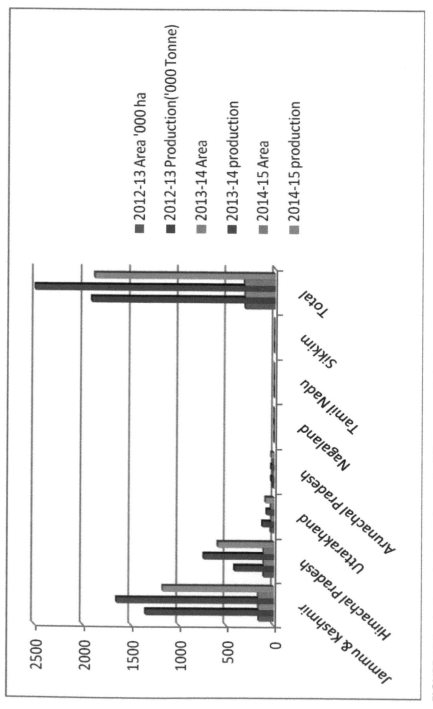

FIGURE 10.1 Apple production per hectare in different states during 2012–2015.

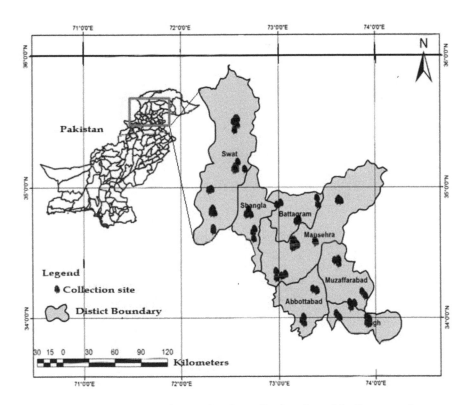

FIGURE 11.1   Map of the study area showing collection sites of the *Pyrus* samples.

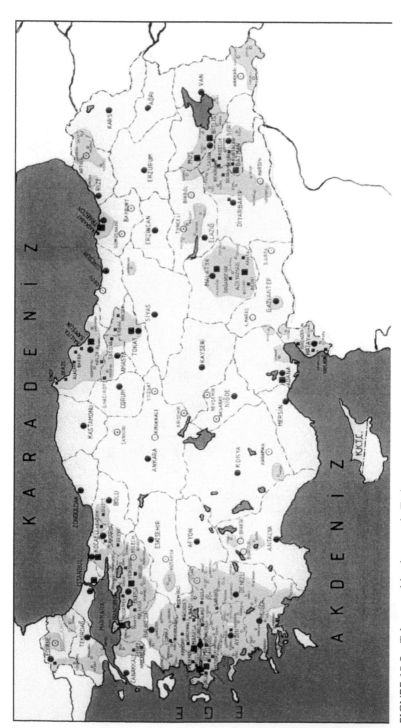

FIGURE 15.3   Tobacco cultivation area in Turkey.

CHAPTER 10

BIOLOGY AND BIOTECHNOLOGICAL ADVANCES IN APPLE: CURRENT AND FUTURE SCENARIO

WAJIDA SHAFI, SUMIRA JAN, SHAFIA ZAFFAR FAKTOO, JAVID IQBAL MIR, and MUNEER SHEIKH

ICAR- Central Institute of Temperate Horticulture, Rangreth, Air Field, Srinagar, Jammu, and Kashmir, India, E-mail: javidiqbal1234@gmail.com

ABSTRACT

Apple is an extremely cherished fruit in various temperate regions across world, and is currently grown vigorously with total world production over 71 million tonnes. Apple is the fourth most significant economical crop following citrus, banana, and grapes. Apart from its consumption as fruit, apple can be processed into diverse products like juice, vinegar, juice, and cider. More than 7,500 cultivars of apple have been discovered across various countries. But due to disease susceptibility and inappropriate quality only few cultivars have market acceptability. This review summarizes biology, production, fruit quality and advancement in biotechnological interventions from classical genetics to more advanced molecular tools in extrication of genetic characterization, development of improved fruit quality as well as development of resistant varieties of apples. Currently, the efforts are directed towards development of apple varieties with tailored traits including fruit quality and nutritional aspects. The review concludes with endnote on prospect guidelines summarizing scope of genetic engineering and advance of transgenic in improving apple quality and productivity globally.

10.1   INTRODUCTION

Apples (Malus × domestica Borkh.) are one of the most commonly consumed fruits in the world and is the fourth most important fruit crop after citrus, grapes, and banana, and one of the commercially most important horticultural crops grown in temperate parts of the world (Ferree and Warrington, 2003). Apple belongs to the Rosaceae family which includes many well-known genera with economically important fruits, particularly edible, temperate-zone fruits and berries such as apple, pear, almond, apricot, cherries, peach, plums, strawberries, and raspberries. In 2013, world apple production was estimated at around 80.8 million tonnes, with China accounting for 49% of the total. Apples can be grown under different climatic conditions, ranging from temperate climates such as southern Siberia or the Mediterranean to subtropical climates such as Brazil or South Africa. Nowadays, it has become increasingly popular to cultivate apples also in subtropical and tropical (high altitudes) areas since they fetch a comparatively high price on the market. Apples are an important agricultural commodity in the global market for fresh products. The quality of an apple depends on its external characteristics, such as color, size, and surface texture, and internal parameters, such as sweetness, acidity, firmness, tissue texture, ascorbic acid, and polyphenolic compounds (Wojdylo et al., 2008). These characteristics, especially internal and external parameters, are similar to a variety. However, each variety has its special characteristics and flavor, which results in different prices and preferences by different people.

10.2   TAXONOMY OF APPLE

The genus *Malus* belongs to the rose family (Rosaceae) which includes over 100 genera and 3000 species distributed worldwide, most commonly in temperate regions (Velasco et al., 2010). Determination of phylogenetic relationships within the Rosaceae is complex and ongoing but regardless of which classification scheme is used, apples, and pears belong to the same subfamily, and along with a handful of other closely related genera, are distinct in having a haploid (x) number of 17 chromosomes (x = 17). They are thought to be of allopolyploid origin, having originated from ancient hybridizations between species in the Prunoideae (x = 8) and the Spiraeoideae (x = 9), followed by chromosome doubling. The original hybrids would have been sterile and only after chromosome doubling would they have formed fertile allopolyploids (Hancock et al., 2008; Luby, 2003; Way et al., 1990; Webster, 2005a).

Taxonomic Classification	
Domain	Eukarya
Kingdom	Plantae
Subkingdom	Tracheobionta
Superdivision	Spermaphyta
Division	Magnoliophyta
Class	Magnoliopsida
Subclass	Rosida
Order Family	Rosaceae
Subfamily	Maloidea
Tribe	Maleae
Subtribe	Malinae
Genus	Malus
Species	*Malus domestica* Borkh

Like the family it belongs to, the genus *Malus* is diverse and complex, and there are considerable challenges in species delimitation due to hybridization, polyploidy, and apomixis (Luby, 2003). The primary center of species richness and diversity is in southwest China and Central Asia, with several species ranging eastwards to Manchuria and Japan, and others ranging westwards to Europe (Ferree and Carlson, 1987; Ignatov and Bodishevskaya, 2011; Luby, 2003). A secondary center is in North America, with four native species. The number of species included in the genus is still a subject of debate, with different treatments recognizing as few as 8 to as many as 78 primary species, depending on the rank given to certain taxa, and the acceptance of reported hybrids (Hancock et al., 2008; Harris et al.2002; Jackson, 2003; Luby, 2003; Rieger, 2006; Robinson et al., 2001). Most species in the genus can be readily hybridized, and many hybrid species, derived naturally or artificially, are recognized (Hancock et al., 2008; Luby, 2003). Most apple species are diploid ($2n = 2x = 34$) but higher somatic numbers exist (e.g., 51, 68, 85) and several cultivated types are triploid ($2n = 3x = 51$) (Hancock et al., 2008).

10.3 ORIGIN OF APPLE

Apple species are distributed throughout very large regions of the world including West Asia, Himalayan, Central Asia, India, Western provinces of China, Europe, and some parts of America and Africa (Juniper et al., 1999).

Historical studies have shown that apple seed transfer by human or animals have probably helped in its distribution from the center of origin (the region where the species originated) to other parts of the world. Central Asia has been reported to contain the greatest diversity of *Malus*, and this area also appears to be the center of origin of the domesticated apple (Janick et al., 1996). This is in accordance with Vavilov's hypothesis about the wild apples in central Asia and their close relatives being the progenitors of the domesticated apple (Harris et al., 2002). Nowadays, *M. sieversii* which grows wild in Kazakhstan and Kyrgyzstan is thought to be the main progenitor species (Pereira-Lorenzo, 2009). *Malus sieversii* has very high similarity with *M.* × *domestica* in morphology and fruit flavor. According to observations made on extensive collection tours, *M. sieversii* has been claimed to incorporate all the fruit qualities which are present in the domesticated apples (Forsline, 1995).

Relationships among apple species have been evaluated by morphological and molecular DNA analysis, and have confirmed that *M. sieversii* is the species from which apple domestication started (Forte et al., 2002). This species may have hybridized with *M. prunifolia*, *M. baccata* and *M. sieboldi* to the East and with *M. turkmenorum* and *M. sylvestris* to the West. Subsequently, well-established apple cultivars were selected and introduced into Europe and especially the Mediterranean regions by the Romans (Juniper et al., 1999). Other *Malus* species are occasionally used for introgression into modern cultivars but this usually requires several generations of backcrossing to reach acceptable fruit size and quality. There are numerous famous apple cultivars with huge commercial prospects (Table 10.1).

10.4   APPLE PRODUCTION AND GEOGRAPHICAL DISTRIBUTION

Global apple production remained stable for a large part of the previous century until China began to expand its apple production in the 1990s. However, no one predicted that the rate of growth would be maintained from the higher base in 1990 so that China's production would exceed 20 million tonnes by the end of the decade (O'Rourke, 2003). Food and Agricultural Organization (FAO) data indicate that China's share of world apple production has gone from 10.7% in 1990 to 36.7% in the year 2000. Currently, China is the largest apple producer in the world with a production of more than 31 million tonnes, which is several times higher than the production of the four countries in the closest positions, e.g., USA, Turkey, Poland, and

TABLE 10.1 Description of Apple Varieties and Their Market Usage

Common Name	Image	Origin	Date	Description	Market usage
Anna		Israel	1965	Color is yellow with a red blush. This variety does not grow well in the cold and prefers heat and humidity	Eating
Antonovka		Kursk, Russia	17th Century	A very old Russian variety, often planted at dachas. Apples are large, yellow-green, and bracingly tart to eat out of hand, but superb for cooking, as they keep their shape. Extremely tolerant of cold weather, and because it produces a single, deep taproot (unusual among apple trees), Antonovka is propagated for use as a rootstock. Antonovka rootstock provides a cold-hardy (to −45°C), well-anchored, vigorous, standard-sized tree.	Cooking, Cider
Ben Davis		Southeastern US		Noted for keeping well prior to refrigerated storage, but flavor has been compared with cork.	Eating
Braeburn		New Zealand	1952	Chance seedling. The fruit is widely sold commercially in the UK	Eating

TABLE 10.1 *(Continued)*

Common Name	Image	Origin	Date	Description	Market usage
Cox's Orange Pippin		England	1829	One of the most celebrated apples in the UK, valued for its aromatic "orange" color and flavor. The fruit is widely sold commercially. Mainly grown in UK, Belgium, and the Netherlands but also grown for export in New Zealand	Eating
Cripps pink (Pink Lady)		Australia	1970's	Crisp, very sweet and slightly tart. Light red, pink, and light yellow-green striped skin	Cooking, Eating
Fuji		Japan	1930's	Red Delicious × Ralls Genet. Dark red, conic apple. Sweet, crisp, dense flesh is very mildly flavored. Keeps very well. One of the most widely grown apple varieties in the world	Eating
Gala (Royal Gala)		New Zealand	1970's	A small to medium-sized conic apple. Thin, tannic skin is yellow-green with a red blush overlaid with reddish-orange streaks. Flesh is yellowish-white, crisp, and grainy with a mild flavor. Cross of three of the world's best known apples: Kidd's Orange Red (a cross of Red Delicious and Cox's Orange Pippin) × Golden Delicious. One of the most widely available commercial fruit	Eating

TABLE 10.1 *(Continued)*

Common Name	Image	Origin	Date	Description	Market usage
Golden Delicious		Clay County, West Virginia, US	1914	One of the most popular varieties in the world. Due to its regular size, even color and storage qualities the fruit is widely sold commercially. Uniform light green-yellow coloration, very sweet. A good pollinator.	Eating
Goldspur					Eating
Granny Smith		Australia	1868	This is the apple once used to represent Apple Records. A favorite variety, widely sold in the UK. Also noted as common pie apple. Lime green coloring. Extremely tart	Cooking, Eating
Green sleeves		Kent, UK	1966	Golden Delicious × James Grieve; good garden apple, with a pleasant but unexceptional flavor. Likely named for famous Renaissance era song	Eating

TABLE 10.1 *(Continued)*

Common Name	Image	Origin	Date	Description	Market usage
Jonagold		New York	1968	Popular in Europe and land of origin. Several highly colored strains are available. Widely sold commercially in the UK	Cooking, Eating
Jonathan		New York	1820's	Tart taste. Mostly red apple with patches of lime green. Does well in cooler areas; some frost resistance.	Cooking (Pie), Eating
Laxton's Fortune			1904	Cox's Orange Pippin × Wealthy	Eating
Liberty		New York	1978	Very disease-resistant. Very similar appearance to McIntosh, relatively short storage life in air	Eating
McIntosh		Ontario, Canada	1811	A popular, cold-tolerant eating apple in North America.	Cooking (applesauce), Eating, Pies
Mollie's Delicious		New Jersey, US	1966	Conical shape, pinkish red color. Lasts long in refrigeration. Good aftertaste.	Eating

TABLE 10.1 *(Continued)*

Common Name	Image	Origin	Date	Description	Market usage
Prima		USA	1958	Resistant to scab and most diseases	Eating
Red Delicious		Iowa, US	1870	Unmistakable for its acutely conic shape, dark red color and telltale bumps on bottom. Flavor is sweet and mild, bordering on bland. Poor choice for cooking or cider. Original seedling known as "Hawkeye." Rights bought by Stark Brothers in 1893. First marketed as "Delicious" or "Stark's Delicious," name changed to "Red Delicious" in 1914 when Stark bought the rights to Mullin's Yellow Seedling, changing that Apple's name to "Yellow Delicious." Red Delicious has many sports and ranks as the world's most prolific apple	Eating
Rome Beauty		Rome, Ohio, United States	19th century	Rounded, deep red, and very glossy. Crisp, juicy white flesh is mild as a dessert apple, but develops an extraordinary depth and richness when cooked. Good keeper	Cooking
Spartan		British Columbia, Canada	1926	Good all-purpose, medium-sized apple. Has a bright red blush and may have background patches of greens and yellows. Popular across border in United States as well.	Cooking, Eating

TABLE 10.1 *(Continued)*

Common Name	Image	Origin	Date	Description	Market usage
Stark Earliest		US	1938	Does nicely in fruit salads. Red striping on light background. Ripens in summer	Eating
Tydeman's Early Worcester		England	1929	McIntosh × Worcester Pearmain. Crimson over yellow background color	Eating
Vista Bella		USA	1956	A very-early season dessert apple.	Eating
June Eating					

TABLE 10.1 *(Continued)*

Common Name	Image	Origin	Date	Description	Market usage
Benoni		Massachusetts, USA		Benoni is a fine dessert apple, very attractive in appearance and excellent in quality, fruit medium to small, roundish inclined to conic, faintly ribbed toward the apex; sides unequal, cavity acute, rather narrow, moderately deep, wavy, greenish-russet. Skin smooth, yellowish orange partly covered with lively deep carmine stripping, dots scattering, minute, whitish, fruit is fine grained juicy and crisp fruit weight (48–50g), fruit length (39–43 mm) and diameter (50–53mm), respectively.	
Fanny					
Hardiman					

TABLE 10.1 *(Continued)*

Common Name	Image	Origin	Date	Description	Market usage
Oregon spur		US	1968	It is a bud support of Delicious with large number of spur bearing branches, conic fruit shape, strong crowning at calyx end, red purple group, dark intensity of over color, solid flesh, medium depth of stalk cavity and large width of stalk cavity aperture of locules is fully open.	
Cooper –IV				It is bud mutant of a Delicious, regular bearing, matures in Mid-Season, fruit general shape is conical, weakly defined flush, strongly defined strips, purple red color, strong crowning at calyx end, medium depth of stalk cavity, large width of stalk cavity.	
Vance Delicious					
Gala Mast				It is precocious, regular bearing mid-maturing, high yielding, conic shape, moderate fruit ribbing, red purple color, flushed, and mottled, aperture of locules on transverse section is closed or slightly open.	

TABLE 10.1 *(Continued)*

Common Name	Image	Origin	Date	Description	Market usage
Red Spur		US	1959	It is bud sport of Red Delicious from complete tree variation, fruits resembles to Rich-a-Red.	
Super Chief		US	1988	This strain starts out as a stripe and fills into a solid red ten days ahead of harvest. Tree is a super-spur and stays compact even on semi-dwarf roots. It is a consistent, annual bearer even when not thinned aggressively.	
Scarlet Spur		US	1982	Scarlet Spur has dark mahogany color, crisp white flesh, and excellent fruit production.	
Red velox		Italy		Mutant of Red Delicious, tree is of moderate vigor, fruit is of medium size, dark red in color, with light stripes, color development is full and 100%, shape typically of red delicious. Particular characteristics of the variety, is the early and homogenous color on the whole tree.	

TABLE 10.1 *(Continued)*

Common Name	Image	Origin	Date	Description	Market usage
Akbar		J&K		Developed from Ambri x Cox'x Orange Pippin.	
Coe Red Fuji				It is precocious regular bearer, late maturing, high yielding cultivar, globose fruit shape, moderate fruit ribbing, weak crowning at calyx end, solid flesh, aperture of locules on transverse section is moderately open.	
Silver Spur		USA	1977	It is spur type, precocious regular bearer, late maturing, high yielding cultivar.	

TABLE 10.1 *(Continued)*

Common Name	Image	Origin	Date	Description	Market usage
Lal Ambri		SKUAST–K Shalimar		This was evolved by SKUAST–K Shalimar by crossing of Ambri X Red Delicious. Fruits are attractive in appearance, conical shape with blushed colored and good self life.	
Red Fuji		USA		This is very productive and annual bearer and late maturing variety, fruit general shape globose,	
Michal				It very low chilling, early maturing variety. Fruits are medium in size slightly conical in shape, with smooth calyx end, stripped red colored skin our green yellow ground,	

TABLE 10.1 *(Continued)*

Common Name	Image	Origin	Date	Description	Market usage
Ambri		J&K		It is only indigenous cultivars of India due to high fragrance and long shelf life, it has high in demand. Fruits typical conical shape with variable shape, size, and color.	
Firdous		J&K		Developed from Golden Delicious x Rome Beauty x Prima. Fruit medium in size, sweet with slight acidic blend, crisp, juicy having resistance to scab and moderate resistance to Alternaria and San jose scale.	
Belle De boskoop		Netherlands.		A popular old dual-purpose apple	

Iran (Figure 10.1). China is currently responsible for approximately half of the world apple productions. Some of the main apple-producing countries and their production volumes are listed in Table 10.2.

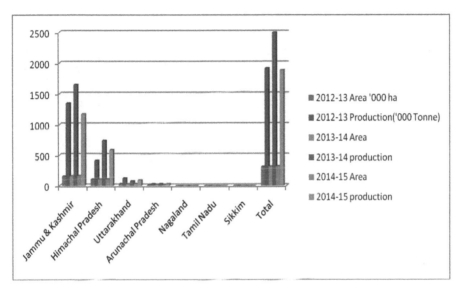

FIGURE 10.1 **(See color insert.)** Apple production per hectare in different states during 2012–2015.

Most temperate-zone woody deciduous trees, including apple, require a certain amount of chilling accumulation during the wintertime to break bud endo-dormancy before active shoot growth in the spring and for normal growth (O'Rourke, 2003). In general, apples are therefore suitable for growing mainly in areas with a temperate climate. However, apples can also be grown in other climates, like subtropical and even tropical areas at high altitudes, where sometimes two crops can be produced per year (Pereira-Lorenzo, 2009). Apple production has thus been reported from countries like India, Mexico, Brazil, Egypt, South Africa, Kenya, Ethiopia, Uganda, and Zimbabwe (Wamocho and Ombwara, 2001; Ashebir et al., 2010). In the subtropical and tropical areas of Asia, India appears to be the largest apple producer. Many different apple cultivars are grown in the northern, mountainous parts of the country, especially in the provinces of Jamma and Kashmir (Verma et al., 2010),the highest productivity of Apples had been noted in J&K (13.1t/ha) followed by H.P. (8.8t/ha) and Uttarakhand (4.1 t/ha) (Table 10.3) (Sharma and Krishna, 2014). In the subtropical areas of the

TABLE 10.2 World Apple Production –All Volumes are in Metric Tons (MT)

	2001	2002	2003	2004	2005	2006	2007	2008	2009	2010	2011	2012	2013
China	20,022,763	19,250,650	21,105,185	23,681,994	24,016,901	26,064,930	27,865,853	29,850,763	31,684,445	33,265,186	35,987,221	37,000,000	39,684,118
U.S.A.	4,276,810	3,866,440	3,947,620	4,735,780	4,408,870	4,568,630	4,122,880	4,369,590	4,402,070	4,210,060	4,272,840	4,110,046	4,081,608
Turkey	2,450,000	2,200,000	2,600,000	2,100,000	2,570,000	2,002,030	2,457,850	2,504,490	2,782,370	2,600,000	2,680,080	2,889,000	3,128,450
Iran	2,353,360	2,334,000	2,400,000	2,178,650	2,661,900	2,700,000	2,660,000	2,718,780	2,000,000	1,662,430	1,651,840	1,700,000	1,693,370
Poland	2,433,940	2,167,520	2,427,750	2,521,510	2,074,950	2,304,890	1,039,970	2,830,660	2,626,270	1,858,970	2,493,080	2,877,336	3,085,074
Italy	2,299,100	2,199,220	1,953,750	2,136,230	2,192,000	2,130,980	2,230,190	2,210,100	2,325,650	2,204,970	2,411,200	1,991,312	2,216,963
France	2,397,000	2,432,230	2,136,890	2,203,650	2,241,480	2,080,920	2,143,670	1,701,750	1,729,650	1,711,230	1,858,880	1,382,901	1,737,482
India	1,230,000	1,160,000	1,470,000	1,521,600	1,739,000	1,814,000	1,624,000	2,001,000	1,985,000	1,777,200	2,891,000	2,203,400	1,915,000
Russian F*	1,640,000	1,950,000	1,700,000	2,026,000	1,786,000	1,626,000	2,342,000	1,122,400	1,441,200	986,000	1,200,000	1,403,000	1,574,000
Chile	1,135,000	1,150,000	1,250,000	1,300,000	1,300,000	1,350,000	1,400,000	1,280,000	1,090,000	1,100,000	1,169,090	1,625,000	1,709,589
Argentina	1,428,800	1,156,830	1,307,460	1,262,440	1,206,210	1,100,000	1,000,000	950,000	1,027,090	1,050,000	1,115,950	1,250,000	1,245,018
Brazil	716,030	857,388	841,821	980,203	850,535	863,019	1,115,380	1,124,160	1,222,890	1,279,030	1,339,000	1,335,478	1,231,472
Germany	922,400	762,800	818,032	979,730	891,402	947,611	1,070,040	1,047,000	1,070,680	834,960	898,848	972,405	803,784
Japan	930,700	925,800	842,100	754,600	818,900	831,800	840,100	910,700	845,600	798,200	655,300	793,800	741,700
Ukraine	474,700	522,300	871,300	716,900	719,800	536,500	754,900	719,300	853,400	897,000	954,100	1,126,800	1,211,400
Spain	917,409	694,822	881,101	690,886	774,210	650,384	721,178	661,724	594,800	596,000	670,566	558,900	546,400
South Africa	562,510	591,432	701,663	765,359	680,426	639,763	708,089	770,741	817,698	724,232	781,124	795,758	811,523
Korea**	660,000	636,902	650,000	665,000	668,000	665,000	635,000	680,564	719,682	752,300	806,718	*785,000	493,701
Romania	507,440	491,500	811,100	1,097,840	637,979	590,413	475,300	459,016	517,491	552,860	620,362	462,935	493,405

TABLE 10.2 (Continued)

	2001	2002	2003	2004	2005	2006	2007	2008	2009	2010	2011	2012	2013
Mexico	442,679	479,613	495,217	572,906	583,994	601,915	505,078	511,988	561,493	584,655	630,533	375,045	858,608
Egypt	468,269	524,948	533,360	546,183	578,250	570,330	557,944	550,743	508,833	493,119	455,817	541,239	546,164
Hungary	605,440	526,865	507,505	700,391	510,361	537,345	170,901	568,600	573,368	496,916	292,810	650,595	552,400
Uzbekistan	454,500	444,600	365,200	352,400	402,000	514,441	502,500	585,000	635,000	712,000	779,000	820,000	937,000
New Zealand	473,726	530,633	501,192	546,000	524,000	354,000	421,000	446,000	431,000	318,900	437,782	448,000	438,952

Russian F*= Russian Federation Korea**=Democratic People's Republic of Korea. Source: The Statistical Division (FAOSTAT) of the Food and Agriculture Organization of the United Nations (FAO).

TABLE 10.3 Area, Production, and Productivity Scenario of Apple in India

S. No.	State	2012–13			2013–14			2014–15		
		Area '000 ha	Production ('000 tonnes)	Productivity (MT/Ha)	Area	Production	Productivity (MT/Ha)	Area	Production	Productivity (MT/Ha)
1	Jammu & Kashmir	157.28	1348.15	8.57	160.87	1647.69	10.24	163.43	1170.31	7.16
2	Himachal Pradesh	106.23	412.40	3.88	107.69	738.72	6.85	107.69	588.97	5.46
3	Uttarakhand	33.76	123.23	3.65	29.97	77.45	2.58	33.91	91.47	2.69
4	Arunachal Pradesh	14.07	30.95	2.19	14.28	31.87	2.23	14.50	32.00	2.20
5	Nagaland	0.10	0.60	6	0.21	1.89	9	0.21	1.89	9
6	Tamil Nadu	0.00	0.03		0.01	0.03	3	0.01	0.07	7
7	Sikkim	0.05	0.03	0.6	0.02	0.03	1.5	0.02	0.03	1.5
	Total	311.50	1915.38	6.14	313.04	2497.68	7.97	319.77	1884.73	5.89

Source: Horticulture Statistics Division, DAC&FW.

America, apples are grown, e.g., in the highlands of the northern regions of Mexico and also in large subtropical areas of Brazil (Leite, 2008; Hauagge, 2010). In Africa, the most important apple-growing country is South Africa, where roughly 20,000 ha of apple are cultivated (Cook, 2010).

10.5   VARIETIES

Apple varieties should have climatic adaptability, attractive fruit size, shape, color, good dessert quality, and long shelf-life, resistance to diseases and pest and tolerance to drought conditions besides high productivity. In the fifties, the green English variety McIntosh, Baldwin, Jonathan, Cox's Orange Pippin, Golden Delicious, Black Ben Davis and Pippins-predominated. Of late, the colored Delicious apples have replaced the English ones. As a result of this phenomenal change, Delicious group occupies more than 83% of the total areas under apple in Himachal Pradesh, 15% in Jammu and Kashmir and 30% in Uttarakhand. In Jammu and Kashmir, the area under Ambri has decreased tb l ess than 1% due to late-bearing of this variety, though the fruits are highly attractive with a long shelf-life. The recommended apple varieties in Jammu, and Kashmir, Himachal Pradesh and hills of Uttarakhand are given in Table 10.4. Recently, a further shift from Delicious group to improved spur type and standard color mutants has been observed; The spur-type mutants produce trees 50–80% of the standard size with 20–50% higher yield potential in addition to early fruit maturity and better fruit coloration. The high yielding apple Scarlet Gala and Red Fuzi have also been introduced in Himachal Pradesh and hills of Uttarakhand. These are being evaluated on size-controlling clonal rootstocks M9, M26, M7, MM 106 and MM 111.

10.6   BIOLOGY OF APPLE

The apple tree has hermaphroditic flowers with a gametophytic type of self-incompatibility controlled by a single multi-allelic locus (Pereira-Lorenzo, 2009). Therefore, at least two different cultivars must be inter-planted in the orchard to ensure high levels of cross-pollination in order to achieve adequate fruit development. Alternatively, trees of undomesticated *Malus* species can be inter-planted among (or top-worked onto) some of the trees in the row. Blooming and ripening time of different cultivars vary considerably and form different categories, e.g., early, middle, and late blooming or ripening cultivars.

TABLE 10.4   Recommended Apple Varieties in Different States

S. No	Seasons	Himachal Pradesh	Jammu & Kashmir	Uttarakhand
1	Early Season	Tydeman's Early (P) Michael Molies Delicious Schlomit Starkrimson	Irish Peach Benoni	Early Shanburry (P) Fenny Benoni Chaubattia Princess
2	Mid-Season	Starking Delicious Red Delicious Richared Vance Delicious Top Red Golden Lord Lambourne (P) Red Chief Oregon Spur Redspur Red Gold (P)	American Mother Razakwar Jpnathan (P) Cox's Orange Pippin (P) Red Gold (P) Queen's Apple Rome Beauty Scarlet Siberian	Rea Delicious Starking Delicious McIntosh (P) Cortland Delicious (P)
3	Late Season	Golden Delicious (P) Yellow Newton (P) Winter Banana Granny Smith (P)	King Pippin American Apirouge Kerry Pippin Lal Ambri Sunhari Chamure Golden Delicious (P) Red Delicious Ambri Baldwin Yellow Newton (P) White Dotted Red	Rymer Buckingham (P)
	Scab-resistant varieties	Florina, Macfree, Nova Easy Grow, Coop12, Coop13 (Redfree), Nova Mac, Liberty and Freedom.	Firdous and Shireen	
	Hybrids	Ambred (Red Delicious x Ambri), Ambrich (Richared x Ambri) and Ambroyal (Starking Delicious x Ambri)	Lal Ambri (Red Delicious x Ambri) Sunehari (Ambri x Golden Delicious)	Chaubattia Princess and Chaubattia Anupam (Early Shanburry x Red Delicious)

TABLE 10.4 *(Continued)*

S. No	Seasons	Himachal Pradesh	Jammu & Kashmir	Uttarakhand
	Low-chilling varieties	Michal, Schlomit, Anna, Tamma., Vered, and Neomi Tropical Beauty and Parlin's Beauty		
	Pollinizing varieties	Tydeman's Early, Red Gold, Golden Delicious, McIntosh, Lord Lambourne, Winter Banana, Granny Smith, Starkspur Golden and Golden Spur		
	A combination of early, mid-season, and late-flowering pollinizers	Tydeman's early, Red Gold Golden Delicious		
	Pollinizers plus scab resistant	Redfree and Liberty		

Flowering in apple is the result of several physiological changes from vegetative to reproductive phase. Like in many other fruit crops, newly initiated apple buds become dormant in late summer or early autumn. Winter chilling (defined as a certain number of hours at or below 7.2°C) is necessary to break bud dormancy. If chilling is not sufficient, both flower buds and vegetative buds (producing leaves only) are delayed. Flowering time of the different cultivars must overlap to a large extent for pollination to be successful. Number of days from pollination to fruit ripening varies considerably due to inherent differences between cultivars and to environmental effects (e.g., weather conditions). During this period, physiological processes like cell division and expansion, starch accumulation, ethylene production, and color changes take place (Janssen et al., 2008).

10.7 BIOCHEMICAL AND BIOPHYSICAL EVALUATION FOR IDENTIFYING SUPERIOR VARIETIES

The attractiveness of fruit to consumers is determined by visual attributes that include appearance, size, uniformity, color, and freshness, as well as non-visual attributes such as taste, aroma, flavor, firmness (texture), nutritional value and healthiness. Among these attributes, firmness, and aroma appear to be the most important for consumers. Sugars, organic acids and phenolic compounds all contribute to the aroma of apples (Mikulic Petkovsek et al., 2009).

Chemical composition of apple fruit is very complex. It consists of numerous organic and inorganic compounds and macro biogenic and micro biogenic elements. Most represented are sugars, acids, pectin, tannins, starch, cellulose, vitamins, enzymes, and phytohormones, while most represented chemical elements are nitrogen, phosphorus, potassium, calcium, sulfur, iron, and magnesium (Table 10.5). Organic compounds present in fruit depend on fruit cultivar, ripeness, physiological condition of a tree as well as soil and weather conditions (Markuszewski and Kopytowski, 2008).

Fruits which in time of optimal ripe have most of the stated compounds in harmony are of great nutritive value, because in our regular diet we have a deficiency in compounds and elements that are in apple (Skendrovic Babojelic et al., 2007). Apples are a part in all food diets and its therapeutic value is well known for different illnesses (determines the absorption of gastric secretions, the elimination of toxins, has diuretic effect). Firmness and sugar content are important quality attributes that directly influence consumers on purchasing fresh apple fruit. Organic acids are an important component of fruit flavor and, together with soluble sugars and aromas, contribute to the overall organoleptic quality of fresh apple fruits. Malic acid is the predominant organic acid in apple fruits (Campeanu et al., 2009). Malic acid is the major component of apple that helps to maintain the liver in a healthy condition and it also help in digestion process. The content of organic acids might be also of interest in that certain acids may lead to a lowering of the postprandial blood glucose and insulin responses.

Apples, and especially apple peels, have been found to have a potent antioxidant activity and can greatly inhibit the growth of liver cancer and colon cancer cells. The total antioxidant activity of apples with the peel was approximately 83 μmol vitamin C equivalents, which means that the antioxidant activity of 100 g apples (about one serving of apple) is equivalent to about 1500 mg of vitamin C. However, the amount of vitamin C in 100 g of apples is only about 5.7 mg. Vitamin C in apples contributed less than 0.4% of total antioxidant activity (Boyer and Liu, 2004). Lee et al. (2003) found that the average concentration of ascorbic acid among six apple cultivars was 12.8 mg/100 g fruit. Apples contain a large concentration of flavonoids, as well as a variety of other phytochemicals, and the concentration of these phytochemicals may depend on many factors, such as cultivar, harvest, storage, and processing of the apples. The concentration of phytochemicals also varies greatly between the apple peels and the apple flesh (Lee et al., 2003). Consumers are becoming more interested in the content of the health-promoting compounds in fruit because of their antioxidant activity (Robards et al., 1999).

The high antioxidant activity of apples has effective functions in inhibiting cancer cell proliferation, decreasing lipid oxidation and lowering cholesterol, which has confirmed their role in potentially reducing the risk of chronic diseases (Boyer & Liu, 2004). In addition to their antioxidant effect, recent studies in humans and animals suggest that intake of apples can positively affect lipid metabolism (Ravn-Haren et al., 2013; Nagasako-Akazome et al., 2007), weight management (De Oliveira et al., 2003) and vascular function (Gasper et al., 2014). Furthermore, evidence of various mechanisms whereby triterpenoid compounds exert an effect on CVD has been found (Waldbauer et al., 2015; Han & Bakovic, 2015). Compared with apple flesh, apple peel shows more potent antioxidant activity and antiproliferative activity (Wolfe & Liu, 2003; Wolfe et al., 2003; Eberhardt et al., 2000). More than 300 volatile molecules have been reported in fresh apples (Nijssen et al., 2011). The total number, identity, and concentration of volatile compounds emitted by ripening apple fruit are cultivar specific (Dixon & Hewett, 2000). The contribution of each compound to the specific aroma profile of each cultivar depends on the activity and substrate specificity of the relevant enzymes in the biosynthetic pathway, the substrate availability, the odor threshold above which the compound can be detected by smell, and the presence of other compounds (Rizzolo et al., 2006). Esters are the most abundant volatile compounds emitted by apple and, together with α-farnesene, have been proposed for cultivar classification [Holland et al., 2005]. Ethyl 2-methyl butanoate, 2-methyl butyl acetate, and hexyl acetate contribute mostly to the characteristic aroma of "Fuji" apples, while ethyl butanoate and ethyl 2-methyl butanoate are the active odor compounds in "Elstar" apples, and ethyl butanoate, acetaldehyde, 2-methyl butanol, and ethyl methyl propanoate in "Cox Orange" (Berger, 2007). For the "Pink Lady" cultivar, hexyl acetate, hexyl 2-methyl butanoate, hexyl hexanoate, hexyl butanoate, 2-methyl butyl acetate and butyl acetate were prominent within the blend of volatiles produced by fruit throughout maturation (Villatoro et al., 2008). Among fruit tissues, it has been shown that epidermal tissue produces a greater amount of volatiles than internal tissues (Rudell et al., 2002). This higher capacity for aroma production by the peel has been attributed to either the abundance of fatty acid substrates or the higher metaboltic activity (Guadagni et al., 2004).

TABLE 10.5 Nutritional Profile of Apple Raw Skin Per 100 Gram Serving

Nutrients	Amount	Vitamins	Amount	Aminoacids	Amount
Calories	52	Vitamin A	3 µg	Tryptophan	1 mg
Water	86%	Vitamin C	4.6 mg	Threonine	6 mg
Protein	0.3 g	Vitamin D	~	Isoleucine	6 mg
Carbs	13.8 g	Vitamin E	0.18 mg	Leucine	13 mg
Sugar	10.4 g	Vitamin K	2.2 µg	Lysine	12 mg
Fiber	2.4 g	Vitamin B1	0.02 mg	Methionine	1 mg
Fat	0.2 g	Vitamin B2	0.03 mg	Cysteine	1 mg
Saturated	0.03 g	Vitamin B3	0.09 mg	Tyrosine	1 mg
Monounsaturated	0.01 g	Vitamin B5	0.06 mg	Valine	12 mg
Polyunsaturated	0.05 g	Vitamin B6	0.04 mg	Arginine	6 mg
Omega–3	0.01 g	Vitamin B12	0 µg	Histidine	5 mg
Omega–6	0.04 g	Folate	3 µg	Alanine	11 mg
Trans fat	0 g	Choline	3.4 mg	Aspartic acid	70 mg
Carbohydrate	13.8 g	Monosaturated fatty acids	0.007 g	Glutamic acid	25 mg
Fiber	2.4 g	18:1	7 mg	Glycine	9 mg
Sugars	10.4 g	Polyunsaturated fatty acids	0.051 g	Proline	6 mg
Sucrose	2.1 g	18:2	43 mg	Serine	10 mg
Glucose	2.4 g	18:3	9 mg		
Fructose	5.9 g	Sterols	12 mg		
Starch	0.05 g	Phytosterols	12 mg		

10.7.1 APPLE ATTRIBUTES INVOLVED IN FRUIT QUALITY

Fruit quality is a major element of grower returns and consequently has been studied widely. There are many factors that are responsible for fruit quality including appearance (size, shape, color, gloss, and freedom from defects and decay), texture (firmness, crispness, juiciness, mealiness, and toughness, depending on the commodity), flavor (sweetness, sourness (acidity), astringency, aroma, and off-flavors), and nutritive value (vitamins, minerals, dietary fiber, phytonutrients). In addition to this there are many factors, which may influence fruit quality, some of which are outside of our control such as weather, site suitability and varietal genetic potential (Table 10.6).

TABLE 10.6 Description of Apple Varieties, Gene, and Gene Products Associated With Fruit Quality

Apple cultivar/variety	Gene Product	Candidate gene	Trait associated with fruit quality	Reference
Royal Gala Golden Delicious McIntosh Cox's Orange Pippin Gala xanoa	Aldehydes acetaldehyde/ra/w–2-hexenal hexanal 2-methyl butyl acetate, butyl acetate, hexyl acetate, butanol, 2-methylbutanol and hexanol	MYB1/MYBA	Coloration Skin color Skin/leaf/flesh Weighted cortical intensity	Young et al., 1996 Ban et al., 2007 Kunihisa et al., 2014 Zhang et al., 2014
'Royal Gala'		MdPG1	Endo-polygalacturonase	Atkinson et al., 2012
Granny Smith' and 'Golden Delicious'		MdPPO	Polyphenol oxidases	
Golden Delicious (winter cv.) and McIntosh (autumn cv.)	Starch degradation Starch amylase starch glucosidase	StA StG	Skin firmness	Morimoto et al., 2013
McIntosh	Sucrose metabolism sucrose phosphatate sucrose phosphate synthase	SP SPS	Skin texture	Kumar et al., 2012, 2013
McIntosh	Cell division	CDKB2		
Golden Delicious	Protein phosphatases	DSPTP1		
Granny Smith' and 'Royal Gala'	Enzyme phytoene synthase	PSY gene PSY1, PSY2, and PSY4 MdPSY1 and MaPSY2,	Caroteniod content	Ampomah-Dwamena et al.,2015
'Lenghaitang' Cultivar	Aluminum-activated malate transporter	MaI gene	Malic acid	Ma et al., 2015

TABLE 10.6 *(Continued)*

Apple cultivar/variety	Gene Product	Candidate gene	Trait associated with fruit quality	Reference
Granny Smith' and 'Royal Gala'	**Esters** butyl acetate pentyl acetate hexyl acetate 2 methyl butyl acetate ethyl butanoate ethyl–2–methyl butanoate 4-methoxyallyl benzene methyl–2-methyl butanoate propyl–2-methyl butanoate butyl–2-methyl butanoate hexyl–2–methyl butanoate butyl hexanoate hexyl propanoate butyl butanoate butyl propanoate hexyl butanoate hexyl hexanoate Putative cinmamyl alcohol dehydrogenase 9 Probable xyloglucan glycosyltransferase 12-1		Fruit firmness	Cai et al., 2006 Kumar et al., 2013, Buti et al., 2015) Kumar et al., 2013 (GS)
McIntosh	Hydroquinone glucosyltransferase		Volatile/polyphenol glycosylation	Louveau et al., 2011
Royal Gala	Flavonoid biosynthesis oxidoreductase protein		Flavanoid synthesis	Flachowsky et al., 2012

Color is important, with preference for clear yellow, bright red, bright green or bicolor fruit. These should be free from russet, although there may be a place for sour completely russet apples. Dull-looking apples are not worthy of further consideration. Shape has become important. Irregularly shaped apples and especially very oblate apples will not be accepted by marketers. Flavor is of prime importance. Many seedlings have fruit with very little acid and in consequence they are very sweet and insipid. This type is unacceptable except in some Asian markets. Fruit may have a strongly aromatic or distinct anise-like flavor and, while these may prove to be good home garden apples, they are unacceptable as commercial apples because such flavors are not universally appreciated. The ideal is sub-acid with a pleasant flavor but there is increasing demand for apples with complex, spicy, and full flavor. Of particular importance is the flavor after storage. Apples that develop an aldehyde or off-flavor will be unacceptable. Apples that are too mild will often taste bland after storage because acidity decreases. The texture and mouthfeel of the flesh should be firm and crisp with plenty of juice. Crisp flesh should persist in storage; soft apples are unacceptable. Overly thick skin can be a problem as well as skin that is too sensitive to bruising, such as 'Sir Prize.' These characteristics are usually attractive to consumers in terms of market acceptability or human health improvement (Kader, 1999). Most postharvest researchers, producers, and handlers are product-oriented in that quality is described by specific attributes of the product itself, such as sugar content, color, or firmness. In contrast, consumers, marketers, and economists are more likely to be consumer-oriented in that quality is described by consumer wants and needs (Shewfelt, 1999). It has been reported that consumers generally prefer apple fruits that are juicy, crisp, and sweet. There are many practices, which can influence fruit quality including crop load management, pruning for tree balance and light interception, maturity assessment for harvesting and nutrient (vitamins, minerals, dietary fiber, phytonutrients) and water management.

Several attempts have been made to develop portable instruments with sensors that detect volatile production by fruits as a way to detect maturity and quality. Other strategies include the removal of a very small amount of fruit tissue and measurement of total sugars or soluble solids content. Near-infrared detectors have great potential for nondestructive estimation of sugar content in fruits (Abbott, 1999). Until such methods become widely available, we will continue to depend on destructive techniques, such as soluble solids determination by a refractometer and titratable acidity measurement by titration, to evaluate flavor quality of fruits. Use of these indices in a quality assurance program must be coupled with tolerances of deviation from the proposed averages because of the large variation among cultivars, production areas and seasons, maturity at harvest and ripeness stage at the time of evaluation.

10.7.2 APPLE CHEMICAL CONTENTS AND DISEASE RESISTANCE

Different compounds in the fruit can probably play a role in resistance to fungal diseases, especially storage diseases. It has been reported that total phenol content is one of the factors affecting the apple storability and disease resistance (Table 10.7). A large number of volatile compounds are important in disease resistance of apple like alcohols, aldehydes, carboxylic esters and ketones. In resistant cultivars, phenolic components accumulate at a higher rate than in susceptible cultivars (Dixon and Hewett, 2000; Usenik et al., 2004; Treutter, 2005; Lattanzio et al., 2006). Among phenolic compounds, the flavonoid quercetin has been considered the most important agent. Sanzani et al. (2009a) has investigated the role of quercetin as an alternative strategy to control blue mold and patulin accumulation in 'Golden Delicious.' By exogenous application of different phenolic compounds, they found that quercetin is effective in controlling blue mold and patulin accumulation. Subsequent studies have demonstrated that this control is achieved through an increased transcription of genes 20 involved in the quercetin biosynthetic pathway (Sanzani et al., 2009b; Sanzani et al., 2010). In a wide variety of plants, organic acids and nutritional compounds such as vitamin C and glutathione are associated with fruit taste and quality. These compounds are apparently related also to the level of disease resistance (Ferguson & Boyd, 2002). The relationship between harvest day and vitamin C content of apple fruits has been investigated by Davey et al. (2007). Low pH can enhance P. expansum colonization, which means that cultivars with a lower pH in their fruits are more susceptible to fungal attack (Prusky et al., 2004).

10.8 APPLICATION OF MOLECULAR MARKERS IN APPLE

Molecular markers have been replacing or complementing traditional morphological and agronomic characterization, since they are virtually unlimited, coverall the genome, are not influenced by the environment and, particularly in the case of fruit trees with long juvenile period, can be less time consuming for the characterization of new hybrids. Several molecular markers studies on apple have been published using techniques such as

TABLE 10.7 Description of Apple Varieties, Candidate Gene and Resistance Associated Against Disease

Apple cultivar/variety	Candidate gene	Disease	Reference
'White Angel'	Vf, Vm, Vb, Vbj, Vr, and Va	Scab resistance genes	Young et al., 1996
	Pl1 on LG12, Pl2 on LG11, Plw on LG8, Pld on LG12, and Plmis on LG11	Powdery mildew resistance	Dunemann et al., 2007, Fernández-Fernández et al., 2008, Gardiner et al., 2003, 2004, James et al., 2004, James, and Evans 2004, Rikkerink et al., 2010
'Fiesta,' 'Prima,' 'Gala,' 'Golden Delicious'	Rvi2, Rvi4, Rvi5, Rvi6, Rvi11, Rvi12, Rvi13, Rvi14, and Rvi15	Scab resistance genes	Patocchi et al. (2009)
Malus floribunda 821	Rvi6/Vf	Scab resistance	Ban et al., 2007
Gala	HcrVf2	Cladosporium fulvum resistance	(Belfanti et al., 2004
	Attacin E gene	Fire blight	(Borejsza-Wysocka et al., 2010
Galaxy and M26	MpNPR1 PR genes (PR2, PR5, and PR8)	Venturia inaequalis and Gymnosporangium juniperi-virginianae.	Malnoy et al., 2007
Malus floribunda 821	HcrVf2	Scab-resistance	Belfanti et al., 2004
M. floribunda 821	HcrVf1	Scab-resistance	Malnoy et al., 2008
Malus domestica Borkh.	MdPG1 (Polygalacturonase)	Changes in cell adhesion/maduration	Atkinson et al., 2002
M. domestica Borkh.	MpNPR1 (pathogenesis-related gene)	Fungal disease resistance	Malnoy et al., 2007a
M. domestica Borkh.	MdMYB10 (MYB transcription factor)	Induce anthocyanin accumulation/red apple fruit color	Espley et al., 2007
Hyalophora cecropia	Attacin E	Fire blight resistance	Norelli et al., 2000
Hyalophora cecropia	Cecropin MB39	Fire blight resistance	Liu et al., 2001
Trichoderma harzianum	ech42 (Endochitinase)	Scab-resistance	Wong et al., 1999

TABLE 10.7 *(Continued)*

Apple cultivar/variety	Candidate gene	Disease	Reference
Trichoderma harzianum	*ech42* (Endochitinase) *Nag70* (Exochitinase)	Scab-resistance	Bolar et al., 2001
Grapevine (*Vitis vinifera* L.)	*Vst1* (Stilbene synthase)	Antifungal activity	Szankowski et al., 2003
PGIP (Polygalacturonase inhibitor)	Kiwi (*Actinidia deliciosa*)	Antifungal activity	Szankowski et al., 2003
Avidin or Streptavidin	*Streptomyces avidinii*	Insect resistance (lightbrown apple moth	Markwick et al., 2003
PinB (puroindoline)	Wheat (*Triticum aestivum*)	Antifungal activity	Faize et al., 2004
'Fiesta,' 'Prima,' 'Gala,' 'Golden Delicious'	*Rvi2, Rvi4, Rvi5, Rvi6, Rvi11, Rvi12, Rvi13, Rvi14, and Rvi15*	scab resistance genes	Patocchi et al. (2009)
Malus floribunda 821	*Rvi6/Vf*	scab resistance	Ban et al., 2007
Gala	*Hcr-Vf2*	*Cladosporium fulvum* resistance	(Belfanti et al., 2004
	Attacin E gene	Fire blight	(Borejsza-Wysocka et al., 2010
Galaxy and M26	MpNPR1 PR genes (PR2, PR5, and PR8)	*Venturia inaequalis* and *Gymnosporangium juniperi-virginianae.*	Malnoy et al., 2007
Malus floribunda 821	*Hcr-Vf2*	Scab-resistance	Belfanti et al., 2004
M. floribunda 821	*Hcr-Vf1*	Scab-resistance	Malnoy et al., 2008
Malus domestica Borkh.	*MdPG1* (Polygalacturonase)	Changes in cell adhesion/ maduration	Atkinson et al., 2002
M. domestica Borkh.	*MpNPR1* (pathogenesis-related gene)	Fungal disease resistance	Malnoy et al., 2007a

TABLE 10.7 (Continued)

Apple cultivar/variety	Candidate gene	Disease	Reference
M. domestica Borkh.	MdMYB10 (MYB transcription factor)	Induce anthocyanin accumulation/ red apple fruit color	Espley et al., 2007
Hyalophora cecropia	Attacin E	Fire blight resistance	Norelli et al., 2000
Hyalophora cecropia	Cecropin MB39	Fire blight resistance	Liu et al., 2001
Trichoderma harzianum	ech42 (Endochitinase)	Scab–resistance	Wong et al., 1999
Trichoderma harzianum	ech42 (Endochitinase) Nag70 (Exochitinase)	Scab–resistance	Bolar et al., 2001
Grapevine (Vitis vinifera L.)	Vst1 (Stilbene synthase)	Antifungal activity	Szankowski et al., 2003
PGIP (Polygalacturonase inhibitor)	Kiwi (Actinidia deliciosa)	Antifungal activity	Szankowski et al., 2003
Avidin or Streptavidin	Streptomyces avidinii	Insect resistance (lightbrown apple moth	Markwick et al., 2003
PinB (puroindoline)	Wheat (Triticum aestivum)	Antifungal activity	Faize et al., 2004

RFLP (Restriction Fragment Length Polymorphism) (Nybom & Schaal, 1990; Watillon et al., 1991), RAPD (Random Amplified Polymorphic DNA) (Koller et al., 1993; Mulcahy et al., 1993; Harada et al., 1993; Dunemann et al., 1994; Gardiner et al., 1996; Goulão et al., 2001), AFLP (Amplified Fragment Length Polymorphism) (Goulão et al., 2001) and SSR (Simple Sequence Repeats) (Guilford et al., 1997; Gianfranceschi et al., 1998). The increasing development and generalized use of a large number of methodologies during the last years, requires comparative studies in order to choose the best DNA marker technology to be used in fingerprinting and in diversity studies, considering reproducibility, costs, sensibility, and level of polymorphisms detection. Molecular technique comparisons have become important because, depending on the objective of the study, one technique can be more appropriate than another, as well as different techniques being informative at different taxonomic levels. Molecular markers are classified into different categories, e.g., biochemical, i.e., isoenzyme and DNA markers. They can be biomolecules related to a genetic trait, or just a difference in the sequence of a piece of DNA.

10.8.1   ISOENZYME MARKERS

Isoenzymes are different forms of an enzyme that vary in size or conformation. Isoenzyme markers have been used for clonal identification of apple (Gardiner et al., 1996), and for developing markers for important genes (Hemmat et al., 1994; Chevreau et al., 1999). Presently their role has, however, been overtaken by DNA-based markers.

10.8.2   DNA MARKERS

Because of the low level of polymorphism in isoenzymes, other groups of molecular markers, i.e., DNA markers, were developed that are able to detect more polymorphism. DNA markers are classified into a wide range of different discriminative techniques that reveal the genetic diversity between or within different species or cultivars. DNA markers are widely used for various purposes like studies of genetic diversity and phylogenetic analyzes (Coart et al., 2003), constructing linkage maps (Liebhard et al., 2002), QTL analysis (Liebhard et al., 2003) and marker-assisted selection (Costa et al., 2004). RAPD (Random amplified polymorphic DNA) fragments have been useful and reliable for genetic mapping and gene tagging in many different plant species including apple (Weeden et al., 1992). The advantage of the RAPD technique is that it can provide numerous genetic markers in a short time, making it the method of choice for mapping economically important plants. A genetic link age map constructed from over 400 markers, mainly RAPDs has been developed for apple based on a cross of 'Rome Beauty' with 'White Angel' (Olemmat et al.1994). RAPD markers can be very useful for cultivar identification. Landry et al. (1994) used RAPD marker data in phylogeny analysis of 25 apple rootstocks and developed a DNA fingerprinting system and identification key for apple rootstocks. Koller et al. (1993) reported 14 RAPD markers resulting from two decamer primers could discriminate 11 apple cultivars A phenetic characterization of forty-one apple cultivars, comparing RAPD and AFLP markers was reported previously (Goulão et al., 2001). In that study the results obtained were positively correlated, and only small differences were observed between the UPGMA dendrograms originated from both methods.

However, these two techniques show some disadvantages regarding reproducibility. In fact, the use of short primers and low annealing temperatures makes RAPD markers extremely sensitive to the reaction conditions and therefore irreproducible among different laboratories. With the AFLP technique, although extremely reproducible in the amplification steps, when most of the widely used DNA extraction protocols are employed, incomplete restriction cuts can occur, resulting in irreproducible band patterns. This fact is particularly important when working with woody species, due to its generally high content in polyphenols, and reinforces the need to develop more reliable markers. Microsatellites are regions of short, tandem repeated DNA sequences of 1 to 6 base pairs, ubiquitous in eukaryotic genomes. Two different marker strategies have been used based on microsatellites: SSR (simple sequence repeats) and ISSR (inter-simple sequence repeats). SSR are highly reproducible co-dominant markers, in which a single pair of PCR primers that flanks the repeated sequences produces polymorphic patterns among alleles, depending on the number of repeat units. Although these markers are generally highly polymorphic, the initial cost of developing them is relatively high. However, once these primers are determined and its sequences published, they can be shared among groups working with material from the same species or related ones, and the method becomes fast and readily employed. Primer pairs designed for SSRs in apple have been published (Guilford et al., 1997; Gianfranceschi et al.,1998) and can be used in further studies. ISSR (Zietkiewicz et al., 1994) is a different microsatellite-based method, which does not need prior knowledge of the genome, cloning or primer design. While the SSR protocol relies on the amplification of the repeated region using two locus-specific primers, in ISSR, a single primer composed of a microsatellite sequence anchored at the 3_ or 5_ end by 2 to 4 arbitrary, often degenerate nucleotides, is used to amplify the DNA between two opposed microsatellites of the same type. Allelic polymorphisms occur whenever one genome is missing the sequence repeated or has a deletion or insertion that modifies the distance between repeats. For five anchored primers, polymorphisms also occur due to differences in the length of the microsatellite. The sequences of repeats and anchored nucleotides are randomly selected. Although ISSR are dominant markers, they have the advantage of analyzing multiple loci in a single reaction.

10.8.3   DNA MARKERS AND APPLE DIVERSITY

The accessibility of reliable genetic markers is essential for variety identification and distinction, for development of efficient breeding methods to create new apple cultivars selected for disease resistance, fruit quality or tree growth characteristics and other traits. The emergence of PCR-based molecular markers has created the opportunity for fine-scale genetic characterizations of germplasm collections, as well. Many studies on molecular markers in apple have been published such as RFLP (Nymbom and Schall, 1990), RAPD (Landry et al., 1994; Gardiner et al., 1996; Dantas et al., 2000; and Goulao and Cristina, 2001), AFLP (Hokanson et al., 2001), SSR (Kenis et al., 2001; and Volk et al., 2008) and ISSR (Hokanson et al., 2001, Kashyap et al. (2010). Microsatellites (SSRs) are the most widely used molecular markers in this field. Repetitive nucleotide sequences are ubiquitously found in eukaryotic nuclear genomes and their flanking regions are highly conserved, which make them suitable for amplifying the usually polymorphic intervening repeat loci. Microsatellites show high polymorphism, co-dominant inheritance and good reproducibility. Since all individuals of the same cultivar originated from a single progenitor by vegetative propagation in apple breeding and cultivation, SSR fingerprinting can be used for cultivar differentiation. Up to now, almost 200 SSR markers are available in apple (Guilford et al., 1997; Gianfranceschi et al., 1998; Hokanson et al., 1998; Liebhard et al., 2002). SSR markers have proven to be highly informative and useful for distinguishing genotypes and for determining genetic relationships among Malus cultivars and species (Goulão and Oliveira, 2001; Hokanson et al., 2001; Laurens et al., 2004; Liebhard et al., 2002). Genetic similarity of 41 apple cultivars was assessed by RAPD (random amplified polymorphic DNA) and AFLP (amplified fragment length polymorphism) markers by Goulao et al. (2001). Oraguzie et al. (2001) used RAPD to evaluate the genetic relationships among four subsets of apple germplasm (including 155 genotypes; modern and old cultivars) in New Zealand. In other studies, genetic diversity of apple genotypes has been evaluated with different markers like RFLP (restriction fragment length polymorphism) (Gardiner et al., 1996), AFLP (Tignon et al., 2000, 2001), SSRs (Oraguzie et al., 2005; Pereira-Lorenzo et al., 2007) and RAPDs (Royo and Itoiz, 2004). The random amplified polymorphic DNA (RAPD) technique determines genetic diversity and relationships among different fruit species and cultivars, including apples (Koc et al., 2009; Erturk and Akcay, 2010; Smolik et al., 2011; Ansari and Khan, 2012). Apple molecular markers are the most appropriate for characterizing apple collections, more than morphological markers (Santesteban et al., 2009), although the later are necessary to complete germplasm descriptions.

Characterization of genetic resource collections has also been greatly facilitated by the availability of a number of molecular marker systems. Morphological traits were among the earliest markers used in germplasm management, but they have a number of limitations, including low polymorphism, low heritability, late expression, and vulnerability to environmental influences (Smith and Smith, 1992). On the other hand, DNA markers do not have such limitations. They can be used to detect variation at the DNA level and have proven to be effective tools for distinguishing between closely related genotypes. Different types of molecular markers have been used to assess the genetic diversity in crop species, but no single technique is universally ideal. Therefore, the choice of the technique depends on the objective of the study, financial constraints, skills, and available facilities (Kafkas et al., 2008; Pavlovic et al., 2012).

10.8.4   DNA MARKERS AND GENE TAGGING

Most of the markers identified so far are linked to monogenic traits, i.e., mainly forms of resistance to the main pathogens and pests. Most of the markers identified so far are linked to the Vf gene for scab resistance, derived from Malus floribunda 821. Starting from this position, a map based gene cloning approach was used to identify putative resistance genes at the Vf locus and at least one of these genes, called HcrVf2, is able to confer resistance to transgenic apple plants (Belfanti et al., 2003). Other sources of monogenic scab resistance have been reported and recently a few markers have been identified for the Va, Vb, Vbj, Vm, and Vr genes.

DNA based markers have been identified also for some other important genes, including, e.g., genes related to ethylene biosynthesis and firmness of the fruits like *Md*-ACS1 (Costa et al., 2005; Oraguzie et al., 2007; Li and Yuan, 2008; Nybom et al., 2008b; Zhu and Baritt, 2008), *Md-ACO1* (Costa et al., 2005), *Md-PG1* (Wakasa et al., 2006) and *Md-Exp7* (Costa et al., 2008), genes related to chilling requirement (*Chr*) (Lawson et al., 1995), genes related with apple fertility (*MADS-box*) (Yao et al., 1999), fruit color (Cheng et al., 1996), resistance to fire blight (Malnoy et al., 2004) and powdery mildew (Markussen et al., 1995). Powdery mildew (Podosphaera leucotricha) for which different resistance sources (Pl1, Pl2, Plw, and Pld genes) have been identified in apple germplasm. The early evaluation of mildew susceptibility in segregating progenies is very difficult therefore reliable markers for these genes would be of great interest. The available markers for the Pl1, Pl2, Plw, and Pld genes are still under testing because sometimes the correlation between molecular and two phenotypic data set proved to be poor. The Sd1 gene conferring resistance to two rosy leaf curling aphid (Dysaphis devecta) biotypes was finely mapped on linkage group 12 of apple and from the closest markers a map-based gene cloning was also started. Some markers have been developed for the Eriosoma lanigerum resistance genes but their reliability is still to be demonstrated. There are also a few markers in apple available for the selection of agronomic and fruit quality traits as the columnar Co gene, the Ma gene controlling fruit acidity and the red or yellow skin color.

10.8.5 MARKER-ASSISTED SELECTION AND QTL MAPPING

Most of the available apple and pear markers can be used in marker-assisted selection but MAS efficiency can widely varying depending on the estimated genetic distance between the marker and the linked gene. Of course, the use of two markers flanking the gene of interest is more advisable, in particular, if the distance between the gene and each marker is not very close. Molecular-based selection in pome fruit breeding progenies can be particularly important for traits that are difficult to evaluate (e.g., mildew resistance in apple) or delayed in time by juvenility (e.g., fruit traits) but today the broadest molecular screening can only be efficiently performed for the Vf scab-resistance gene. The availability of a number of markers linked to the Vf gene made it possible to optimize MAS and to investigate in detail its advantages with respect to phenotypic selection methods. MAS is estimated to be more precise or less time-consuming than the available phenotypic selection protocols and can also make it possible to distinguish heterozygous and homozygous genotypes since two reliable flanking codominant markers are available. Therefore, "positive" MAS selection in favor to a specific allele can even be very informative even for easy-to-score phenotypic traits, as the Vf scab-resistance. Another MAS advantage is the possibility of performing an efficient "negative" selection against the "donor" chromosomal regions in the vicinity of the introgressed gene and this type of selection cannot be performed by using a standard phenotypic selection. In fact it has been demonstrated that most of the advanced apple Vf-selections chosen only on a phenotypic basis are still carrying a large portion of the floribunda genome in the Vf-chromosome even after 5–6 generations. Of course the elimination of "wild" chromosomal regions in pseudo-test cross progenies can also be speeded up at the whole-genome level using a map with few but well-distributed SSR markers. Availability of molecular markers linked to different resistance genes against the same pathogen and their map position can also be used to estimate the possible relationship among various, apparently unrelated resistance sources. In fact, it has been demonstrated that markers linked to a specific gene (i.e., Vf or Vm; Sd1) are not present in selection carrying other resistance genes. This marker-specificity can be used to easily select plants carrying multiple resistances against the same pathogen. (*S. Tartarini*)

The identification of QTLs in fruit trees is rather difficult as the two paren-tals are highly heterozygous and the segregation analysis of each quantitative trait requires large progenies to increase its reliability. The first example of QTL mapping in apple led to the identification of two main loci involved in fruit firmness and other QTLs were also identified for sensory assessments of fruit texture. Various QTLs related to different multigenic sources of scab resistance (i.e., Discovery, TN10–8, Durello di Forlì) have been found with respect to various inocula (Durel et al., 2003). Fire blight resistance disease in apple has been investigated thoroughly in which detection of prime QTL was revealed which was later advanced to development of SCAR marker appropriate for MAS (Khan et al., 2006; 2007). Durel et al., 2009 revealed remarkable QTL effect on resistance on LG12. Parravicini et al., 2011 devel-oped novel microsatellite through identification and cloning of QTL within 189kb sequence. This sequence exhibits numerous homolog genes of *Pto/Prf* complex that confer resistance against bacterial disease in tomato. The prime QTL on LG3 and three fire blight resistance QTLs in *Malus x robusta* 5 was also developed by assessing three populations originating from 'Robusta 5' accessions (Peil et al., 2007; Gardiner et al., 2012). Identification of QTL on LG7 has been recognized along with heat shock protein gene (HSP90) and a

WRKY transcription factor gene. Similarly QTL on LG3 parallel to leucine rich repeat family receptor like protein gene (MxdRLP1) was also detected in apple cultivar from New Zealand. Khan et al., 2012 illustrated comprehensive data of 27 QTLs constructed on diverse genetic resources, strains, and used them for breeding resistance against fire blight in apple cultivars by using three remarkably stable QTLs (LG3/*Malus* × *robusta* 5, LG12/*M. floribunda* clone 821, 'Evereste,' LG7/'Fiesta'). Genomic wide association research report discovered QTLs on LG2, LG6 and LG15 illustrating 34 remarkable associations for resistance against fire blight (Khan et al., 2013). Fahrentrapp et al., 2013 discovered the potential resistance candidate gene Fb_MR5 from CC-NBS-LRR resistance gene family within QTL region on LG3 of *Malus* × *robusta* 5. Broggini et al., 2014 demonstrated this gene as prime factor for conferring resistance in Gala against fire blight disease. Apple breeding programs mainly target for improved disease resistance and enhanced fruit quality for which MAS proves to be appropriate alternative. Multiple resistance gene was developed in apple seedlings against scab by crossing two scab resistant genes (*Rvi6* (*vf*), *Rvi2* (*Vh2*) or *Rvi4* (*Vh4*) and a mildew resistance gene (*Pl1* or *Pl2*) using DNA markers precise to the genes (Kellerhals et al., 2008). They also developed MAS system combining the QTLs for fire blight resistance with scab resistance.

Jänsch et al. (2015) ascertained SNPs associated to eight disease resistance genes (scab: *Rvi2*, *Rvi 4*, *Rvi 6*, *Rvi 11*, *Rvi 15*, powdery mildew: *Pl2*, fire blight: *FB_E*, *FB_MR5*), and advanced the locus of *Rvi2*, *Rvi4*, and *Rvi11*. They then authenticate specificity of their alleles in linking with resistance by resolving the allele composition in eight apple genotypes ('Golden Delicious,' 'Delicious,' 'Cox's Orange Pippin,' 'Jonathan, McIntosh,' 'Granny Smith,' 'Braeburn,' and selection F2–26829–2–2 originated from *M. floribunda 821*) via efficient high throughput analysis in marker-assisted breeding (MAB). Baumgartner et al. (2015) crossed homozygous lines for resistance genes for scab (*Rvi2*, *Rvi4*, *Rvi6*), powdery mildew (*Pl1*, *Pl2*) and fire blight (*FBF7*). Even though incidence of fire blight has not been yet discovered in Japan, Alternaria leaf-blotch has been challenging for apple growers and researchers since the 1960's. Comprehensive genetic evaluation of F1 populations originated from crosses between cultivars susceptibility to Alternaria blotch was resolved by prime dominant gene (*Alt*) and was inherited as dominant trait (Saito and Takeda, 1984). Fukasawa- Akada et al., 2003 detected RAPD markers associated to vulnerability to "Kaori" originating from Delicious. This study established strong positive correlation between susceptibility symptoms and incidence of markers via analysis of 108 cultivars originating from 'Delicious' and 'Indo.' Moriya et al. (2011) ascertained the close association between LG11 of Starking Delicious' and DR033892 which were mapped between two SSRs (Moriya et al., 2013). These SSRs were utilized in MAS system and associated with "Alt" blotch resistance and fruit color (Moriya et al., 2012b). This region was further curtailed to 102 kb with ten gene and used for developing new SSR (Moriya et al., 2013). Tabira and Otani (2004) authenticated the utilization of SNPs in the alpha sub-unit of chloroplast chaperon (cpn-alpha) gene in screening the seedling resistance to Alternaria blotch. Though Okada et al., 2011 illustrated that cpn-alpha is associated to *Alt*, but not *Alt* itself. Abe et al. (2012a, 2012b) demonstrated inheritance of restrained susceptibility in 'Sekai-ichi,' 'Golden Delicious' and 'Orin' and indicated incidence of dominant gene (*Alt–2*) dissimilar to *Alt* which was widespread among cultivars. Though Morriya et al., 2012a presented QTL map for normal susceptibility on Orin LG11 within similar area of *Alt*. However, co-location of Alt and Alt- 2 is yet to be determined. On the contrary, there is study demonstrating association of LGs to susceptibility of 'Golden Delicious' on SSR (Li et al., 2011). Crown gall is most prevalent disease in apples. The resistant gene 'Cg' against the strain Peach CG8331 is reported from Japanese wild apple *Malus sieboldii* Sanashi 63 which confers resistance against crown gall and develops crown gall resistant rootstock (Moriya et al., 2008). Moriya et al. (2010) recorded Cg to LG2 of wild apple and developed selectable markers for MAS.

10.8.6 *COMMON APPLE DISEASE AND DISEASE RESISTANCE GENES*

Apple scab is main prevalent disease that immensely deters apple production leading to high fruit damage. The main scab resistance genes (*Vf, Vm, Vb, Vbj, Vr,* and *Va*) have been detected from low productivity Malus species (Williams and Kuc, 1969; Biggs, 1990). Among all the scab resistant gene, *Vf* gene originating from Malus floribunda 821 Siebold ex Van Houtte has been extensively introgressed into susceptible high market value apple cultivars (Korban, 1998). The Vf genes confers resistance against identified races of *Venturia inequalis* and has been employed successfully in the development of resistance in orchards for over 80 years. Sansavini et al., 2004 demonstrated discreet approach to develop scab resistant plants which provides effective linkage map and molecular markers mapped in the region of scab resistance and initiation of cloning in *Vf* gene. This study provided contig of BAC clones across the regions flanking between two Vf molecular markers, M18 and AM19 which lead to detection of four genes namely *HcrVf1* to

HcrVf4. These gene transcript receptor like proteins having homology to the *Cladosporium fulvum* (Cf) resistance gene family of tomato. These genes have an extracellular leucine rich repeat domain and a transmembrane domain (Vinatzer et al., 2001). This information was utilized in developing resistance in susceptible cv. Gala via introduction of *HcrVf2* under control of *Ca MV 35S* promoter via *nptII* gene for selection. The primary step involves the scab infection indicated by penetration and stroma formation by the fungus (Barbieri et al., 2003; Sansavini et al., 2003) as estimated in vitro, trailed by a greenhouse scab inoculations of the experimental lines comprising of a single functional copy of *HcrVf2*. These investigations illustrated explicitly that the four lines having HcrVf2 were at least resistant to scab (Belfanti et al., 2004a; 2004b) as the usually bred Vf resistant cv 'Florina.' Plants having both *HcrVf2* and *nptII*, and those with nptII only, and wild-type '*Gala*' and '*Florina*' (Vf) were infected with a field inoculum (assortment of genotypes) identified to have the capability to cause scab on 'Gala,' but not 'Florina,' and with an inoculum procured from *M. floribunda 821*, the prime donor of *Vf*. These results signify that the *HcrVf2* line was, as usual, resistant to the field inoculum likewise to Florina, while all the other plant exhibited general scab lesions with profuse sporulation. *M. floribunda 821* inoculated with race 7 lead to sporulation lesions on all plants though inoculums was less virulent and sporulation was less profuse than the field inoculums. Furthermore, HcrVf2 line still preserved some resistance, it is reveals relatively higher resistance than line transformed only with nptII and the original 'Gala' (Gessler and Patocchi, 2007). This experiment illustrated that resistance in *HcrVf2* transformed lines comprised the *Vf* gene. Furthermore investigation detected the promoter sequence of *HcrVf1*, 2 and 4 and revealed their functionality (Silfverberg-Dilworth et al., 2005b). Malnoy et al., 2008 demonstrated regulation of Vfa1 (HcrVf1) and Vfa2 (HcrVf2) gene under its own promoter can confer resistance against *V. inaequalis* (Malnoy et al., 2008).

At present, numerous research teams have illustrated that the plant generate defense associated proteins for instance pathogenesis-related proteins (PR) and antimicrobial peptides. Constitutive expression of these molecules might augment plant resistance such as puroindolines (*PinA* and *PinB*), antifungal cystein rich proteins found in wheat seeds. Both these genes are thoroughly characterized and has been in introduced in transformed rice (Krishnamurthy et al., 2001). Same gene PinB was introduced in apple and results signified the Strain 104, race 1 representing the common *V. inaequalis* population on the commercial cultivars is unaffected by the PinB at any expression level; however, the strain EU D42, race 6, is repressed gradually with the rising levels of PinB. Hence two strains exhibited discrepancy in tolerance against PinB (Faize et al., 2004). On the contrary, various research studies illustrated that constitutively high-level expression of PR proteins could avert any infection by diverse pathogens. Resistant apple varieties namely cv. Remo demonstrated up-regulation of proteins in contrast to susceptible cultivars cv. Elstar. These proteins include β–1,3-glucanase, ribonuclease-like PR10, cysteine protease inhibitor, endochitinase, ferrochelatase, and ADP-ribosylation factor and assists in dismutation of superoxides as revealed by transcriptional analysis. Further, resistant cultivar Remo exhibited enormous number of clones originating from mRNA for Type 3 metallothioneins. On the contrary, susceptible cultivar cv. Elstar revealed relatively lower levels of transcripts in young control leaves as compared to leaves inoculated with *V. inaequalis* (Degenhardt et al., 2005). Further research reports demonstrated that over expression of *MpNPR1* gene confers disease resistance in *M. domestica*. The *NPR1* gene has prime function in systemic acquired resistance in plants. An NPR1 homolog, MpNPR1–1, was cloned from *M. domestica* and over-expressed into two chief apple genotypes, Galaxy, and M26. Over expression of this gene leads to higher expression of pathogenesis-related (PR) genes. Resistant varieties of Galaxy exhibit high expression of MpNPR1confering resistance against two fungal pathogens of apple *V. inaequalis* and Gymnosporangium juniperi-virginianae. These cultivar line were then propagated in fields and investigated for disease resistance and fruit quality (Malnoy et al., 2007a; 2007b).

In apple, scab, powdery mildew (caused by the fungal pathogen *Podosphaera leucotricha*), and fire blight (caused by *Erwinia amylovora*) are the major diseases affecting commercial apple production in many countries. For breeding of resistant apple cultivars, genes, and QTLs related to disease resistance, and the linked DNA markers, have been successively identified. Development of DNA markers for scab resistance has preceded that of markers for other diseases. Seventeen genes for apple scab resistance have been identified, and their global positions have been located on the apple genetic map (Bus et al., 2011). Among them, the most intensively studied has been the *Rvi6* (*Vf*) gene from *M. floribunda* 821.This was the first fine-mapped scab resistance gene, and defined as a receptor-like gene showing homology to candidate tomato genes for *Cladosporium fulvum* resistance (Vinatzer et al., 2001). The *Rvi15* (*Vr2*) locus was found to contain three candidate genes (of the Toll and mammalian interleukin–1 receptor protein nucleotide-binding site leucine-rich repeat structure resistance gene family) (Galli et al., 2010a, 2010b, Schouten et al., 2014), and the *Rvi1 Vg*) locus was shown to contain 6 ORFs of four putative TIR-NBS-LRR (TNL) genes, a TNL pseudogene, and a serine/threonine protein phosphatase 2A gene (Cova et al., 2015). Furthermore, Soriano et al. (2014) have developed SSR markers linked to the broad-spectrum resistance of the selection 1980–015–025 (*V25*), and fine-mapped them on LG11 as *Rvi18*. This region contains a lectin-like receptor kinase (LRK) as a candidate gene for resistance. Clark et al. (2014) have also identified two novel scab resistance loci in 'Honeycrisp,' and mapped the loci as *Rvi19* and *Rvi20* on LG1 and LG15, respectively. They suggest that genes containing a leucine rich repeat region (LRR), a motif common in R genes, would be the prime candidate at each locus. Bastiaanse et al. (2015) have reported that resistance in 'Geneva' is conditioned by at least five NBS-LRR candidate genes clustered on LG4. Padmarasu et al. (2014) have mapped *Rvi12* (*Vb*) on LG12 of *Malus baccata* Hansen's baccata #2, and developed 16 SNP markers for resistance selection. Among the identified scab-resistance genes, *Rvi15* (*Vr2*) and *Rvi6* (*Vf*) have been proven to be practical for transformation of common susceptible cultivars (*Vr2*: Shouten et al., 2014, *Vf*: Belfanti et al., 2004, Joshi et al., 2011, Würdig et al., 2015). Many apple cultivars with the *Rvi6* (*Vf*) gene have been developed by MAS, and are now in commercial use. In Japan, a scab-resistant cultivar 'Aori 25' has been developed, and the presence of the *Rvi6* (*Vf*) gene has been identified on the basis of DNA markers (Kudo et al., 2013). However, as breakdown of resistance conferred by a single gene has been observed at several experimental farms (Bénaouf and Parisi 2000, Parisi et al., 2006), accumulation of multiple resistance genes has become an essential strategy. Although each of the developed molecular markers is a powerful tool for the pyramiding of resistance genes, it is necessary to include reference cultivars or strains derived from original studies for the appropriate use of such markers. The most serious diseases that hamper apple cultivation are scab and fire blight. Over the last few decades, apple varieties have been bred for production of better resistance. For more than half a century (1914–1970), the scab resistance gene *Rvi6/Vf*, originating from the wild species *Malus floribunda* 821, was incorporated into a wide variety of apple cultivars through crossing. However, the creation of a variety possessing the *Vf* resistance gene, but with commercially sufficient fruit quality, was not easy (Dayton et al., 1970). In order to improve the scab resistance of apples, genes encoding chitinolytic enzymes from a bio-control organism *Trichoderma harzianum* were introduced into apple (Bolar et al., 2000, 2001). The resulting transgenic lines expressing the genes were more resistant than nontransformed controls. Analysis of the *Vf* region led to the identification of a cluster of genes homologous to the tomato *Cladosporium fulvum* resistance gene family (Vinatzer et al., 2001). One of these genes,

HcrVf2, was used to transform the susceptible apple cultivar 'Gala' (Belfanti et al., 2004). As a candidate gene conferring resistance to fire blight disease, the attacin E gene derived from *Hyalophora cecropia* (the North America silkworm moth) has been used. Attacin E exhibits substantial lytic activity against many important plant pathogenic bacteria, and apple trees transformed with this gene exhibited resistance to fire blight (Borejsza-Wysocka et al., 2010; Ko et al., 2002). Apples incorporating the *Lc* gene, a bHLH transcription factor of maize, exhibited resistance to both scab and fire blight (Li et al., 2007). It is considered that the effect is likely related to enhancement of biosynthesis of a specific flavonoid, which plays important roles in the plant response to pathogens. Moreover, Krens et al. (2011) have reported that transgenic apple lines carrying the barley hordothionin gene (*hth*), which inhibits *in vitro* growth of a number of fungi and bacteria (Terras et al., 1993), were significantly less susceptible to scab disease.

10.9 MOLECULAR MECHANISM UNDERLYING PHENOL AND FLAVANOID BIOSYNTHETIC PATHWAY AMONG DIVERSE APPLE CULTIVARS

Apple is a valuable dietary fruit because of its phenolic compounds pool (Henriquez et al., 2010). Plant phenolics are high-valued natural metabolites biosynthesized via the shikimate/phenylpropanoids pathways (Lattanzio et al., 2001, Cheynier, 2005). This group of compounds encompasses a wide array of structurally diverse constituents with some more biological activities (Lattanzio et al., 2001). Plants employ phenolic compounds for pigmentation, growth, reproduction, pollination, protection against UV radiation, resistance to pathogens and defense mechanisms under different environmental stressful conditions such as wounding, infection, and for many other biological functions (Lattanzio et al., 2001, Cheynier, 2005). Furthermore, these compounds are responsible for the organoleptic characteristic such as taste and color of fruits during per and post harvest stages (Cheynier, 2005). Strong antioxidant capacity of phenolic compounds made them excellent natural products in coping with cardiac disorders and cancer (Lattanzio et al., 2001, Cheynier 2005, Lata et al., 2009). Flavonoids have long been recognized to possess antiallergenic, anti-inflammatory, antiviral, and antiproliferative activities (Lattanzio et al., 2001, Cheynier 2005, Lata et al., 2009). Undoubtedly, different parts of apple fruit owe various concentrations of phenolic compounds. Noticeably, apple peel has the major affinity for the biosynthesis of aforementioned compounds compared with the other parts of the fruit and also has a high bioactivity potential (Henriquez et al., 2010).

10.9.1   PHENOL BIOSYNTHETIC PATHWAY

Plant phenolic compounds include a wide range of secondary metabolites that are synthesized from carbohydrates via the shikimate pathway and the acetate pathway (Bravo, 1998). Therefore the distribution of phenolic compounds is almost ubiquitous in the plant kingdom (Pereira et al., 2009). It has been estimated that among 100,000–200,000 existing second metabolites (Metcalf, 1987), 20% of the carbon fixed by photosynthesis is channeled into the phenylpropanoid pathway, generating the majority of natural phenolics (Weisshaar & Jenkins, 1998). The phenolic compounds include more than 8000 currently known phenolic structures (Harborne, 1989), and they have one common structural feature, a phenol, which is an aromatic ring bearing at least one 'acidic' hydroxyl substituent (Harborne, 1989). Phenolic compounds are essential to the physiology of plants, mainly as regards: their contribution to plant morphology; providing plants with resistance to pathogens and predators; and protecting crops from pre-harvest seed germination (Bravo, 1998). Recent epidemiological studies have shown important antioxidant properties and other physiological effects of polyphenols and their probable role in the prevention of various diseases, such as cancer and CVD (Dai & Mumper, 2010; Cai et al., 2004; Morton et al., 2000). The antioxidant activities are related to the structure of phenolic compounds (Rice-Evans et al., 1996), in particular the reactivity of the phenol moiety, which depends on the number and position of hydroxyl groups and other substituents, and glycosylation of flavonoid molecules (Cai et al., 2004; Robbins, 2003). Phenolic compounds can be divided into 16 different classes depending on their basic chemical structure, which ranges from simple phenols (C6) to highly polymerized compounds such as lignins ((C6-C3)n) (Harborne, 1989). The two major groups of phenolic compounds present in apple fruit are: (1) phenolic acids and related compounds and (2) flavonoids (Spanos & Wrolstad, 1992).

10.9.2   PHENOLIC ACIDS AND RELATED COMPOUNDS

Hydroxycinnamic acid esters constitute one of the major subgroups of phenolic acids in various apple varieties (Tsao et al., 2003). Chlorogenic acid, the ester of caffeic with quinic acid, and p-coumaroylquinic acid are two classic compounds in the hydroxycinnamic acid ester group (Tsao et al., 2003). The browning occurring in apple juice and cider is mainly due to oxidation of chlorogenic acid by oxidative enzymes (Nicolas et al., 1994). Chlorogenic acid is present in relatively high concentrations in both the peel and flesh of most apple cultivars (Marks et al., 2007; Pérez-Ilzarbe et al., 1991).

10.9.3 FLAVONOID BIOSYNTHETIC PATHWAY

Flavonoids are the most common and widely distributed group of plant phenolic compounds (Bravo, 1998). Over 4,000 flavonoids have been identified and they are widely distributed in different tissues of plants, such as leaves, seeds, bark, and flowers (Heim et al., 2002). Flavonoids have a diphenylpropane skeleton (C6-C3-C6) and the three-carbon bridge between the phenyl groups is usually cyclised with oxygen (Spanos & Wrolstad, 1992). The flavonoids can be divided into 13 sub-groups, according to differences in the number of substituent hydroxyl groups, degree of unsaturation and degree of oxidation of the three-carbon bridge (Bravo, 1998; Spanos & Wrolstad, 1992). Flavonoids are referred to as glycosides when they contain one or more sugar groups (or glucosides in the case of a glucose moiety), and as aglycones when no sugar group is present (de Rijke et al., 2006). The 3 position on flavonoids is the preferred glycosylation site and the 7 position is less frequent (Rice-Evans et al., 1996). In apple, anthocyanidin, flavanol (also named flavan–3-ols) flavo-nols (mainly quercetin glycosides), and dihydrochalcones are the major sub-groups of flavonoids (Chinnici et al., 2004; Guyot et al., 1998; Spanos & Wrolstad, 1992).

Flavonoids and other flavonoid-related compounds, such as stilbenes, are synthesized via a complex network of routes based primarily on the shiki-mate, phenylpropanoid, and malonate pathways (Winkel-Shirley, 2001a; Tanaka et al., 2008). Flavonoids, including anthocyanins and condensed tannins (CTs), as well as stibenes are synthesized from malonyl-CoA, derived from the malonyl pathway, and p-coumaroyl-CoA, derived from the phenylpropanoid pathway

Apple flavor, aroma, and causal agents and candidate gene: In apples, the profile of volatile compounds changes with maturation; aldehydes predominate at the beginning, then the content of alcohols starts to increase considerably, and finally the profile is dominated by esters. Therefore, it is important to discuss exactly how the cultivar and biotic and abiotic factors affect the profile of aldehydes, alcohols, and esters in apples. There are reports of more than 25 aldehydes in apple, mostly hexanal, *trans*–2-hexenal and butanal. Aldehydes are abundant in pre-climacteric apples, but after ripening, the content of some aldehydes becomes almost imperceptible. However, when volatile content is determined in homogenized tissue (juice), high concentrations of some aldehydes are found, mostly hexanal and hexenals. Under hypoxic conditions apples can also produce acetaldehyde, which can be reduced to ethanol. Nonetheless, there are apple varieties such as Royal Gala and Golden Delicious that are very resistant to extremely low oxygen concentrations The volatile flavor constituents of apple have been the studied for over 50 years and have been reviewed by Dimick and Hoskin (1983). Although over 200 volatile compounds have been reported to be in various cultivars of apple there is little information on the contribution these chemicals have on the sensory perception consumers have of apple flavor. Flath et al. (1967) recognized the importance of sensory evaluation and the influence of varietal differences. The importance of cultivar was reinforced by an investigation on cider apples (Williams et a1 1980). The flavor volatiles of some of the older commercial cultivars such as Delicious, Golden Delicious, Cox's Orange Pippin and McIntosh has been the subject of several studies (Flath et al., 1967; Sapers et a1 1977; Williams and Knee, 1977; Dirinck et al., 1983). The early studies clearly show that the characteristic apple aroma/flavor results from a complex mixture of alcohols, esters, aldehydes, and ketones. Volatile flavor compounds produced by Royal Gala apple have been identified by GC-MS. Major components were 2-methyl butyl acetate, butyl acetate, hexyl acetate, butanol, 2-methylbutanol and hexanol. Odor-port evaluation of the components separated by GC indicated that the first four compounds were important contributors to the aroma and flavor. Use of analytical sensory panels revealed that 2-methyl butyl acetate, butanol, and hexyl acetate had the greatest causal effect on those aroma and flavor attributes considered important for Royal Gala apples. Although there is a great range of compounds in the volatile profile of apples, the majority are esters (78–92%) and alcohols (6–16%) (Paillard, 1990). The most abundant compounds are even numbered carbon chains including combinations of acetic, butanoic, and hexanoic acids with ethyl, butyl, and hexyl alcohols (Paillard, 1990). Higher molecular weight volatiles, often with one or two hydrophobic aliphatic chains, are likely to be trapped by skin waxes and are generally not found in the headspace (Paillard, 1990). The knowledge of the molecular mechanisms underlying fruit quality traits seems to be fundamental for apple ripening physiological status description and for improving apple breeding. Since the fruit development and ripening are multi-affected biological processes (Soglio et al., 2009), identification of the particular genes/sequences, and potential functional/physiological molecular markers is difficult. Our analysis shows that the StG amplifying gene sequence encoding starch glucosidase seems to be a good candidate for such functional molecular marker. It could be useful for breeders for monitoring the fruit ripening processes and predicting the collective maturity and storage ability of fruits of cvs Golden Delicious and McIntosh. However, its utilization for testing the other apple varieties has to be validated. Climacteric fruit maturation is polygenic, complex process. Gene activity has a significant effect on the quality characteristics of the fruit for harvest and storage. Existing methods generally allow determining the degree of ripeness at harvest (point '0'). Since there is no method defining onset of the climacteric stage of the fruits, an attempt to identify the functional molecular marker that would determine a physiological ripeness of the fruit several days in advance before harvest was conducted. The analysis of changes in transcript of ten selected genes, and evaluation of the correlation of these genes with changes in fruits quality of two apple varieties Golden Delicious (winter cv.) and McIntosh (autumn cv.), allowed to identify a potential marker, activated a few days before harvesting the fruits. Over expression of the starch glucosidase gene (StG) in the late fruits has been observed prior to the onset of ethylene production. The results confirm that it could be a potential functional marker useful for assessment of physiological ripeness status of cvs Golden Delicious and McIntosh. An important group of genes playing a key role in fruit formation and growth are genes regulating cells division processes. Transition from cell production to cell expansion in fruits occurs between 3–8 weeks after full bloom and is strictly associated with meiotic cell production. This is facilitated by the core group of cell cycling genes such as cyclin dependent kinases (CDK) expressed mainly at the early stages of fruit development. Our data correspond to the research conducted by the other authors, who also indicated that these genes are early induced and negatively correlated with the fruit weight (Janssen et al., 2008; Malladi, Johson, 2011).

Expression of genes related to ethylene Biosynthesis and ethylene perception: The ethylene biosynthesis pathway has been well established in higher plants (Yang and Hoffman, 1984). The biosynthesis of ethylene begins with the production of S-adenosyl methionine (SAM) from the amino acid methionine. The conversions of SAM to 1-aminocyclopropane–1-carboxylate (ACC) and ACC to ethylene are the rate-limiting steps in ethylene biosynthesis and are catalyzed by ACC synthase (ACS) and ACC oxidase

(ACO), respectively (Alexander and Grierson, 2002; Wang et al., 2002). Genes encoding ACS and ACO are members of multi-gene families, and their expression is differentially regulated by various developmental, environmental, and hormonal signals (Kende, 1993; Wang et al., 2002). After synthesis, ethylene is perceived by a family of membrane-localized receptors that are similar to bacterial two-component histidine kinase receptors (Bleecker and Kende, 2000; Klee, 2004; Wang et al., 2002). Expression of the ethylene receptor genes, MdETR1, MdETR2, MdERS1, and MdERS2, increased in the fruit cortex for both cultivars, but only MdETR2 and MdERS2 increased in the FAZ of 'Golden Delicious' apples. The transcript levels of MdPG2, a polygalacturonase gene (PG), and MdEG1,ab–1,4-glucanase gene, markedly increased only in the FAZ of 'Golden Delicious' apples, whereas only MdPG1 rapidly increased in the fruit cortex of 'Golden Delicious' apples. Our results suggested that MdACS5A, MdACO1, MdPG2, and MdEG1 in the FAZ might be related to the difference in PFA between these two cultivars, whereas MdACS1 and MdPG1 were associated with fruit softening.

Apples are typical climacteric fruit characterized by a drastic increase in ethylene production and respiration at ripening (Blanpied, 1972; Yuan and Carbaugh, 2007). It has been reported that the occurrence of the respiratory climacteric and high levels of endogenous ethylene are correlated with PFA and fruit ripening (Blanpied, 1972; Yuan and Carbaugh, 2007). Application of ethephon, an ethylene-releasing compound, effectively promoted PFA and ripening in apples (Edgerton and Blanpied, 1970), whereas aminoethoxyvinylglycine (AVG), an inhibitor of ethylene biosynthesis, or 1-methylcyclopropene (1-MCP), an inhibitor of ethylene action, reduced fruit ethylene production and delayed PFA and ripening (Schupp and Greene, 2004; Yuan and Carbaugh, 2007). In addition, MdACS1–2, an allele of MdACS1, andMdACO1–2, an allele of MdACO1, genes have been implicated in differences in fruit ethylene production, fruit softening, and storage life among apple cultivars (Costa et al., 2005; Harada et al., 2000; Sunako et al., 1999). In contrast, some reports showed that apple fruit softening is not related to theMdACS1allelotype and ethylene production (Oraguzie et al., 2004; Wakasa et al., 2006) but related toMdPG1 (Wakasa et al., 2006). It has been found that fruit abscission-related EG and PG genes are significantly different from those associated with fruit softening in tomato (Solanum lycopersicum L.) (Brummell et al., 1999; Roberts et al., 2002; Taylor et al., 1990; Taylor and Whitelaw, 2001), whereas the same EG gene (cel1) is responsible for both fruit softening and mature fruit abscission in avocado

(Persea americana Mill.) (Tonutti et al., 1995). However, no information is available about whether the difference in the expression of fruit softening-related PG genes such asMdPG1 is also responsible for the difference in PFA among apple cultivars (Table 10.8).

TABLE 10.8 Description of Apple Varieties, Candidate Gene and Gene Class Associated With Ethylene Perception

Apple cultivar/variety	Gene Class	Gene Name	Reference
Golden Delicious	SAMS	*MdSAMS1*	Young et al., 1996
		MdSAMS2	
		MdSAMS3	
		MdSAMS4	
		MdSAMS5	
		MdSAMS5	
		MdSAMS8	
		MdSAMS9	
'White Angel'	*ETR*	*MdETR1*	Rikkerink et al., 2010
		MdETR2	
		MdETR3	
		MdETR4	
		MdETR5	
		MdCTR1	
Royal Gala	ERS	*ERS1,*	Ireland et al., 2012
		ERS2	
Golden Delicious	*EIN*	*MdEIN2–1*	Liu and Yuan, 2008
		MdEIN2–2	
		MdEIN3–1	
		MdEIN3–2	
		MdEIN3–2	
Golden Delicious	*ERF*	*ERF1/2*	Gu et al., 2017
		ERF1/2	
		EBF1/2	
		MdMPK6–1	
		MdMPK6–2	

TABLE 10.8 *(Continued)*

Apple cultivar/variety	Gene Class	Gene Name	Reference
Prima × Fiesta and Fuji × Mondial Gala	ACS	*Md-ACS1*	Costa et al., 2005
		Md-ACS1–2	Li et al., 2013
Golden Delicious		*MdACS3a*	
		MdACS3b	
		MdACS3c	
		MdACS4	
		MdACS5A	
		MdACS5B	
Prima × Fiesta and Fuji × Mondial Gala	ACO	*Md-ACO1*	Costa et al., 2005
		Md-ACO1–1	

10.10 CONCLUSIONS AND FUTURE PROSPECTS

Following developments in molecular biology, genomics, and bioinformatics, new breeding technologies are being developed rapidly. Traditional apple breeding involves the deliberate crossing of closely or distantly related individuals to produce new varieties with desirable properties. For such breeding, MAS is being steadily applied. Simultaneously, for the improvement of existing trusted cultivars, new transgenic technologies can be applied in order to quickly eliminate any foreign genes. Furthermore, genome editing, by which only the target gene can be accurately modified, is emerging as a novel breeding technology. These new technologies will undoubtedly facilitate apple breeding, and yield novel and attractive apple cultivars. In Japan, techniques, such as MAS, trans-grafting, and reduction of generation time by virus vectors, are being studied for practical use. On the other hand, technologies utilizing a large volume of genomic information and molecular markers, such as GS and Meta QTL, are yet to be acquired. It will also be necessary to adopt information on markers to Japanese apple cultivars, which have emerged via unique evolution. The importance of individual diseases and fruit characteristics may differ among countries, as growth conditions and consumer preference vary internationally. Development of apple cultivars that satisfy consumers and related industries will be accelerated by the integration of new genomic information, new technologies and existing breeding programs.

KEYWORDS

- apple
- aroma
- disease resistance
- fruit quality
- molecular markers

REFERENCES

Abbott, J. A., (1999). Quality measurement of fruits and vegetables. *Postharvest Biol. Technol., 15,* 207–225.

Ansari, I. A., & Khan, M. S., (2012). An efficient protocol for isolation of high-quality genomic DNA from seeds of apple cultivars (Malus × Domestica) for random amplified polymorphic DNA (RAPD) analysis. *Pharmaceutical Crops, 3,* 78–83.

Ashebir, D., Deckers, T., Nyssen, J., Bihon, W., Tsegay, A., Tekie, H., et al. (2010). Growing apple *(Malus domestica*) under tropical mountain climate conditions in northern Ethiopia. *Experimental Agriculture, 46,* 53–65.

Boyer, J., & Liu, R. H., (2004). Apple phytochemicals and their health benefits. *Nutrition Journal, 3* (5), pp. 1–45.

Bravo, L., (1998). Polyphenols: Chemistry, dietary sources, metabolism, and nutritional significance. *Nutrition Reviews, 56* (11), 317–333.

Cai, Y., Luo, Q., Sun, M., & Corke, H., (2004). Antioxidant activity and phenolic compounds of 112 traditional Chinese medicinal plants associated with anticancer. *Life Sciences, 74* (17), pp. 0022157–2184.

Campeanu, G., Neata, G., & Darjanschi, G., (2009). Chemical Composition of the Fruits of Several Apple Cultivars Growth, *Not. Bot. Hort. Agrobot. Cluj 37*(2), 161–164.

Cheng, F., Weeden, N., & Brown, S., (1996). Identification of co-dominant RAPD markers tightly linked to fruit skin color in apple. *Theoretical and Applied Genetics, 93,* 222–227.

Chevreau, E., Manganaris, A. G., & Gallet, M., (1999). Isozyme segregation in five apple progenies and potential use for map construction. *Theoretical and Applied Genetics, 98,* 329–336.

Cheynier, V., (2005). Polyphenols in foods are more complex than often thought. *Am. J. Clin. Nutr., 81* (1), 2235–2295.

Chinnici, F., Bendini, A., Gaiani, A., & Riponi, C., (2004). Radical scavenging activities of peels and pulps from cv. Golden Delicious apples as related to their phenolic composition. *Journal of Agricultural and Food Chemistry, 52* (15), 4684–4689.

Coart, E., Vekemans, X., Smulders, M. J. M., Wagner, I., Van Huylenbroeck J., Bockstaele, V. E., & Roldan-Ruiz, I., (2003). Genetic variation in the endangered wild apple (M. sylvestris (L.) Mill.) in Belgium as revealed by amplified fragment length polymorphism and microsatellite markers. *Molecular Ecology, 12,* 845–857.

Cook, N. C., (2010). Apple production under conditions of sub-optimal winter chilling in South Africa. *Acta Horticulturae, 872*, 199–204.

Costa, F., Stella, S., Magnani, R., & Sansavini, S., (2004). Characterization of apple expansion sequences for the development of SSR markers associated with fruit firmness. *Acta Horticulturae, 663*, 341–344.

Costa, F., Stella, S., Van de Weg, W. E., Guerra, W., Cecchinel, M., Dallavia, J., Koller, B., & Sansavini, S., (2005). Role of the genes *Md-ACO1* and *Md-ACS1* in ethylene production and shelf life of apple (*Malus domestica* Borkh). *Euphytica, 141*, 181–190.

Costa. F., Van de Weg. W. E., Stella. S., Dondini. L., Pratesi. D., Musacchi. S., & Sansavini, S., (2008). Map position and functional allelic diversity of *Md-Exp7*, a new putative expansin gene associated with fruit softening in apple (*Malus domestica* Borkh.) and pear (*Pyrus communis*). *Tree Genetics and Genome, 4*, 575–586.

Dai, J., & Mumper, R. J., (2010). Plant phenolics: extraction, analysis, and their antioxidant and anticancer properties. *Molecules, 15* (10), pp. 7313–7352.

Dantas, A., Silva de, M., Fortes, J. A., & Rombaldi, C., (2000). "RAPD in somaclones of apple rootstocks cultivar M. *9 Regenerated from Aluminum Medium,"* *Revista-Brasileira-de-Fruticultura (Brazil)* (Vol. 22, pp. 303–305).

Davey, M. W., Auwerkerken, A., & Keulemans, J., (2007). Relationship of apple vitamin C and antioxidant contents to harvest date and postharvest pathogen infection. *Journal of the Science of Food and Agriculture, 87*, 802–813.

De Oliveira, M. C., Sichieri, R., & Moura, A. S., (2003). Weight loss associated with a daily intake of three apples or three pears among overweight women. *Nutrition, 19* (3), 253–256.

De Rijke, E., Out, P., Niessen, W. M., Ariese, F., Gooijer, C., & Udo, A. T., (2006). Analytical separation and detection methods for flavonoids. *Journal of Chromatography A, 1112* (1), 31–63.

Defilippi, B. G., Dandekar, A. M., & Kader, A. A., (2005). Relationship of ethylene biosynthesis to volatile production, related enzymes and precursor availability in apple peel and flesh tissues. *J. Agric. Food Chem., 53*, 3133–3141.

Dixon, J., & Hewett, E. W., (2000). Factors affecting apple aroma/flavor volatile concentration: a review. *New Zealand Journal of Crop and Horticultural Science, 28*, 155–173.

Dunemann, F., Kahnau, R., & Shmidt, H., (1994). Genetic relationships in *Malus* evaluated by RAPD 'fingerprinting' of cultivars and wild species. *Plant Breeding, 113*, 150–159.

Durel, C. E., Calenge, F., Parisi, L., Van de Weg, W. E., Kodde, L. P., Liebhard, R., et al. (2003). Stability of scab resistance QTLs in several mapped progenies. *Eucarpia Symposium in Fruit Breeding and Genetics*, Angers, France.

Eberhardt, M. V., Lee, C. Y., & Liu, R. H., (2000). Nutrition: antioxidant activity of fresh apples. *Nature, 405* (6789), pp. 903–904.

Erturk, U., & Akcay, M. E., (2010). Genetic variability in accessions of 'Amasya' apple cultivar using RAPD markers. *Not Bot Horti Agrobot Cluj-Napoca, 38*, 239–245.

Ferguson, I. B., & Boyd, L. M., (2002). Inorganic nutrients and fruit quality. In: Knee, M., (ed.), *Fruit Quality and its Biological Basis* (pp. 17–45). Sheffield Academic Press, UK.

Ferree, D. C., & Carlson, R. F., (1987). Apple rootstocks. In: Rom, R. C., Carlson, R. F., (eds.), *Rootstocks for Fruit Crops* (p. 107143). John Wiley & Sons, New York, NY (New York).

Ferree, D. C., & Warrington, I. J., (2003). *Apples: Botany, Production, and Uses*. CABI publishing, CAB international, UK. pp. 672.

Forsline, P. L., (1995). Adding diversity to the national apple germplasm collection: Collecting wild apples in Kazakstan. *New York Fruit Quarterly, 3* (3), 3–6.

Forte, A. V., Ignatov, A. N., Ponomarenko, V. V., Dorokhov, D. B., & Savelyev, N. I., (2002). Phylogeny of the *Malus* (apple tree) species, inferred from the morphological traits and molecular DNA analysis. *Russian Journal of Genetics, 38*, 1150–1160.

Gardiner, S. E., Bassett, H. C. M., & Madie, C., (1996). Isozyme, randomly amplified polymorphic DNA (RAPD), and restriction fragment-length polymorphism (RFLP) markers used to deduce a putative parent for the 'Braeburn' apple. *Journal of the American Society for Horticultural Science, 121*, 996–1001.

Gasper, A., Hollands, W., Casgrain, A., Saha, S., Teucher, B., Dainty, J. R., et al. (2014). Consumption of both low and high (−)-epicatechin apple puree attenuates platelet reactivity and increases plasma concentrations of nitric oxide metabolites: A randomized controlled trial. *Archives of Biochemistry and Biophysics, 559*, pp. 29–37.

Gianfranceschi, L., Seglias, N., Tarchini, R. K., & Gessler, C., (1998). Simple sequence repeats for genetic analysis of apple. *Theor. Appl. Genet., 96*, 1069–1076.

Goulão, L., & Cristina, M., (2001). "Molecular characterization of apple using microsatellite (SSR and ISSR) markers." *Euphytica, 122*, pp. 81–89.

Goulão, L., & Oliveira, C. M., (2001). Molecular characterization of cultivars of apple (*Malus* ×domestica Borkh.) using microsatellite (SSR and ISSR) markers. *Euphytica, 122*, 81–89.

Goulão, L., Cabrita, L., Oliveira, C. M., & Leitao, J. M., (2001). Comparing RAPD and AFLPTM analysis in discrimination and estimation of genetic similarities among apple (*Malus domestica* Borkh.) cultivars. *Euphytica, 119*, 259–270.

Guadagni, D. G., Bomben, J. L., & Hudson, J. S., (1971). Factors influencing the development of aroma in apple peel. *J. Sci. Food Agric., 22*, 110–115.

Guilford, P., Prakash, S., Zhu, J. M., Rikkerink, E., Gardiner, S., Bassett, H., & Forster, R., (1997). Microsatellites in *Malus* ×domestica (apple): abundance, polymorphism, and cultivar identification. *Theor. Appl. Genet., 94*, 249–254.

Guyot, S., Marnet, N., Laraba, D., Sanoner, P., & Drilleau, J. F., (1998). Reversed-phase HPLC following thiolysis for quantitative estimation and characterization of the four main classes of phenolic compounds in different tissue zones of a French cider apple variety (Malus domestica var. Kermerrien). *Journal of Agricultural and Food Chemistry, 46* (5), pp. 1698–1705.

Han, N., & Bakovic, M., (2015). Biologically active triterpenoids and their cardio protective and anti-inflammatory effects. *J. Bioanal. Biomed. S., 12*, pp. 2.

Hancock, J. F., Luby, J. J., Brown, S. K., & Lobos, G. A., (2008). *Apples*. p. 137

Harada, T., Matsukawa, K., Sato, T., Ishikawa, R., Niizeki, M., & Saito, K., (1993). DNA-RAPD detect genetic variation and paternity in *Malus. Euphytica, 65*, 87–91.

Harborne, J. B., (1989). Methods in Plant Biochemistry (Vol. 1). Plant phenolics. Academic Press Ltd.

Harris, S. A., Robinson, J. P., & Juniper, B. E., (2002). Genetic clues to the origin of the apple. *Trends in Genetics, 18* (8), 426430.

Hauagge, R., (2010). 'IPR Julieta,' A new early low chill requirement apple cultivar. *Proc. 8th IS on Temperate Zone Fruits. Acta Horticulturae, 872*, 193–196.

Heim, K. E., Tagliaferro, A. R., & Bobilya, D. J., (2002). Flavonoid antioxidants: chemistry, metabolism, and structure-activity relationships. *The Journal of Nutritional Biochemistry, 13* (10), pp. 572–584.

Hemmat, M., Weeden, N. F., Manganaris, A. G., & Lawson, D. M., (1994). Molecular marker linkage map of apple. *Journal of Heredity, 85*, 4–11.

Henriquez, C., Almonacid, S., Chiffelle, I., Valenzuela, T., Araya, M., Cabezas, L., Simpson, R., & Speisky, H., (2010). *Determination of Antioxidant Capacity, Total Phenolic Content and Mineral Composition of Different Fruit Tissue*

Hokanson, S. C., Lamboy, W. F., Szewc-McFadden, A. K., & McFerson, J. R., (2001). Microsatellite (SSR) variation in a collection of Malus (apple) species and hybrids. *Euphytica, 118*, 281–294.

Hokanson, S. C., Szewc-McFadden, A. K., Lamboy, W. F., & McFerson, J. R., (1998). Microsatellite (SSR) markers reveal genetic identities, genetic diversity, and relationships in a Malus × domestica Borkh. core subset collection. *Theor. Appl. Genet., 97*, 671–683.

Holland, D., Larkov, O., Bar-Yaákov, I., Bar, E., Zax, A., & Brandeis, E., (2005). Developmental and varietal differences in volatile ester formation and acetyl-CoA: Alcohol acetyl transferase activities in apple (*Malus domestica* Borkh.) fruit. *J. Agric. Food Chem., 53*, 7198–7203.

Ignatov, A., & Bodishevskaya, A., (2011). *Malus.* In: Kole, C., (ed.), *Wild crop Relatives: Genomic and Breeding Resources Temperate Fruits* (p. 4564). SpringerVerlag, Berlin, Heidelberg.

Jackson, J. E., (2003). *Biology of Apples and Pears.* Cambridge University Press, Cambridge.

Janick, J., & Moore, J. N., (1996). *Fruit Breeding, Tree, and Tropical Fruits.* John Wiley & Sons, Inc. Oxford, UK. p. 77.

Janssen, B., Thodey, K., Schaffer, R. J., Alba, R., Balakrishnan, L., Bishop, R., et al. (2008). Global gene expression analysis of apple fruit development from the floral bud to ripe fruit. *BMC Plant Biology, 8* (16), p. 29.

Juniper, B. E., Watkins, R., & Harris, S. A., (1999). The origin of the apple. *Act Horticulturae. 484*, 27–33.

Kader, A. A., (1999). Fruit maturity, ripening, and quality relationships. *Acta Horticulturae, 485*, 203–207.

Kafkas, S., Özgen, M., Doğan, Y., Özcan, B., Ercişli, S., & Serçe, S., (2008). Molecular characterization of mulberry accessions in Turkey by AFLP markers. *J. Am. Soc. Hortic. Sci., 4*, 593–597.

Kashyap, P., Singh, A. K., Singh, S. K., & Deshmukh, R., (2010). "Genetic diversity analysis of indigenous and exotic apple genotypes using inter simple sequence repeat markers." *Indian J. Hort., 67*, pp. 15–20.

Kenis, K., Pauwels, E., Houlvinck, N. V., & Keulemans, J., (2001). "The use of microsatellites to establish unique fingerprints for apple cultivars and some of their descendents," *Acta Hort., 546*, pp. 427–431.

Koc, A., Akbulut, M., Orhan, E., Celik, Z., Bilgener, S., & Ercisli, S., (2009). Identification of Turkish and standard apple rootstocks by morphological and molecular markers. *Genet. Mol. Res., 2*, 420–425.

Koller, B., Lehmann, A., McDermot, J. M., & Gessler, C., (1993). Identification of apple cultivars using RAPD markers. *Theor. Appl. Genet., 85*, 901–904.

Landry, B. S., Li, R. Q., Cheung, W. Y., & Granger, R. L., (1994), "Phylogeny Analysis of 25 Apple Rootstocks Using RAPD Markers and Tactical Gene Tagging," *Theor. Appl. Genet., 89*, pp. 847–852.

Landry, B. S., Li, R. Q., Cheung, W. Y., & Granger, R. L., (1994). Phylogeny analysis of 25 apple rootstocks using RAPD markers and tactical gene tagging. *Theor. Appl. Genet., 89*, 847–852.

Łata, B., Trąmpczyńska, A., & Pacześna, J., (2009). Cultivar variation in apple peel and whole fruit phenolic composition. *Sci. Horticult., 121*, 176–181.

Lattanzio, V., Di-Venere, D., Linsalata, V., Bertolini, P., Ippolito, A., & Salerno, M., (2001). Low-temperature metabolismof apple phenolics and quiescence of Phlyctaena vagabunda. *J. Agric. Food Chem., 49* (12), 5817–5821.

Lattanzio, V., Lattanzio, V. M. T., & Cardinali, A., (2006). Role of phenolics in the resistance mechanisms of plants against fungal pathogens and insects. *Phytochemistry: Advances in Research, 661,* 23–67.

Laurens, F., Durel, C. E., & Lascostes, M., (2004). Molecular characterization of French local apple cultivars using SSRs. *Acta Hort., 663,* 639–642.

Lawson, D. M., Hemmat, M., & Weeden, N. F., (1995). The use of molecular markers to analyze the inheritance of morphological and developmental traits in apple. *Journal of the American Society for Horticultural Science, 120,* 532–537.

Lee, K., Kim, Y., Kim, D., Lee, H., & Lee, C., (2003). Major phenolics in apple and their contribution to the total antioxidant capacity. *Journal of Agriculture and Food Chemistry, 51,* 6516–6520.

Leite, G. B., Denardi, F., & Raseira, M. C. B., (2008). Breeding of temperate zone fruits for sub-tropical conditions. *Acta Horticulturae, 772,* 507–512.

Li, J., & Yuan, A. R., (2008). NAA and ethylene regulate of genes related to ethylene biosynthesis, perception, and cell wall degradation during fruit abscission and ripening in 'Delicious' apples. *Journal of Plant Growth and Regulation, 27,* 283–295.

Liebhard, R., Gianfranceschi, L., Koller, B., Ryder, C. D., Tarchini, R., Van De Weg E., & Gessler, C., (2002). Development and characterization of 140 new microsatellites in apple (Malus × domestica Borkh.). *Molecular Breeding, 10,* 217–241.

Liebhard, R., Koller, B., Gianfranceschi, L., & Gessler, C., (2003). Creating a saturated reference map for the apple (Malus domestica Borkh.) genome. *Theoretical and Applied Genetics, 106,* 1497–1508.

Luby, J. J., (2003). Taxonomic classification and brief history. In: Ferree, D. C., & Warrington, I. J., (eds.), *Apples: Botany, Production, and Uses* (p. 114). CABI International, Cambridge, UK.

Malnoy, M., Borejsza-Wysocka, E. E., Jin, L. Q., He, S. Y., & Aldwinckle, II. S., (2004). Over-expression of the apple gene *MpNPR1* causes increased disease resistance in *Malus domestica. Acta Horticulturae, 663,* 463–468.

Marks, S. C., Mullen, W., & Crozier, A., (2007). Flavonoid and chlorogenic acid profiles of English cider apples. *Journal of the Science of Food and Agriculture, 87* (4), pp. 719–728.

Markussen, T., Kruger, J., Schmidt, H., & Dunemann, F., (1995). Identification of PCR-based markers linked to the powdery-mildew-resistance gene *Pl–1* from *Malus robusta* in cultivated apple. *Plant Breeding, 114,* 530–534.

Markuszewski, B., & Kopytowski, J., (2008). Transformations of chemical compounds during apple storage. Scientific Works of the Lithuanian Institute of Horticulture and Lithuanian University of Agriculture. *Sodininkyste Ir Darzininkyste, 27* (2), 329–338.

Metcalf, R. L., (1987). Plant volatiles as insect attractants. *Critical Reviews in Plant Sciences, 5* (3), pp. 251–301.

Mikulic, P. M., Stampar, F., & Veberic, R., (2009). Changes in the inner quality parameters of apple fruit from technological to edible maturity. *Acta Agriculturae Slovenica, 93* (1), 17–29.

Morton, L. W., Caccetta, R. A. A., Puddey, I. B., & Croft, K. D., (2000). Chemistry and biological effects of dietary phenolic compounds: relevance to cardiovascular disease. *Clinical and Experimental Pharmacology and Physiology, 27* (3), pp. 152–159.

Mulcahy, D. L., Cresti, M., Sansavini, S., Douglas, G. C., Linskens, H. F., Mulcahy, G. B., Vignani, R., & Pancaldi, M., (1993). The use of random amplified polymorphic DNAs to fingerprint apple genomes. *Sci Horticulturae, 54,* 89–96.

Nagasako-Akazome, Y., Kanda, T., Ohtake, Y., Shimasaki, H., & Kobayashi, T., (2007). Apple polyphenols influence cholesterol metabolism in healthy subjects with relatively high body mass index. *Journal of Oleo Science, 56* (8), pp. 417–428.

Nicolas, J. J., Richard-Forget, F. C., Goupy, P. M., Amiot, M. J., & Aubert, S. Y., (1994). Enzymatic browning reaction in apple and apple products. *Critical Reviews in Food Science & Nutrition, 34* (2), pp. 109–157.

Nijssen, L. M., Van Ingen-Visscher, C. A., & Donders, J. J. H., (2011). *VCF Volatile Compounds in Food: Database (Version 13. 1.).* Zeist (The Netherlands): TNO Triskelion Recuperato da.

Nybom, H., & Schaal, B. A., (1990). DNA 'fingerprints' applied to paternity analysis in apples (Malus × domestica). *Theor. Appl. Genet., 79,* 763–768.

Nybom, H., Sehic, J., & Garkava-Gustavsson, L., (2008b). Modern apple breeding is associated with a significant change in allelic ratio of the ethylene production gene *Md-ACS1. Journal of Horticultural Science and Biotechnology, 83,* 673–677.

O'Rourke, D., (2003). World production, trade, consumption, and economic outlook for apples. In: Ferree D. C., & Warrington, I. J., (eds.), *Apples: Botany, Production, and Uses* (pp. 15–28). CABI publishing, CAB international, UK.

Oraguzie, N. C., Gardiner, S. E., Basset, H. C. M., Stefanati, M., Ball, R. D., Bus, V. G. M., & White, A. G., (2001). Genetic diversity and relationships in *Malus* sp. germplasm collections as determined by randomly amplified polymorphic DNA. *Journal of the American Society for Horticultural Science, 126,* 318–328.

Oraguzie, N. C., Volz, R. K., Whitworth, C. J., Bassett, H. C. M., Hall, A. J., & Gardiner, S. E., (2007). Influence of *Md-ACS1* allelotype and harvest season within an apple germplasm collection on fruit softening during cold air storage. *Postharvest Biology Technology, 44,* 212–219.

Oraguzie, N. C., Yamamoto, T., Soejima, J., Suzuki, T., & De Silva, H. N., (2005). DNA fingerprinting of apple (*Malus* spp.) rootstocks using Simple Sequence Repeats. *Plant Breeding, 124,* 197–202.

Pavlovic, N., Zdravkovic, J., Cvikic, D., Zdravkovic, M., Adzic, S., Pavlovic, S., & Surlan-Momirovic, G., (2012). Characterization of onion genotypes by use of RAPD markers. *Genetika, 2,* 269–278.

Pereira-Lorenzo, S., Ramos-Cabrer, A. M., & Fischer, M., (2009). Breeding apple (Malus x domestica Borkh). In: *Breeding Plantation Tree Crops: Temperate Species* (pp. 33–81). Springer New York.

Pérez-Ilzarbe, J., Hernández, T., & Estrella, I., (1991). Phenolic compounds in apples: varietal differences. *Zeitschrift für Lebensmittel-Untersuchung und Forschung, 192* (6), pp. 551–554.

Prusky, D., McEvoy, J. L., Saftner, R., Conway, W. S., & Jones, R., (2004). Relationship between host acidification and virulence of Penicillium spp. on apple and citrus fruit. *Phytopatology, 94,* 44–51.

Ravn-Haren, G., Dragsted, L. O., Buch-Andersen, T., Jensen, E. N., Jensen, R. I., Németh-Balogh, M., et al. (2013). Intake of whole apples or clear apple juice has contrasting effects on plasma lipids in healthy volunteers. *European Journal of Nutrition, 52* (8), pp. 1875–1889.

Rice-Evans, C. A., Miller, N. J., & Paganga, G., (1996). Structure–antioxidant activity relationships of flavonoids and phenolic acids. *Free Radical Biology and Medicine, 20* (7), pp. 933–956.

Rieger, M., (2006). *Introduction to Fruit Crops*. Food Products Press, Binghamton.

Robards, K., Prenzler, P. D., Tucker, G., Swatsitang, P., & Glover, W., (1999). Phenolic compounds and their role in oxidative processes in fruits. *Food Chem., 66,* 401–436.

Robbins, R. J., (2003). Phenolic acids in foods: an overview of analytical methodology. *Journal of Agricultural and Food Chemistry, 51* (10), pp. 2866–2887.

Robinson, J. P., Harris, S. A., & Juniper, B. E., (2001). Taxonomy of the genus *Malus* Mill. (Rosaceae) with emphasis on the cultivated apple, *Malus domestica* Borkh. *Plant Systematics and Evolution 226,* 3558.

Rudell, D. R., Mattinson, D. S., Mattheis, J. P., Wyllie, S. G., & Fellman, J. K., (2002). Investigations of aroma volatile biosynthesis under anoxic conditions and in different tissues of "Redchief Dilicious"apple fruit (*Malus domestica* Borkh.) *J. Agric. Food Chem., 50,* 2627–2632.

Santesteban, L. G., Miranda, C., & Royo, B. J., (2009). Assessment of the genetic and phenotypic diversity maintained in apple core collections constructed by using either agro-morphologic or molecular marker data. *Span. J. Agric. Res., 7,* 572–584.

Sanzani, S. M., De Girolamo, A., Schena, L., Solfrizzo, M., Ippolito, A., & Visconti, A., (2009a). Control of Penicillium expansum and patulin accumulation on apples by quercetin and umbelliferone. *European Food Research and Technology, 228,* 381–389.

Sanzani, S. M., Schena, L., Nigro, F., De Girolamo, A., & Ippolito, A., (2009b). Effect of quercetin and umbelliferone on the transcript level of Penicillium expansum genes involved in patulin biosynthesis. *European Journal of Plant Pathology, 125,* 223–233.

Sanzani, S. M., Schenab, L., De Girolamo, A., Ippolito, A., & González-Candelas, L., (2010). Characterization of genes associated withinduced resistance against Penicillium expansum in apple fruit treated with quercetin. *Postharvest Biology and Technology, 56,* 1–11.

Sharma, R., & Krishna, H., (2014). Apple. In: *Fruit Production (Major fruits)* (pp. 190–254). Daya Publishing House, Astral International, New Delhi.

Shewfelt, R. L., (1999). What is quality? *Postharvest Biol. Technol., 15,* 197–200.

Shulaev, V., Korban, S. S., Sosinski, B., Abbott, A. G., Aldwinckle, H. S., Folta, K. M., et al. (2008). Multiple models for Rosaceae genomics. *Plant Physiology, 147,* 985–1003.

Smith, J. S. C., & Smith, O. S., (1992). Fingerprinting crop varieties. *Adv. Agron., 47,* 85–140.

Smolik, M., Malinowska, K., Smolik, B., & Pacewicz, K., (2011). Polymorphism in random amplified and nuclear rDNA sequences assessed in certain apple (Malus × domestica Borkh.) cultivars. *Not Bot Horti Agrobot Cluj-Napoca, 39,* 264–270.

Spanos, G. A., & Wrolstad, R. E., (1992). Phenolics of apple, pear, and white grape juices and their changes with processing and storage. A review. *Journal of Agricultural and Food Chemistry, 40* (9), pp. 1478–1487.

Tanaka, Y., Sasaki, N., & Ohmiya, A., (2008) Biosynthesis of plant pigments: Anthocyanins, betalains, and carotenoids. *Plant J., 54,* 733–749

Tartarini, S., (1992). Marker-assisted selection in pome fruit breeding. Session I: MAS in plants, Dipartimento Colture Arboree, Bologna University, Italy.

Tignon, M., Kettmann, R., & Watillon, B., (2000). AFLP: Use for the identification of apple cultivars and mutants. *Acta Horticulturae, 521,* 219–226.

Tignon, M., Lateur, M., Kettmann, R., & Watillon, B., (2001). Distinction between closely related apple cultivars of the belle-fleur family using RFLP and AFLP markers. *Acta Horticulturae, 546*, 509–513.

Treutter, D., (2005). Significance of flavanoids in plant resistance and enhancement of their biosynthesis. *Plant Biology, 7*, 581–591.

Tsao, R., Yang, R., Young, J. C., & Zhu, H., (2003). Polyphenolic profiles in eight apple cultivars using high-performance liquid chromatography (HPLC). *Journal of Agricultural and Food Chemistry, 51* (21), pp. 6347–6353.

Usenik, V., Mikulic, P. M., Solar, A., & Stampar, F., (2004). Flavanols of leaves in relation to apple scab resistance. *Journal of Plant Diseases and Protection, 111*, 137–144.

Velasco, R., Zharkikh, A., Affourtit, J., Dhingra, A., Cestaro, A., Kalyanaraman, A., et al. (2010). The genome of the domesticatedapple (*Malus* x *domestica* Borkh.). *Nature Genetics: 833839.*

Verma, M. K., Ahmed, N., Singh, A. K., & Awasthi, O. P., (2010). Temperate tree fruits and nuts in India. *Chronica Horticulture, 50*, 43–48.

Villatoro, C., Altisent, R., Echeverria, G., Graell, J., Lopez, M. L., & Lara, I., (2008). Changes in biosynthesis of aroma volatile compounds during on-tree maturation of "Pink Lady" apples. *Postharvest Biol. Technol., 47*, 286–295.

Wakasa, Y., Kudo, H., Ishikawa, R., Akada, S., Senda, M., Niizeki, M., & Harada, T., (2006). Low expression of an endopolygalacturonase gene in apple fruit with long-term storage potential. *Postharvest Biology and Technology, 39*, 193–198.

Waldbauer, K., Seiringer, G., Nguyen, D. L., Winkler, J., Blaschke, M., McKinnon, R., et al. (2015). Triterpenoic acids from apple pomace enhance the activity of the endothelial nitric oxide synthase (eNOS). *Journal of Agricultural and Food Chemistry. 13; 64*(1), 185–194. doi: 10.1021/acs.jafc.5b05061

Wamocho, L. S., & Ombwara, F. K., (2001). Deciduous fruit tree germplasm in Kenya. *Acta Horticulturae, 565*, 45–47.

Watillon, B., Druart, P., Du Jardin, P., Kettmann, R., Boxus, P., & Burny, A., (1991). Use of a random cDNA probes to detect restriction fragment length polymorphisms among apple clones. *Sci. Horticulturae, 46*, 235–243.

Way, R. D., Aldwinckle, H. S., Lamb, R. C., Rejman, A., Sansavini, S., Shen, T., et al. (1990). Apples (*Malus*). In: Moore, J. N., & Ballington, R., (eds.), *Genetic Resources of Temperate Fruit and Nut Crops* (pp. 1–62). International Society of Horticultural Science, Wageningen, Netherlands.

Webster, A. D., (2005a). The origin, distribution, and genetic diversity of temperate tree fruits. In: Tromp, J., Webster, A. D., & Wertheim, S. J., (eds.), *Fundamentals of Temperate Zone Tree Fruit Production Backhuys Publishers* (p. 111), Leiden, The Netherlands.

Weeden, N. F., Timmerman, G. M., Hemmat, M., Kneen, B. E., & Lodhi, M. A. (1992). Inheritance and reliability of RAPD markers. In: *Application of RAPD Technology to Plant Breeding* (pp. 12–17). *Joint Plant Breeding Symp. series. Crop Sci. Soc Am./Am. Soc. Hort. Sci./Am. Genet. Assoc.* pp. 12–17.

Weisshaar, B., & Jenkins, G. I., (1998). Phenylpropanoid biosynthesis and its regulation. *Current Opinion in Plant Biology, 1* (3), pp. 251–257.

Winkel-Shirley, B., (2001a). Flavonoid biosynthesis. A colorful model for genetics, biochemistry, cell biology, and biotechnology. *Plant Physiol, 126*, 485–493.

Wojdylo, A., et al. (2008). Polyphenolic compounds and antioxidant activity of new and old apple varieties. *Journal of Agriculture Food C, 13, 56*(15), 6520–6530. doi: 10.1021/ jf800510j.

Wolfe, K. L., & Liu, R. H., (2003). Apple peels as a value-added food ingredient. *Journal of Agricultural and Food Chemistry*, *51* (6), pp. 1676–1683.

Wolfe, K., Wu, X., & Liu, R. H., (2003). Antioxidant activity of apple peels. *Journal of Agricultural and Food Chemistry*, *51* (3), 609–614.

Yao, J. L., Dong, Y. H., Kvarnheden, A., & Morris, B., (1999). Seven *MADS-box* genes in apple are expressed in different parts of the fruit. *Journal of the American Society for Horticultural Science, 124*, 8–13.

Zhu, Y., & Barritt, B. H., (2008). *Md-ACS1* and *Md-ACO1* genotyping of apple (*Malus domestica* Borkh.) breeding parents and suitability for the marker-assisted selection. *Tree Genetics and Genomes, 4*, 555–562.

Zietkiewicz, E., Rafalski, A., & Labuda, D., (1994). Genome fingerprinting by simple sequence repeat (SSR)-anchored polymerase chain reaction amplification. *Genomics, 20*, 176–183.

CHAPTER 11

PEARS (*PYRUS*) OF NORTHERN PAKISTAN

MOHAMMAD ISLAM[1] and HABIB AHMAD[2]

[1]*Department of Genetics, Hazara University, Mansehra, KPK, Pakistan*

[2]*Vice Chancellor, Islamia College University, Peshawar, Pakistan,*
Tel.: +92-(0)997-414131
E-mail: drhahmad@gmail.com; vc@icp.edu.pk

ABSTRACT

Biosystematic information about 110 specimens of the landraces of *Pyrus* trees available in nature and traditional farms of the moist temperate region of Northern Pakistan is presented here. The landraces were surveyed in 100,565 km^2 area and best representative plants of different types were selected for morphological, DNA, and ribosomal gene analyzes. For morphological analyzes the numerical parameters viz. petiole length, leaf area, pedicel length, fruit length, fruit width, and fruit weight were considered. The morphological analyzes sorted out all the collected specimens into 14 species viz., *Pyrus Pashia, Pyrus calleryana, Pyrus bretschneideri, Pyrus pyrifolia, Pyrus pseudopashiae, Pyrus Communis, Pyrus sinkiangensis, Pyrus hopeienses, Pyrus serrulata, Pyrus ovoidea, Pyrus turcomanica, Pyrus ussuriensis, Pyrus xerophila, and Pyrus armenicefolia.* Only two species, i.e., *P. pashia* and *P. communis* were previously known from this area. The result further revealed that the landraces had sufficient variability with respect to all the parameters, except the leaf area. Mean values shows that the landraces Kushbago Batang (Kbb), Atti Bating (Ab) and Shardi Tanchi (Srt) had longer petioles with mean of 56 mm, 49 mm and 48 mm, respectively while the landraces Batangi and Glass Batang (Gb) had minimum values of 30.60 mm and 30.67 mm of petiole length, respectively. For pedicel length, landraces Batangi, Ghata Zira Tangai (Gzt), Klak Nak (Kn) Shardi Tanchi (Srt) had a minimum value of 15–20 mm whereas in Kushbago Batang (Kbb) a maximum pedicel length of 65.5 mm was recorded. For fruit length, Glass Batang (Gb) had the highest means, 95.2 mm followed by Kado Batang (Kb) and China Batang with a mean of 73.1 mm and 72.2 mm, respectively. The landraces Batangi, Gzt, and Srt fruit length was the minimum. For fruit with the landrace Cb had highest (59.3 mm) value and the landraces Batangi, Srtand Kzt had the lowest value, 23 mm, 23 mm, and 25 mm of fruit width, respectively, while landraces Kadobatang (Kb) and Glass batang (Gb) proved similar in fruit width, 50.50 mm and 50.37 mm, respectively. Fruit weight was maximum in the landraces Cb and Gb ranges from 148.0–163.7 g followed by Kb, Kbb, and Nhs while the minimum values showed by landrace Batangi, Srt, Kb, and Kzt ranges from 8.7–11.8 g. The numerical treatment of the analyzed traits concluded that the parameters like petiole length, pedicel length, fruit length, fruit width and fruit weight provides strong basis for the identification of *Pyrus* species and should be kept under special consideration in taxonomic studies. For molecular characterizations of *Pyrus* l and races a reliable protocol of DNA isolation was optimized and tested on herbarium specimens using bark, wood, and leaves was yielding 100µg/µl, 68µg/µl and 53µg/µl quantity of DNA, respectively. The obtained DNA was used both for the marker-assisted elaboration of the specimens and nucleotide sequencing of 18S RNA. PCR amplification of 36 landraces with 60 RAPD primers showed that only 28 primers were successfully generated 304 reproducible bands with band sizes ranging from 150–2600 bp. The average bands per primer were 10.85 with 100% polymorphism. Fourteen among the primers showed landrace specificity by producing 35 different size bands ranging from 150–2100 bp. Out of the 14 landraces specific primers, 8 primers showed specificity to single landraces with 1–2 loci. The primers D–16, K–09, J–05 and F–13 were specific to three different groups of landraces in the range of 4, 5 and 6 respectively. The homology tree based upon the reproducible bands sorted out all the 36 *Pyrus* landraces into 6 major groups with 62%–100% homology. The clustering showed that most of the landraces share 80%–100% phylogeny and lineage similarity with each other. Results regarding the 24 landraces analyzed for 18S rRNA. The results showed that Ktt was closely related to *P. pyrifolia* cv. Shinil whereas Gtt occupied and intermediate position between *P. pyrifolia* cvs. Nijisseiki and Okusankichi. The accession Gzt showed its close relationship with *P. pyrifolia* cv. Mansoo, Zm showed its close relationship with *P. communis* cv. Clap's Favorite. Khan Tango (Kt) was closely related to *P. pyrifolia* cv. Nijiseeki and *P. pyrifolia* cv. Okusankichi. The Pakistani Nashpati had close affinities with *P. pyrifolia* cv. Minibae. Parawoo Tango (Pt) occupied an independent position in *Pyrus* sub clade I and Pekhawry Tango showed its affinity with *P. pyrifolia* cv. Mansoo whereas the Pakistani landrace Nak Tango (Nk) showed its close resemblance with *P. communis cv.* Favorite. Asmasy Tango (At) showed close relation with *P. communis* cv. Beurre. Mamosay Batal–8 showed close relation with *P. pyrifolia* cv. Shinsui. Mamosay and Batal–12 occupied an intermediate position between *P. pyrifolia* cv. Miwang and *P. communis cv.* Conference, "Mamosay Batal–14" showed the landrace lies in between *P. pyrifolia cvs.* Shinsuiand Niitaka. Mamosay-B15 showed close relation with *P. communis* cv. Clapps-Favorite. The Pakistani landrace Gultar Tango (Gt) was similar to *P. communis cv.* Clapp's Favorite. Hary Tango-Batal (Ht) was closely related to *P. communis cv.* Clapp's Favorite. Kado Batang (Kb) showed close resemblance with *P. communis cv.* Clapp's Favorite. The Pakistani landrace of Malyzay Tango (Mt) showed close relationship with *P. communis* cv. And Mamosranga showed close relationship with *P. pyrifolia*cv. Gamcheonbae. Guraky Tango (Gkt) showed close relationship with *P. pyrifolia* cv. Nijisseiki. Shaker Batang (Sb) occupied an intermediate position in between *Pyrus pyrifolia cvs.* Nijisseiki and Okusankichi. Similarly, locally unidentified landrace Pak–24 was in between the *P. pyrifolia* cv. Nijisseiki and Okusankichi. Shaker Tango proved to be closely related with *P. communis* cultivar Pachkan's Triumph. The biosystematics treatment of all the 110 specimens collected from northern Pakistan established 12 new records to the species list of *Pyrus* from Pakistan. The study established phylogenetic relationship with the Pakistani landraces with the recorded cultivars available in different parts of the world. We conclude the landraces Ktt, At, St, and Pak–24 were hybrid in nature and their origin can easily be traced from their potential progenitors. It is also concluded that the same local name can be referred to landraces of different taxonomic or phylogenetic origins.

11.1   INTRODUCTION

Pears belong to the genus *Pyrus* (subfamily Pomoideae of family Rosaceae) with basic chromosome number x= 17 (Challice and Westwood, 1973). Most of the plants belonging to this genus are shrubs or trees. The trees are 4- 20 m tall, pyramidal in outlook, umbrella-like or narrow and straight in shape. On the bases of morphological and chemical characters, the genus is divided into four groups, the East Asian Pea Pears, the larger fruited East Asian Pears, the North African Pears and the European & West Asian Pears (Challice and Westwood, 2008). While on the basis of morphological and geographical features, the genus *Pyrus* divided into two groups, the Oriental and Occidental pears (Zhukovsky and Zeelinski, 1965). Oriental Pears are characterized by deciduous sepals, woody pedicels and subglobose fruits reported from Eastern Asia, Hindu Kush mountain range to Japan throughout Himalaya. While Occidental pears have persistent sepals, fleshy pedicels and fruit pyriform, reported from Europe, North Africa, Central Asia, Turkey, Afghanistan, Iran, and Kashmir.

The northern part of Pakistan is blessed with wide variety of scientifically unknown resources of pear and pear allies, traditionally grown in kitchen gardens and field boundaries of the subsistence farms. The scientific exploration of these resources is not only of academic interest but could also contribute to livelihood of the marginal communities. Keeping in view, both the scientific and commercial important of *Pyrus* resources scientific endeavor was undertaken for taxonomic exploration and characterization of the available biodiversity through DNA and protein technology. The Districts of northern Pakistan included in this study were Swat, Shangla, Battagram, Mansehra, and Abbottabad of Khyber Pakhtunkhwa (KP) province, Muzafar Abad, Bagh, and Palandri of Azad Jammu & Kashmir (AJK). Line sketch of the study area is given in Figure 11.1.

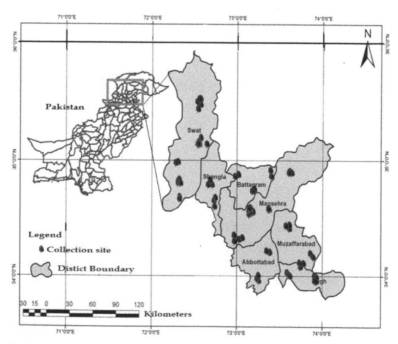

FIGURE 11.1   (See color insert.) Map of the study area showing collection sites of the *Pyrus* samples.

Pear and pear allies are of special importance in northern Pakistan. Historically, the area have been visited by a large number of invaders, visitors, traders, and pilgrims, etc. of different origins. Archeological remains, ruins, inscriptions, and petroglyphs are available everywhere in the area. Information about the famous people who remained in the area in the recent past is presented in Table 11.1. Besides cultural and other biological resources, these people had introduced a variety of crops into the area, some of which like the pears are still available in the traditional agricultural farmlands of the area.

TABLE 11.1   Historic Profile of the Northern Pakistan

S. No	Period	Duration
1	Pre-historic	40,000–4000 before present (B.P)
		40,000–10,000
		40,000–10,000 upper, lower, middle Palaeolithic
		6000–4000 Neolithic
2	Protohistoric	3280–1750 BC
3	Aryan	Middle of the 2nd millennium B.C
4	Buddhism Gandhara Civilization	Indo-Greek: 190 B.C–90 B.C
		Scythians: 90 B.C–25 A.D
		Parthians: 25 A.D–49 A.D
		Kushans:49 A.D–458 A.D
		White Huns (Epthalite): 458 A.D
5	Turk Shahi	666–822 A.D
6	Hindu Shahi	822–977 A.D
7	Islam	977 A.D

Pears are one of the most important fruit plants worldwide, cultivated in Europe and Asia for two to three thousand years. They are commercially cultivated in temperate regions of more than 50 countries in the world (Bell, 1990; Bell et al., 1996). According to (Cuizhi and Spongberg, 2003), there are 25 species in Asia while 14 species in China including 8 species endemic species. Fourteen (14) landraces belonging to the genus *Pyrus* has been reported from northern Pakistan (Islam, 2012). On the basis of morphological characteristics, 14 species of *Pyrus* had been identified and reported from northern Pakistan and AJK (Islam, 2014). The genus *Pyrus* contains at least 22–26 widely recognized primary species, all indigenous to Europe, Asia, and the mountainous area of North America (Bailey, 1917). Among *Pyrus species*, only a few species have been domesticated for commercial production (Bailey, 1917; Bell and Itai, 2011). On the bases of geographical distribution and morphological characters, most cultivated pears are native to East Asia (Teng and Tanabe, 2002). In China *Pyrus pashia* Buchanan-Hamilton ex D. Don, Prodr, *Pyrus bretschneideri* Rehder, *Pyrus pyrifolia* (N. L. Burman) Nakai, *Pyrus betulifolia* Bunge, *Pyrus phaeocarpa* Rehder, *Pyrus calleryana* Decaisne, *Pyrus communis* L., *Pyrus serrulata* Rehder, *Pyrus sinkiangensis* T. T. Yu, *Pyrus armeniacifolia* T. T. Yu, *Pyrus pseudopashia* T. T. Yu, and *Pyrus xerophila* T. T. Yu are very common (Burman, 1926).

11.2   THE GENUS *PYRUS*

Pyrus L. Sp. Pl. 1: 479. 1753; *Hook, Fl. Brit. Ind.* 2: 387, 1878; Rehder, *Proc. Am. Acad. Arts. Sci.* 50: 223, 1915; Terpo and Franco, *Fl. Europaea* 2: 65, 1968; Schonbeck-Temesy, *Fl. Iran.* 66: 27, 1969; Malvee, *Fl. of U.S.S.R.* 9: 259, 1985; Cuizhi, *Fl. China* 9: 173, 2003.

The term *Pyrus* is derived from a Latin root referring to pears. Pears are mostly shrubs or trees; trees may be 4- 20 m height, pyramidal, umbrella or cylindrical in shape. The members of the genus may either be deciduous or semi-evergreen, sometimes thorny at the lower parts in young stage. Young leaves are mostly tomentose or pubescent which become glabrous at later stages. Leaves are simple, petiolate, stipulate, base ovate-elliptic or ovate orbicular, rounded or elliptic. Leafmarginare serrate, serrulate, crenulated, dentate to entire. Dorsal surfaces of leaves are generally dark green, ventral surfaces are light green. Inflorescence corymbose raceme, flower appearing either later or before leaves and born on a short lateral spurs. Hypanthum copular, bell shape or somewhat flattened. There are 5 sepals, green, mostly triangular, sometimes hooded, spread at mature flowers, forming crown at fruit, apex acute–acuminate, margin glandular or dentate, mostly tomentose inside and slightly outside, sometimes glabrous, free or united at base, persistent or deciduous in fruit and villous on both sides. Petals white or pinkish in bud condition, sub-sessile, narrow at the base, rounded, broad oblong or lobed at apex. Stamens 15–35, filaments usually white, anthers radish to pinkish. Styles 2–5, free, united together with the hypanthum. Ovary inferior, 2–5 loculed, 2 ovules per locule. Fruitisa pome, succulent, usually pyriform, subglobose, globose, turbinate, depressed globose, or oval, flesh may be with sclerenchymatous cells, rich in stone cells, 2–5 celled, with cartilaginous endocarp. Seeds are black or nearly so, testa cartilaginous, cotyledons plano-convex. Flowering time ranges from March, 20 to April

30, depending on altitudinal variation while fruits maturation time ranges from July to September, varies from species to species, variety to variety and altitudinal variation too.

Basic chromosome no (X) = 17; some triploid and tetraploids; most species require cross pollination due to self-incompatibility. From North Africa, Asia, Europe about 25 species are reported. On the basis of morphological and geographical features, Zhukovsky, and Zeelinski, 1965, divided the genus *Pyrus into* two groups, the Oriental, and Occidental Pears.

11.2.1   ORIENTAL PEARS

Oriental Pears are characterized by deciduous sepals, woody pedicels and subglobose fruits reported from Eastern Asia, Hindu Kush mountain range to Japan through Himalaya.

11.2.2   OCCIDENTAL PEARS

Occidental pears has persistent sepals, pedicels fleshy and fruit pyriform, reported from Europe, North Africa, Central Asia, Turkey, Afghanistan, Iran, and Kashmir.

11.3   ORIGIN AND DISTRIBUTION

It is generally accepted opinion that pears originated in the Caucasus and spread to Europe and Asia. They were first cultivated more than 4000 years ago. Different species belonging to the genus *Pyrus* are present in most of the countries in the world. In the Corvallis (Oregon USA) gene bank maintains 2031 clonal *Pyrus* accessions, 327 seedlots of 36 taxa from 53 countries. The clonal accessions include 844 European cultivars, 144 Asian cultivars, 87 hybrid cultivars and 159 rootstock selections of *Pyrus* species are assorted (Postman, 2008).

In Britain, its cultivation started during Roman occupation and by the 13[th]century many varieties of pears had been imported from France and the fruit was used mainly for cooking rather than eating raw. At the end of the 14th century, the Warden pear had been bred and became famous for its use in pies.

By 1640, at least 64 varieties of pears were under cultivation in England. In the 18th century, new and improved strains were introduced. However, the majority of pears continued to be used for cooking. Dessert pears were grown mainly in private gardens but were unsuitable for commercial cultivation. One exception was the William's Pear, which became very popular by 1770 and is still produced on a limited scale http://www.usapears.com. It is reported that the Royal Horticultural Society encouraged pear growing and in 1826 there were 622 varieties in their gardens at Chiswick. The first significant English Pear to be produced by controlled breeding was Fertility in 1875, although this variety is no longer produced commercially. In early 19th century, the renowned horticulturist Thomas Andrew Knight began to develop pear varieties. It has been reported that by the middle of the 20th century the scale and production of Comice cultivar decreased while that of Conference increased both in scale and production and now a day's represents more than 90% commercial production in UK http://www.englishapplesandpears.co.uk/.

Regarding the origin of Japanese pear cultivars, it is generally accepted that, this group of cultivars has been domesticated from wild *P. pyrifolia* occurring in Japan (Kikuchi, 1948). The genus *Pyrus* is believed to have arisen during the Tertiary period in the mountainous regions of western China. Dispersal and speciation is believed to have followed the mountain chains both east and west (Yamamoto and Chevreau, 2009). European pear cultivars was well established in Greece and cultivars with distinct names were propagated as early as 300 B.C. Oriental pears, which arose independently, were also grown in China for more than 2000 years (Kikuchi, 1946). Pears have been cultivated in Europe since as early as 1000 B. C., when Homer wrote of the garden of Alcinous (Hedrick et al., 1921). *Pyrus sinkangenses* is a natural hybrid between *Pyrus communis* and *P. bretschneideri* while *Pyrus pashia, Pyrus betulifolia, Pyrus phaeocarpa, Pyrus calleryana, Pyrus xerophila* are often used as stock for grafting pear cultivars and *Pyrus bretschneideri, Pyrus pyrifolia, Pyrus communis, Pyrus serrulata, Pyrus sinkiangensis, Pyrus armeniacifolia, Pyrus pseudopashia* are produces excellent fruits (Cuizhi and Spongberg, 2003).

11.4 CYTOGENETICS

Mitotic and meiotic analysis in *Pyrus* species has revealed that all the species are diploid, with 2n = 34, though several cultivars, variants are however triploid in nature. Speciation within the genus *Pyrus*, as in other genera of the Pomoideae, has occurred without a change in chromosome number (Zielinski and Thompson, 1967). Majority of cultivated apple and pear are diploid (2n = 2x = 34). The whole tribe of Maloideae is characterized by a basic chromosomes number (x = 17), compared to the three other tribes of family Rosacéae, where basic chromosome number is 7, 8, 9 for Rosoideae, Prunoideae, and Spiraeoideae, respectively. This number of 17 chromosomes has stress discussion about the genetic origin of Maloideae. Several hypotheses were put forward on auto or allopolyploid origin of Maloideae. Apple and pear chromosomes are difficult to distinguish because of their small size from 0.5 to 1.5μm (Lespinasse and Salesses, 1973). The result shows that *Pyrus armeniacaefolia* is triploid, *P. ussuriensis* has diploid and triploid cultivars; *P. bretschneideri* has diploid, triploid, and tetraploid cultivars. Diploid and triploid cultivars of *P. pyrifolia* also exist. Diploid, triploid, and tetraploid cultivars exist in *P. sinkiangensis; P. phaeocarpa, P. betulaefolia, P. hopeiensis, P. serrulata, P. pashia, P. xerophilus* and *P. calleryana.* The *Pyrus* in China provides a change in chromosome number in the evolutionary process. It will enable to enrich the polyploid composition of *Pyrus* in the world and may be used in the production, genetics, and for the breeding purpose in pear (Shenghua and Chengquan, 1993).

11.5   TRADE AND COMMERCE

In 2005–2006, the World pear exports were 1.6 million tons, up 6% from last year. Despite a slight decline in pear exports from Argentina and China's, 20% increase in export levels. In the Northern Hemisphere, exports were increased 12%, supported by growth in most reporting countries. Export volume from the US for 2005/06 was decreased 6% with lower production level.

In 2010, the share of pear production with respect to Argentina, Australia, Austria, Belgium, Brazil, Canada, Chile, China, France, Germany, Greece, Hungry, India, Iran, Italy, Japan, Mexico, Netherland, New Zeeland, Poland, Portugal, Romania, Russia, South Africa, Spain, Turkey, Ukraine, United Kingdom, United States and Uruguay was 704200, 95111,48400, 260000, 16367, 7830, 180000, 15231858, 173746, 38895, 93600, 24176, 382000, 160000, 736646, 284900, 24986, 274000, 25100, 57514, 176900, 60375, 41000, 366216, 473400, 380003, 141700, 32800, 732642, and 18072 metric tons, respectively, while the net production was 22,644,756 metric tons. Similarly, in 2011 and 2012 the net production of pears of Europion countries was 2.629 x1000 and 2.060 x 1000 metric tons, respectively.

China produce more Asian pear as compared to other countries in the world, having an annual production is 8 million tons and the area under cultivation is 0.94 million hectares. The prominent cultivars are mainly produced from *Pyrus bretschneideri* Rehd., *P. ussuriensis* Maxim. and *P. serotina* Rehd and many imported from Japan (Jun and Hongsheng, 2001).

In China, the pear production during 2005–06 was 11.2 million metric tons. China produced 10.6 million MT of pears in 2004–05. In China, pear production increasing due to planting area and improved tree management. According to "World Pear Export Production," the share of Argentina, China, Belgium, Netherland, USA, South Africa, Chile, and other countries 18%, 17%, 13%, 12%, 10%, 8%, 7%, 16%, respectively, to the world export. In Pakistan, the total area under cultivation was 2115 ha, 97 ha, 44 ha in Khyber Pakhtunkhwa, Baluchistan, and Punjab, respectively while the net production was 28343 tones, in which 27596, 431 and 316 tons produced by Khyber Pakhtunkhwa, Baluchistan, and Punjab, respectively (Anonymous, 2010) and only Shawar Valley, Swat- K.P, like other fruit cops, pears contributed about 10412 boxes to market and the net income of the valley during 2004 was Rs. 0.72 million (Hussain et al., 2006; Islam, 2010).

11.6   MOLECULAR BIOLOGY

Molecular genetics analyzes can resolve many aspects of species that are critical in conservation (Frankham et al., 2004). However, the so-called wild types are only found near human habitation and supposed to be escapes (Shimura, 1988). In addition, genetics resources have not been fully identified due the low morphological diversity, lack of prominent differentiating characteristics among species and widespread cross-ability. It is unfortunate that a large number of pear cultivars available in the conventional farms, is neither expressed taxonomically nor genetically tagged, due the low morphological diversity, lack of prominent differentiating characteristics among species and widespread cross-ability. Classical morphological characters are very useful in taxonomy but sometimes these characters are very poor or influenced by environmental conditions which do not fulfill the purpose of genotypes identification (Bailey, 1917). Morphological characterization is the first step in the description and characterization and classification of plants germplasm (Smith and Smith, 1989; Singh and Tripathi, 1985). Morphological characteristics and isozyme analysis have been the major approaches to assess the genetic variation within *Pyrus spp.* or cultivated Pears, (Chevreau et al., 1985; Jang et al., 1991). Nevertheless, their exact variation still remains unclear because of isozyme markers and morphological characteristics are limited in number. Identification of *Pyrus* is mainly carried out through morphological and geographical distribution but identification to species level is difficult work because 1) very poor wild population 2) very poor morphological characteristics and diversity 3) widespread cross-ability and interspecific hybridization and introgression among different species. Therefore presently many DNA markers have been used for Taxonomical relationships and evolution in *Pyrus* (Yamamoto and Chevreau, 2009).

Estimation of genetic diversity among *Pyrus spp.*, however, is often difficult because it is considered that there are at least 9 interspecific hybrids (Bell et al., 1996). Estimation of genetic diversity and relationship between germplasm collections are very important for facilitating efficient germplasm collection, evaluation, and utilization. Many tools are now available for identifying desirable variation in the germplasm including total seed protein, isozymes, and various types of DNA markers.

During last few years, many attempts using different research methods, especially DNA markers have been tried to solve the disputing and the origin of pear cultivars native to East Asia. Several different types of DNA markers have been successfully applied for *Pyrus* cultivars identification and the analysis of genetic relationship (Kim et al., 1997; Kimura et al., 2003). The identification and localization of red color trait on morphological marker in the linkage maps 4 of Max Red Bartlef for the first time has been mapped out of linkage group–9 in a species of Maloideae (Dondini et al., 2008). For this purposes, we need to isolate high-quality DNA whether from leaves, bark, and wood in the normal season and even in the off-season for molecular characterization of Pear to identify genotypes.

11.7 DNA TECHNOLOGY

Before starting the detailed biosystematics endeavor of pears, it was imperative to optimize protocol for its DNA isolation of plant parts available in different parts of the year, which can be used for genetic characterization of pears, which have not been identified due to low morphological diversity and overlapping traits (Jang et al., 1991; Chevreau et al., 1997; Frankham et al., 2004). The recent application of molecular biology have revolutionized the use of DNA for better understanding of genomestructure, evolution, and identification of species, which needs high-quality genomic DNA (Kim et al., 1997; Kimura et al., 2003). The isolation of high-quality genomic DNA is essential for many molecular biological applications such as PCR (Chakraborti et al., 2006) but DNA isolation from mature trees of high altitude is difficult job due to the presence of large number of phenolic compounds and polysaccharids etc. (Gupta et al., 2011).

Various protocols have been reported for isolation of high yielding quality DNA (Saghai-Maroof et al., 1984; Doyle, 1991) each having their own limitations and scope. The longer processing time, need of expensive equipment, costly kits/chemicals and low-quality yield generally hampers their application in fruit trees (Adams and Graver, 1991; John, 1992; Kim et al., 1997; Shepherd et al., 2002; Aganga and Tshwenyane, 2003; Varma et al., 2007). The low-quality yield is due to variety of chemical constituents in trees (Scott and Playford, 1996; Shepherd and McLay, 2011) which needs modification in the protocol accordingly (Barzegari et al., 2010; Li et al., 2010; Smyth et al., 2010).

First of all DNA isolation was developed by (Saghai-Maroof et al., 1984) when the isolation of nuclei were required, from a lengthy, expensive, and low yielding cesium chloride-ethedium bromide ultracentrifugation procedure which obtained only degraded DNA from soybean leaves. Doyl and Doyl (1987) applied exactly the same procedure of Saghai-Maroof et al. (1984) but failed with soybean leaves. So they modified the said procedure by doubling the concentration of extraction buffer and the modified protocol work beautifully. This procedure was published in different time (Taylor and Powell, 1982; Doyle and Doyle, 1987; Doyle and Dickson, 1987; Doyle, 1991). Similarly, a modified CTAB was adopted for extraction of high-quality DNA which is suitable for RAPD. In such procedure, high salt concentration was used for removal of polysaccharide and PVP for polyphenolic compounds (Porebski et al., 1997). DNA degradation is mediated by secondary plant products such as phenolic terpenoids bind to DNA after cell lysis (John, 1992). The isolation of High-quality DNA from plants containing a high content of polyphenolics such as pear, apples, grape, and conifers was a difficult problem (John, 1992; Pich and Schubert, 1993; Kim et al., 1997) tried to isolate DNA from materials containing highly polyphenolic compounds by modifying several existing methods. (Kimura et al., 2003) isolated DNA from dry root of *Berberis lyceum* by modifying CTAB procedure using 1% PVP to remove polysaccharides and purification using low melting temperature agarose which work beautifully for RAPD.

11.7.1   RANDOMLY AMPLIFIED POLYMORPHIC DNA

RAPD stands for random amplification of polymorphic DNA. It is a PCR reaction in which segments of DNA are amplified randomly by using many arbitrary, short primers (8–10 nucleotides), hoping that fragment will amplify. In RAPD, primers will bind somewhere in the sequence but not sure exactly where and no knowledge of sequence for the genomic DNA is required (Oliveira et al., 1999; Teng and Tanabe, 2002) (Srivastava and Mishra, 2009). Presently, RAPD has been used to evaluate, trace, and determine the phylogenetic relationship of plants and animals species. RAPD markers are 10bp DNA fragments and able to differentiate genetically distinct individuals.

Many molecular markers are uses for genetic diversity, DNA fingerprinting and plant identification, etc. Among these, random amplified polymorphic DNA, RAPD generate DNA fingerprints with a single oligonucleotide primer. RAPD are dominant markers and inherited in a simple Mendlian fashion, less expensive, faster, and require a small amount of DNA, reproducible, require less skill for operation and do not need radioisotopes so therefore RAPD have proven to be used in genotype identification, gene mapping and have wide application in developing countries (Demeke and Adams, 1994), used for taxonomic identification of *Allium*, cauliflower, cabbage, and *Brassica* at cultivar and species level (Wilkie et al., 1993), used for authentication and identification of different species of medicinal valued plants (Monte-Corvo et al., 2000; Arif et al., 2010; Khan et al., 2010). (Monte-Corvo et al., 2000) used RAPD, AFLP for genetic similarity of *Pyrus* cultivars and varieties of *Pyrus* (Botta et al., 1997). Similarly, different cultivars of *P. pyrifolia* and *P. communis* were identified by RAPD and sequenced characterized amplified region (SCAR) primers (Lee et al., 2004), RAPD is convenient in performance and does not require any information about the DNA sequence to be amplified. Molecular and phenotypic characterization among *P. communis*, *P. pyrifolia*, *P. cordata*, *P. bourgaeana*, and *P. pyraster* were evaluated through RAPD markers and it was pointed out that some primers are genotype specific which can be used for cultivar identification (Oliveira et al., 1999). Similarly, RAPD markers were also used for classification and identification of *P. pyrifolia* and *P. communis* cultivars (Lee et al., 2004). RAPD marker was linked to major genes that controlling skin color in Japanese pear, important with respect to market value and external pressures. The RAPD marker (OPH–19$_{425}$) was specific to green fruit color genotypes with 92% probability, useful in breeding program (Inoue et al., 2006).

A genetic diversity of 56 accessions, 8 reference cultivars belong to the genus *Pyrus* and 12 microsatellite markers, among 12 primers, 9 primer pairs revealed 106 putative alleles that ranged from 7 to 19, average value was 11.8 alleles per locus. Coefficients varies from 0.00 to 1.00 was observed in KT53

(Btung) and between BG21 and MZ26 andKT53 (Btung), representing the highest genetic diversity among all genotypes (Ahmed et al., 2010). *Pyrus ussuriensis var. aromatica* is a wild *Pyrus* species of Japan was studied for conservation and evaluation. Over 500 accessions of *Pyrus* species, five SSR markers, were examined for 86 *Pyrus* individuals including 58 accessions from Iwate. Due to high allelic frequency, Iwate accessions were genetically more different than the Japanese pear varieties. Most Japanese pears possessed a 219 bp deletion at a spacer region between the *accD* and *psaI* genes in the chloroplast DNA (cpDNA), but other *Pyrus* species and other did not. A combined analysis of SSR and cpDNA showed high genetic diversity in Iwateyamanashi and coexistence of Iwateyamanashi and hybrid progeny with *P. pyrifolia* (Katayama et al., 2007). Seven SSRs markers derived from apple were successfully transferred to 25 local Tunisian pear genotypes and 6 common varieties of *Pyrus communis* for low chilling requirements and adaptation to dry conditions. All the microsatellites except one are amplify more than one locus in some of the genotypes (Brini et al., 2008). Similarly the genetic diversity and relationship of *Pyrus* cultivars for 168 putative alleles studied through SSR markers which were generated from six primer-pairs, markers showed a high level of genetic polymorphism with a mean of 28 putative alleles per locus and the heterozygosity of 0.63 while the Dice's similarity coefficient between cultivars ranged from 0.02 to 0.98 and Occidental pears generally had low affinities to Asian pears. Similarly Chinese white pears as a variety or an ecotype of Chinese sand pears (*P. pyrifolia var. sinensis*) and the progenitor of Japanese pears coming from China (Bao et al., 2007). A seven microsatellite loci (SSR) developed in apple, used in the identification of 63 European pear and a total of 46 fragments were amplified and its average was 6.6 alleles per SSR. A single microsatellite amplified more than one locus. The He and H$o$heterozygosities over the six single-locus SSRs averaged 0.68 and 0.44, respectively, and the number of effective alleles per loci was 3.43 (Wünsch and Hormaza, 2007).

Apple and pear have complex genetic constitution with respect to reproductive cycle and total self-sterility is a complicating factor in the genetic knowledge of *Pyrus* species. Most agronomic features show in lineage a continued variation allowing a polygenic heredity conclusion (Janick and Moore, 1996). Observed segregation on several lineage shows a bigenic disomic heredity as well as fixed heterozygosity. These species should have been considered secondary polyploid with disomic behavior (Challice, 1981; Chevreau et al., 1985; Chevreau and Laurens, 1987). Perennial fruit trees such pear require a long-term effort for breeding because of their long generation time and high level of heterozygosity (Bouvier et al., 2002). Fragments of *copia*-like retrotransposons were obtained from pear, peach, and apple and fifty-one non-redundant sequences derived from Japanese pear were classified into 15 groups by 80% nucleotide identity. Phylogenetic survey revealed a high degree of heterogeneity among the groups. Southern bloting demonstrated that several types of retrotransposon-like sequences existed in the genomes of *Pyrus* species and polymorphisms were detected among *Pyrus* species as well as within the species. Retrotransposons contributes to the understanding of the genome structures and the principles of mutation in pear as well as other fruit tree species (Shi et al., 2001).

11.7.2 SMALL SUBUNIT RRNA SEQUENCES

Ribosomal ribonucleic acid (rRNA) is the ribonucleic acid (RNA) component of ribosome, play an important role in protein synthesis. It is a major component of ribosome, 60% rRNA and 40% protein. Ribosomes contain two major rRNAs, large subunit (LSU) and small subunit (SSU), present in large and small ribosomal subunits, respectively. In eukaryotes, total size of ribosome is 80S, large subunit (LSU) consist 60S (5S, 5.8S nt, 256, 28S: 5070 nt sequences (Wuyts et al., 2001) and small subunit (SSU) consist of 40S (18S: 1869 nt sequences) (Wuyts et al., 2004). Eukaryotes possess many copies of RNA genes, organized in tandem repeats and adopted special structure and transcription behavior. The rRNA genes clusters are commonly known as ribosomal DNA but the actual DNA is not present in them. Ribosomal rRNAs are ancient in origin and present in all forms of life. Therefore, in phylogenetic relationship and taxonomic identification, its nucleotides sequences are excessively used. Ribosomal RNA (rRNA) is the result of few gene product which is available in types of cells (Smit et al., 2007) and the genes that encode the ribosomal RNA (rRNA) are sequenced for taxonomy, calculation of related groups and for the estimation of species divergence rates. Therefore, thousands of rRNA sequences are available in different databases such as Ribosomal Database project-II (RDP) and SILVA (Cole et al., 2003; Pruesse et al., 2007).

A revolution is occurring in biological Sciences on the bases of rapid sequencing of nucleic acids which is basically transformed by different molecular approaches and concepts and its great impact on genetics, medicine, and biotechnology and ever-lasting effect on evolution and relationship among organisms. The cell is basically an historical document, getting the capacity to read it through gene sequencing (Woese, 1987).

The traditional classification of plants into classes, order, family, genus, and species is based on morphological, cytological, biochemical, and ecological characteristics. The invention and advancement of molecular techniques in molecular biology such as molecular hybridization, cloning, nucleotides, and proteins sequencing, and restriction endonucleas digestion have provided new tools for investigation and phylogentics relationship of plants, animals, and microbes. The most possible comparisons of primary nucleotides sequences of homologous genes in different populations or species in molecular biology.

The strategies adopted for comparative based identification are the new window for prokaryotic and eukaryotic organisms (Summerbell et al., 2005). In this method, a target region of the genomic DNA is amplified through polymerase Chain Reaction (PCR), isolating the amplicon and sequenced and sequencing the selected amplicons. After this, developing a consensus sequence through database library, software, and species are identified by generating dendrogram, evaluating presence of similarity and absent of dissimilarity or through advanced phylogenetic analysis. In interpreting sequence comparison data by creating a % identity score, single numeric score determined for each aligned pair of sequences and the number of similar matched nucleotide in relation to the length of the alignment. A point of termination scores are arbitrary in species identification and the score can vary depending on many factors, quality of the sequence, number, and authenticity of existing database library records from the same species and locus, fragment length of the sequence and the software program that use researcher for analysis.

Identification of different species depends on the success of comparative sequencing strategy in the choice of target gene or fragment of DNA. The target gene should be evolved by a common descent (orthologous), high level of interspecific variation combined with low levels variation on intraspecific bases and should not consider recombination. Except these factors, the target gene must be easily amplify and a standard universal primer sets. The size of the amplicon should be within the range of DNA sequencers using for sequencing and capable of easily alignment with a sequence database for comparison.

The genome of all living organisms consist of DNA sequences that code for ribosomal RNAs which is an essential components during cellular proteins synthesis. In plants, ribosomal DNA (rDNA) is present in chloroplast, nuclear, and mitochondrial genomes. The presence of rRNA throughout in nature and development of techniques for the rapid determination of primary nucleotides sequences rRNA molecules make rRNA a good tool for leading conclusion to evolutionary relationship (Hamby and Zimmer, 1992). In flowering plants, determination of relationship in different species is difficult task because of poor, less prominent morphological characteristics and a small number of molecular markers to be useful for phylogenetic relationship in angiosperms. Therefore the use of small subunit 18S rRNA sequences techniques solves the above problems, It has been reported 18S rRNA sequences from 3 representative genera of three families i.e., Olacaceae, Santalaceae, and Viscaceae within the Santalales and 6 dicot outgroup families i.e., Celastraceae, Comaceae, Nyssaceae, Buxaceae, Apiaceae, and Araliaceae and observed that Santalales was holophyletic taxon, most closely related to Euronymus (Celastraceae) and the Santalales has shown approximately 13% transitional mutations than the other 7 seven group belonging to dicotyledinous species. Ro et al. (1997) evaluated 31 and 4 species belonging to family Ranunculaceae and Berberidaceae, respectively to demonstrate phylogentic relationships through 26S rDNA sequences. The result showed that the phylogeny strongly support the concept of *Thalictrum* chromosome group is not monophyletic but comprises of three independent lineages, i.e., Hydrastis, Xanthorhizaand, and *Thalictrum, Aquilegia*, and *Enemion*. The results of 26S rDNA topology were compared with results from two previously published DNA sequences of *rbc* L, *atp* B, 18S rDNA genes and RFLP data of cpDNA. The 3 topologies are highly similar and match with karyological characters. In eukaryotes, small subunit rRNA sequences open new window with respect to evolution. With regard to evolution, animals, plants, and fungi form monophyletic groups which seem be originated at the same time, approximately. While in contrast to eukaryotes, the dissimilarities in small ribosomal subunit RNA sequences among protoctists is large and exceed in prokaryotes and the bases of r RNA data, Protoctista branch off very soon in eukaryotic evolution while other diverge later and protoctista consider to be a collection of independent evolutionary lineages. Generally, eukaryotes are recently diverged lineage while on rRNA sequences they are seem to be as ancient as prokaryotes (van de Peer et al., 1993). An evolutionary distances measured from comparison of sequences of small subunit (16S rRNA) of *Giardia lamblia* and other eukaryotes, showing similar measurements of evolutionary diversity between archae bacteria and eubacteria and ensure the phylogenetic significance of multiple eukaryotic kingdoms. The *Giardia lamblia*16S like RNA possess many of the properties that might have been occurred in the common ancestor of eukaryotic and prokaryotic organisms (Sogin et al., 1989). The gene sequences of nuclear encoded small-subunit 18S rRNA were determined for the *Mantoniellasquanata, Charafoetidaetc, and Mougeotiascalaris, Marchantiapolymorpha, Fossombroniapusilla*, and

Funariahygrometrica and *Selaginellagalleottii* for better understanding of sequential evolution from green algae to land Plants. After sequential alignments and conclusion of maximum parsimony and maximum likelihood analysis, Charophyceae was continuously placed on the branch going to the land plants (Kranz et al., 1995). Similarly a coding region sequences of SSU, 16S-like rRNA were determined for 8 species belonging to Chlorophyceae. After constructing phylogenetic tree, showed evolutionary relationships between several plant and green algae (Huss and Sogin, 1990). The Ginseng drugs were identified through PCR-RFLP and mutant allele specific amplification (MASA) on the differences bases of 18S rRNA gene sequence among three *Panax* species. The 18S ribosomal RNA gene were amplified and digested with restriction enzymes Ban-II and Dde-I. Each piece give unique electrophrotic profiles for each species (PCR-RFLP analysis). The genomic DNA of each species were amplified with specific primer and the expected size of the fragments of corresponding species were determined with help of PCR conditions. With the help of these two molecular techniques, Ginseng drugs were identified. To insure identification, partial sequence of plastid gene (matK) was determined in addition to the 18S ribosomal RNA gene (Fushimi et al., 1997).

First time DNA sequences obtained from arbuscularendomycorrhizal fungi are reported and these sequences were achieved by direct sequencing overlapping amplified fragments of the nuclear genes coding for SSURNAa. The said sequences were used to develop PCR-based primer which amplifies a portion of the vesicular- arbuscularendomycorrhizal fungus SSUrRNA directly from a mixture of plant and fungal tissues (Simon et al., 1992).

In biodiversity and taxonomic research, DNA bar-coding provide and effective techniques, tools for species identifications and the sequences of target genes are collected and will provide a horizontal genomic view with vast implications, i.e., comparing the goals and methods of DNA bar-coding with population genetics and molecular phylogeny and assumed that DNA bar-coding can complement in the above areas and provided basic information in the identification of species for further evaluation (Hajibabaei et al., 2007). Identification of species on morphological parameters/characteristics is very difficult task for non-taxonomists and taxonomist too because of poor morphological traits, environmental effects, and anthropogenic activities. Therefore DNA bar-coding, DNA taxonomy or molecular taxonomy in combination with morphological data generates an integrated taxonomy. DNA bar-coding generate a universal molecular identification key on the bases of taxonomic knowledge that is assembled in logical sequences in reference library. The DNA bar-coding extensively strengthen the field of molecular identification (Teletchea, 2010).

The molecular trees have the capacity to enclose both minimum and maximum periods of time based on the observation that rate of evolution of genes are different and the DNA specifying ribosomal RNA (rRNA) changes relatively slow as compared to compared to mitochondrial DNA (mtDNA) that evolve relatively rapidly. Therefore rRNA is useful for investigating relationships between taxa that diverged hundreds of millions of years ago. The procedure adopted in DNA bar-coding (a unique sequence of nucleotides that characterized by species) is simple. Sequences of the bar-coding region are obtained from different individuals and the resulting sequence data are then used to construct a phylogenetic tree using a distance-based 'neighbor-joining' method. In such a tree, similar, putatively related individuals are clustered together and each cluster is assumed to represent a separate species (Hebert et al., 2003).

11.8   SCOPE OF THE STUDY

Updated knowledge of genetic diversity is a prerequisite for the improvement of crop plants, which can be elaborated through using morphological markers. Usually limited in number, overlapping in expression and generally influenced by environmental changes. Unfortunately, no serious effort has been made with reference to important of pears and their allies in Pakistan. Pears which are traditionally cultivated in temperate parts of the country have been introduced and highly adopted in traditional farming systems of northern parts of Pakistan. During the present research, recently developed molecular techniques were used to study genome structure and to document amount of existing genetic variability in pear genotypes commonly grown in Northern parts of Pakistan. Techniques for the isolation of total genomic DNA from selected genotypes were developed/optimized. Genetic variation, in terms of Genetic Distance, present in various taxons of pears was studied at the DNA level through Polymerase Chain Reaction using Randomly Amplified Polymorphic DNA (RAPD) primers and 18S rRNA. DNA profiles were statistically analyzed to find out phylogenetic relationship among the landraces using cluster analysis and dendrogram. Results of the present research will be helpful for developing better strategies for exploration of plant natural resources with respect to pears, improvement of pear genotypes in Pakistan and their conservation. Successful completion of the project will result tagging the pear genotypes in Pakistan and on the other hand it will contribute more to collection of pear germplasm and pear allies.

KEYWORDS

- **genetic diversity**
- **molecular markers**
- **pears**
- *Pyrus*

REFERENCES

Adams, R. P., Graver, T., (1991). A simple technique for removing plant polysaccharide contaminants from DNA. *BioTechniques*, *10* (2), 162–164.

Aganga, A., & Tshwenyane, S., (2003). Feeding values and anti-nutritive factors of forage tree legumes. *Pak. J. Nutr.*, *2* (3), 170–177.

Ahmed, M., Anjum, M. A., Khan, M. Q., Ahmed, M., & Pearce, S., (2010). Evaluation of genetic diversity in *Pyrus* germplasm native to Azad Jammu and Kashmir (Northern Pakistan) revealed by microsatellite markers. *Afri. J. Biotechnol.*, *9* (49), 8323–8333.

Anonymous, (2010). *Agricultural Statistics, Finance Division*, Islamabad, Government of Pakistan. Annual. Islamabad.

Arif, I., Bakir, M., Khan, H., Al Farhan, A., Al Homaidan, A., Bahkali, A., et al. (2010). Application of RAPD for molecular characterization of plant species of medicinal value from an arid environment. *Genet. Mol. Rese.*, *9* (4), 2191–9198.

Bailey, L., (1917). *Pyrus. Stand. Cyclop. Horti.*, *5,* 2865–2878.

Bao, L., Chen, K., Zhang, D., Cao, Y., Yamamoto, T., & Teng, Y., (2007). Genetic diversity and similarity of pear (*Pyrus* L.) cultivars native to East Asia revealed by SSR (simple sequence repeat) markers. *Genet. Res. Crop Evol.*, *54* (5), 959–971.

Barzegari, A., Vahed, S. Z., Atashpaz, S., Khani, S., & Omidi, Y., (2010). Rapid and simple methodology for isolation of high-quality genomic DNA from coniferous tissues (*Taxus baccata*). *Molecul. Biol. Rep.*, *37* (2), 833–837.

Bell, R. L., & Itai, A., (2011). *Pyrus*. Wild Crop Relatives: *Pyrus. " Wild Crop Relatives: Genomic and Breeding Resources* (pp. 147–177.). Springer Berlin Heidelberg.

Bell, R., (1991). Pears (*Pyrus*). *Genetic Resources of Temperate Fruit and Nut Crops*, *290*, 655–697.

Bell, R., Quamme, H., Layne, R., & Skirvin, R., (1996). Pears. *Fruit Breed.*, *1,* 441–514.

Botta, R., Akkak, A., Me, G., Radicati, L., & Casavecchia, V., (1997). *Identification of Pear Cultivars by Molecular Markers*. Symposium on Plant Biotechnology as a tool for the Exploitation of Mountain Lands, *457*.

Bouvier, L., Guerif, P., Djulbic, M., & Lespinasse, Y., (2002). First doubled haploid plants of pear *(Pyrus communis)*. *Acta Hort., 1,* 173–176.

Brini, W., Mars, M., & Hormaza, J., (2008). Genetic diversity in local Tunisian pears *(Pyrus communis* L.) studied with SSR markers. *Sci. Hort., 115* (4), 337–341.

Burman, N. L., (1926). *Pyrus pyrifolia* (N. L. Burman) Nakai. *Bot. Mag., 40,* 564.

Chakraborti, D., Sarkar, A., Gupta, S., & Das, S., (2006). Small and large scale genomic DNA isolation protocol for chickpea *(Cicer arietinum* L.), suitable for molecular marker and transgenic analyzes. *Afr. J. Biotechnol., 5,* 585–589.

Challice, J., & Westwood, M., (1973). Numerical taxonomic studies of the genus *Pyrus* using both chemical and botanical characters. *Bot. J. Linn. Soci., 67* (2), 121–148.

Challice, J., (1981). Chemotaxonomic studies in the family Rosaceae and the evolutionary origins of the subfamily Maloideae. *Preslia, 53,* 289–304.

Chevreau, E., & Laurens, F., (1987). The pattern of inheritance in apple *(Malus x domestica* Borkh.): further results from leaf isozyme analysis. *Theor. Appl. Genet., 75* (1), 90–95.

Chevreau, E., Lespinasse, Y., & Gallet, M., (1985). Inheritance of pollen enzymes and polyploid origin of apple *(Malus domestica). Theor. Appl. Genet., 71* (2), 268–277.

Chevreau, E., Leuliette, S., & Gallet, M., (1997). Inheritance and linkage of isozyme loci in pear *(Pyrus communis* L.). *Theor. Appl. Genet., 94* (3), 498–506.

Cole, J. R., Chai, B., Marsh, T. L., Farris, R. J., Wang, Q., Kulam, S., et al. (2003). The Ribosomal Database Project (RDP-II): Previewing a new autoaligner that allows regular updates and the new prokaryotic taxonomy. *Nucl. Acids Res., 31* (1), 442–443.

Cuizhi, G., & Spongberg, S., (2003). *Pyrus. Flora China, 9,* 173–179.

Demeke, T., & Adams, R. P., (1994). The use of PCR-RAPD analysis in plant taxonomy and evolution. *PCR Technology: Current Innovations,* 179–191.

Dondini, L., Pierantoni, L., Ancarani, V., D'Angelo, M., Cho, K. H., Shin, I. S., et al. (2008). The inheritance of the red color character in European pear *(Pyrus communis)* and its map position in the mutated cultivar 'Max Red Bartlett.' *Plant Breed., 127* (5), 524–526.

Doyle, J. J., & Dickson, E. E., (1987). Preservation of plant samples for DNA restriction endonuclease analysis. *Taxon, 1,* 715–722.

Doyle, J., & Doyle, J., (1987). A rapid DNA isolation procedure for small quantities of fresh leaf tissue. *Phytochem. Bull., 19,* 11–15.

Doyle, J., (1991). DNA protocols for plants-CTAB total DNA isolation. In: Hewitt, G. M., & Johnston, A., (eds.), *Molecular Techniques in Taxonomy* (pp. 283–293).

Frankham, R., Ballou, J. D., & Briscoe, D. A., (2004). *A Primer of Conservation Genetics.* Cambridge University Press.

Fushimi, H., Komatsu, K., Isobe, M., & Namba, T., (1997). Application of PCR-RFLP and MASA analyzes on 18S ribosomal RNA gene sequence for the identification of three Ginseng drugs. *Biol. Pharm. Bull., 20* (7), 765.

Gupta, V., Chaudhary, N., Srivastava, R., Sharma, G. D., Bhardwaj, R., & Chand, S., (2011). Luminscent graphene quantum dots for organic photovoltaic devices. *J. Amer. Chem. Soc., 133* (26), 9960–9963.

Hajibabaei, M., Singer, G. A., Hebert, P. D., & Hickey, D. A., (2007). DNA barcoding: how it complements taxonomy, molecular phylogenetics, and population genetics. *Trends Genet., 23* (4), 167–172.

Hamby, R. K., & Zimmer, E. A., (1992). Ribosomal RNA as a phylogenetic tool in plant systematics. *Mol. Syst. Plants, 50,* 91.

Hebert, P. D., Cywinska, A., & Ball, S. L., (2003). Biological identifications through DNA barcodes. *Proceedings of the Royal Society of London. Series B: Biological Sciences, 270* (1512), 313–321.

Hedrick, U. P., Howe, G. H., Taylor, O. M., Francis, E. H., & Tukey, H. B., (1921). *The Pears of New York.* JB Lyon Company.

Huss, V., & Sogin, M., (1990). Phylogenetic position of some *Chlorella* species within the chlorococcales based upon complete small-subunit ribosomal RNA sequences. *J. Mol. Eevol., 31* (5), 432–442.

Hussain, F., Islam, M., & Zaman, A., (2006). Ethnobotanical profile of plants of Shawar Valley, District Swat, Pakistan. *Int. J. Biol. Biotecnol., 3* (2), 301–307.

Inoue, E., Kasumi, M., Sakuma, F., Anzai, H., Amano, K., & Hara, H., (2006). Identification of RAPD marker linked to fruit skin color in Japanese pear (*Pyrus pyrifolia* Nakai). *Scien. Horti., 107* (3), 254–258.

Islam, M., & Ahmad, H., (2012). Morphological evaluation of *Pyrus* genotypes of Kaghan Valley, Pakistan through quantitative parameters. *Int. J. Biosci., 2* (12), 1–6.

Islam, M., & Ahmad, H., (2014). Distriburtion of *Pyrus* species across the Hindu Kush-Himalayn region of northern Pakistan. *The Third Int. Symposium (BIORARE–2014), Turkey, 65.*

Islam, M., & Razziq, A., (2010). *Economic and Fruit Plants of Shawar Vallely, District Swat, Pakistan.* Economic value and conservation status of plants of Shawar valley, District Swat, Pakistan. Germany, VDM Verlag Dr. Müller Aktiengesellschaft & Co. KG Dudweiler Landstr, 99 D –66123 43.

Jang, J., Tanabe, K., Tamura, F., & Banno, K., (1991). Identification of *Pyrus* species by peroxidase isozyme phenotypes of flower buds. *J. Japanese Soci. Horti. Sci., 60* (3), 513–519.

Janick, J., & Moore, J. N., (1996). *Fruit Breeding* (Vol. I.). Tree and tropical fruits, John Wiley and Sons.

John, M. E., (1992). An efficient method for isolation of RNA and DNA from plants containing polyphenolics. *Nucleic Acids Res., 20* (9), 2381.

Jun, W., & Hongsheng, G., (2001). *The Production of Asian Pears in China.* International Symposium on Asian Pears, Commemorating the 100th Anniversary of Nijisseiki Pear 587.

Katayama, H., Adachi, S., Yamamoto, T., & Uematsu, C., (2007). A wide range of genetic diversity in pear (*Pyrus ussuriensis* var. *aromatica*) genetic resources from Iwate, Japan revealed by SSR and chloroplast DNA markers. *Genet. Res. Crop Evol., 54* (7), 1573–1585.

Khan, S., Mirza, K. J., & Abdin, M. Z., (2010). Development of RAPD markers for authentication of medicinal plant Cuscuta reflexa. *Eur. Asia J. Biol. Sci., 4* (1), 1–7.

Kikuchi, A., (1946). *Speciation and Taxonomy of Chinese Pears.* Unpublished.

Kikuchi, A., (1948). *Horticulture of Fruit Trees.* Yokendo, Tokyo (in Japanese).

Kim, C., Lee, C., Shin, J., Chung, Y., & Hyung, N., (1997). A simple and rapid method for isolation of high-quality genomic DNA from fruit trees and conifers using PVP. *Nucleic Acids Res., 25* (5), 1085–1086.

Kimura, T., Sawamura, Y., Kotobuki, K., Matsuta, N., Hayashi, T., Ban, Y., & Yamamoto, T., (2003). Parentage analysis in pear [*Pyrus*] cultivars characterized by SSR markers. *J. Japanese Soci. Horti. Sci., 72,* 182–189.

Kranz, H. D., Mikš, D., Siegler, M. L., Capesius, I., Sensen, C. W., & Huss, V. A., (1995). The origin of land plants: phylogenetic relationships among charophytes, bryophytes, and vascular plants inferred from complete small-subunit ribosomal RNA gene sequences. *J. Mol. Evol., 41* (1), 74–84.

Lee, G., Lee, C., & Kim, C., (2004). Molecular markers derived from RAPD, SCAR, and the conserved 18S rDNA sequences for classification and identification in *Pyrus pyrifolia* and *P. communis*. *Theor. Appl. Genet.*, *108* (8), 1487–1491.

Lespinasse, Y., & Salesses, G., (1973). Application de techniques nouvelles à l'observation des chromosomes chez les genres *Malus* et *Pyrus*. *Ann. Amelior. Plant*, *23*, 381–386.

Li, J. F., Li, L., & Sheen, J., (2010). Protocol: a rapid and economical procedure for purification of plasmid or plant DNA with diverse applications in plant biology. *Plant Methods*, *6* (1), 1.

Monte-Corvo, L., Cabrita, L., Oliveira, C., & Leitão, J., (2000). Assessment of genetic relationships among *Pyrus* species and cultivars using AFLP and RAPD markers. *Genet. Res. Crop Evol.*, *47* (3), 257–265.

Oliveira, C., Mota, M., Monte-Corvo, L., Goulao, L., & Silva, D., (1999). Molecular typing of *Pyrus* based on RAPD markers. *Sci. Hort.*, *79* (3), 163–174.

Pich, U., & Schubert, I., (1993). Midiprep method for isolation of DNA from plants with a high content of polyphenolics. *Nucl. Acids Res.*, *21* (14), 3328–3330.

Porebski, S., Bailey, L. G., & Baum, B. R., (1997). Modification of a CTAB DNA extraction protocol for plants containing high polysaccharide and polyphenol components. *Plant Mol. Biol. Report.*, *15* (1), 8–15.

Postman, J. D., (2008). *World Pyrus Collection at USD Agenebank in Corvallis, Oregon*. InX International Pear Symposium 800 2007 May 22, (pp. 527–534).

Pruesse, E., Quast, C., Knittel, K., Fuchs, B. M., Ludwig, W., Peplies, J., & Glöckner, F. O., (2007). SILVA: A comprehensive online resource for quality checked and aligned ribosomal RNA sequence data compatible with ARB. *Nucl. Acids Res.*, *35* (21), 7188–7196.

Ro, K. -E., Keener, C. S., & McPheron, B. A., (1997). Molecular phylogenetic study of the Ranunculaceae: Utility of the nuclear 26S ribosomal DNA in inferring intrafamilial relationships. *Mol. Phylogenet. Evol.*, *8* (2), 117–127.

Saghai-Maroof, M., Soliman, K., Jorgensen, R. A., & Allard, R., (1984). Ribosomal DNA spacer-length polymorphisms in barley: Mendelian inheritance, chromosomal location, and population dynamics. *Proceed. Nat. Acad. Sci.*, *81* (24), 8014–8018.

Scott, K., & Playford, J., (1996). DNA extraction technique for PCR in rain forest plant species. *BioTechniques*, *20* (6), 974, 977, 979.

Shenghua, L., & Chengquan, F., (1993). Studies on chromosome of *Pyrus* in China. *VI International Symposium on Pear Growing*, *367*.

Shepherd, L. D., & McLay, T. G., (2011). Two micro-scale protocols for the isolation of DNA from polysaccharide-rich plant tissue. *J. Plant Res.*, *124* (2), 311–314.

Shepherd, M., Cross, M. J., Stokoe, R. L., Scott, L. J., & Jones, M. E., (2002). High-throughput DNA extraction from forest trees. *Plant Mol. Biol. Report.*, *20* (4), 425.

Shi, Y., Yamamoto, T., & Hayashi, T., (2001). *The Japanese Pear Genome Program III*. Copialike retrotransposons in pear. International Symposium on Asian Pears, Commemorating the 100th Anniversary of Nijisseiki Pear, 587.

Shimura, I., (1988). Nashi (Pear). *Heibonsha's World Encyclopedia*, *36*, 354–372.

Simon, L., Lalonde, M., & Bruns, T., (1992). Specific amplification of 18S fungal ribosomal genes from vesicular-arbuscular endomycorrhizal fungi colonizing roots. *Appl. Environ. Microbiol.*, *58* (1), 291–295.

Singh, S., & Tripathi, B., (1985). Genetic divergence in pea. *Indian J. Genet. Plant Breed.*, *45* (2), 389–393.

Smit, S., Widmann, J., & Knight, R., (2007). Evolutionary rates vary among rRNA structural elements. *Nucl. Acids Res.*, *35* (10), 3339–3354.

Smith, J., & Smith, O., (1989). The description and assessment of distance between inbred lines of maize. 2, The utility of morphological-biochemical-and genetic descriptors and a scheme for the testing of distinctiveness between inbred lines. *Maydica*, 34151–34161.

Smyth, R., Schlub, T., Grimm, A., Venturi, V., Chopra, A., Mallal, S., Davenport, M., & Mak, J., (2010). Reducing chimera formation during PCR amplification to ensure accurate genotyping. *Gene*, *469* (1), 45–51.

Sogin, M. L., Gunderson, J. H., Elwood, H. J., Alonso, R. A., & Peattie, D. A., (1989). Phylogenetic meaning of the kingdom concept: an unusual ribosomal RNA from *Giardia lamblia*. *Science*, *243* (4887), 75.

Srivastava, S., & Mishra, N., (2009). Genetic markers-A cutting edge technology in herbal drug research. *J. Chem. Pharm. Res.*, *1* (1), 1–18.

Summerbell, R., Levesque, C., Seifert, K., Bovers, M., Fell, J., Diaz, M., et al. (2005). Microcoding: the second step in DNA barcoding Philos. *Tran. R. Soc. B-Biol. Sci.*, *360* (1897), 1903.

Taylor, B., & Powell, A., (1982). Isolation of plant DNA and RNA. *Focus*, *4* (4), 6.

Teletchea, F., (2010). After 7 years and 1000 citations: Xomparative assessment of the DNA barcoding and the DNA taxonomy proposals for taxonomists and non-taxonomists. *Mitochondrial DNA*, *21* (6), 206–226.

Teng, Y., & Tanabe, K., (2002). Reconsideration on the origin of cultivated pears native to East Asia. *XXVI International Horticultural Congress: IV International Symposium on Taxonomy of Cultivated Plants, 634*.

Van de Peer, Y., Neefs, J. -M., de Rijk, P., & De Wachter, R., (1993). Evolution of eukaryotes as deduced from small ribosomal subunit RNA sequences. *Biochem. Syst. Ecol.*, *21* (1), 43–55.

Varma, A., Padh, H., & Shrivastava, N., (2007). Plant genomic DNA isolation: An art or a science. *Biotechnol. J.*, *2* (3), 386–392.

Wilkie, S. E., Isaac, P. G., & Slater, R. J., (1993). Random amplified polymorphic DNA (RAPD) markers for genetic analysis in *Allium*. *Theor. Appl. Genet.*, *86* (4), 497–504.

Woese, C. R., (1987). Bacterial evolution. *Microbiol. Rev.*, *51* (2), 221.

Wünsch, A., & Hormaza, J., (2007). Characterization of variability and genetic similarity of European pear using microsatellite loci developed in apple. *Sci. Hort.*, *113* (1), 37–43.

Wuyts, J., De Rijk, P., Van de Peer, Y., Winkelmans, T., & De Wachter, R., (2001). The European large subunit ribosomal RNA database. *Nucl. Acids Res.*, *29* (1), 175–177.

Wuyts, J., Perrière, G., & Van de Peer, Y., (2004). The European ribosomal RNA database. *Nucleic Acids Research 32* (1), D101–D103.

Yamamoto, T., & Chevreau, E., (2009). Pear genomics. *Genetics and Genomics of Rosaceae*, 163–186.

Zielinski, Q. B., & Thompson, M. M., (1967). Speciation in *Pyrus:* Chromosome number and meiotic behavior. *Botanical Gazette*, 109–112.

CHAPTER 12

ENHANCING THE DROUGHT TOLERANCE IN EGGPLANT BY EXOGENOUS APPLICATION OF OSMOLYTES IN WATER-DEFICIT ENVIRONMENTS

MAHAM SADDIQUE and MUHAMMAD SHAHBAZ

Department of Botany, University of Agriculture, Faisalabad, Pakistan, E-mail: shahbazmuaf@yahoo.com

12.1 INTRODUCTION

Various adverse environmental conditions like as salt stress, cold stress, heat stress and drought are major threats to agricultural lands and plant production (Cheeseman, 2013). Crop plants are subjected to these stresses when grown in various parts of the world (Atkinson and Urwin, 2012). High-temperatures conditions are frequent with water deficit conditions (Pitman and Läuchli, 2004). More than 100 countries are under the shock of variable water deficit conditions, which are categorized through different characteristics and kinds (Rengasamy, 2006). Salt stress adversely affects growth of the various plant parts by reducing cell elongation due to reduction in water relation attributes of plants (Munns and Tester, 2008).

12.2 DROUGHT STRESS AN OVERVIEW

The specific time periods when rate of precipitation is significantly decreased are termed as drought. The main causes of water deficit conditions are evaporation and transpiration (Jaleel et al., 2009). Drought tolerance ability is variable among various plant species (Reddy et al., 2004). Drought stress is one of major environmental stresses, which minimizes yield of crops (Zhao and Running, 2010).

12.3 HARMFUL EFFECTS OF DROUGHT STRESS ON PLANTS

12.3.1 CROP GROWTH AND YIELD

Primary effects of drought stress are reduction in seedlings growth (Zheng et al., 2016), germination rates, elongation of hypocotyl, shoot, and root biomass. In some species, like alfalfa root length is enhanced under drought stress conditions (Zeid and Shedeed, 2006). In rice, adverse effects of drought stress have also been observed (Manikavelu et al., 2006). Water deficit conditions decreased various yield attributes like grain yield, number of grains per plant and number of spikes or spikelet in *Hordeum vulgare* (Li et al., 2006).

Adverse effect of water deficit conditions is relatively high at post-anthesis stage as compared to other growth stages (Samarah, 2005). Low yield in *Zea mays* occurred due to delay in silking under water deficit conditions (Cattivelli et al., 2008). Similarly, total harvest of seeds and branches reduced in soybean under drought stress (Frederick et al., 2001). Water deficit conditions cause barrenness at floral stage (Anjum et al., 2011). It also decreased the cotton production due to bolls' abortion and less flowering at reproduction stage (Pettigrew, 2004).

12.3.2 WATER RELATIONS

Relative water contents (RWC) of *Triticum aestivum* increased with increase in plant growth but reduced at maturity (Siddique et al., 2001). When plants were exposed to water deficit conditions, water potential of leaves, RWC, and transpiration rates were reduced due to rise in leaves' temperature (Siddiqui et al., 2001). Water stress also caused reduction in RWC, transpiration, WUE, turgor potential and stomatal conductance in shoe flower (Egilla et al., 2005). Wheat plants showed more water use efficiencies underwater deficit environments than the well-irrigated environments (Abbate et al., 2004). Luecern plants also showed high WUE underwater deficit condition (Miyashita et al., 2005) while, in *Solanum tuberosum* limited water supply caused significant reduction in WUE and plants development (Costa et al., 1997).

12.3.3   NUTRIENT RELATIONS

Drought stress reduces uptake of nutrients and hindered in tissues formation in various crops (Waraich et al., 2011). Transpiration activity is reduced due to disturbed uptake of nutrients (McWilliams, 2003). Plant production can be increased with increasing efficiencies of nutrients under low availability of water (Garg, 2003). Uptake of sodium and potassium inhibited in *Gossypium hirsutum* under limited supply of water (McWilliams, 2003). Water stress reduces translocation and metabolic rates of nutrients and also causes reduction in transpiration process (Giri et al., 2003).

12.3.4   PHOTOSYNTHESIS

Water stress impaired photosynthetic apparatus and caused senescence of leaves and lesser yield thus reduced photosynthetic activity (Wahid and Rasul, 2005). Water stress altered the photosynthetic pigments' accumulation (Anjum et al., 2008). Drought produced damaging effects on photosynthetic machinery (Fu and Huang, 2001). Drought stress led to limited plant production due to limited activity of enzymes of Calvin cycle (Monakhova and Chernyadèv, 2002). The imbalance among generation of enzymatic antioxidants and reactive oxygen species (ROS) reduced the photosynthetic activity of plants (Reddy et al., 2004). Production of ROS caused oxidative burst in lipids, proteins, and various cellular structures, however closing of stomata was major factor for reduced photosynthetic activities under moderately water deficit conditions (Yokota et al., 2002). Stomatal closure is the main priority of plants under moderate or severe water scarcity (Cornic and Massacci, 1996). It is evident from different reports that soil humidity is the main factor that brings about diverse responses in stomata (Turner et al., 2001). Under severe water deficit condition, photosynthetic activities are reduced because of the reduction in activities of Rubisco (Bota et al., 2004). The activities of photosynthetic electron transport chain (ETC) are linked with supply of carbon dioxide in (chloroplast and changes in PS II activity) under water stress conditions. Drought stress caused reduction in cell volume and as a result, cell shrinking took place (Guerfel et al., 2009). There are many reasons for limited photosynthetic activity underwater deficit conditions. Of various, one is the low production of ATP in photophosphorylation (Tezara et al., 1999). In fact, the activities of the enzymes of carbon assimilation and those involved in adenosine triphosphate synthesis are retarded and sometimes inhibited depending on the severity of drought stress conditions. Of these enzymes, Rubisco, which is dual functions, acts as oxygenase underwater-limiting conditions and therefore, limited CO_2 fixation has been observed (Munné-Bosch et al., 2005).

12.3.5   PHOTO-ASSIMILATE PARTITIONING

Transport of photo-assimilates towards the reproductive parts of plants is an important phenomenon for newly developing seeds. Seed formation and seed set are reduced due to low supply and further use of these photo-assimilates (Asch et al., 2005). Translocation of dry matter to root is enhanced underwater deficit conditions which increases water potential (Leport et al., 2006). Translocation of sucrose is dependent on rates of photosynthesis and sucrose concentrations in leaf (Komor, 2000). Water deficit conditions reduced photosynthesis, disturbed metabolic rates of carbohydrates and reduced translocation of sucrose (Kim et al., 2000). Reduced translocation of sucrose and photosynthetic activities hamper the transport of sucrose from (source to sink) and effects the organs of reproduction (Zinselmeier et al., 1999). When phloem unloading is impropriated, activities of acid invertase are increased causing limited development of various tissues (reproductive tissues) (Goetz et al., 2001).

12.3.6   RESPIRATION

The overall metabolic efficiency of the plant is determined by the loss of carbohydrate fraction (Davidson et al., 2000). The root is chief consumer of the photosynthetically fixed carbon. This fixation is further used for growth, maintenance, and generation of dry matter (Lambers et al., 1996). Austere drought led to reduced shoot and root biomass and photosynthetic and respiration rates of roots (Liu and Li, 2005). When exposed to drought stress, plants produced reactive oxygen species (ROS), which damage membrane components (Blokhina et al., 2003). In short, water shortage in the rhizosphere causes an increase in root respiration rate that leads to an unequal use of carbon resources, reduction in adenosine triphosphate production and improved reactive oxygen species generation.

12.3.7 OXIDATIVE DAMAGE

Exposure of plants to some ecological stresses lead to the generation of ROS, including superoxide anion radicals (O^{2-}), hydroxyl radicals (OH^-), hydrogen peroxide (H_2O_2), alkoxy radicals (RO) and singlet oxygen (Munné-Bosch and Penuelas, 2003). Reactive oxygen species can react with lipids, proteins, and DNA, causing oxidative damage and impairing the cell's normal functioning (Foyer and Fletcher, 2001). Many cellular compartments produce ROS; most importantly chloroplasts, because excited pigments in thylakoid membranes may interact with O_2 to form strong oxidants such as O^{2-} (Reddy et al., 2004). Additional reactions produce other reactive oxygen species such as H_2O_2 and OH^- (Fazeli et al., 2007). Mechanism for ROS generation in biological systems are characterized to both non-enzymatic and enzymatic reactions (Apel and Hirt, 2004). Peroxidases and catalases play a significant role in the regulation of reactive oxygen species in the cells by activating and deactivating H_2O_2 (Sairam et al., 2005). Mitochondria produce reactive oxygen species because of the electron leakage at the ubiquinone site, the ubiquinone: cytochrome b region (Gille and Nohl, 2001) and at the matrix side of complex I (NADH dehydrogenase) under drought stress (Möller, 2001). ROS are involved in lipid peroxidation, and destruction of membranes and degradation of proteins and inactivation of enzymes (Sairam et al., 2005).

12.4 DROUGHT RESISTANCE MECHANISMS

Plants show various biochemical and physiological adaptation for their survival underwater deficit environments. Plants are categorized as drought tolerant when they show development, flowering, and economical production under limited water availability. Water relations of plants are affected by arid conditions (cellular, tissue, and organ level) which damages adaptive processes of plants (Beck et al., 2007). Plants show drought tolerance by initiating defensive mechanisms (Chaves and Oliveira, 2004).

12.4.1 MORPHOLOGICAL MECHANISMS

Drought tolerance in plants is variable in plants at whole-plant, tissue, physiological, and molecular levels. Plant tolerance toward drought is determined through expression of various plant responses underwater deficit conditions.

12.4.1.1   ESCAPE

Plants escape from water stress by shortening their life cycles, which allows plant reproduction in arid environments. Shortening in life cycles is an important adaptive response towards stress tolerance (Araus et al., 2002). Genotype and environmental changes determine durations for crops and its abilities to survive under stress. Development of crops with shortened durations is effective tool to minimize production losses due to water stress (Turner et al., 2001).

12.4.1.2   LEAF AREA

Drought inhibits cell growth and leaves' development, transpirational processes are reduced due to decreased water uptake from roots (Escalona et al., 2015). Leaf surfaces have various restrictions, which are helpful in defensive mechanisms (Fathi and Tari, 2016).

12.4.1.3   LEAF ABSCISSION

Older leaves are fallen after exposure to drought, and this is essential adaptation of plant survival in arid environments (Maleki et al., 2013). Leaf abscission due to drought stress is due to enhanced activities of ethylene hormone (Kabiri and Naghizadeh, 2015).

12.4.2   PLANT OSMOLYTES

The main effects of osmotic stresses are addition of increased number of solutes with lower molecular weights and high solubility and non-interference with plant metabolism. They are not toxic for plants even if accumulated in high concentrations, these are termed as compatible solutes (Yancey, 2005). Accumulation of compatible solutes results in maintenance of water potential through osmotic adjustment (Hasegawa *et al.*, 2000). Compatible solutes maintain solute potentials; enhance thermodynamic stabilities of protein and protection of macromolecules (Yancey, 2005). These organic osmolytes acts as osmoprotectants by protecting cellular structures from water stress (Gagneul *et al.*, 2007). Between these compatible solutes few are subjected to changes (short and long-term) as degradation and accumulation of proline, whereas some other as betaines are accumulated for long durations (Szabados and Savouré, 2010). Cellular level of compatible solutes is changed in accordance to developmental periods and stages, and various environmental conditions (Murakeözy *et al.*, 2003).

12.4.2.1 ENDOGENOUS PRODUCTION OF OSMOLYTES UNDER DROUGHT STRESS

Proline is synthesized by an enzyme pyrroline–5-carboxylate synthetase (P5CS) which reduces glutamate to glutamate semialdehyde and instantly transformed into pyrroline–5-carboxylate (P5C) (Szabados and Savouré, 2010). Prolineis formed by the further reduction of P5C reductase (P5CR) to P5C intermediate. Ornithine is the other precursor for proline biosynthesis other than glutamate which is transaminated by mitochondrial ornithine-δ-aminotransferase to P5C (Verbruggen and Hermans, 2008). Under the dehydration caused by heat, cold, drought or salinity, plant produced glycine betaine in addition to the various quaternary ammonium compounds (Guo et al., 2009). Non-halophytic plants such as *Madicago* and *Citrus* species accumulates prolinebetaine which is a dimethyl proline also known as stachydrine (Trinchant et al., 2004). Under abiotic stresses, plants accumulate various sugars like trehalose, sucrose (Yuanyuan et al., 2009) etc. These sugars are involved in membranes stabilization and osmotic adjustment underwater stress (Lokhande and Suprasanna, 2012). These compounds play very crucial regulatory roles in plants under drought stress (Lunn et al., 2014). In plant tissues, the most common inositol six isomers also known as Myo-inositol found and their all methylated isomers are also isolated which have important functions as signaling molecule and performed protective functions (Sureshan et al., 2009). The primary form of vitamin E such as Tocotrienolsis to increase antioxidant contents. These vitamins are present in seed of most of the monocot plants as well as in corn, wheat, and rice (Cahoon et al., 2003).

12.4.2.2 EXOGENOUS APPLICATION OF OSMOLYTES UNDERWATER DEFICIT CONDITION

Sun et al. (2016) conducted an experiment on 3-week-old cucumber seedlings applied with foliar-spray of 1.5 mM H$_2$O$_2$ and exposed to different soil water deficit condition. Exogenous application of H_2O_2 extensively increased relative water content (RWC), leaf biomass and chlorophyll content, activities of antioxidant enzymes such as peroxidase (POD), superoxide dismutase (SOD), proline contents and soluble protein contents and net photosynthetic rate underwater deficit condition. In stressful environment, exogenous application of proline and GB enhance the production of many crops like wheat, rice, and corn (Ashraf and Foolad, 2005; Arfan et al., 2006). Underwater deficit condition, exogenous applied salicylic acid (SA) and glycinebetaine (GB) has been found very efficient in alleviating the harmful effects of drought stress by improving the yield of hybrid sunflower (Hussain et al., 2008). Moreover, the exogenous application of various enzymes and Polyamines such as citrulline help to reduce detrimental effects of water deficit stress (Farooq et al., 2009). It was also studied that exogenous application as foliar spray of trehalose drastically elevate the biomass accumulation, plant water relation and photosynthetic attributes (Nounjan and Theerakulpisut, 2012). It was also studied that under drought stress condition, trehalose application helps to increase the non-enzymatic compounds such as phenolics and tocopherols and some of the enzymatic antioxidants such as catalase (CAT) and peroxidase (POD) (Ali and Ashraf, 2011). Moreover, in another experiment, the exogenous spray of 24-epibrassinolide @ 0.01 and 1 µMwas applied to the tomato leaves underwater stress condition for about 3 days with one-day interval. Application of 24-epibrassinolide increased the activity of antioxidants such as POD, CAT, MDA, and H_2O_2 and decreased the contents of H_2O_2 and MDA (Behnamnia et al., 2009). In the same way chlormequat chloride (CCC) and salicylic acid (SA) application on wheat enhanced the level of antioxidants enzymes such as peroxidase, superoxide dismutase and catalase and upregulated free proline and total soluble proteins by alleviating the adverse effects of stress (Anosheh et al., 2012). Similarly, in onion plants the exogenous application of potassium dihydrogen phosphate (KH_2PO_4) and alpha-tocopherol also called Vitamin E alleviated the harmful effects of salt stress (Mohamed and Aly, 2008). Moreover, it was also documented that underwater stress condition, exogenously applied trehalose increased the level of superoxide dismutase (SOD), catalase (CAT) and peroxidase (POD) enzymes and upgraded the concentration of ascorbic acid (AsA), carotenoids, malondialdehyde (MDA), total soluble proteins (TSP) and glycine betaine (GB) in the roots of *Raphanussativus* L. (radish) plants (Shafiq et al., 2015). In another experiment, it was observed that under high-temperature (35 °C) plant treated with α-Tocopherol (Figures 12.1–12.6) enhanced activity of non-enzymatic and enzymatic antioxidants (Kumar et al., 2013).

FIGURE 12.1 Differential effect of drought on two eggplant F1 lines (a) Sultan (tolerant) and (b) Jhanak (sensitive) under exogenously application of a-tocopherol.

(a) Sultan F1 (Foliar spray)   (b) Sultan F1 (Pre-sowing)

FIGURE 12.2 Effect of various methods of a-tocopherol application (a) as foliar spray or (b) pre-sowing treatment on drought-stressed eggplant line Sultan F1 (tolerant).

(a) Janak F1 (Foliar spray)   (b) Janak F1 (Pre-sowing)

FIGURE 12.3   Effect of various methods of a-tocopherol application (a) as foliar spray or (b) pre-sowing treatment on drought-stressed eggplant line Janak F1 (sensitive).

FIGURE 12.4   Effect of various levels of a-tocopherol applied as foliar spray (a) or pre-sowing (b) treatment on drought-stressed eggplant line Sultan F1.

FIGURE 12.5 Effect of various levels of a-tocopherol applied as foliar spray (a) or pre-sowing (b) treatment on drought-stressed eggplant line Janak F1.\

FIGURE 12.6 A comparison of the effect of a-tocopherol on vegetative growth of two F1 lines of eggplant (a) sultan (tolerant) and (b) Janak (sensitive).

12.4.3 EGGPLANT A MAJOR VEGETABLE CROP

Eggplant is among popular vegetables in Indonesia and available in market with lowest prices. Eggplant consists of following vitamins (B1, C, A, B2 and P) and used as medicine to cure diseases like hemorrhoids, high blood pressure, drug itching of the skin and toothache (Siaga et al., 2016).

12.4.3.1 WATER REQUIREMENTS OF THE CROP

It is investigated through various researches that at (67% ETc) plants exhibited no adverse effect on developmental and photosynthetic parameters, yield of fruits was alike to plant with irrigation at (100% ETc). Eggplant belonging to *Solanaceae* family like pepper and tomato but it has more sensitivity to water stress (Ertek et al., 2006)

12.4.3.2 EFFECTS OF DROUGHT STRESS ON EGGPLANT

Eggplant show sensitivity to drought stresses during whole growth periods. Large amounts of water are required for their growth periods. Plant responses to drought are variable on the basis of duration of drought and the severity of drought (Song et al., 2011). Stomatal opening and closing determine the carbon and water cycles (Miura and Tada, 2014). Stomatal characters as (density, size, position) occupies main positions in gaseous exchange processes (Zou et al., 2010). Resistance of stomata depends on numbers and size of stomata (Yu et al., 2016).

12.4.3.3 ROLE OF OSMOLYTES IN EGGPLANT

Exposure of plants to drought causes physical and chemical changes. Plants synthesize large amounts of compatible solutes as an adaptive mechanism for stress tolerance (Cazzonelli, 2011). They are essential for human health (Rao and Rao, 2007). These solutes detoxify free radicals as antioxidative enzymes (Grassmann et al., 2002). Carotenoids are in extensive studies in eggplants because they are essential for human nutritional and health advantages, whereas, effects of proline in maintaining osmotic balance are also important under stress conditions (Meloni et al., 2004).

12.4.3.4   ROLE OF EXOGENOUS APPLICATION OF OSMOLYTES ON EGGPLANT

Treatment of eggplants with 6-benzyladenine showed positive impact under salt stress conditions (Wu et al., 2014). Treatment of 6-benzyladenine increased content of chlorophyll, whereas gaseous exchange parameters like stomatal conductance, intercellular CO_2 concentration, net CO_2 assimilation rate and transpiration rate also increased (Wu et al., 2014). These findings showed that exogenously applied 6-benzyladenine enhanced the salinity resistance of eggplants (Wu et al., 2014). Likewise, exogenously applied Betaine (50 and 25 mgL^{-1}) on eggplants and wheat enhanced the growth rates and photosynthetic intensities underwater deficit conditions (Zhang et al., 2003). Exogenously applied proline showed positive effects on photosynthetic parameters, germination rate and mineral ions in eggplants growing in salt stress. Proline treatment deteriorated the toxicities of saline conditions on water use efficiency and shoot biomass of eggplants (Shahbaz et al., 2013). Treatment of brassinosteroids to plants under cold stress enhanced photosynthetic activities and photosynthetic pigments. Application of brassinosteroids also enhanced activity of various enzymatic antioxidants as SOD, POD, APX, and GPX, whereas ascorbic acid and proline contents were also enhanced in sprouts, facing cold stresses (Wu et al., 2015). Sugar beet extracts when applied to cultivars of eggplants (Bemisal and Dilnasheen) improved germination and transpirational rates, activity of photosynthesis, thus increased overall production (Abbas et al., 2010). Abul-Soudand Abd-Elrahman (2016) observed that Se supplement with 20 µM showed the good effects on vegetative growth and yield of eggplants under different salinity levels. On the other hand, 24-Epibrassinolide (EBL), a plant steroid hormone, plays a pivotal role in regulating plant resistance to various stresses. Study suggested that exogenous EBL confers resistance against Zn-induced oxidative damage by enhancing the ASA-GSH cycle by up-regulating key gene expression and key enzyme activities (Wu et al., 2015).

12.5   CONCLUSION AND PROSPECTS

Water deficit reduces plant growth, development, flowers, and fruits production. In plants, facing drought stomata close gradually with a parallel decline in net photosynthesis and water use efficiency. Water deficit conditions reduced or inhibited the activities of some enzymes. One of the major factors, the reactive oxygen species targeted the peroxidation of cellular membrane lipids and degradation of enzyme proteins and nucleic acids. Being very complex, the drought tolerance mechanism involves many morphological, physiological, and biochemical processes at cell, tissue, organ, and whole-plant levels, when activated at various plant developmental stages. These mechanisms caused reduction in water loss by closing stomata, improved water uptake by emerging deep root systems, accumulation of osmolytes in different plant parts. Osmolytes have been reported to play an important role during drought tolerance in eggplants and others include, salicylic acid, cytokinin, and abscisic acid. These osmolytes produced endogenously or applied exogenously have been proved to be effective for decreasing the ameliorative effects of drought due to scavenging of reactive oxygen species by enzymatic and non-enzymatic systems, cell membrane stability and stress proteins under drought stress. Drought stress can be managed in eggplants and other vegetable crops by exogenous application of osmolytes (glycinebetaine, proline, and other compatible solutes). It is crucial to determine the mechanisms of action for various osmolytes, the most optimal concentrations, and appropriate plant developmental stages for better growth of plants.

ACKNOWLEDGMENT

This chapter has been extracted from the "Review of Literature" section of PhD Thesis of Maham Saddique (2014-ag-21) submitted to Department of Botany, University of Agriculture, Faisalabad.

KEYWORDS

- glycinebetaine
- osmolytes
- photosynthesis
- proline
- water use efficiency

REFERENCES

Abbas, W., Ashraf, M., & Akram, N. A., (2010). Alleviation of salt-induced adverse effects in eggplant (*Solanummelongena* L.) by glycinebetaine and sugar beet extracts. *Sci. Hort. 125*, 188–95.

Abbate, P. E., Dardanelli, J. L., Cantarero, M. G., Maturano, M., Melchiori, R. J., & Suero, E. E., (2004). Climatic and water availability effects on water-use efficiency in wheat. *Crop Sci., 44*, 474–83.

Abul-Soud, M. A., & Abd-Elrahman, S. H., (2016). Foliar selenium application to improve the tolerance of eggplant grown under salt stress conditions. *Int. J. Plant Soil Sci.*, 1–10.

Ali, Q., & Ashraf, M., (2011). Induction of drought tolerance in maize (*Zea mays* L.) due to exogenous application of trehalose: growth, photosynthesis, water relations and oxidative defence mechanism. *J. Agron. Crop Sci., 197*, 258–271.

Anjum, N. A., Umar, S., Ahmad, A., Iqbal, M., & Khan, N. A., (2008). Sulphur protects mustard (*Brassica campestris* L.) from cadmium toxicity by improving leaf ascorbate and glutathione. *Plant Growth Regul., 54*, 271–279.

Anjum, S. A., Xie, X. Y., Wang, L. C., Saleem, M. F., Man, C., & Lei, W., (2011). Morphological, physiological, and biochemical responses of plants to drought stress. *Afr. J. Agric. Res., 6*, 2026–2032.

Anosheh, H. P., Emam, Y., Ashraf, M., & Foolad, M. R., (2012). Exogenous application of salicylic acid and chlormequat chloride alleviates negative effects of drought stress in wheat. *Adv. Stud. Biol., 4*, 501–520.

Apel, K., & Hirt, H., (2004). Reactive oxygen species: metabolism, oxidative stress, and signal transduction. *Annu. Rev. Plant Biol., 2*, 373–399.

Araus, J. L., Slafer, G. A., Reynolds, M. P., & Royo, C., (2002). Plant breeding and drought in C3 cereals: what should we breed for? *Ann. Bot., 89*, 925–940.

Asch, F., Dingkuhn, M., Sow, A., & Audebert, A., (2005). Drought-induced changes in rooting patterns and assimilate partitioning between root and shoot in upland rice. *Field Crops Res., 93*, 223–236.

Ashraf, M., & Foolad, M. R., (2005). Pre-sowing seed treatment –A shotgun approach to improve germination, plant growth, and crop yield under saline and non-saline conditions. *Adv. Agron., 88*, 223–271.

Atkinson, N. J., & Urwin, P. E., (2012). The interaction of plant biotic and abiotic stresses: From genes to the field. *J. Exp. Bot., 63*, 3523–3543.

Beck, E. H., Fettig, S., Knake, C., Hartig, K., & Bhattarai, T., (2007). Specific and unspecific responses of plants to cold and drought stress. *J. Biosci., 32*, 501.

Behnamnia, M., Kalantari, K. M., & Ziaie, J., (2009). The effects of brassinosteroid on the induction of biochemical changes in *Lycopersiconesculentum* under drought stress. *Turk. J. Bot., 33*, 417–428.

Blokhina, O., Virolainen, E., & Fagerstedt, K. V., (2003). Antioxidants, oxidative damage and oxygen deprivation stress: A review. *Ann. Bot., 91*, 179–194.

Bota, J., Medrano, H., & Flexas, J., (2004). Is photosynthesis limited by decreased Rubisco activity and RuBP content under progressive water stress? *New Phytol., 162*, 671–681.

Cahoon, E. B., Hall, S. E., Ripp, K. G., Ganzke, T. S., Hitz, W. D., & Coughlan, S. J., (2003). Metabolic redesign of vitamin E biosynthesis in plants for tocotrienol production and increased antioxidant content. *Nat. Biotechnol., 21*, 1082–1087.

Cattivelli, L., Rizza, F., Badeck, F. W., Mazzucotelli, E., Mastrangelo, A. M., Francia, E., et al. (2008). Drought tolerance improvement in crop plants: an integrated view from breeding to genomics. *Field Crop Res.*, *105,* 1–4.

Cazzonelli, C. I., (2011). Carotenoids in nature: Insights from plants and beyond. *Funct. Plant Biol.*, *38,* 833–847.

Chaves, M. M., & Oliveira, M. M., (2004). Mechanisms underlying plant resilience to water deficits: prospects for water-saving agriculture. *J. Exp. Bot.*, *55,* 2365–2384.

Cheeseman, J. M., (2013). The integration of activity in saline environments: Problems and perspectives. *Funct. Plant Biol.*, *40,* 759–774.

Cornic, G., & Massacci, A., (1996). Leaf photosynthesis under drought stress. *Photosynthesis Environ.*, *30,* 347–366.

Costa, D. L., Delle, V. G., Gianquinto, G., Giovanardi, R., & Peressotti, A., (1997). Yield, water use efficiency and nitrogen uptake in potato: influence of drought stress. *Potato Res.*, *40,* 19–34.

Davidson, E. A., Verchot, L. V., Cattanio, J. H., Ackerman, I. L., & Carvalho, J. E., (2000). Effects of soil water content on soil respiration in forests and cattle pastures of eastern Amazonia. *Biogeochemistry*, *48,* 53–69.

Egilla, J. N., Davies, F. T., & Boutton, T. W., (2005). Drought stress influences leaf water content, photosynthesis, and water-use efficiency of Hibiscus rosa-sinensis at three potassium concentrations. *Photosynthetica.*, *31,* *43* (1), 135–40.

Ertek, A., Şensoy, S., Küçükyumuk, C., & Gedik, İ., (2006). Determination of plant-pan coefficients for field-grown eggplant (*Solanummelongena* L.) using class a pan evaporation values. *Agric. Water Manage.*, *85,* 58–66.

Escalona, J. M., Bota, J., & Medrano, H., (2015). Distribution of leaf photosynthesis and transpiration within grapevine canopies under different drought conditions. *VITIS J. Grapevine Res.*, *42,* 57.

Farooq, M., Wahid, A., Kobayashi, N., Fujita., & Basra, S. M., (2009). Plant drought stress: effects, mechanisms, and management. *Agron. Sus. Dev.*, *29,* 185–212.

Fathi, A., & Tari, D. B., (2016). Effect of drought stress and its mechanism in plants. *Int. J. Life Sci.*, *10,* 1–6.

Fazeli, F., Ghorbanli, M., & Niknam, V., (2007). Effect of drought on biomass, protein content, lipid peroxidation and antioxidant enzymes in two sesame cultivars. *Biol. Plant.*, *51,* 98–103.

Foyer, C. H., & Fletcher, J. M., (2001). Plant antioxidants: Color me healthy. *Biologist (London, England)*, *48,* 115–120.

Frederick, J. R., Camp, C. R., & Bauer, P. J., (2001). Drought-stress effects on branch and mainstem seed yield and yield components of determinate soybean. *Crop Sci.*, *41,* 759–763.

Fu, J., & Huang, B., (2001). Involvement of antioxidants and lipid peroxidation in the adaptation of two cool-season grasses to localized drought stress. *Environ Exp. Bot.*, *45,* 105–114.

Gagneul, D., Aïnouche, A., Duhazé, C., Lugan, R., Larher, F. R., & Bouchereau, A., (2007). A reassessment of the function of the so-called compatible solutes in the halophytic Plumbaginaceae *Limoniumlatifolium*. *Plant Physiol.*, *144,* 1598–1611.

Garg, B. K., (2003). Nutrient uptake and management under drought: Nutrient-moisture interaction. *Curr. Agric.*, *27,* 1–8.

Gille, L., & Nohl, H., (2001). The ubiquinol/bc1 redox couple regulates mitochondrial oxygen radical formation. *Arch. Biochem. Biophys.*, *388,* 34–38.

Giri, G. S., & Schillinger, W. F., (2003). Seed priming winter wheat for germination, emergence, and yield. *Crop Sci., 43,* 2135–2141.

Goetz, M., Godt, D. E., Guivarc'h, A., Kahmann, U., Chriqui, D., & Roitsch, T., (2001). Induction of male sterility in plants by metabolic engineering of the carbohydrate supply. *Proc. Natl. Acad. Sci., 98,* 6522–6527.

Grossman, S. R., Deato, M. E., Brignone, C., Chan, H. M., Kung, A. L., Tagami, H., et al. (2003). Polyubiquitination of p53 by a ubiquitin ligase activity of p300. *Sci., 300,* 342–344.

Guerfel, M., Baccouri, O., Boujnah, D., Chaïbi, W., & Zarrouk, M., (2009). Impacts of water stress on gas exchange, water relations, chlorophyll content and leaf structure in the two main Tunisian olive (*Oleaeuropaea* L.) cultivars. *Sci. Hortic., 119,* 257–263.

Guo, W. S., Schaefer, D. M., Guo, X. X., Ren, L. P., & Meng, Q. X., (2009). Use of nitrate-nitrogen as a sole dietary nitrogen source to inhibit ruminalmethanogenesis and to improve microbial nitrogen synthesis in vitro. *Asian-Aust J. Anim. Sci., 22,* 542–549.

Hasegawa, P. M., Bressan, R. A., Zhu, J. K., & Bohnert, H. J., (2000). Plant cellular and molecular responses to high salinity. *Ann. Rev. Plant Biol., 51,* 463–499.

Hussain, T. M., Hazara, M., Sultan, Z., Saleh, B. K., & Gopal, G. R., (2008). Recent advances in salt stress biology a review. *Biotech. Mol. Biol. Rev., 3,* 8–13.

Jaleel, C. A., Manivannan, P. A., Wahid, A., Farooq, M., Al-Juburi, H. J., Somasundaram, R. A., & Panneerselvam, R., (2009). Drought stress in plants: A review on morphological characteristics and pigments composition. *Int. J. Agric. Biol., 11,* 100–105.

Kabiri, R., & Naghizadeh, M., (2015). Exogenous acetylsalicylic acid stimulates physiological changes to improve growth, yield, and yield components of barley underwater stress condition. *J. Plant Physiol. Breed., 5,* 35–45.

Kim, J. Y., Mahé, A., Brangeon, J., & Prioul, J. L., (2000). A maize vacuolar invertase, IVR2, is induced by water stress. Organ/tissue specificity and diurnal modulation of expression. *Plant Physiol., 124,* 71–84.

Komor, E., (2000). Source physiology and assimilate transport: The interaction of sucrose metabolism, starch storage and phloem export in source leaves and the effects on sugar status in phloem. *Funct. Biol., 27,* 497–505.

Kumar, D., Yusuf, M. A., Singh, P., Sardar, M., & Sarin, N. B., (2013). Modulation of antioxidant machinery in [alpha]-tocopherol-enriched transgenic *Brassica juncea* plants tolerant to abiotic stress conditions. *Protoplasma., 250,* 1079.

Lambers, H., Atkin, O. K., & Millenaar, F. F., (1996). Respiratory patterns in roots in relation to their functioning. Plant Roots. *Hidden Half,* 323–362.

Leport, L., Turner, N. C., Davies, S. L., & Siddique, K. H., (2006). Variation in pod production and abortion among chickpea cultivars under terminal drought. *Eur. J. Agron., 24,* 236–246.

Li, R. H., Guo, P. G., Michael, B., Stefaniam G., & Salvatore, C., (2006). Evaluation of chlorophyll content and fluorescence parameters as indicators of drought tolerance in barley. *Agric. Sci., 5,* 751–757.

Liu, H. S., & Li, F. M., (2005). Root respiration, photosynthesis, and grain yield of two spring wheat in response to soil drying. *Plant Growth Reg., 46,* 233–240.

Lokhande, V. H., & Suprasanna, P., (2012). Prospects of halophytes in understanding and managing abiotic stress tolerance. In: *Environmental Adaptations and Stress Tolerance of Plants in the Era of Climate Change,* (pp. 29–56). Springer New York.

Lunn, J. E., Delorge, I., Figueroa, C. M., Van, D. P., & Stitt, M., (2014). Trehalose metabolism in plants. *Plant J., 79,* 544–567.

Maleki, A., Naderi, A., Naseri, R., Fathi, A., Bahamin, S., & Maleki, R., (2013). Physiological performance of soybean cultivars under drought stress. *Bull. Env. Pharmacol. Life Sci., 2,* 38–44.

Manickavelu, A., Nadarajan, N., Ganesh, S. K., Gnanamalar, R. P., & Chandra, B. R., (2006). Drought tolerance in rice: Morphological and molecular genetics consideration. *Plant Growth Reg., 50,* 121–138.

McWilliam, E. L., Barry, T. N., Lopez-Villalobos, N., Cameron, P. N., & Kemp, P. D., (2004). The effect of different levels of poplar (Populus) supplementation on the reproductive performance of ewes grazing low-quality drought pasture during mating. *Anim. Feed Sci. Technol., 115,* 1–8.

Meloni, D. A., Gulotta, M. R., Martínez, C. A., & Oliva, M. A., (2004). The effects of salt stress on growth, nitrate reduction and proline and glycinebetaine accumulation in Prosopis alba. *Braz. J. Plant Physiol., 16,* 39–46.

Miura, K., & Tada, Y., (2014). Regulation of water, salinity, and cold stress responses by salicylic acid. *Front. Plant Sci., 5.*

Miyashita, K., Tanakamaru, S., Maitani, T., & Kimura, K., (2005). Recovery responses of photosynthesis, transpiration, and stomatal conductance in kidney bean following drought stress. *Environ. Exp. Bot., 53,* 205–214.

Mohamed, A. A., & Aly, A. A., (2008). Alterations of some secondary metabolites and enzymes activity by using exogenous antioxidant compound in onion plants grown under seawater salt stress. *American-Eurasian J. Sci. Res., 3,* 139–146.

Møller, I. M., (2001). Plant mitochondria and oxidative stress: electron transport, NADPH turnover, and metabolism of reactive oxygen species. *Ann. Rev. Plant Biol., 52,* 561–591.

Monakhova, O. F., Chernyadèv, I. I., Munne-Bosch, S., & Penuelas, J., (2003). Protective role of kartolin–4 in wheat plants exposed to Photo-and antioxidative protection, and a role for salicylic acid during drought and recovery in field-grown Phillyreaangustifolia plants. *Planta., 217,* 758–766.

Munné-Bosch, S., Shikanai, T. l., & Asada, K., (2005). Enhanced ferredoxin-dependent cyclic electron flow around photosystem I and α-tocopherolquinone accumulation in water-stressed ndhB-inactivated tobacco mutants. *Planta., 222,* 502–511.

Munns, R., & Tester, M., (2008). Mechanisms of salinity tolerance. *Annu. Rev. Plant Biol., 59,* 651–681.

Murakeözy, É. P., Nagy, Z., Duhazé, C., Bouchereau, A., & Tuba, Z., (2003). Seasonal changes in the levels of compatible osmolytes in three halophytic species of inland saline vegetation in Hungary. *J. Plant Physiol., 160,* 395–401.

Nounjan, N., & Theerakulpisut, P., (2012). Effects of exogenous proline and trehalose on physiological responses in rice seedlings during salt-stress and after recovery. *Plant Soil Environ., 58,* 309–315.

Pettigrew, W. T., (2004). Physiological consequences of moisture deficit stress in cotton. *Crop Sci., 44,* 1265–1272.

Pitman, M., & Läuchli, A., (2004). Global impact of salinity and agricultural ecosystems. In: Läuchli, A., & Lüttge, U., (eds.), *Salinity: Environment–Plants–Molecules* (pp. 3–20). Springer Verlag, Netherlands.

Rao, A. V., & Rao, L. G., (2007). Carotenoids and human health. *Pharmacol. Res., 55,* 207–216.

Reddy, A. R., Chaitanya, K. V., & Vivekanandan, M., (2004). Drought-induced responses of photosynthesis and antioxidant metabolism in higher plants. *J. Plant Physiol., 161,* 1189–202.

Rengasamy, P., (2006). World salinization with emphasis on Australia. *J. Exp. Bot. 57,* 1017–1023.

Sairam, R. K., Srivastava, G. C., Agarwal, S., & Meena, R. C., (2005). Differences in antioxidant activity in response to salinity stress in tolerant and susceptible wheat genotypes. *Biol. Plant., 49,* 85–91.

Samarah, N. H., (2005). Effects of drought stress on growth and yield of barley. *Agron. Sustain. Dev., 25* (1), 145–149.

Shafiq, S., Akram, N. A., & Ashraf, M., (2015). Does exogenously-applied trehalose alter oxidative defense system in the edible part of radish (*Raphanussativus* L.) underwater-deficit conditions?. *Sci. Hortic., 185,* 68–75.

Shahbaz, M., Mushtaq, Z., Andaz, F., & Masood, A., (2013). Does proline application ameliorate adverse effects of salt stress on growth, ions, and photosynthetic ability of eggplant (*Solanummelongena* L.)?. *Sci. Hortic., 164,* 507–511.

Siaga, E., Maharijaya, A., & Rahayu, M. S., (2016). Plant growth of eggplant (*Solanummelongena*L.) in vitro in drought stress polyethylene glycol (Peg). Biovalentia: *Biol. Res. J., 2,* 218–235.

Siddiqui, M. H., Al-Whaibi, M. H., & Basalah, M. O., (2011). Interactive effect of calcium and gibberellin on nickel tolerance in relation to antioxidant systems in *Triticum aestivum* L. *Protoplasma., 248,* 503–511.

Song, T., Pranovich, A., & Holmbom, B., (2011). Effects of pH control with phthalate buffers on hot-water extraction of hemicelluloses from spruce wood. *Bioresour. Technol., 102,* 10518–10523.

Sun, J., Lu, N., Xu, H., Maruo, T., & Guo, S., (2016). Root zone cooling and exogenous spermidine root-pretreatment promoting *Lactuca sativa* L. Growth and photosynthesis in the high-temperature season. *Front. Plant Sci., 7.*

Sureshan, K. M., Murakami, T., & Watanabe, Y., (2009). Total syntheses of cyclitol based natural products from myo-inositol: brahol and pinpollitol. *Tetrahedron., 65,* 3998–4006.

Szabados, L., & Savouré, A., (2010). Proline: a multifunctional amino acid. *Trends Plant Sci., 15,* 89–97.

Tezara, W., Mitchell, V. J., Driscoll, S. D., & Lawlor, D. W., (1999). Water stress inhibits plant photosynthesis by decreasing coupling factor and ATP. *Nature, 401,* 914–917.

Trinchant, J. C., Boscari, A., Spennato, G., Van-de-Sype, G., & Le-Rudulier, D., (2004). Prolinebetaine accumulation and metabolism in alfalfa plants under sodium chloride stress. Exploring its compartmentalization in nodules. *Plant Physiol., 135,* 1583–1594.

Turner, N. C., Wright, G. C., & Siddique, K. H., (2001). Adaptation of grain legumes (pulses) to water-limited environments. *Adv. Agron., 71,* 193–231.

Verbruggen, N., & Hermans, C., (2008). Proline accumulation in plants: A review. *Amino Acids, 35,* 753–759.

Wahid, A., & Rasul, E., (1997). Identification of salt tolerance traits in sugarcane lines. *Field Crops Res., 54,* 9–17.

Waraich, E. A., Ahmad, R., & Ashraf, M. Y., (2011). Role of mineral nutrition in alleviation of drought stress in plants. *Australian J. Crop Sci., 5,* 764.

Wu, X., He, J., Chen, J., Yang, S., & Zha, D., (2014). Alleviation of exogenous 6-benzyladenine on two genotypes of eggplant (*Solanummelongena* Mill.) growth under salt stress. *Protoplasma., 251,* 169.

Wu, X., He, J., Ding, H., Zhu, Z., Chen, J., Xu, S., & Zha, D., (2015). Modulation of zinc-induced oxidative damage in *Solanummelongena* by 6-benzylaminopurine involves ascorbate–glutathione cycle metabolism. *Environ. Exp. Bot., 116,* 1–1.

Yancey, P. H., (2005). Organic osmolytes as compatible, metabolic, and counteracting cytoprotectants in high osmolarity and other stresses. *J. Exp. Biol.*, *208*, 2819–2830.

Yokota, A., Kawasaki, S., Iwano, M., Nakamura, C., Miyake, C., & Akashi, K., (2002). Citrulline and DRIP-1 protein (ArgE homolog) in drought tolerance of wild watermelon. *Ann. Bot.*, *89*, 825–832.

Yu, L. H., Wu, S. J., Peng, Y. S., Liu, R. N., Chen, X., Zhao, P., et al. (2016). Arabidopsis EDT1/HDG11 improves drought and salt tolerance in cotton and poplar and increases cotton yield in the field. *Plant Biotechnol. J.*, *14*, 72–84.

Yuanyuan, M., Yali, Z., Jiang, L., & Hongbo, S., (2009). Roles of plant soluble sugars and their responses to plant cold stress. *Afri. J. Biotechnol.*, *8*, 101–112.

Zeid, I. M., & Shedeed, Z. A., (2006). Response of alfalfa to putrescine treatment under drought stress. *Biol. Plant.*, *50*, 635–640.

Zhang, J. X., Xu, F. L., Lu, J. L., Li, Y. L., & Li, D., (2003). Effect of exogenous betaine on drought resistance of different crops. *Agric. Res. Arid Areas*, *2*, 1–9.

Zhao, M., & Running, S. W., (2010). Drought-induced reduction in global terrestrial net primary production from 2000 through 2009. *Science*, *329*, 940–943.

Zheng, M., Tao, Y., Hussain, S., Jiang, Q., Peng, S., Huang, J., Cui, K., & Nie, L., (2016). Seed priming in dry direct-seeded rice: Consequences for emergence, seedling growth and associated metabolic events under drought stress. *Plant Growth Reg.*, *78*, 167.

Zinselmeier, C., Jeong, B. R., & Boyer, J. S., (1999). Starch and the control of kernel number in maize at low water potentials. *Plant Physiol.*, *121*, 25–36.

Zou, J. J., Wei, F. J., Wang, C., Wu, J. J., Ratnasekera, D., Liu, W. X., & Wu, W. H., (2010). *Arabidopsis* calcium-dependent protein kinase CPK10 functions in abscisic acid-and Ca^{2+}-mediated stomatal regulation in response to drought stress. *Plant Physiol.*, *154*, 1232–1243.

CHAPTER 13

ENHANCING MORPHO-PHYSIOLOGICAL RESPONSES AND YIELD POTENTIAL IN OKRA (*ABELMOSCHUS ESCULENTUS* L. MOENCH) UNDER SALINITY STRESS

MARIA NAQVE and MUHAMMAD SHAHBAZ

Department of Botany, University of Agriculture, Faisalabad, Pakistan,
E-mail: shahbazmuaf@yahoo.com

13.1 INTRODUCTION

Vegetables are considered as important food crops due to their nutritional gradient (Noreen and Ashraf, 2009). Okra (*Abelmoschus esculentus* L. Moench) is the major edible crop belongs to Malvaceae, the crop is believed to be originated in tropics of Africa and Mediterranean regions (Abbas et al., 2014; Hameed et al., 2017) (Figure 13.1). Okra contains high amounts of fats, carbohydrates, minerals including calcium, phosphorus, iron, and vitamins including ascorbic acid, β-carotene, niacin, riboflavin, and thiamin (Adewale et al., 2017).

Seeds of okra serve as good source of linoleic acid (48%) and rich in edible oil (Gemede et al., 2015). On the other hand, Okra is also a rich source of mucilage which is mainly found in its fresh and dry fruits. This crop is also a great source of mucilage, which has important medicinal and industrial values (Olympia et al., 2014). Approximately, 4.8 million tons of okra fruits have been produced globally per year but only 0.116 million tons are contributed by Pakistan (Shahid et al., 2013). Reduction in yield is associated with antagonistic effects of abiotic stresses primarily due to salinization in the soil and groundwater mainly (Khan et al., 2001).

Unfavorable environmental conditions like salinization of soil and water, extreme temperatures, heavy metals, UV, and drought which adversely affect productivity of crops are collectively termed as "abiotic stress" (Ahmad and Prasad 2012a). Among these unpredictable environmental conditions, salinization mainly restrained the productivity of vegetable crops. It is difficult to evaluate salt stress precisely and the area of saline soils is increasing annually posing greater risk to irrigated soils (Shanker and Venkateswarlu, 2011). Salinization mainly reduces the growth and yield of plants by 50% worldwide (Zhu, 2001).

Salt stress imposes adverse effects overall plant by disturbing its cellular levels via ionic and osmotic stress. Salinization pressurizes the plants in two ways, primarily by the accumulation of higher levels of salts in root zone which interrupts the normal metabolism of plant by inducing osmotic stress, and larger quantities of salt inside the plant tissues cause ion toxicity (Munns, 2004).

The adverse physiological responses of a plant to salt stress are often complex and multi-faceted, which are mainly associated with increased levels of NaCl (Tavakkoli et al., 2010). Vegetable crops can be demised by increased NaCl levels which affects the crops through osmotic stress and ionic toxicity which ultimately causes devastating effects on plant metabolism, photosynthetic apparatus, enzymes dysfunction and cellular injuries (Hasanuzzaman et al., 2012a).

FIGURE 13.1   A general view of okra fruits.

13.2   RESPONSES OF CROP PLANTS TO SALT STRESS

13.2.1   GERMINATION AND GROWTH

Germination process plays a central part in the plant growth and higher productivity of plants that is badly influenced by salinity stress (Xu et al., 2011). Salinity imposes harmful effects on plant development. Growth attributes show reduction with increasing levels of Na^+ and Cl^- ions (Bazihizina et al., 2012). Increasing salinity levels show adverse impacts on photosynthesis, height, and weight of plants and productivity yield in wheat (Chaabane et al., 2011) and rice (Rad et al., 2012). Elevated levels of salts in the rhizosphere are one of the most brutal factors, which limit the seed germination altering hormonal balance and enzymes function reducing crop yield (Khan and Rizvi, 1994; Dantas et al., 2007). Abbas et al. (2014) directed a research to examine the effect of salt (NaCl) tress on twenty genotypes of okra. Various levels of salt stress markedly suppressed plant growth and germination parameters including percentage index as well as length of embryonic axis. The response of all okra genotypes to salinity stress varied significantly for all the morpho-physiological attributes investigated. Overall, genotypes OH–713, OH–139, OH–138, OH–2324 and OH–001 proved to be superior while OH–597, MD–02, OH–152, PMF-Beauty, PusaSwani, JKOH- 456, OH–809 and MF–04 were inferior under saline conditions. Genotypes Kiran, Okra–1548, Ikra–3, Sabzpari, Okra–7100, Sitara–9101 and Okra–7080 have shown susceptibility to salinity stress based on germination characteristics.

13.2.2 NUTRIENT IMBALANCE

Salinization results abnormalities in ionic status of plants. Higher levels of salts cause nutrients imbalance and stunted growth through osmotic and ionic stresses (Hasegawa et al., 2000). Accumulation of Na^+ and Cl^- ions in rhizosphere interferes with plants efficiency to acquire water that finally causes water stress inducing reduction in growth and physiological damages (Munns, 2002a). Vegetable crops suffer from severe nutritional ailments due to competition among Cl^-, Na^+ Ca^{2+} K^+ Mg^{2+} and NO^-_3 ions (Grattan et al., 1992; Hu and Schmidhalter, 2005) Higher concentrations of toxic ions induce numerous inhibitory effects on biomass and morphological characteristics of various vegetable crops grown in saline soils (Giuffrida et al., 2013).

13.2.3 PHOTOSYNTHESIS

Increased amount of Cl^- and Na^+ ions adversely affects photosynthetic apparatus of crop plants, physiological attributes and ultimately yield (Nawaz et al., 2010; Flowers et al., 2010). Elevated levels of NaCl seems to augment stomatal conductance inhibiting the assimilation rate of CO_2 (Flexas et al., 2007; Maxwell and Johnson, 2000). The increase in NaCl concentration in chloroplast is commonly reported phenomenon, which reduces the photosynthetic rate causing poor yield (Sudhir and Murthy, 2004). Shahid et al. (2011) determined that salt stress significantly affect physio-biochemical attributes of okra cultivar Chinese red. It was reported that ion toxicity in the leaves and root zone disturb germination process photosynthesis and ambient CO_2 concentration. Saleem et al. (2011) investigated the physiological response of okra (*Abelmoschus esculentus* L. Moench) cultivars Nirali and Posa Sawni under the influence of salinity stress. The results revealed reduction in growth traits, transpiration rate and chlorophyll *b* content and enhancement in the levels of toxic ions. Moreover, it was noted that N level in the root, chlorophyll *b* content and effectiveness of PS-II was significantly improved in salinity tolerant Nirali cultivar of okra. While Na^+ and Cl^- ion contents were enhanced in the root of Posa Sawni cultivar of okra with reduced growth traits under saline condition.

13.2.4　PLANT WATER RELATIONS

Plants lose their ability to uptake water and water relation characteristics negatively affect under varying saline regimes (Parida and Das, 2005). Higher concentrations of salts in the rhizosphere reduce water relations of crop plants such as water potential thus, plants become unable to maintain their turgor potential inducing physiological damages (Romero-Aranda et al., 2001). When plants suffer water deficit conditions RWC (relative water content) have shown reduction due to increased salt (NaCl) content in sugar beet (Ghoulam et al., 2002).

Ion toxicity disturbs the water relation attributes of plants such as solute and water potentials (Parida and Das, 2005). Plants can resist ion toxicity through adjusting themselves osmotically by enhancing biosynthesis of proline and trehalose (Serrano et al., 1999).

13.2.5　YIELD

Salt stress reduced physiological attributes, which ultimately caused yield losses. Plants have to face troubles in maintaining growth sequence and yield production (Yokoi et al., 2002). Salinity stress significantly induced reduction in seed yield and seed quality in faba bean (Orabi et al., 2014). Different yield components of okra cultivar Chinese red were significantly affected by salinity stress due to physio-biochemical damages including gas exchange parameters (Shahid et al., 2011).

13.2.6 OXIDATIVE STRESS

Oxygen-containing free radicals are defined as reactive oxygen species (ROS) and mainly ROS generated in crop plants are superoxide anion radical, hydroxyl radical, singlet oxygen and hydrogen peroxide. Level of these species increased under salt stress (Ahmad and Umar, 2011). Assaha et al. (2015) observed the adverse effect of salinization on the role of antioxidant enzymes and cell wall peroxidase enzymes to compare the adaptations between eggplant and huckleberry to overcome saline stress. Higher accumulation of Na$^+$, malondialdehyde (MDA) content and reduced growth was observed in eggplant than huckleberry by the influence of salt stress. Activities of glutathione reductase (GR) and cell wall Peroxidase (cwPOD) and catalase (CAT) were increased during salinity stress, which are main scavengers of reactive oxygen species. Enhanced antioxidant properties are used as mechanism to resist salinity. Significant damages have been reported on the structure and activities of enzymes, lipids, DNA, RNA, and membrane dysfunction by salt stress in vegetable crops including pea and tomato (Hasegawa et al., 2000; Mittova et al., 2004).

13.3 TOLERANCE OF VEGETABLE CROPS TO SALT STRESS

Salt stress tolerance can be defined as the ability of crop plants to combat adverse effects of surplus amounts of salts in the rhizosphere and vegetable crops respond differently in tolerance of salinity stress (Genuchten and Hoffman, 1984). The salinity tolerance of plants depends on several aspects e.g., initial and final growth stages, forms of accumulated salts, Characteristics of soil and absorption of ambient CO_2 (Shannon et al., 1998; Maggio et al., 2002). Vegetables show more sensitivity to salinization during initial level of growth such as germination and seedling emergence rather than during final phase of growth (Läuchli et al., 2007). Plants mainly respond to ion toxicity either by sequestrating at cell vacuoles or by preventing their accumulation via salt exclusion mechanism (Munns et al., 2006). Salt stress can shunt the oxidative burst by producing reactive oxygen species. ROS-mediated cellular damages occur by peroxidation of lipids, while proteins dysfunction have been observed due to elevated levels of salt stress in different crops including rice, tomato, pea, and turnip. Exogenous application of various osmolytes, phytohormones, and antioxidants such as enzymatic and non-enzymatic offer remarkable protection against these damages in plants by increasing their rate of germination, growth, and developmental processes, photosynthetic rate, and ultimately yield (Ahmad et al., 2010b; Hasanuzzaman et al., 2011a).

13.4   ROLE OF EXOGENOUS AND ENDOGENOUS PROTECTANTS TO ALLEVIATE HARMFUL EFFECTS OF SALINIZATION

It has been reported by various researchers that applications of antioxidants, osmolytes, and various phytohormones are the substances that scavenge ROS and prevent the damages of salinity on crop plants by increasing germination and growth rates, physiological characteristics and crop productivity (Hasanuzzaman et al., 2012b).

13.4.1   ANTIOXIDANTS

13.4.1.1   GLUTATHIONE

Glutathione (GSH) plays vital role as antioxidant in plant tissues by protecting physiological processes of plants as a reducing agent for many free radicals. Glutathione reduced oxidative damages primarily by quenching the hydroxyl radical and singlet oxygen species. GSH also helps the crop plants to minimize salinity stress by stabilizing redox potential. In addition, GSH helps in maintaining protein structure and enzymes activities by providing binding sites to (GST) glutathione-S-transferases and (GPX) glutathione peroxidase (Noctor et al., 2002). GSH accumulation has been reported under saline regimes inducing tolerance by acting as a signaling molecule. GSH can also directly scavenge ROS by enhancing the initiation of multiple genes which are involved in the biosynthesis of (GR) glutathione reductase. Foliar application of glutathione combats the brutal impacts of salinity on growth of *Brassica napus*, *Allium cepa* and cotton seedlings. These plants showed enhancement in growth coupled with an increase in synthesis of cysteine and sulfur under varying saline regimes (Gossett et al., 1996; Kattab, 2007; Salama and Al-Mutawa, 2009). Exogenous applications of GSH have been found to overcome salt-induced damages by enhancing morpho-physiological characteristics such as fresh and dry weight, tallness of plants, weight, and number of flowers, carbohydrates content, photosynthetic pigments and ionic content (Rawia et al., 2011).

13.4.1.2   ASCORBIC ACID

The accumulation of antioxidants such as ascorbic acid (AsA) is a well adaptive strategy of plants against salinity. Ascorbic acid is also known as vitamin C. Its biosynthesis occurs in cytosol from D-glucose. Great experimental work has been conducted about role of vitamin C as antioxidant to combat salt stress as ROS scavenging agent. It has been reported that AsA along with other antioxidants such as CAT and POD combat adverse effects of ROS produced by salinity stress in crop plants (Foyer and Noctor, 2005). Findings of numerous researches have revealed that exogenous applications of ascorbic acid induced the regulation of many enzymes to manage salt stress along with other antioxidants and act as a signaling molecule (Athar et al., 2008). Pre-seed treatment of C. annuum and Glycine max with ascorbic acid speed up the ability of crop plants by accelerating the activities of catalase and peroxide to combat the negative impacts of salt stress (Beltagi, 2008; Munir and Aftab, 2011). Significant synergistic effect between NaCl and AsA application have been observed where AsA remarkably enhanced the efficiency of photosynthetic pigments by scavenging the H_2O_2 activities (Azzedine et al., 2011).

13.4.1.3   TOCOPHEROLS

Tocopherols are essentially present in almost all higher plants and their biosynthesis mainly occurs in plastids (Figure 13.2). They are also known as vitamin E and have a sub-family of tocotrienols possessing four isomeric forms. Vitamin E also plays vital role in protecting membranes by breaking chain reaction of lipid peroxyl radicals (Maeda et al., 2005; Garg and Manchanda, 2009). In addition, tocopherols play vital role in scavenging free radicals and by protecting plant photosystems and membranes from damages. Photosystem II is protected from photoinhibition by the antioxidative role of alpha tocopherol (Havaux et al., 2005).

Salt stress activated the inhibition in the morpho-physiological characteristics of Vicia faba lines Giza 40 and Giza 429 Semida et al., 2014). Significant response of various levels of tocopherol isomer was observed in enhancing growth and yield potential of these lines. Farouk (2011) reported the pre-seed treatment of H. annuus with α-tocopherol which is effective to tolerate salinity due to increased levels of catalase and peroxidase enzymes and by enhancing proline concentration, photosynthesis, and reducing ion toxicity.

(a) Control	(b) 100 mM NaCl; 200 mg L-1 α-Toco	(c) 100 mM NaCl; 300 mg L-1 α-Toco

FIGURE 13.2   Effect of various levels of α-tocophrol on vegetative growth and fruit size of Okra plants exposed to 100 mM NaCl in sand culture.

13.4.2   PLANT OSMOPROTECTANTS

13.4.2.1   GLYCINEBETAINE

Glycinebetaine (GB) is an important natural compound which is water soluble. Glycinebetaine is non-toxic even at higher levels and provides potential protective effect against salinity (Ashraf and Foolad, 2007; Chen and Murata, 2008). Glycinebetaine serves as compatible solute and protects the cells from salinity effects (Gadallah, 1999) by membrane stability (Makela et al., 2000), reduction of active oxygen species (Ashraf and Foolad, 2007), and protecting the photosynthetic pigments (Cha-Um and Kirdmanee, 2010).

Exogenous application of glycinebetaine enhances chlorophyll content, photosynthetic rate and water use efficiency of rice plants under varying saline regimes (Cha-Um and Kirdmanee, 2010). A study on rice seedling suggested that glycinebetaine has important role in the absorption and translocation of positive ions under saline conditions. Application of glycinebetaine along nutrient solution had no damaging effect on plants growing under normal conditions but it improves the salinity tolerance of stressed plants. The effective role of glycinebetaine is due to its ability to reduce the sodium content (Na^+) and accumulate potassium ions (K^+) under saline conditions (Lutts, 2000).

13.4.2.2 PROLINE

Proline is an important compatible osmolyte which provides an adaptive mechanism for salt stress in plants. Under salinity stress conditions, accumulation of proline act as selection criterion for salinity tolerance in several plant species (Parida and Das, 2005). It has been reported in various studies that proline has antioxidant properties and it serves as scavenger of active oxygen species especially singlet oxygen ($^1O^2$) (Matysik et al., 2002). Salt stress disturbs the mechanism of antioxidants such as peroxidase and catalase but proline application significantly enhances their activities by introducing genes of salt tolerant proteins (Khedr et al., 2003). Proline enhances the tolerance ability of plants to quench salinity by activating the Rubisco enzyme, photosynthetic machinery and membrane structure (Hamilton and Heckathorn, 2001). Foliar application of proline in combination with glycinebetaine reduces the apoplastic flow of sodium ions in rice seedlings. Proline application also improves the chlorophyll pigments, photosynthetic rate and *Fv/Fm* ratio (Sobahan et al., 2009).

13.4.2.3 TREHALOSE

Trehalose is a sugar which acts as compatible solute and its concentration increases in plants during stress conditions. It provides the resistance against salt stress conditions by acting as an osmoprotectant in physiological responses of plants (Zeid, 2009). A study on maize plants revealed that pre-soaking treatment with trehalose releases the negative impacts of salt stress by improving chlorophyll and nucleic acid content. Application of trehalose imparts the salinity tolerance by protecting membranes, by lowering electrolyte leakage and enhancing the sodium-potassium ratio in maize leaves (Cortina and Culianez-Macia, 2005).

13.4.3   PLANT HORMONES

13.4.3.1   ABSCISIC ACID (ABA)

Abscisic acid is major plant hormone which has effective role under stress conditions. It has an important role in stress signaling. There are many biological and physiological processes such as seed dormancy, germination delay, seed development, closing of stomata, leaf falling and lipoprotein synthesis which are controlled by abscisic acid. Another important function of abscisic acid is the control of water balance and osmotic stress tolerance (Tuteja, 2007). Abscisic acid mitigates the salt stress by increasing sodium chloride ratio, sodium-potassium ratio and increasing the concentration of proline, soluble sugars and calcium and potassium ratio (Gurmani et al., 2011).

High accumulation of abscisic acid in plant tissue under salinity were reported in several studies (Babu et al., 2012). Jeschke et al. (1997) studied that the enhanced concentration of abscisic acid in the vascular tissues is associated with non-significant stomatal conductance. Salt stress enhances the ABA synthesis in roots and its transportation through xylem is also coordinated with stomatal conductance.

13.4.3.2   INDOLE ACETIC ACID (IAA)

Rare reports are available about the association between auxin accumulation and salinity tolerance in plants and its role in relieving salinity stress damages. Some studies have shown that IAA has role in salt tolerance in plants. The alterations in IAA accumulation under saline regimes are like the ABA content (Ribaut and Pilet, 1991). Sakhabutdinova et al. (2003) reported that salinity declines the indole acetic acid level in wheat plants. Under salt stress conditions, decline in indole acetic acid concentration was also observed in rice plants (Nilsen and Orcutt, 1996).

Salt stress lowers the germination rate of wheat but pre-sowing treatment of seeds with IAA lowers the growth reducing effects of NaCl salt (Afzal et al., 2005). There is an evidence that salt stress enhances the concentration of indole acetic acid which works as signaling molecule in the growing zone of leaf (Akhiyarova et al., 2005). Akbari et al. (2007) described that exogenously applied auxin promotes the lengths of hypocotyls, plant weights of wheat seedling under higher levels of NaCl.

13.4.3.3 GIBBERELLIC ACID

Gibberellic acid or Gibberellin is mostly involved in growth and development, i.e., it controls leaf expansion, seed germination, flowering, and stem elongation (Kim and Park, 2008). Moreover, gibberellic acid interacts with other plant growth regulators to enhance metabolism of plants. Gibberellins show their beneficial effects under saline condition (Kaya et al., 2009). Maggio et al. (2010) reported that application of GA3 helps to reduce stomatal resistance in tomato (*Lycopersicum esculentum*). Grain yield in *Triticum aestivum* improved when pre-sowing grain treatment was made by GA3 under saline regimes. Application of GA3 (10 mM) in *Brassica juncea* significantly increased dry mass, leaf area, chlorophyll contents and photosynthesis under varying levels of NaCl (Shah, 2007). Moreover, GA3 proved to cope with salinity by counteracted on water content level, chlorophyll level and electrolytic leakage (Ahmad et al., 2009).

It has sufficient stimulatory effect on proline and GB biosynthesis (Ahmad, 2010). Foliar-applied gibberellic acid (100 ppm) sugar plays significant role in enhancing sugar contents under salt stress by enhancing uptake of nutrients, morphological, and physiological parameters as well. GA increases soluble protein, sugar, while chlorophyll remained uniform (Shomeili et al., 2011). Final percentage germination in *Beta vulgaris* increased by the seed priming with GA3. Root-shoot length and fresh mass of plant also showed considerable increase by priming method (Jamil and Rha, 2007). GA regulates phytochrome level in saline condition so plant resumes its normal growth. Phytochromonal analysis of soybean showed that GA3 treated plants along with jasmonates showed higher values while salicylic acid and ABA contents showed lower values under same treatment (Hamayun et al., 2010).

13.4.3.4 JASMONATES

Jasmonates (JA) are pervasive in plants and play vital role as hormone. They are involved in several regulatory processes in cell-like stomatal closure, germination process of seeds, fertility, fruit metabolism and fruit ripening (Hossain et al., 2011). Jasmonic acid is involved in various stress responses in plants by playing unique role as hormone. Foliar application of JA under salt stress is coped with the production of abundant proteins synthesis called JIPs. Jasmonic acid priming has vital role in increasing photosynthesis, relative water content and protein contents under salt stress (Rohwer and Erwin, 2008).

Foliar application of JA modulates the level of various plant hormones like ABA that is thought to be very important in understanding the protection mechanism in varying saline regimes (Kang et al., 2005). Accumulation of GA increases with increase in JA content under salinity (Seo et al., 2005). Pre-seed treatment with MeJA stabilizes the plant metabolism to tolerate salinity by enhancing morphological attributes, chl. contents, photosynthesis, transpiration, and proline content (Yoon et al., 2009).

13.5 CONCLUSION AND FUTURE PERSPECTIVE

Overall, it is concluded that salt stress imposes adverse effects on the whole plant by disturbing its cellular levels via ionic and osmotic stress. Crop species mainly experience yield losses because of these physiological level injuries. It is useful to work on tolerance level enhancements of crop plants and application of exogenous protectants. Among endogenous natural substances, plant hormones play vital role in amelioration of salt-induced damages. Although, their role as signaling molecules is not clearly understood yet. However, there is a need to understand the role of key molecular players involved in tolerance of salinity stress mechanism.

ACKNOWLEDGMENTS

This chapter is extracted from PhD thesis review of literature by Maria Naqve, Reg. No. 2012-ag–73 from The Department of Botany, University of Agriculture Faisalabad, Pakistan.

KEYWORDS

- abscisic acid
- gibberellic acid
- indole acetic acid
- jasmonates

REFERENCES

Abbas, T., Pervez, M. A., Ayyub, C. M., Shaheen, M. R., Tahseen, S., Shahid, M. A., et al. (2014). Evaluation of different okra genotypes for salt tolerance. *Int. J. Plant Anim. Environ. Sci.*, *4*, 23–30.

Afzal, I., Basra, S., & Iqbal, A., (2005). The effect of seed soaking with plant growth regulators on seedling vigor of wheat under salinity stress. *J. Stress Physiol. Biochem.*, *1*, 6–14.

Ahmad, P., & Prasad, M. N. V., (2012a). *Abiotic Stress Responses in Plants: Metabolism, Productivity, and Sustainability*. Springer, New York.

Ahmad, P., & Umar, S., (2011). *Oxidative Stress: Role of Antioxidants in Plants*. Studium Press, New Delhi.

Ahmad, P., Jaleel, C. A., & Sharma, S., (2010b). Antioxidative defense system, lipid peroxidation, proline metabolizing enzymes and Biochemical activity in two genotypes of *Morus alba* L. subjected to NaCl stress. *Russ. J. Plant Physiol.*, *57*, 509–517.

Ahmad, P., Jaleel, C. A., Salem, M. A., Nabi, G., & Sharma, S., (2010). Roles of Enzymatic and non–enzymatic antioxidants in plants during abiotic stress. *Crit. Rev. Biotechnol.*, *30*, 161–175.

Ahmad, P., Jeleel, C. A., Azooz, M. M., & Nabi, G., (2009). Generation of ROS and non–enzymatic antioxidants during abiotic stress in plants. *Bot. Res. Int.*, *2*, 11–20.

Akbari, G., Sanavy, S. A., & Yousefzadeh, S., (2007). Effect of auxin and salt stress (NaCl) on seed germination of wheat cultivars (*Triticum aestivum* L.). *Pak. J. Biol. Sci.*, *10*, 2557–2561.

Akhiyarova, G. R., Sabirzhanova, I. B., Veselov, D. S., & Frike, V., (2005). Participation of plant hormones in growth resumption of wheat shoots following short-term NaCl treatment. *Rus. J. Plant Physiol.*, *52*, 788–792.

Ashraf, M., & Foolad, M. R., (2007). Roles of glycine betaine and proline in improving plant abiotic stress resistance. *Environ. Exp. Bot.*, *59*, 206–216.

Assaha, D. V. M., L. Liu, Mekawy, A. M. M., Ueda, A., Nagaoka, T., & Saneoka, H., (2015). Effect of salt stress on Na accumulation, antioxidant enzyme activities and activity of cell wall peroxidase of huckleberry (*Solanum scabrum*) and eggplant (*Solanum melongena*) *Int. J. Agric. Biol.*, *17*, 1149–1156.

Athar, H. R., Khan, A., & Ashraf, M., (2008). Exogenously applied ascorbic acid alleviates salt-induced oxidative stress in wheat. *Environ. Exp. Bot.*, *63*, 224–231.

Azzedine, F., Gherroucha, H., & Baka, M., (2011). Improvement of salt tolerance in durum wheat by ascorbic acid application. *J. Stress Physiol. Biochem.*, *7*, 27–37.

Babu, M. A., Singh, D., & Gothandam, K. M., (2012). The effect of salinity on growth, hormone, and mineral elements in leaf and fruit of tomato cultivar PKM1. *J. Anim. Plant Sci.*, *22*, 159–164.

Bazihizina, N., Barrett-Lennard, E. G., & Colmer, T. D., (2012). Plant growth and physiology under heterogeneous salinity. *Plant Soil*, *354*, 1–19.

Bebelib, H., & Passama, C., (2014). Genetic and morphological diversity of okra (*Abelmoschus esculentus* [L.] Moench.) genotypes and their possible relationships, with particular reference to Greek landraces. *Hort. J.*, *171*, 58–70.

Beltagi, M. S., (2008). Exogenous ascorbic acid (Vitamin C) induced anabolic changes for salt tolerance in chick pea (*Cicer arietinum* L.) plants. *Afr. J. Plant Sci.*, *2*, 118–123.

Chaabane, R., Khoui, S., Khamassi, K., Teixeira da Silva, J. A., Ben Naceur, M., Bchini, H., et al. (2012). Molecular and agro-physiological approach for parental selection before intercrossing in salt tolerance breeding programs of durum wheat. *Int. J. Plant Breed.*, *6*, 100–105.

Cha–Um, S., & Kirdmanee, C., (2010). Effect of glycinebetaine on proline, water use, and photosynthetic efficiencies, and growth of rice seedlings under salt stress. *Turk. J. Agric. For.*, *34*, 517–527.

Chen, T. H. H., & Murata, N., (2008). Glycinebetaine: An effective protectant against abiotic stress in plants. *Trends Plant Sci.*, *13*, 499–505.

Cortina, C., Culianez–Maciá, F. A., (2005). Tomato abiotic stress enhanced tolerance by trehalose biosynthesis. *Plant Sci.*, *169*, 75–82.

Dantas, B. F., De Sá Ribeiro, L., & Aragao, C. A., (2007). Germination, initial growth and cotyledon protein content of bean cultivars under salinity stress. *Rev. Bras. de Sementes.*, *29*, 106–110.

Esan, A. M., Masisi, K., Dada, F. A., & Olaiyaa, C. O., (2017). Comparative effects of indole acetic acid and salicylic acid on oxidative stress marker and antioxidant potential of okra (*Abelmoschus esculentus*) fruit under salinity stress. *Sci. Hortic.*, *216*, 278–283.

Farouk, S., (2011). Ascorbic acid and α –Tocopherol minimize salt-induced wheat leaf senescence. *J. Stress Physiol. Biochem.*, *7*, 58–79.

Flexas, J., Diaz–Espejo, A., Galmés, J., Kaldenhoff, R., Medrano, H., & Ribas–Carbo, M., (2007). Rapid variations of mesophyll conductance in response to changes in CO_2 concentration around leaves. *Plant Cell Environ.*, *30*, 1284–1298.

Flowers, T. S., Gaur, P. M., Laxmipath, I., Gawda, C. L., Krishnamurithy, L., Saminen, S., et al. (2010). Salt sensitivity in chick pea. *Plant Cell Environ.*, *33*, 490–509.

Foyer, C. H., & Noctor, G., (2005). Redox homeostasis and antioxidant signaling: a metabolic interface between stress perception and physiological responses. *Plant Cell*, *17*, 1866–1875.

Gadallah, M. A. A., (1999). Effect of proline and glycinebetaine on *Vicia faba* responses to salt stress. *Biol. Plant.*, *42*, 249–257.

Garg, N., & Manchanda, G., (2009). ROS generation in plants: Boon or bane? *Plant Biosys.*, *143*, 81–96.

Gemede, H. F., Ratta, N., Haki, G. D., Woldegiorgis, A. Z., & Beyene, F., (2015). Nutritional quality and health benefits of okra (*Abelmoschus esculentus*): A Review. *J. Food Process. Technol.*, *6*, 458–463.

Genuchten, M. T., & Hoffman, G. J., (1984). Analysis of crop salt tolerance data. In: Shainberg, I., & Shalhevet, J., (eds.), *Soil Salinity under Irrigation, Processes, and Management, Ecological Studies* (Vol. 3, pp. 258–271). Springer: New York, NY, USA.

Ghoulam, C., Foursy, A., & Fares, K., (2002). Effects of salt stress on growth, inorganic ions and proline accumulation in relation to osmotic adjustment in five sugar beet cultivars. *Environ. Exp. Bot.*, *47*, 39–50.

Giuffrida, F., Scuderi, D., Giurato, R., & Leonardi, C., (2013). Physiological response of broccoli and cauliflower as affected by NaCl salinity. *Acta Hortic.*, *1005*, 435–441.

Gossett, D. R., Banks, S. W., Millhollon, E. P., & Lucas, M. C., (1996). Antioxidant response to NaCl stress in a control and a NaCl–tolerant cotton cell line grown in the presence of paraquat, buthionine sulfoximine and exogenous glutathione. *Plant Physiol.*, *112*, 803–809.

Grattan, S. R., & Grieve, C. M., (1992). Mineral element acquisition and growth response of plants grown in saline environments. *Agric. Ecosyst. Environ.*, *38*, 275–300.

Gurmani, A. R., Bano, A., Khan, S. U., Din, J., & Zhang, J. L., (2011). Alleviation of salt stress by seed treatment with abscisic acid (ABA), 6–benzylaminopurine (BA) and chlormequat chloride (CCC) optimizes ion and organic matter accumulation and increases yield of rice (*Oryza sativa* L.). *Aust. J. Crop Sci.*, *5*, 1278–1285.

Hamayun, M., Khan, S. A., Khan, A. L., Shin, J. H., Ahmad, B., Shin, D. H., & Lee, I. J., (2010). Exogenous gibberellic acid reprograms soybean to higher growth and salt stress tolerance. *J. Agric. Food Chem.*, *58*, 7226–7232.

Hameed, A., Qadri, N. T., Mahmooduzzafar, S. T. M., Ozturk, M., Altay, V., & Ahmad, P., (2017). Physio-biochemical and nutritional responses of *Abelmoschus esculentus* (L.) Moench (Okra) under mercury contamination. *Fresenius Environmental Bulletin*, (In Press).

Hamilton, E. W., & Heckathorn, S. A., (2001). Mitochondrial adaptation to NaCl. Complex I is protected by antioxidants and small heat shock proteins, whereas complex II is protected by proline and betaine. *Plant Physiol.*, *126*, 1266–1274.

Hasanuzzaman, M., Hossain, M. A., & Fujita, M., (2011a). Nitric oxide modulates antioxidant defense and the methylglyoxal detoxification system and reduces salinity–induced damage of wheat seedlings. *Plant Biotechnol. Rep.*, *5*, 353–365.

Hasanuzzaman, M., Hossain, M. A., & Fujita, M., (2012b). Exogenous selenium pretreatment protects rapeseed seedlings from cadmium–induced oxidative stress by up-regulating the antioxidant defense and methylglyoxal detoxification systems. *Biol. Trace Elem. Res.*, *49*, 248–261.

Hasanuzzaman, M., Hossain, M. A., Da Silva, J. A. T., & Fujita, M., (2012a). Plant responses and tolerance to abiotic oxidative stress: Antioxidant defenses is a key factors. In: Bandi, V., Shanker, A. K., Shanker, C., & Mandapaka, M., (eds.), *Crop Stress and its Management: Perspectives and Strategies* (pp. 261–316). Springer, Berlin.

Hasegawa, P., Bressan, R. A., Zhu, J. K., & Bohnert, H. J., (2000). Plant cellular and molecular responses to high salinity. *Annu. Rev. Plant Physiol. Plant Mol. Biol.*, *51*, 463–499.

Havaux, M., Eymery, F., Porfirova, S., Rey, P., & Dormann, P., (2005). Vitamin E protects against photoinhibition and photooxidative stress in *Arabidopsis thaliana*. *Plant Cell*, *17*, 3451–3469.

Hossain, M. A., Munemasa, S., Uraji, M., Nakamura, Y., Mori, I. C., & Murata, Y., (2011). Involvement of endogenous abscisic acid in methyl jasmonate–induced stomatal closure in *Arabidopsis*. *Plant Physiol.*, *156*, 430–438.

Hu, Y., & Schmidhalter, U., (2005). Drought and salinity: a comparison of their effects on mineral nutrition of Plants. *J. Plant Nutr. Soil Sci.*, *168*, 541–549.

Jamil, M., & Rha, E. S., (2007). Gibberellic acid (GA 3) enhance seed water uptake, germination, and early seedling growth in sugar beet under salt stress. *Pak. J. Biol. Sci.*, *10*, 654–658.

Jeschke, W. D., Peuke, A. D., Pate, J. S., & Hartung, W., (1997). Transport, synthesis, and catabolism of abscisic acid (ABA) in intact plants of castor bean (*Ricinus communis* L.) under phosphate deficiency and moderate salinity. *J. Exp. Bot.*, *48*, 1737–1747.

Kang, D. J., Seo, Y. J., Lee, J. D., Ishii, R., Kim, K. U., Shin, D. H., Park, S. K., Jang, S. W., & Lee, I. J., (2005). Jasmonic acid differentially affects growth, ion uptake and abscisic acid concentration in salt-tolerant and salt-sensitive rice cultivars. *J. Agron. Crop Sci.*, *191*, 273–282.

Kattab, H., (2007). Role of glutathione and polyadenylic acid on the oxidative defense systems of two different cultivars of canola seedlings grown under saline condition. *Aust. J. Basic Appl. Sci.*, *1*, 323–334.

Kaya, C., Tuna, A. L., & Yokas, I., (2009). The role of plant hormones in plants under salinity stress. In: Ashraf, M., Ozturk, M., & Athar, H. R., (eds.), *Salinity, and Water Stress: Improving Crop Efficiency* (pp. 45–50). Springer, Berlin.

Khan, M. A., & Rizvi, Y., (1994). Effect of salinity, temperature, and growth regulators on the germination and early seedling growth of *Atriplex griffithii* var. Stocksii *Can. J. Bot.*, *72*, 475–479.

Khan, M. A., Ungar, I. A., & Showalter, A. M., (2001). Effects of salinity on growth water relations and ion accumulation in the subtropical perennial halophyte, *Atriplex griffithii*, vr. Stocksii. *Annu. Bot.*, *85*, 225–232.

Khedr, A. H. A., Abbas, M. A., Wahid, A. A. A., Quick, W. P., & Abogadallah, G. M., (2003). Proline induces the expression of salt-stress-responsive proteins and may improve the adaptation of *Pancratium maritimum* L. to salt-stress. *J. Exp. Bot.*, *54*, 2553–2562.

Kim, S. G., & Park, C. M., (2008). Gibberellic acid–mediated salt signaling in seed germination. *Plant Signal. Behav.*, *3*, 877–879.

Läuchli, A., & Grattan, S. R., (2007). Plant growth and development under salinity stress. In: *Advances in Molecular Breeding toward Drought and Salt Tolerant Crops* (pp. 1–32). Springer: Dordrecht, The Netherlands.

Lutts, S., (2000). Exogenous glycinebetaine reduces sodium accumulation in salt-stressed rice plants. *Int. Rice Res. Notes*, *25*, 39–40

Maeda, H., Sakuragi, Y., Bryant, D. A., & Della-Penna, D., (2005). Tocopherols protect *Synechocystis* sp. strain PCC 6803 from lipid peroxidation. *Plant Physiol.*, *138*, 1422–1435.

Maggio, A., Barbieri, G., Raimondi, G., & De Pascale, S., (2010). Contrasting effects of GA 3 treatments on tomato plants exposed to increasing salinity. *J. Plant Growth Regul.*, *29*, 63–72.

Maggio, A., Dalton, F. N., & Piccinni, G., (2002). The effects of elevated carbon dioxide on static and dynamic indices for tomato salt tolerance. *Eur. J. Agron.*, *16*, 197–206.

Mäkelä, P., Kärkkäinen, J., & Somersalo, S., (2000). Effect of glycine betaine on chloroplast ultrastructure, chlorophyll, and protein content, and RuBPCO activity in tomato grown under drought or salinity. *Biol. Plant.*, *43*, 471–475.

Matysik, J., Alia, Bhalu, B., & Mohanty, P., (2002). Molecular mechanisms of quenching of reactive oxygen species by proline under stress in plants. *Curr. Sci.*, *82*, 525–532.

Maxwell, K., & Johnson, G. N., (2000). Chlorophyll fluorescence – A practical guide. *J. Exp. Bot.*, *51*, 659–668.

Mittova, V., Guy, M., Tal, M., & Volokita, M., (2004). Salinity upregulates the antioxidative system in root mitochondria and peroxisomes of the wild salt-tolerant tomato species *Lycopersicon pennellii*. *J. Exp. Bot.*, *55*, 1105–1113.

Munir, N., & Aftab, F., (2011). Enhancement of salt tolerance in sugarcane by ascorbic acid pretreatment. *Afr. J. Biotechnol.*, *10*, 18362–18370.

Munns, R., (2002a/2004). Comparative physiology of salt and water stress. *Plant Cell Environ.*, *25*, 239–250.

Munns, R., James, R. A., & Lauchli, A., (2006). Approaches to increasing the salt tolerance of wheat and other cereals. *J. Exp. Bot.*, *57*, 1025–1043.

Murphy, K. S. T., & Durako, M. J., (2003). Physiological effects of short-term salinity changes on *Ruppia maritima*. *Aquat. Bot.*, *75*, 293–309.

Nawaz, K., Hussain, K., Majeed, A., Khan, F., Afghan, S., & Ali, K., (2010). Fatality of salt stress to plants: Morphological, physiological, and biochemical aspects. *Afr. J. Biotechnol.*, *9*, 5475–5480.

Nilsen, E., & Orcutt, D. M., (1996). *The Physiology of Plants Under Stress – Abiotic Factors*. Wiley, New York, pp. 118–130.

Noctor, G., Gomez, L., Vanacker, H., & Foyer, C. H., (2002). Interactions between biosynthesis, compartmentation, and transport in the control of glutathione homeostasis and signaling. *J. Exp. Bot.*, *53*, 1283–1304.

Noreen, Z., & Ashraf, M., (2009). Assessment of variation in antioxidative defense system in salt-treated pea (*Pisum sativum*) cultivars and its putative use as salinity tolerance markers. *J. Plant Physiol.*, *166*, 1764–1774.

Olympia, G., K., Arensc, P., Koen, T. B. P., Karapanosa, I. P., Shahid, M. A., Pervez, M. A., et al. (2011). Salt stress effects on some morphological and physiological characteristics of okra (*Abelmoschus esculentus* L.). *Soil Environ.*, *30*, 66–73.

Orabi, S. A., & Abdelhamid, M. T., (2016). Protective role of α–tocopherol on two *Vicia faba* cultivars against seawater–induced lipid peroxidation by enhancing capacity of antioxidative system. *J. Saudi Soc. Agric. Sci.*, *15*, 145–154.

Parida, A. K., & Das, A. B., (2005). Salt tolerance and salinity effects on plants. *Rev. Ecol. Environ. Saf.*, *60*, 324–349.

Rad, H. E., Aref, F., & Rezaei, M., (2012). Response of rice to different salinity levels during different growth stages. *Res. J. Appl. Sci. Eng. Technol.*, *4*, 3040–3047.

Rawia, A. E., Taha, L. S., & Ibrahiem, S. M. M., (2011). Alleviation of adverse effects of salinity on growth, and chemical constituents of marigold plants by using glutathione and ascorbate. *J. Appl. Sci. Res.*, *7*, 714–721.

Ribaut, J. M., & Pilet, P. E., (1991). Effect of water stress on growth, osmotic potential and abscisic acid content of maize roots. *Physiol. Plant.*, *81*, 156–162.

Rohwer, C. L., & Erwin, J. E., (2008). Horticultural applications of jasmonates: A review. *J. Hortic. Sci. Biotechnol.*, *83*, 283–304.

Romero–Aranda R., Soria, T., & Cuartero, S., (2001). Tomato plant–water uptake and plant–water relationships under saline growth conditions. *Plant Sci.*, *160*, 265–272.

Sakhabutdinova, A. R., Fatkhutdinova, D. R., Bezrukova, M. V., & Shakirova, F. M., (2003). Salicylic acid prevents the damaging action of stress factors on wheat plants. *Bulg. J. Plant Physiol.*, *37*, 314–319.

Salama, K. H. A., & Al–Mutawa, M. M., (2009). Glutathione–triggered mitigation in salt-induced alterations in plasmalemma of onion epidermal cells. *Int. J. Agric. Biol.*, *11*, 639–642.

Saleem, M., Ashraf, M., & Akram, N. A., (2011). Salt (NaCl)–induced modulation in some key physio–biochemical attributes in okra (*Abelmoschus esculentus* L.). *J. Agron. Crop Sci.*, *197*, 202–213.

Sattler, S. E., Gilliland, L. U., Magallanes–Lundback, M., Pollard, M., & Penna, D. D., (2004). Vitamin E is essential for seed longevity and for preventing lipid peroxidation during germination. *Plant Cell*, *16*, 1419–1432.

Semida, W. M., Taha, R. S., Abdelhamid, M. T., & Rady, M. M., (2014). Foliar-applied α–tocopherol enhances salt-tolerance in *Vicia faba* L. plants grown under saline conditions. *South Afr. J. Bot.*, *95*, 24–31.

Seo, H. S., Kim, S. K., Jang, S. W., Choo, Y. S., Sohn, E. Y., & Lee, I. J., (2005). Effect of jasmonic acid on endogenous gibberellins and abscisic acid in rice under NaCl stress. *Biol. Plant.*, *49*, 447–450.

Serrano, R., Mulet, J. M., Rios, G., Marquez, J. A., Larriona, I. F., Leube, M. P., et al. (1999). A glimpse of the mechanism of ion homeostasis during salt stress. *J. Exp. Bot.*, *50*, 1023–1036.

Shah, S. H., (2007). Stress on mustard as affected by gibberellic acid application. *Gen. Appl. Plant Physiol.*, *33*, 97–106.

Shahid, M. A., Pervez, M. A., Balal, R. M., Ahmad, R., Ayyub, C. M., Abbas, T., & Akhtar, N., (2011). Salt stress effects on some morphological and physiological characteristics of okra (*Abelmoschus esculentus* L.). *Soil Environ.*, *30*, 66–73.

Shannon, M. C., & Grieve, C. M., (1998). Tolerance of vegetable crops to salinity. *Sci. Hortic.*, *78*, 5–38.

Shomeili, M., Nabipour, M., Meskarbashee, M., & Memari, H. R., (2011). Effects of gibberellic acid on sugarcane plants exposed to salinity under a hydroponic system. *Afr. J. Plant Sci.*, *5*, 609–616.

Sobahan, M. A., Arias, C. R., Okuma, E., Shimoishi, Y., Nakamura, Y., Hirai. Y., Mori, I. C., & Murata, Y., (2009). Exogenous proline and glycinebetaine suppress apoplastic flow to reduce Na$^+$ uptake in rice seedlings. *Biosci. Biotechnol. Biochem.*, *73*, 2037–2042.

Sudhir, P., & Murthy, S. D. S., (2004). Effects of salt stress on basic processes of photosynthesis. *Photosynthetica*, *42*, 481–486.

Tavakkoli, E., Rengasamy, P., & McDonald, G. K., (2010). High concentrations of Na$^+$ and Cl$^-$ ions in soil solution have simultaneous detrimental effects on growth of faba bean under salinity stress. *J. Exp. Bot.*, *61*, 4449–4459.

Tuteja, N., (2007). Abscisic acid and abiotic stress signaling. *Plant Signal. Behav.*, *2*, 135–138.

Uchida, A., Jagendorf, A. T., Hibino, T., & Takabe, T., (2002). Effects of hydrogen peroxide and nitric oxide on both salt and heat stress tolerance in rice. *Plant Sci.*, *163*, 515–523.

Xu, S., Hu, B., He, Z., Ma, F., Feng, J., Shen, W., & Yan, J., (2011). Enhancement of salinity tolerance during rice seed germination by presoaking with hemoglobin. *Int. J. Mol. Sci.*, *12*, 2488–2501.

Yokoi, S., Bressan, R. A., & Hasegawa, P. M., (2002). Salt stress tolerance of plants. *JIRCAS Working Report*, pp. 25–33.

Yoon, J. Y., Hamayun, M., Lee, S. K., & Lee, I. J., (2009). Methyl jasmonate alleviated salinity stress in soybean. *J. Crop Sci. Biotechnol.*, *12*, 63–68.

Zeid, I. M., (2009). Trehalose as osmoprotectant for maize under salinity–induced stress. *Res. J. Agric. Biol. Sci.*, *5*, 613–622.

Zhu, J. K., (2001). Plant salt tolerance. *Trends Plant Sci.*, *6*, 66–71.

CHAPTER 14

SUGAR BEET: AN OVERUTILIZED ANCIENT CROP

KADRIYE TAŞPINAR[1], MÜNIR ÖZTÜRK[2], VOLKAN ALTAY[3], and HALIL POLAT[1]

[1]Transitional Zone Agricultural Research Institute, Soil & Water Resources Department, Plant Nutrient and Soil Unit, Eskişehir, Turkey

[2]Department of Botany and Center for Environmental Studies, Ege University, Izmir, Turkey, E-mail: munirozturk@gmail.com

[3]Biology Department, Faculty of Science and Arts, Hatay Mustafa Kemal University, Hatay, Turkey

ABSTRACT

A sustainable conservation and knowledge of plant resources from our natural wealth are very important for future research. In the light of this statement, sugar beet is one of the over-utilized industrial crops in the world. It is widely distributed in the northern hemisphere. The crop is usually sown in late winter or early spring. The harvesting of the crop is realized 5–9 months of growth depending on the climatic and soil conditions. The crop is sown in autumn in the Mediterranean with spring/summer/fall harvests. Tremendous work is being done on the improvement of sugar beet crops for economic benefits. Research work is also being conducted on the sustainability of beet growing as well as minimization of all environmental threats. Its evaluation as a source of other income-gaining crop is also carried out on a large scale in different countries. This review highlights the latest information about the sugar beet, with an emphasis on its taxonomic status, ecological, and ecophysiological characteristics, and potential alternative economical uses such as; compost, biosorption, biochar, bioenergy, and molasses.

14.1   INTRODUCTION

The annual global sugar production is around 160 Mt, with about 23 kg per capita consumption. Its use is going up by nearly 1.5% on annual basis, particularly in densely populated nations China and India (Biancardi et al., 2010). Nearly ¼ of the global sugar production comes from beets (*Beta vulgaris*), and the rest from sugar cane (*Saccharum officinarum*). Sucrose in the major component of these products (over 99.5% in white crystalline sugar). The climatic requirements and photosynthetic pathways differ in these two crops. The yield of beets is much better under temperate climatic conditions, especially France, Germany, and North America. The sugarcane prefers tropical to subtropical climate and is therefore found largely in India, Australia, Cuba, and Brazil (Biancardi et al., 2010).

The sugar production from sugar beet and sugarcane has been on the market since the earliest sugar beet factories established in the early 1800s (Biancardi et al., 2010). ASs against the beet sugarcane enjoys is that these factories are energy sufficient due to the burning of bagasse. On the other hand, the power for processing beets generally relies on fossil fuels (Biancardi et al., 2010). The cost of sugarcane is generally lower, however, the price of sugar from beets and cane generally follows the price of crude oil (Biancardi et al., 2010).

The countries mainly involved in the sugar production from the sugar beet are: France, USA, Germany, Russian Federation, Turkey, Poland, Ukraine, UK, and China (FAO, 2011). In 2004, sugar beet cultivation was common in 51 countries, with a total production of around 240 million metric tons, and the major producers in 2004 were France, USA, Germany, Russian Federation, Turkey, Ukraine, and Poland (AGBIOS, 2005).

The sugar industry has played an important technological, economical, and social developmental role in rural communities dependent on sugar beet in Turkey. It has at the same time played a significant role in Turkey's agriculture and agro-industry (Erdal et al., 2007; Ozturk et al., 2017). Currently the area is around 330,000 ha and about 13,965,000 tons of sugar is produced annually from the sugar beet. This constitutes more than 5.5% of world total sugar beet production (FAO, 2005; Erdal et al., 2007).

14.2   TAXONOMIC STATUS OF SUGAR BEET

The genus *Beta* belongs to the family Amaranthaceae (formerly Chenopodiaceae). It is subdivided into four sections, but there are some problems due to the existence of quite distinct and diverse cultivated forms (Letschert, 1993; Letschert et al., 1994; Lange et al., 1999; Ford-Lloyd, 2005). First one is; the affinity between the wild species and cultivated forms originating from them. The second one is the evolutionary and taxonomic relations of the cultivated forms. *B. vulgaris* includes all cultivated forms. The selections for agricultural and horticultural purposes has resulted in four crops with a greater number of cultivars. The beets used for sugar extraction, fodder beet, red, and yellow beets and foliage beet. The fodder group is used for feeding cattle, garden beet and beetroot are consumed as vegetable, particularly around the Mediterranean basin. The leaves are also consumed as a vegetable. The spinach beet-leaves are used for salads or as spinach, the swiss chard or seakale for the swollen midribs and petioles, used in salads or as a vegetable along with the green parts.

The origin of beet names in ancient languages reveals that the origin is from a Greek word Sicula both in the case of silga, selgand silg the Babylonian, Arabic, and Nabatean words referring to beet. However, earliest Greek word for beet is teutlon which has persisted in modern Greek as seutlon, referring to maritime beet, and teutlorizon. The earliest description in this connection refers to the foliage beets (or chards). Theophrastus around 300 BC mentions two different beets-white (or Sicilian) and black (or dark green)-the colors referring to the light and dark green appearance of the leaves, whereas Aristotle has written something about a red chard around 350 BC. Again in the 2nd century AD, a description of sessile beet, white beet, common beet and dark or swarthy beets has been given, all related to the leaf characters, meaning thereby that plants were used as leaf beets. In this sense beets of earlier ages were from the leaf forms group. These have been mentioned much by Romans in the 1st century AD. Similarly, the mention has been made of this leafy group in Chinese since the 7th century AD. Undoubtedly, Roman writers have mentioned about the beetroots for both culinary and medicinal purposes. The word "fleshy" used earlier is affiliated to the nonwoody roots. From 15th century onwards several types have been described such; a beet with fleshy sweet root and a single, long straight shoot. This was followed by some information on the red beet with a swollen, turnip-like root which was eaten, accepted by some researchers as the ancestor of the present day long-rooted beets; followed by a shorter, thick-rooted form, similar to modern half-long blood beets. During the following centuries descriptions of many other types of wild and cultivated beets has been published. These include two forms namely white beet (Sicula) and black beet (Nigra). In another work three forms have been mentioned such as; white (Sicula or Candida), common red (Beta nigru) and a "Strange red beet " (*Beta nigra* Romana), with a thick short root which was very sweet, and this has been suggested as the ancestor of globular beets. Later on five different morphological forms are mentioned; common white, common red, common green, Roman red beet (*Beta raposa* or *rupa*) and the Italian beet with green leaves having large white midribs and used either for boiling or uncooked in salads. The existence of considerable earlier variation could mean a common ancestry for all red-rooted types and more recent selection could have produced the distinct cultivars. In the eighteenth century 8 varieties of red and yellow garden beets are mentioned (Ford-Lloyd and Williams, 1975).

In the botanical evaluation by Linnaeus (1753), *B. vulgaris:* var. *perennh* (wild type), var. *rubra* (garden beet) and var. *cicla* (foliage beet) have been described, later on a wild maritime species *B. maritima* was added to this by him. The cultivated beets all fall within section *vulgares*, as *B. vulgaris* and the wild species *B. atriplicifolia, B. macrocarpa* and *B. patula*, includes *B. maritima* and two other wild species more recently described, *B. adanensis* and *B. trojana,* both endemic in Turkey. Other sections are: Corollinae including 5 perennial species, *B.macrorhiza, B. lomatogona, B. trigyna, B. intermedia* and *B. corollijlora*; Patellares with 3 species: *B. patellaris, B. webbiana* and *B. procumbens.* The biotypes have been reported which are intermediate between *B. webbiana* and *B. procumbens;* Nanae including only 1 species, *B. nana* not studied at length (Ford-Lloyd and Williams, 1975).

The morphological studies on *B. vulgaris* prove that the variations are continuous between two extremes wild aff. *B. maritima* and cultivated sugar beet. Both the maritime and foliage beets seem to be similar to the original primitive types from which all other cultivated beets have originated as seen in the comparative ranges of variation in the differences between these two extremes. The morphological features reveal that, greatest range of variation is found in the maritime and foliage beets; whereas small ranges of variation are visible in the highly cultivated types. However, a large diversity in root type is observed; because of the original selections carried out by humans. A continuous variation in *B. vulgaris* sensu lato complex has been reported after working on the material of highly selected cultivated origin. An analysis of the data from Turkey clearly depicts that continuous variation is even stronger. As such, relationships between beetroot (red beet), primitive leaf beets and *B. maritima,* are still open, probably gene exchange has occurred more readily in the more primitive farming conditions on a large scale, therefore identity of modern forms is quite often lost. Undoubtedly, some rare characters are common to both maritime and cultivated beets. This could be an indication of their close relationship and genetic similarity.

The ecotypic differentiation has probably resulted in microevolution with the products like *B. adanensis* and *B. trojana*. Probably the seed of both these endemics have been evaluated in the past by local farmers for spacing purposes in the crops. The degree of microevolution within wild forms and existence of hybridization between maritime beets and cultivars and among cultivated types of beet, the relegation of *B. mantima, B. adanensis* and also *B. patula* to the rank of subspecies within *B. vulgaris* has been proposed. In the subspecies *maritima,* a large number of varieties exist such as; var. *maritima,* var. *trojana,* var. *macrocarpa,* var. *prostrata,* var. *erecta,* and var. *atriplicifolia.* A continuous variation has been observed amongst cultivars within the *B. vulgaris* complex. The field observations from the Turkish heterogeneous complex of cultivated beets reveals that, a form considered to be the remains of an ancestral subspecies *provulgaris* comb. nov. is possible, which is a weedy or primitively cultivated and appears to be somewhat inter-mediate between ssp. *maritima* and ssp. *cicla.* It has been subjected to human selection. Other 2 subspecies could be the one representing swollen-rooted types (ssp. *vulgaris),* other non-swollen-rooted types (ssp. *cicla).* A recogni-tion of the ancestral ssp. *provulgaris* with its central relationship within the variable gene pool of beets existing in Turkey depict that, it is largely the remains of a primitive gene pool within which the broadly-based genetic variation is becoming progressively narrower due to increased communica-tions between originally closed farming communities. It seems very much plausable to identify 2 varieties within ssp. *cicla,* one representing leaf beets with swollen midribs (var. *flavescens),* the other of ordinary foliage beets with abundant leaf material (var. *cicla).*

Beta vulgaris ssp. *vulgaris* var. *boissieri* conv.' *vulgaris,* var. *vulgaris,* var. *flavescens,* conv. *crassa,* var. *crassa,* var. *altissima,* var. *iurea,* var. *condifiva* classification looks like to pose a confusion according to the International Code of Botanical Nomenclature (Ford-Lloyd and Williams, 1975).

A general evaluation of the information published by different workers shows that the plant list of beets includes 50 scientific names of species placed under the genus *Beta.* Of these nine are accepted species (www. theplantlist.org):

1. *Beta corolliflora* Zosimovic ex Buttler;
2. *Beta lomatogona* Fisch. & C.A.Mey;
3. *Beta macrocarpa* Guss;
4. *Beta macrorhiza* Steven;
5. *Beta nana* Boiss. & Heldr;
6. *Beta palonga* R.K.Basu & K.K.Mukh;

7. *Beta patula* Aiton;
8. *Beta trigyna* Waldst. & Kit; and
9. *Beta vulgaris* L.

Until today, although many scientific names have been given to sugar beet (*Beta vulgaris*), of these names, only *Beta vulgaris* (Figure 14.1) has been accepted. The synonyms of *Beta vulgaris* listed in different research publications are as follows:

FIGURE 14.1   General appearance of sugar beet in a cultivated field.

Synonyms: *Beta alba* DC.; *B. altissima* Steud.; *B. atriplicifolia* Rouy; *B. bengalensis* Roxb.; *B. brasiliensis* Voss; *B. carnulosa* Gren.; *B. cicla* (L.) L.; *B. cicla* (L.) Pers.; *B. cicla* var. *argentea* Krassochkin & Burenin; *B. cicla* var. *viridis* Krassochkin & Burenin; *B. crispa* Tratt.; *B. decumbens* Moench; *B. esculenta* Salisb.; *B. foliosa* Ehrenb. ex Steud.; *B. hortensis* Mill.; *B. hybrida* Andrz.; *B. incarnata* Steud.; *B. lutea* Steud.; *B. marina* Crantz; *B. maritima* L.; *B. maritima* var. *atriplicifolia* Krassochkin; *B. maritima* subsp. *atriplicifolia* (Rouy) Burenin; *B. maritima* subsp. *danica* Krassochkin; *B. maritima* var. *erecta* Krassochkin; *B. maritima* var. *glabra* Delile; *B. maritima* subsp. *marcosii* (O.Bolòs & Vigo) Juan & M.B.Crespo; *B. maritima* subsp. *orientalis* (Roth) Burenin; *B. maritima* var. *pilosa* Delile; *B. maritima* var. *prostrata* Krassochkin; *B.noeana* Bunge ex Boiss.; *B. orientalis* Roth; *B. orientalis* L.; *B. purpurea* Steud.; *B. rapa* Dumort.; *B. rapacea* Hegetschw.; *B. rosea* Steud.; *B. sativa* Bernh.; *B. stricta* K.Koch; *B. sulcata* Gasp.; *B. triflora* Salisb.; *B. vulgaris* var. *altissima* Döll, *B. vulgaris* var. *annua*

Asch. & Graebn.; *B. vulgaris* subsp. *asiatica* Krassochkin ex Burenin; *B. vulgaris* var. *asiatica* Burenin; *B. vulgaris* var. *atriplicifolia* (Rouy) Krassochkin; *B. vulgaris* var. *aurantia* Burenin; *B. vulgaris* subsp. *cicla* (L.) Schübl. & G.Martens; *B. vulgaris* var. *cicla* L.; *B. vulgaris* subsp. *cicla* (L.) W.D.J. Koch; *B. vulgaris* var. *coniciformis* Burenin; *B. vulgaris* var. *crassa* Alef.; *B. vulgaris* var. *debeauxii* Clary; *B. vulgaris* var. *foliosa* (Asch. & Schweinf.) Aellen; *B. vulgaris* subsp. *foliosa* Asch. & Schweinf.; *B. vulgaris* var. *glabra* (Delile) Aellen; *B. vulgaris* var. *grisea* Aellen; *B. vulgaris* subsp. *lomatogonoides* Aellen; *B. vulgaris* var. *lutea* DC.; *B. vulgaris* var. *marcosii* O.Bolòs & Vigo; *B. vulgaris* var. *maritima* (L.) Moq.; *B. vulgaris* subsp. *maritima* (L.) Arcang.; *B. vulgaris* subsp. *maritima* (L.) Thell.; *B. vulgaris* var. *mediasiatica* Burenin; *B. vulgaris* var. *orientalis* (Roth) Moq.; *B. vulgaris* subsp. *orientalis* (Roth) Aellen; *B. vulgaris* var. *ovaliformis* Burenin; *B. vulgaris* var. *perennis* L.; *B. vulgaris* var. *pilosa* (Delile) Moq.; *B. vulgaris* subsp. *provulgaris* Ford-Lloyd & J.T. Williams; *B. vulgaris* var. *rapacea* W.D.J.Koch; *B. vulgaris* var. *rosea* Moq.; *B. vulgaris* var. *rubidus* Burenin; *B. vulgaris* var. *rubra* L.; *B. vulgaris* var. *rubrifolia* Krassochkin ex Burenin; *B. vulgaris* var. *saccharifera* Alef.; *B. vulgaris* var. *virescens* Burenin; *B. vulgaris* var. *viridifolia* Krassochkin ex Burenin; *B. vulgaris* var. *vulgaris*; *B. vulgaris* subsp. *vulgaris* (www.theplantlist.org).

The two taxa names accepted by the taxonomist and confirmed as to belonging to *Beta vulgaris* are (www.theplantlist.org):

- *Beta vulgaris* subsp. *adanensis* (Pamukç.) Ford-Lloyd & J.T. Williams (synonyms: *Beta adenensis* Pamukç.).
- *Beta vulgaris* var. *trojana* (Pamukç.) Ford-Lloyd & J.T. Williams (synonyms: *Beta trojana* Pamukç.).

14.3 ECOLOGICAL AND ECOPHYSIOLOGICAL ASPECTS

The *vulgares* section is well spread in the Mediterranean basin. *B. maritima* is widely distributed as far as the outer Asiatic steppes and East India in the east, to the Cape Verde and Canary Islands in the west, and along the Atlantic Coast to the North Sea including the coast of the British Isles, Denmark, and Sweden towards the north. Other taxa show a restricted distribution, however all occur within the wide range of *B. maritima*. The taxa included under *B. vulgaris* are found in cultivation throughout Europe, Turkey, and Asia. The corollinae section is grouped within Turkey with an extention towards Iran in the east as well as extending to the west towards eastern

Europe. The only species of section name is *B. nana,* recorded only from the high altitudes of Olympos, Parnassos, and Taiyetos mountains in Greece. The section patellares is found mainly in the western Mediterranean area, including S.E. Spain, as well as Cape Verde, Canary, Salvage, and Madeira Islands (Ford-Lloyd and Williams, 1975).

The taxa included under beets are either maritime or inland taxa, with corollinae section members exclusively distributed in the inlands and always above 300 m altitudes and above. *B. lomatogona* and *B. intermedia* have been recorded as weeds of cultivated fields, mainly occupying the edges of wheat and barley fields in Turkey. *B. trigyna* and *B. corolliflora* too have been reported to grow as weeds, often within cultivated fields. *B. macrorkiza* is a typical ruderal commonly found alongside the roads. The taxa grouped under section patellares occupy both inlands as well as maritime sites. The species of *B. patellaris* flourishes at altitudes of 200 m on inland mountain ranges and within coastal spray zones. The cultivated forms from section vulgares are found on inland lowland agricultural areas. The two Turkish endemics *B. adanensis* and *B. trojana* are distributed on the fields only just inland from saline soils. *B. adanensis* is found on saline fields very close to the sea coast. *B. trojana* mainly occupies the silted-up delta between ancient Troy and the sea. *B. maritima* apparently has a limited inland distribution, occurring on inland saline halophytic habitats together with *Arthrocnemum glaucum* and *Tamarix tetragyna,* as well as in the fields and along the roads in Palestine. The names *B. maritima* and *B. vulgaris* have been used synonymously. The native habitat for a true wild *B. vulgaris* is the saline steppe of Asia Anterior. No such form are reported from the inland saline habitats in Turkey. On the coastal areas only *B. maritima* is found because it is ecologically wholly or partially maritime, commonly occupying the soil banks near the sea with very high salinities. This taxon also occurs on coastal cliffs mainly under the impact of salt spray. In some areas it grows in between the rocks along the seashore, with a wide degree of salt tolerance.

Accessions of the sect. *vulgares* collected by Williams & Ford-Lloyd (1974) from all over Turkey during the summer and grown after full vernalization at an approximate temperature of 20°C with 18 hours of supplemental mercury-vapor lighting. These studies have revealed that Turkish collections are extremely variable. The observations have shown that sometimes yellow beets, red beets and non-pigmented beets are all produced from a single seed collection. No natural groupings were possible with all types of characters used. Only one grouping of individuals was possible. *B. adanensis* accessions formed a clear unit, clearly distinct from

B. maritima. Former formed a very distinct group, separated from the culti-vated types and *B. maritima. B. trojana* appears as a fairly homogeneous group but has been included within the general variation exhibited by *B. maritima* (Ford-Lloyd and Williams, 1975).

14.3.1 CULTIVATION OF SUGAR BEET

The sugar beet has been cultivated since last 200 years. It has originated from a limited range of beet types and was domesticated in the Mediter-ranean basin as a leaf beet from wild forms without hypocotyls or swollen roots. Later were developed through cultivation practices all through ages. The origin of diverse types of sugar beet, fodder beet and garden beet lies mainly in northwestern Europe. For a future breeding of this economically important crop it is a must that interspecific hybridization will play great role in broadening the genetic base (De Bock, 1985).

The data published in 2016 reveals that the total number of villages involved in sugar beet production in Turkey was 2324. The number of villages on the basis of the provinces in Turkey are; 190 for Tokat, 173 for Konya, 161 for Eskişehir, 158 for Afyonkarahisar, 164 for Yozgat, 118 for Ankara, 109 for Kastamonu, 85 for Çorum, 83 for Erzincan, 67 for Sivas, 66 each for Muş, Denizli, Kahramanmaraş, 65 for Burdur, 62 for Kırşehir, 60 for Elazığ, 45 for Aksaray, 43 each for Erzurum and Bursa, 40 each for Karaman and Nevşehir, 33 for Van, 30 for Antalya, 27 each for Malatya, Uşak and Edirne, 24 each for Ağrı and Kırıklareli, 23 for Kırıkkale, 22 for Tekirdağ, 18 for Niğde, 16 for Bayburt, 15 each for Bitlis, Kars, Gaziantep, Gümüşhane, 11 for Amasya, 10 each for Isparta, Çankırı, Iğdır, Kütahya, 9 for Balıkesir, 8 for Şanlıurfa, 5 for Samsun, 4 for Adana, 2 each for Diyarbakır, Adıyaman, Muğla, one each for Kayseri, Bilecik, Istanbul, Bingöl, Manisa, and Tunceli. The statewise sugar beet cultivated area (decares) in 2016 in Turkey is presented in the Figure 14.2a, b, c.

In 2016, Turkey's sugar beet production on average has been recorded as 5.96 tons per decare. The sugar beet production in these provinces as tonnes per decare is as follows:

10.01 for Bursa, 9.32 for Manisa, 7.35 each for Diyarbakır and Gaziantep, 7.12 for Afyonkarahisar, 6.93 for Bilecik, 6.77 for Tekirdağ, 6.73 for Kırıkkale, 6.71 each for Niğde and Uşak, 6.55 for Aksaray, 6.51 for Konya, 6.49 for Ankara, 6.47 for Balıkesir, 6.42 for Eskişehir, 6.41 for Karaman and Edirne, 6.31 for Isparta, 6.17 for Antalya, 6.00 each for Yozgat and Elazığ, 5.99 for Kahramanmaraş, 5.95 for Nevşehir, 5.73 for Van, 5.70

a.

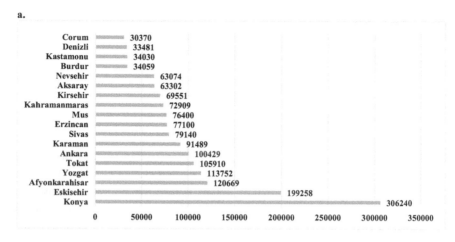

b.

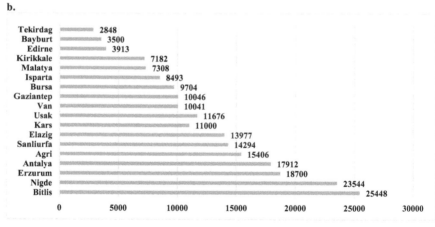

c.

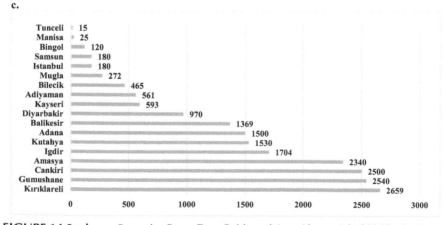

FIGURE 14.2a, b, c   Statewise Sugar Beet Cultivated Area (decares) in 2016 in Turkey.

for Kırşehir, 5.67 for Şanlıurfa, 5.64 for Denizli, 5.58 for Çorum, 5.50 for Kayseri, 5.46 for Adana, 5.41 for Amasya, 5.37 for Kütahya, 5.34 for Burdur, 5.32 for Çankırı, 5.23 each for Tokat and Kırıklareli, 5.16 for Bingöl, 5.14 for Adıyaman, 5.13 for Muğla, 5.12 for Erzincan, 5.06 for Kastamonu, 4.88 for Iğdır, 4.71 for Istanbul, 4.62 for Muş, 4.53 for Tunceli, 4.47 for Sivas, 4.23 for Ağrı, 4.22 for Erzurum, 4.20 for Bitlis, 4.16 for Bayburt, 4.12 for Gümüşhane, 4.10 for Kars, 3.28 for Malatya, and 3.12 for Samsun.

The crop is sown in late winter or early spring in the northern hemisphere, depending upon climate and soil features, and harvesting is done after 5–9 months of growth. In the areas with a Mediterranean climate, sowing is done in autumn and harvesting in spring/summer/fall (Biancardi et al., 2010). The harvested material is transported immediately to the factories or placed in storage chambers. All this depends on the prevalent temperatures and weather conditions as well as throughput of the factory. The crowns, petioles, and leaves are cut as at the start as these are low in sugar content and decrease the value due to high concentration of processing impurities. The beetroots are then washed followed by sugar extraction using hot water diffusion from thinly sliced roots. The lime and carbon dioxide are applied repeatedly to the "raw juice" for its purification, followed by filtration and finally evaporation to get the concentrated "thin juice." The sugar crystallization is initiated in "thick juice" under partial vacuum and high-temperature conditions, when concentration of sucrose is above becomes 60% of the sugar Molasses. The end product is a brown and heavy syrup with 45% sugar. The separation of to produce nearly pure, commercial, white sugar (McGinnis, 1982; Biancardi et al., 2010). The left out material called molasses is evaluated as an animal feed and also used to produce alcohol, glutamate, yeasts, etc. In certain places molecular sieving and ion exchange extractions are used after returning the left out material to the factory for further sugar removal and separation of sucrose. After sugar extraction, the non-soluble portions of the sliced roots; commonly known as pulp; are used mainly for producing food for animal and other pets (Biancardi et al., 2010). Sugar beet pulp has also been evaluated for human consumption as a palatable, fibrous food ingredient (Harland et al., 2006). Both postprandial plasma glucose as well as blood cholesterol have been reported to be reduced by sugar beet fiber inclusion in human diets, as a result of significant physiological changes (Morgan et al., 1988). This has been reviewed at length by Harland (1989, 1993).

The beets cultivated all over the world are grouped under the taxon subspecies *vulgaris*, which is affiliated to the species *vulgaris* and to the genus *Beta* (Letschert, 1993; Letschert et al., 1994; Lange et al., 1999; Ford-Lloyd,

2005). In all the cultivar groups described for the proposed *Beta vulgaris* are listed below (Lange et al., 1999):

1. Leaf Beet Group;
2. Garden Beet Group;
3. Fodder Beet Group;
4. Sugar Beet Group.

A summarized information of these cultivar groups is given below (Lange et al., 1999):

Leaf Beet Group: It includes the cultivars where leaves and petioles are consumed as a vegetable. The size and shape of the leaf blade generally varies, together with thickening of midrib and petiole. The swollen hypocotyls and/or roots are absent or are very small in these. The cultivars known as Mangold, Spinach beet, Chard or Swiss chard are all included in this group. It has been proposed that the use of the UPOV denomination classes Mangold Group and Spinach Beet Group should be discontinued and the cultivars of these groups should be included in the new Leaf Beet Group (Lange et al., 1999). The cultivars under fodder beet group too are used sometimes as leafy vegetables; but these have a swollen hypocotyl and root and cannot be confused in agricultural practices (Lange et al., 1999).

Garden Beet Group: This group consists of cultivars mostly used as a vegetable as they have swollen hypocotyls. These are also used in the canning industry and for making pickles. Adense red coloration is observed due to the presence of betacyanidines in the majority of the cultivars in this group; however, yellow, and pinkish cultivars are also seen. Therefore the color is not a valid character for the description of cultivars in this group. The epithet is this group has been chosen instead of red garden beet group, although red beets are sometimes also used to produce a red natural dye. No information is available about the fact if there are special cultivars bred in this connection. Accordingly one has to see whether this group would be distinct enough to be treated as a separate group. The beetroot group is another group, albeit ambiguous, name for this cultivar group. Although, one word beetroot in English linguistically refers to the vegetable, in American slang the vegetable is named beet, while beet root (two words) means root of any beet. It is proposed to discontinue the UPOV denomination class Beetroot Group, and to include the cultivars of this group in the new Garden Beet Group (Lange et al., 1999).

Fodder Beet Group: This is a large group of cultivars, used as fodder for livestock. The plants have large swollen hypocotyl and root parts. Much variation in shape and color of the swollen parts and skin color is observed here. This variation is not generally used to distinguish subgroups. An alternative name suggested for this group is forage beet group, but currently preference is given to the fodder beet group name. The mangold name too has been used in earlier publications (Knapp, 1958). This name was also used by UPOV for naming a cultivar group containing leaf beet cultivars. Therefore, this epithet has been ambiguously applied and use is recommended to be discontinued. The cultivars of the UPOV denomination class fodder beet group are included in the new fodder beet group (Lange et al., 1999).

Sugar Beet Group: The cultivars included in this group are used for the production of sugar, and are bred for a high sucrose content and good extraction qualities. This group shows less variation in shape and color than the former group. The swollen parts of the cultivars are characterized here by high sucrose content and the swollen part consists mainly or wholly of root tissue. Historically, this cultivar group is the most recently developed (Lange et al., 1999). The cultivars included here can be evaluated for the production of ethanol via fermentation. Still there is controversy on the fact if the cultivars grouped here are distinct enough to be treated separately, when new beet types are developed, e.g., so-called fructan beets, to be used for the production of polymerized fructose instead of sucrose. Such types might be grouped in a separate new denomination class. It seems more plausible that to include the cultivars of the UPOV denomination class sugar beet group in the new sugar beet group (Lange et al., 1999).

14.3.2 ECOLOGICAL AND ECOPHYSIOLOGICAL STUDIES ON SUGAR BEET

Many of the physiological processes are effected by leaf tissue water deficits, these ultimately lead to a reduction in the yield. One of the major factors leading to profit loss in the sugar beet crop is the drought stress (Pidgeon et al., 2001; Tognetti et al., 2003). In south Mediterranean as well as North African countries with subtropical conditions, sowing of sugar beet is undertaken in autumn and harvesting done in summer. The crop reaches maturity for harvest when temperatures and evaporative demands are at the highest (Monreal et al., 2007). These have great impact on sugar production, decreasing yield and increasing the accumulation of soluble non-sucrose compounds. General lyproline and glucose impair the crystallization of sugar in the factory process and lead to the formation of colored components, thereby reducing the industrial quality of the root (Campbell, 2002; Coca et al., 2004).

Under drought conditions sugar beet plants show high osmotic adaptation (Gzik, 1996; Ober et al., 2005; Chołuj et al., 2008). The osmoprotectants such as: potassium or sodium, soluble sugars and their derivatives as well as nitrogen-containing substances such as glycine betaine, proline or other free amino acids end up with the lowering of the osmotic potential (s) in the cells of sugar beet (Shaw et al., 2002; Bloch and Hoffman, 2005; Bloch et al., 2006; Chołuj et al., 2008). Although beneficial for the acclimation response of the plant to dry conditions, the osmoprotectants are melassigenic and influence the quality of sugar beet root negatively (Clarke et al., 1993; Campbell, 2002). These substances at the same time may be rapidly reduced to ammonia, which causes juice alkalization (Chołuj et al., 2014).

Monreal et al. (2007) have focused on the proline quantification in storage roots from field trials under irrigation, sugar beet variety and nitrogen fertilization. The proline accumulation in sugar beetroots has mainly been observed underwater deficit situations. An increase in the proline levels due to excess N supply and partial increase in leaf area index (LAI) and exacerbating drought stress produce an additive effect. Claudia and Ramona beet varieties have been studied and results have shown different responses to water shortage and to nitrogen, with maximum proline levels in Claudia roots subjected to a combination of water shortage and excess N. A significant and positive correlation has been reported between proline and glucose levels in the roots of sugar beet, depicting a relationship between stress responses, carbohydrate catabolism, proline, and glucose accumulation (Monreal et al., 2007). This is further supported by the effects of treatments with di–1-p-menthene (anti-transpirant) and with DMDP (2,5-dihydroxymethyl–3,4-di-hydroxypyrrolidina, a glycosidase inhibitor), which lead to decreased level of proline in non-irrigated Claudia sugar beetroots (Monreal et al., 2007).

Pilon-Smits et al. (1999) have used the same SacB gene from *Bacillus subtilis* to produce bacterial fructans in *Beta vulgaris*. They report that transgenic sugar beets accumulate fructans to low levels (max. 0.5% of dry weight) in both roots and shoots. Two independent transgenic lines of fructan-producing sugar beets have shown significantly better growth under drought stress than untransformed beets. The total dry weights were higher (+25–35%) in drought-stressed fructan-producing plants than wild-type, due to higher biomass production of leaves (+30–33%), storage roots (+16–33%) and fibrous roots (+37–60%) (Pilon-Smits et al., 1999). These researchers have concluded that, the introduction of fructan biosynthesis in transgenic plants is a promising approach to improve crop productivity under drought stress.

Sadeghian and Yavari (2004) have studied sugar beet progeny lines under drought stress conditions in the field, and screened for both high water use efficiency and high sugar yield, together with the germination rate and early seedling growth underwater deficit stress induced by mannitol solutions. They used seeds of 9 different sugar beet progeny lines, grown under 3 experimental conditions using filter paper, perlite, and water agar as substrate, with 3 levels of 0.0, 0.2 and 0.3 m mannitol concentrations, germination percentage and seedling growth parameters such as cotyledon fresh weight, cotyledon dry weight, root fresh weight, root dry weight (RDW) and root length (RL) were measured experimentally. Abnormality has been recorded only in the filter paper experiment. Their results further show that drought stress could be simulated by mannitol solution and significant differences result between stress levels for seedling characteristics. Distinct genetic variances have been found among progeny lines with respect to germination and early seedling growth characteristics, except for cotyledons and RDW. Both the characters severely decline at highest concentrations of mannitol. The rate of abnormality increases progressively at the germination stage with an increase in mannitol concentration but is more pronounced in the drought-susceptible progeny lines. The highest values of relative germination percentage and relative growth percentage of RL have been attained for the most tolerant line (Sadeghian and Yavari, 2004). Sadeghian and Yavari (2004) have also reported that seedling characteristics, in addition to other physiological components involved in the seed germination process under specific stress conditions, may be considered for breeding purposes.

Sugar beet is a glycophytic member shows a high ability for osmotical adaptation (Katerij et al., 1997; Choluj et al., 2008). During osmotic adjustment inorganic ions act as an osmolyte and are sequestered in the vacuole where their high concentration becomes detrimental to cell metabolism. These plants accumulate more potassium, sodium, and chloride in the leaves than in the roots, which are effective in osmotic adjustment induced by salinity or drought (Ghoulam et al., 2002; Choluj et al., 2008).

The biosynthesis and accumulation of compatible solutes under unfavorable conditions such salt, osmotic, and water stress in sugar beet has been reported earlier (Hanson and Wyse, 1982; Hanson and Rhodes, 1983; Gzik, 1996; Rover and Buttner, 1999; Ghoulam et al., 2002; Shaw et al., 2002; Choluj et al., 2008). In majority of the studies either tap-root or leaves are investigated; nevertheless, the role of the selected compounds as compatible and protective solutes has been investigated. Very little information is about organ-specific accumulation of various solutes under drought conditions in sugar beet plants (Mack and Hoffman, 2006; Choluj et al., 2008).

The soluble non-sucrose constituents of sugar beet storage root, accumulating due to water stress generally reduce sucrose crystallization (Choluj et al., 2008). The ratio of sucrose to total soluble solids (sucrose plus impurities) determines the processing quality of sugar beetroots (Campbell, 2002). The changes in the composition of beet organ solutes under drought may be due largely to the affect by severity and duration of stress as well as regeneration processes. Field-grown sugar beets frequently suffer from water stress during summer and recover after rainfall; however, there is no detailed information about the extent to which drought-induced changes are reversible (Choluj et al., 2008).

In an earlier study, Milford, and Lawlor (1976) have indicated different wilting patterns in the young and mature sugar beet leaves. Mature leaves loose the turgor when the plant is water-stressed in spite of a low stomatal conductance and reduction in transpiration, but young leaves usually remain upright although their stomata are open (Milford and Lawlor, 1976). Therese authors suggest that young leaves are better supplied with water and a hydrodynamic equilibrium does not exist within these plants. According to Chołuj et al. (2004, 2008) previous data concerning the different water status in individual organs of sugar beet plants subjected to water stress are in agreement with these findings. We can explain these observations with differential ability of organs in sugar beet with decrease in osmotic potential, which affects water potential gradient and passive water movement within the whole plant.

According to Choluj et al. (2008), studies on the sugar beet cv. Janus reveals an accumulation of various osmolytes in plants grown under 2 soil water treatments: control (60% of the field water capacity; FWC) and drought (30–35% FWC). The water shortage started on the 61st day after emergence (DAE), at the stage of the beginning of tap-roots development and imposed for 35 days reveals that, osmotic potential of sugar beet plant organs particularly tap-roots, decreases significantly as a consequence of a long-term drought. Water shortage reduces univalent (K^+, Na^+) cation concentrations in the petioles and divalent (Ca^{2+}, Mg^{2+}) ion levels in the mature and old leaves. In the taproots cation concentrations are not affected by water shortage. The ratio of univalent to divalent cations significantly increases in young leaves and petioles as a consequence of drought. Long-term water deficit causes a significant reduction of inorganic phosphorus (Pi) concentration in the two types of leaves. Underwater stress condition, proline concentration increases in all individual plant organs, except youngest leaves. Drought treatment causes a significant increase of glycine betaine content in shoots without any change in tap-roots. Glucose concentrations increase significantly only in tap-roots due to drought. Underwater shortage sucrose accumulation is observed in all leaves and tap-roots examined (Choluj et al., 2008). These authors also report that, long-term drought activates an effective mechanism for osmotic adjustment both in the shoot and in the root tissues which may be critical for the survival rather than to maintain plant growth but sugar beet organs accumulate different solutes as a response to water cessation (Choluj et al., 2008).

Kaffka and Hembree (1999) have studied the behavior of sugar beet in saline soils, in particular the effect of seed priming on emergence rate and seedling growth. They have reported a similar rate of germination for primed seeds in saline plots and non-primed seeds in plots with low electric conductivity of soil (ECe). This practice is certain to become a seed enhancing requirement for crop production under stressed conditions (Sadeghian and Yavari, 2004).

In 1982, Hanson and Wyse evaluated sugar beet, fodder beet and *Beta vulgaris* (Synonyms: *Beta maritima*) under increasing salinity conditions. They concluded that salinization increases betaine levels of roots and shoots 2 to 3 fold. *B. maritima* accessions accumulated 40% less betaine in shoots than other progeny lines. The investigations undertaken on the identification of physiological and environmental factors under abiotic stress tolerance in sugar beet have attracted the attention of many researchers. Productivity and germination are linked to the varieties used by beet growers. They tend to note the ultimate crop yield (Tugnoli and Bettini, 2001). Significant variations have been found in the total dry matter, root yield, sugar yield, juice purity and impurities of roots among sugar beet, fodder beet and red beet when exposed to various periods of drought stresses (Sadeghian et al., 2000).

Mahmoud and Hill (1981) have reported that experiments performed on the germination and seedling emergence of sugar beet under various temperatures and NaCl concentrations indicate that the effect of salinity on seedling emergence was increasingly inhibitory as temperature increased from 10–15°C to 25 and 35°C. Moreover, Durrant et al. (1974) have reported that small differences in NaCl concentrations do not change the rate of germination of seeds, but greatly affect the water uptake and seedling growth.

There is strong evidence that N plays a great role in the generation of the foliage canopy as a central mechanism governing the yield of healthy and disease-free sugar beet crops (Malnou et al., 2006). After emergence, sugar beet crops with adequate nutrients and water need approximately 900°C days above a base temperature of 3°C to achieve 85% canopy cover, almost a closed canopy (Malnou et al., 2006; Werker and Jaggard, 1997).

This canopy cover is equivalent to a leaf area index (LAI) of 3. The sugar beet needs to contain 0.04 g N cm^{-2} of leaf, which corresponds to an uptake of 120 kg N ha^{-1} to reach 85% cover. N needs to be available for rapid uptake in order to produce a closed canopy as early in the growing season (Andrieu et al., 1997; Jaggard and Qi, 2006; Malnou et al., 2008).

The study carried out by Malnou et al. (2008) on whether nitrogen (N) fertilizer is necessary to maintain the efficiency of sugar beet foliage in late summer or whether the crop can continue to operate effectively on N mineralized from the soil. Three field experiments have been carried out in 2000, 2001 and 2002. The treatments used as early doses of N fertilizer (0, 80 or 160 kg N ha^{-1}) without and with a late N fertilizer application (60 kg N ha^{-1}) were made as soon as the foliage reached 85 cover percentage of the ground. The late N fertilizer dose increases the N concentration in the plants and canopy size in late summer, but canopy size still declines throughout autumn. The late N application increases chlorophyll concentration in the leaves but has no significant effect on radiation use efficiency in late summer and autumn. Late N application increases foliage dry weight at final harvest but fails to have a positive effect on sugar yield. This occurs on a soil which should not mineralize more N during summer than any other soil used for growing beet. Nitrogen from fertilizer is not necessary to maintain canopy efficiency and seems to be necessary solely to stimulate the rapid growth of the canopy in early summer (Malnou et al., 2008).

The studies carried out by Terry (1968), Milford et al. (1985a), Abdollahian-Noghabi, and Froud-Williams (1998) under controlled conditions in order to quantify the influence of temperature, solar radiation or water input on the growth and development of sugar beet. Other studies include modifying temperature, incident radiation or water supply in the field, using soil covers and/or irrigation (Watson et al., 1972; Sibma, 1983; Schittenhelm, 1999; Kenter et al., 2006).

The early growth of sugar beet in Central Europe is strongly influenced by temperature and crop growth rate is proportional to absorbed solar radiation after canopy closure (Scott and Jaggard, 1993; Durr and Boiffin, 1995). Unlike temperature and radiation, the water supply to plants is not affected by weather conditions alone, it is a function of rainfall and temperature together with evaporative demand and the soil water holding capacity. On sandy soils growth of sugar beet in summer may be limited by rainfall, but on loamy soils it can retain water for the crop to use during drought (Freckleton et al., 1999). In autumn, the crop growth rate goes down due to temperature lowering or solar radiation together with leaf senescence (Milford and Thorne, 1973; Scott et al., 1973). Some studies have focused on individual weather variables and their influence on the final yield at individual sites, but the influence of different weather variables throughout the vegetation period under a wide range of environmental conditions is lacking (Kenter et al., 2006).

The studies on the interactions between weather conditions and growth of sugar beet have shown that there are many differences in the soil and weather conditions in the beet growing areas of continental Europe (Scott and Jaggard, 1993; Freckleton et al., 1999; Clover et al., 2001; Kenter et al., 2006).

According to Durr and Boiffin (1995) there is a close relationship between the early growth of sugar beet and its maximum field emergence or the first leaf stage related to thermal time. Generally sugar beet requires 360 to 750°C d to form a closed canopy (Milford et al., 1985b). However, data published by Kenter et al. (2006) demonstrates that the period during which thermal time controls the dry matter accumulation of sugar beet extends up to 1200°C d above 3°C from sowing. This period starts from end of June or beginning of July. Different patterns have been found in the growth of leaves and taproot. Although leaf dry matter increases linearly from sowing to 1200°C d, the taproot needed up to 700°C d, followed by exponential increase. Thus canopy closure together with maximum radiation interception and assimilate supply are the prerequisites for taproot development (Kenter et al., 2006).

The optimum temperature for leaf growth is 24°C and for taproot growth 17°C (Terry, 1968). The values are 19°C for maximum leaf growth and 12°C for maximum taproot growth (Milford and Riley, 1980). These results undertaken under controlled conditions are difficult to extrapolate to the field. It is possible to estimate the optimum temperature for sugar beet growth under field conditions. Optimum mean daily air temperatures for taproot growth are approximately 18°C, corresponding to maximum daily temperatures between 22–26°C. Maximum leaf growth rates are also observed at 18°C, but no optimum has been identified. During the maximum leaf growth period (66–90 dps), a low-temperature range has been experienced, which coincides with the optimum temperature for net CO_2 assimilation of sugar beet (Schafer et al., 1979). Mean daily temperatures above 18°C during the season lead to root yield losses, which can be attributed to a reduction in both photosynthesis and dry matter accumulation (Kenter et al., 2006).

The dry matter accumulation of sugar beet is influenced by solar radiation which is similar to the temperature. The taproot and leaf growth rates increase strongly with increasing solar radiation during the first 65 dps. Since both these variables are closely correlated, effect cannot be clearly distinguished from that of temperature (Kenter et al., 2006). Despite low leaf area and radiation interception, increase in solar radiation may accelerate growth of beet during spring, because foliage reaches its maximum photo-synthetic capacity per unit leaf area within 64 days of sowing (Hodanova, 1975; Kenter et al., 2006).

From June to September solar radiation did not affect the growth rates of sugar beet leaves and taproot at 66 to 175 dps. When it decreases176–200 dps before October below 10 MJ m^{-2} d^{-1}, it becomes limiting in the case of dry matter accumulation in taproot (Kenter et al., 2006). In spring, the effect is clearly due to radiation, it is not correlated with temperature in October (Kenter et al., 2006). The growth rate of leaves is very low in autumn and is independent of incident radiation (Milford and Thorne, 1973).

One of the major factor affecting sugar beet growth is water supply (Jaggard et al., 1998). For a quantification of its influence, correlation of growth rates of leaves and taproot with different parameters of water balance provide a clue. Observations have revealed that rainfall only slightly affected leaf growth and did not show any significant influence on the growth rate of taproot (Kenter et al., 2006). Similar observations were made by Stockfisch et al. (2002), who attributed this to the water-retentive characteristic of the soil at the experimental plot. This probably was also the cause of poor correlations found by others. It is just possible that too long harvest intervals are responsible for the significant results. A low summer rainfall at the beet harvest site every other week demonstrates that from July to September the daily increment in sugar yield increases linearly with water input with a time lag of 14 days Kenter and Hoffmann (2003).

Wright et al. (1997) has published more significant results where sugar yield regressed with crop transpiration instead of rainfall/irrigation as demonstrated for evapotranspiration by Werker and Jaggard (1998) and Clover et al. (2001) as well. Kenter et al. (2006) has reported that neither actual evapotranspiration nor climatic water balance give better estimates of dry matter accumulation in sugar beet than rainfall, not even in July and August (91–120 and 121–145 dps). In sandy soils with low water holding capacity, drought induces yield loss in sugar beet as a function of the balance of evapotranspiration and rainfall from June to August (Jaggard et al., 1998).

The water supply effects have also been demonstrated by relating the leaf and taproot growth rates with the available soil water content (Kenter et al., 2006). The dry matter accumulation of the taproot increased with increasing soil water content up to 50 to 60% of available field capacity to a depth of 0.9 m in July and August, although maximum rooting depth of sugar beet may exceed 2.5 m (Windt and Marlander, 1994). This is also supported by the findings of Kretschmer and Hoffmann (1985). They stated that on sandy loam soil the available soil water content to a depth of 0.6 m became limiting for sugar beet growth when it fell below 50%. The leaf growth was much more susceptible to water shortage than taproot growth, which goes against the findings of Kenter et al. (2006). They found that from July onwards sugar beet leaf growth is hardly influenced by weather. It is just possible that, different nutrient availability at the experimental plots may be more important, because leaf growth of sugar beet until autumn is enhanced by high nitrogen supply (Scott and Jaggard, 1993). The water supply does not exert a distinct impact on sugar beet yield as per the information of single trials (Kenter et al., 2006).

The sequential addition of maximum growth rates observed in the experiments of Kenter et al. (2006) during each of the 6 time intervals, a theoretical yield of around 32 t taproot dry matter ha^{-1} after circa 200 vegetation days was recorded, which equals 24 t sugar ha^{-1}, accounting for 76% of taproot dry matter (Kenter, 2003). This theoretically achieved maximum yield is possible in a temperate climate, provided that the weather conditions and all agronomic measures are ideal during the entire season. Growth rate of taproots is thus maximized during all developmental stages of the crop (Kenter et al., 2006).

Sugar beets are cultivated over a wide range of climatic conditions and possess a good tolerance to the salinity in soils (Tognetti et al., 2003; Sakellariou-Makrantonaki et al., 2002). However, the major cause of yield loss in sugar beets; under arid and semiarid conditions; is drought stress (Pidgeon et al., 2001). The irrigation therefore plays an important role in sugar beet cultivation (Hassanli et al., 2010). The crop is drought resistant with ability to produce economical good yield even with low irrigation practices (Winter, 1980; Faberio et al., 2003). The water requirement for its cultivation is highly dependent on weather as well as genotype, nitrogen application, growth period, plant density and irrigation management (Kuchaki and Soltani, 1995). Since the growth period of is very long they are the highest water consuming plants; seasonal water consumption lies around 350–1150 mm in different regions of world (Allen et al., 1998). One of methods to maximize water use efficiency and increase yield per unit of water, a low irrigation regime can prove beneficial, particularly when the plants have faced water stress in a special growth step or in a whole season (Kirda, 2002).

According to Ucan and Gencoglan (2004); depending upon the level of irrigation water; the greatest values for WUE and IWUE are found in the treatments with the highest yields. Hassanli et al. (2010) has reported in

Southern Iran, the highest IWUE for sugar is 1.26 kg m^{-3}, using surface drip irrigation. According to Topak et al. (2010) the values for WUE and IWUE range from 7.46 to 8.32 kg m^{-3} and 7.91 to 11.5 kg m^{-3} respectively in the Central Anatolian geographical division of Turkey. These findings support the results of Kiymaz and Ertek (2015). Baclin and Celik (1994) have studied the effect of deficit irrigation on sugar beets under five different irrigation regimes at Tokat Research Institute experimental site during 1983–1986. They have reported that the highest root and sugar yields are observed under the treatments of 20 and 40% of irrigations so as to complete (full) irrigation treatment. They also stated that no significant yield loss will occur at 50% water deficit.

Elverenli (1985) has studies the effects of different levels of water and nitrogen on sugar beet root yield and quality. They have reported that the highest water consumption was 975 mm and the highest root yield was 49.70 t ha^{-1}. They recommended that sugar beets should be given 100 kg ha^{-1} of nitrogen and watered when 50% of the available water in root zone is consumed.

Sharmasarkar et al. (2001) undertook experiments on the effect of surface drip and flood irrigation on water and fertilizer use efficiency of sugar beets, and concluded that drip system irrigation water application uses less water and fertilizer as compared to the flood irrigation (Kiymaz and Ertek, 2015). Similarly Tognetti et al. (2003) have reported root yields of 78.7 and 63.1 kg ha^{-1} for sugar beet using the drip system under full and deficit (50% full irrigation) irrigation in Southern Italy. The largest root yield of 91.5 and 68.0 kg ha^{-1} in 2005 and 2006 respectively has been reported by Suheri et al. (2008) using drip irrigation under the conditions of Konya Plain in Central Anatolia in Turkey. According to the report published by Hassanli et al. (2010), the highest sugar beet root yield has been 79.7 kg ha^{-1} and the lowest 58.6 kg ha^{-1} for surface drip irrigation. In the same way a study carried out in Iran by Abyaneh et al. (2012) shows that the highest and the lowest root yields of 116.8 t ha^{-1} and 52.21 t ha^{-1} respectively are obtained under similar conditions discussed above. Demirer et al. (1994), Kandil et al. (2002), and Esmaeili (2010) have reported similar results. According to Kiymaz and Ertek (2015) deficit irrigation is important for an increase in the root yield of sugar beet

Kiymaz and Ertek (2015) investigated the relationship between water-nitrogen-yield in sugar beets, and reported highest yields from treatments that received of the highest water and nitrogen. As against this, the maximum yield per unit of water and nitrogen is reported for the treatments applied at lowest water and nitrogen. The highest WUE and IWUE values have been determined at the level of the lowest water and nitrogen application (Kiymaz and Ertek, 2015). They also report that the lowest water and nitrogen levels has greater effect on the quality parameters (sugar rate, refined digestion rate, and refined sugar yield) of sugar beets. I1N1 treatment (Kcp1: 0.5 and 30 kg ha^{-1} of N) with the highest WUE value has therefore been recommended for the cultivation of sugar beet under similar climatic and soil conditions in terms of water and fertilizer saving (Kiymaz and Ertek, 2015). The influence of stress factors on the activity of the primary photosynthetic reaction can be determined using Chlorophyll a fluorescence and these parameters can be used in the sugar beet. These measurements are used to detect the photoinhibitory effects of drought (Clarke et al., 1993; Clover et al.,1999; Chołuj et al., 2001, 2014; Ober et al., 2005; Bloch et al., 2006; Mohammadian et al., 2003).

Not much information is available on characterizing the reactions of sugar beet to drought with currently grown varieties (Chołuj et al., 2014). A recent study demonstrates, there is considerable genotypic variability in drought tolerance in the available sugar beet germplasm (Ober and Rajabi, 2010). The sugar beet genotypes investigated under drought conditions have exhibited clear differences in the root yield as well as yield stability, harvest index and radiation use efficiency (RUE) (Ober et al., 2004). The accumulation of sodium cations as well as variation in chlorophyll fluorescence parameters depending on water deficiency tolerance covering genotypic diversity has been demonstrated well (Mohammadian et al., 2003; Sadeghian et al., 2004). Ober et al. (2005) have observed genotypic diversity in drought-related physiological traits such as stomatal conductance, succulence index, SLW, and osmotic adjustment. The results of Romano et al. (2013) too show a wide genetic variation in root morphological parameters in sugar beet genotypes.

Chołuj et al. (2014) have demonstrated significant variation in the response of different sugar beet genotypes to drought. The greatest variation has been observed for wilting score, membrane damage, leave area index, PAR absorption, sum of P, Na, and α-amino N concentration in storage roots, as well as for root and sugar yield, while smaller variability has been observed for the succulence index, chlorophyll a fluorescence parameters and the sucrose concentration in tap-roots. Using PCA, they also demonstrated a strong relationship between some of these traits. A positive relationship has been identified between genetic similarity and root and sugar yield. Also LAI, PAR absorption, chlorophyll a fluorescence parameters and osmotic potential too are reported to be positively correlated. However, these traits are negatively correlated with wilting score, SLW, and succulence index (Chołuj et al., 2014). Their results suggest that the most tolerant genotypes with the highest DTI value for root and sugar yield are characterized by maintenance of leaf area and the activity of the primary photosynthetic light reaction during drought, and also by an effective osmotic adjustment response. LAI, PAR absorption, ϕPSII and s, which seem to be closely correlated with root yield, should also be used in physiological-associated breeding strategies during the selection of sugar beet genotypes, substantially differing in their response to water stress. According to Chołuj et al. (2014) despite the development of molecular techniques for the estimation of genetic distance, prediction of the heterosis effect based on molecular markers remains difficult, and is strongly dependent on the nature of the analyzed genotypes, species, and traits.

Apart from these ecophysiological studies, two detailed studies on yield and quality characteristics of sugar beet in Turkey have been undertaken by Çakmakçı and Oral (1998). They have conducted this study to follow the effect of field emergence levels with and without thinning on the root yield and quality of sugar beet (cv. Eva). The study has been carried out at the Experimental Farm of Turkish Sugar Factories Corporation in Pasinler Plateau (Erzurum-Turkey) in 1992 and 1993. These workers compared two intra-row spacings (8 cm and 15 cm within rows), thinning (thinned and unthinned) and three emergency levels (60, 50 and 35%). Emergence levels have been obtained by mixing seed with 0, 20 and 40% dead seed material. Row spacing was 45 cm apart in all treatments. Laboratory seed germination has been recorded as 90, 72 and 54%, respectively, with corresponding field emergence rates of 60, 50 and 35%. When field emergence percentage decreased to 35%, the differences between nonthinning sown in 15 cm intervals and thinning sown in 8 cm intervals, yield, and quality has varied significantly (P<0.01). White sugar content and white sugar yield are reported to have reached maximum values of 9000, 11000 and 10,000 plant/da. According to these workers, sowing into 8 cm intervals within rows in the presence of thinning may increase seed and labor costs. They have concluded that the 2-year trials indicate non-thinned sowing can be employed without sacrificing root and white sugar yield, by using 45 cm inter, 15 cm intra row spacing's and maintaining minimum 50–60% field emergence. This could eliminate thinning costs and reduce seed use by 50% (Çakmakçı and Oral, 1998).

Çalışkan et al. (1999) have studied the effects of different sowing dates (1st February and 1st March) and row spacing's (30, 45 and 60 cm) on yield and quality characteristics of four sugar beet cvs. (Evita, Sonja, Fiona, and Anadolumono) in the Hatay province (Turkey) under same ecological conditions in 1996 and 1997. Their findings report that delayed sowings negatively affect all the yield and quality characteristics. Moreover, 30 days delaying of sowing root leads to decrease in the sucrose yield around 8.06 and 8.64% in, 1996; 37.7and 40.4% in 1997, respectively. The highest root and sucrose yields have been obtained from 45 cm row spacing in 1996 (47.77 and 8.14 t/ha, resp.) and from 30 cm row spacing in 1997 (73.00 and 12.21 t/ha, resp.). The highest root yield has been obtained from Fiona in 1996 (49.71 t/ha) and from Anadolumono in 1997 (61.57 t/ha). The highest sucrose yield has been obtained from Fiona in 1996 (8.48 t/ha), from Evita in 1997 (10.22 t/ha), the root and sucrose yield differences of cvs were found statistically significant only in 1996 (Çalışkan et al., 1999).

14.4 EVALUATION OF SUGAR BEET WASTES AS AN ALTERNATIVE SOURCE

14.4.1 SUGAR BEET AND COMPOST

Earth does not possess unlimited soil resources. A population increase brings earth to critical stage, various measures have to be taken to avert serious situation. For sustainability, correct soil tillage practices, crop rotation, use of composts from farm, municipal, and industrial wastes are implemented during the last two decades. In the selection of appropriate soil amendment material economic factors are to be considered (Polat and Almaca, 2006).

Recently, compost is widely used with some organic wastes. Their contents and amounts can change according to socio-economic conditions of the area and the season of the year. One of the methods to deal with agricultural wastes is to convert it to waste compost. Composting agricultural waste accelerates organic components. The result is an organic product with potential benefits for agricultural soils. However, agricultural use of compost remains low. The nutrient value of compost is low compared to fertilizers. This concern is mitigated if compost is applied well in advance of planting (Paolo, 1996; Ouédraogo et al., 2001; Hargreaves et al., 2008).

Soil organic matter is one of the most important factors of the soil which positively effects physical, chemical, and biological properties of the soils. Soil chemical factors can be modified withinorganic fertilizers however physical factors of soils are degraded. To obtain maximum yield, soil physical factors must be considered as much as soil chemical factors because, physically degraded soils are not very productive. Generally organic matter contents of soil organic matter of Turkey are considerably low. About 21% of the soils are very low in soil organic matter and 54.6% of the soils have low soil organic matter content. It can be concluded that 75.6% of Turkish soils have inadequate soil organic matter and only 4.3% have adequate soil organic matter (Aydeniz, 1985). An important potential use of compost in the agricultural industry is its application as a soil amendment to eroded soils. Compost is a cheap source of organic matter that can enrich soil and add to the biological diversity (Biala and Wynen, 1998; Clark et al., 2000; Khalilian et al., 2002)

The studies carried out at the Eskişehir Soil and Water Resources Research Institute substation (Turkey) dealt with the supply of and management of potential agricultural wastes compost to get high yield and high polar value form sugar beet. The compost used for this study was composed of a mixture of wastes of wheat, sugar beet, corn, sunflower, and animal wastes (horse). This mixture was turned into compost after 16 weeks and applied to the field @ 0, 20, 40, and 60 tons compost/ha. The compost was not applied to the control treatment and only recommended dosage of fertilizer was applied. All compost treatments had no additional fertilizer application. A randomized block design with 3 replications was followed during the studies.

Compost analysis results are presented in Table 14.1. It indicates that, main determinant, i.e., C/N value was 8.8 and organic matter was 49.2%.

TABLE 14.1   The Composition of Compost

pH (SC)	EC (mmhos/m)	N (%)	OC (%)	C/N	OM (%)
7.42	4.27	3.1	27.2	8.8	49.2

The sugar yields were 7,844 tons/ha (for 0 tons compost/ha treatment), 8,129 tons/ha (for 20 tons compost/ha treatment), 8,585 tons/ha for 40 tons compost/ha treatment) and 9,061 tons/ha (for 60 tons compost/ha treatment). The polar values were 14.80 (for 0 tons compost/ha treatment), 15.30 (for 0 tons compost/ha treatment), 15.20 (for 0 tons compost/ha treatment) and 16.15% (for 0 tons compost/ha treatment) respectively (Table 14.2). Variance analysis gave us 5% significant statistical variance. Similar results have already been obtained by Raviv et al. (1998), Roe (1998), and Evanylo et al. (2008).

Are petition of the experiment following the same parameters revealed that, sugar yields were 6.840 tons/ha (for 0 tons compost/ha treatment), 7.620 tons/ha (for 20 tons compost/ha treatment), 8.244 tons/ha (for 40 tons compost/ha treatment), 8.676 tons/ha (for 60 tons compost/ha treatment) (Table 14.2). The polar value for 2008 form was 16.01 (for 0 tons compost/ha treatment), 16.20 (for 0 tons compost/ha treatment), 16.58 (for 0 tons compost/ha treatment) and %16.69% respectively (for 0 tons compost/ha treatment) (Table 14.2). Variance analysis gave us a 1% significant statistical variance. Duncan grouping placed the treatments in three groups for yields (Table 14.2). These findings are inline with earlier findings of Madejón et al. (2001) and Viator et al. (2002).

TABLE 14.2 Sugar Beet Yield and Polar Values Obtained During the Experiment

2007	A_0	A_1	A_2	A_3
Treatment average (tons/plot)	0.423	0.439	0.463	0.489
Yield (tons/plot)	7.844 B	8.129 B	8.585 AB	9.061 A
Polar value (%)	14.80 C	15.30 B	15.20 B	16.15 A
2008	A_0	A_2	A_4	A_6
Treatment average (tons/plot)	0.369	0.411	0.445	0.468
Yield (tons/plot)	6.840 D	7.620 C	8.244 B	8.676 A
Polar value (%)	16.01 D	16.20 C	16.58 B	16.69 A

The alphabets placed in rows show significance of treatment means based on Duncans's Test.

14.4.2 IMPORTANCE OF SUGAR BEET AS BIOSORPTION

Sugar beet pulp; a by-product of the sugar refining industry; is very cheap, mainly used as animal feed and at the same time one of the low-cost sorbents suited to biosorption, exhibiting a large capacity to bind metals. It is composed of 20% natural polysaccharide and over 40% of cellulosic and pectic substances. Later accounts for more than 40% of the dry matter, being complex heteropolysaccharides containing major sugar constituents galacturonic acid, arabinose, galactose, and rhamnose. The pectic substances strongly bind metal cations in aqueous solution, because of the carboxyl functions of galacturonic acid (Dronnet et al., 1997; Kartel et al., 1999; Gerente et al., 2000; Reddad et al., 2002; Aksu and Işoğlu, 2005).

The heavy metal biosorption by sugar beet pulp has been investigated by several workers. However, noy much attention has been paid to the temperature dependence of biosorption process and evaluating equilibrium, kinetic, and thermodynamic parameters of the system, which are important in the design of treatment systems (Aksu and Işoğlu, 2005). Aksu and Işoğlu (2005) have worked on possible removal of copper (II) from aqueous solution using the agricultural solid waste by-product dry sugar beet pulp as a biosorbent. Their results indicated that dried sugar beet pulp exhibited the highest copper (II) uptake capacity of 28.5 mg g^{-1} at 25°C and at an initial pH value of 4.0 at 250 mg l^{-1} initial copper (II) concentration. They also found that the intraparticle diffusion played an important role in the biosorption mechanisms of copper (II), and biosorption kinetics followed pseudo first- and pseudo second-order kinetic models rather than the saturation type kinetic model for all temperatures. Furthermore, they also evaluated the thermodynamic constants of biosorption using thermodynamic equilibrium coefficients obtained at different temperatures.

In the same way Pehlivan et al. (2008) carried out studies on the adsorption of Cd^{2+} and Pb^{2+} using sugar beet pulp (SBP), a low-cost material, and a comparison made on the ability of native SBP to remove cadmium (Cd^{2+}) and lead (Pb^{2+}) ions from aqueous solutions. SBP was used for the removal of Cd^{2+} and Pb^{2+} ions from aqueous water. The results point out that sorption process is relatively fast and equilibrium is reached after about 70 min of contact. SBP has removed nearly 70–75% of Cd^{2+} and Pb^{2+} ions in about 70 min. SBP uptake of Cd^{2+} and Pb^{2+} ions reveals that this process is pH-dependent profile. Overall uptake by SBP is maximum at pH 5.3 and is around 46.1 mg g^{-1} for Cd^{2+}, but at pH 5.0 it is 43.5 mg g^{-1} for Pb^{2+}, probably removed exclusively by ion exchange, physical sorption and chelation. A dose of 8 g L^{-1} is enough for optimum removal of metal ions. In the presence of 0.1 M NaNO$_3$ the level of metal ion uptake has been found to reach its maximum value very rapidly with the speed increasing both with SPB concentration as well as with increasing initial pH of the suspension (Pehlivan et al., 2008). Moreover, reversibility of the process has shown that desorption of Cd^{2+} and Pb^{2+} ions, previously deposited on SBP, back into the deionized water under acidic pH values during 1-day study and is generally rather low. The extent of adsorption for both metals increases along with an increase of SBP dosage. SBP as a cheap and highly selective seems to be a promising substrate to entrap heavy metals in aqueous solutions (Pehlivan et al., 2008).

14.4.3 IMPORTANCE OF SUGAR BEET AS BIOCHAR

According to Yao et al. (2011a, b) biochar converted from anaerobically digested sugar beet tailings (DSTC) shows superior ability to remove phosphate from water under a range of pH and competitive ion conditions. The phosphate removal is mainly controlled by adsorption onto colloidal and nano-sized MgO particles on the DSTC surface. Both the original and anaerobically prepared DSTC are waste materials, the cost to make DSTC is low. However, there is a benefit of additional energy generation and more efficient production with less CO_2 release by using pre-digested sugar beet tailings. DSTC is thus considered a promising alternative water treatment or environmental remediation technology for phosphate removal. The exhausted biochar can be directly applied to the agricultural fields as a fertilizer, when used as an adsorbent to reclaim phosphate from water. This will improve the soil fertility because the P-loaded biochar contains abundance of valuable nutrients. Other environmental benefits in this connection include fuel or energy produced during both the anaerobic digestion and pyrolysis as well as carbon sequestration due to biochar's refractory nature. Arsenate and molybdate are phosphate analogs, it is expected that the digested sugar beet tailing biochar can be an effective adsorbent for these as well (Manning and Goldberg, 1996; Yao et al., 2011a, b). The biochar physicochemical properties and the preliminary phosphate sorption assessment features show that residue from the anaerobic digestion of sugar beet tailings can be used as a feedstock for biochar production, some of the physicochemical properties (pH and surface functional groups) of the two biochars are similar, but only the anaerobically digested sugar beet tailing biochar has colloidal and nano-sized periclase (MgO) on its surface, and anaerobic digestion enhances the phosphate adsorption ability of biochar produced from digested sugar beet tailings relative to undigested ones (Yao et al., 2011a, b).

14.4.4 IMPORTANCE OF SUGAR BEET IN BIOENERGY

Sugar beet tailings have been evaluated as waste by-products and managed mainly by landfill disposal or direct land applications till now. The beet sugar accounts for a large percentage of all refined sugar consume by humans, every day significant amount of sugar beet tailings are generated as solid waste. These tailings can be anaerobically digested to generate bioenergy (biogas) (Liu et al., 2008; Yao et al., 2011a, b). This practice may reduce the sugar beet tailing waste volume, but disposal of residue materials from the anaerobic digestion still poses significant economic and environmental problem (Yao et al., 2011a, b).

Yaldiz et al. (1993) has reported that sugar beet yield, total energy input and output, and output/input ratio show the values as; 44.6 tons ha^{-1}, 20 567 MJ ha^{-1}, 103 862 MJ ha^{-1} and 5.04, respectively, however the yield has been reported to be 61.8 tons ha^{-1}by Altıntaş (2004). Total energy input and output, and output/input ratio in sugar beet production has also been given as 29 700 MJ ha^{-1}, 239 300–336 300 MJ ha^{-1} and 9.43–17.83 respectively by Hülsbergen et al. (2001). As against this Stout (1999) reports that total energy input used in sugar beet production is 32 900 MJ ha^{-1}. Hacıseferogulları et al. (2003) have calculated energy output/input ratio as 19.15, and Stephen and Jackson (1994) too has reported this ratio as 20.0.

Sugar beet seems to be one of the most efficient sources of ethanol on a per hectare basis, (Panella, 2010). It is reported to produce 100 and 120 l/t (fw) of ethanol through the fermentation process (110 l/t (FAO, 2008), 103.5 l/t (Shapouri et al., 2006), 117 l/t (Panella and Kaffka, 2010). In general the dry weight equivalent of one tonne of sugar beet (fw) is said to contain about 3.89 GJ of energy (Tzilivakis et al., 2005). Its ethanol has an energy content of 21.2 MJ/l (Schmer et al., 2008), which would be given an energy value of 2.44 GJ/t sugar beet (fw), assuming production of 115 l/t when converted to ethanol. A value around 5,060 l/ha yield for sugar beet has been calculated and compared with a 1,960 l/ha yield for maize or 952 l/ha for wheat, using a global estimate of average yield (46 t/ha for beet, 4.9 t/ha for maize, 2.8 t/ha for wheat) FAO (2008). Estimations on anaerobic digestion methods for whole beets to produce bio-methane can produce 137% more energy than would fermentation of sugar beet to ethanol (Von Felde, 2008). Yield estimates in FAO (2008) study are low compared to yields in some industrially advanced countries. Sugar beet production in Europe The studies undertaken during 2003–2004 on the sugar beet production variations in root yield of 9 genotypes in 5 European regions has revealed that, Hungary, Czech Republic, Austria, Poland come under the lowest average yielding, being over 60 t/ha (Hoffmann et al., 2009). In Turkey, Spain, and Italy this averaged over 110 t/ha (Kaffka, 2009; Panella and Kaffka, 2010).

According to Panella (2010) sugar beet will play an important role as a feedstock in the production of bioethanol, particularly in France, Germany, and Spain. There is interest in exploring or expanding sugar beet evaluation as a bioethanol feedstock in Ireland as well (Power et al., 2008), but it will be more in Slovenia and Serbia (Kondili and Kaldellis, 2007; Krajnc et al., 2007; Dodic et al., 2009).

Globally, there is a strong interest in Asia in the use of sugar beet as one of the potential bioethanol feedstocks-Turkey (Icöz et al., 2009; Ozturk et al., 2017): Similarly countries like Japan (Hatano et al., 2009; Koga et al., 2009; Koga, 2008), China (Tian et al., 2009), and India (Srivastava et al., 2008) are also striving hard in this direction. Most attractive areas for biofuel production will be the areas with climates where sugar beet can be cultivated both as spring or fall sowed crops, and it can be harvested daily most of the year, instead of storing the harvested roots, a large impediment to using sugar beet as a bioethanol feedstock (Panella, 2010). If biofuel production were via anaerobic digestion, sugar could be ensiled to extend the processing (Klocke et al., 2007).

A key factor effecting the use of sugar beet is the development rate of advanced technologies, which will offer very sustainable alternatives to some of the food crops now being used as feedstocks (de Wit et al., 2010; de Wit and Faaij, 2010; Panella, 2010). Another resource requirement, which is receiving increased scrutiny is the water footprint for bioenergy crop production. Sugar beet appers to be more efficient than other crops as a source for biofuel at global level (Gerbens-Leenes et al., 2009; Hoogeveen et al., 2009). However, once sugar beet is irrigated, its water footprint also grows, which may impact regions with fast-growing populations dependent on irrigated food production (Morillo-Velarde et al., 2001; Morillo-Velarde and Ober, 2006; Gerbens-Leenes et al., 2009; Hoogeveen et al., 2009).

14.4.5   IMPORTANCE AS SUGAR BEET MOLASSES

The amount of sucrose; which is easily fermented by many microbes; extracted per hectare of sugar beet is dependent on three factors, the weight of harvested beets, sucrose percentage in the beets, and the extractable amount of sucrose. Sugar beetroots may contain 20% of sucrose by fresh weight. But the amount extracted is approximately 15.3% on average basis, because Na^+ and K^+ or betaine and glutamine interfere with its extraction (McGinnis, 1982; Panella, 2010). After sucrose extraction, the remaining juice is molasses. Later weighs about 20 kg from one tonne (t = Mkg) of sugar beet (fw) and is approximately 50% sucrose (Shapouri et al., 2006). The leftover pulp constitutes around 22–28% of the dry mass of the sugar beet root, not solubilized during extraction and is fermentable (Scott and Jaggard, 1993). The amount of beet tops ranges from 4.6 to 7.5 t/ha, they have feed value but are usually left in the field at harvest (Scott and Jaggard, 1993).

Following the processing of sugar beet from which no more sugar can be crystallized by conventional means the residual syrup is beet molasses. Further processing of this through ionic separation technology can produce additional sugar. This technology is found around the many beet sugar factories in many countries. This viscous black liquid is primarily used as an animal feed or for fermented, being 80–85% Brix and of high viscosity. Later makes handling at ambient temperatures difficult. Therefore, it is normally diluted to 72–75% DM for sale and used on farm or in feed mills (Harland et al., 2006).

The protein composition of molasses has 3 fractions; 27% betaine, 33% amino acids and 35% uncharacterized. These nitrogenous components are believed to give beet molasses the characteristic earthy flavor and smell.

Their major component of dry matter is sucrose, with around 50% as sold. In addition there are small quantities of vitamins, reducing sugars and raffinose, ash apart from nitrogen as well as the minerals potassium and sodium. Their relatively high contents (11–55 g/kg and 11 g/kg) of minerals may cause diarrhea in livestock when fed in large amounts (Harland et al., 2006).

In short, beet molasses primarily is used as feed ingredient or a fermentation substrate. The reason for this is energy value and palatability, which usually respond positively to a 10% inclusion of beet molasses (Harland et al., 2006). Traditionally, its direct use was limited by difficulties in the physical handling, however the widespread availability of blends has overcome this problem.

The fermentation industry also uses beet molasses because the remaining fermentable sugars in it are an excellent source of growth in baker's yeast production. It is typically used together with cane molasses, but is preferred more singly due to its lower ash content, which reduces the waste matter flow. Other products are monosodium glutamate, lysine, and citric acid. It is also a good source of betaine which acts as a methyl donor by replacing part of the animal requirement for the amino acid methionine. The betaine obtained in this way is also used in monogastric and fish diets (Harland et al., 2006). Its role in metabolic processes has been recently reviewed by Craig (2004). The desugaring of beet molasses also produces a liquid co-product. Chromatographic separation permits industrial complexes to extract the majority of the sucrose remaining in beet molasses and creates another liquid co-product, desugared molasses. Later ranges in dry matter from 50 to 70% with 7–20% crude protein and 13–20% total sugars; extensively used in liquid feed production as a low-cost substitute for cane molasses. It is also added to beet pulp as a binding agent (Harland et al., 2006).

14.4.6   IMPORTANCE OF SUGAR BEET AS ANIMAL FEED

The sugar beet crop after harvested and processed for sugar production, yields a number of co-products which can be used for animal feeding as well as a forage for livestock. The processing of roots results in the production of two more valuable feeds: sugar beet pulp and beet molasses. The beet molasses is further processed by fermentation to alcohol or ion exclusion to yield additional potential feedstuffs, which can be used separately or combined and dried or otherwise processed in a variety of ways to produce a range of high-quality animal feeds (Harland et al., 2006).

Fresh beet tops and top silage are both very palatable and may be fed to all ruminants. On an energy basis 10 kg beet top silage is equivalent to 1.5 kg barley. The washed and sliced strips of "beet root" called cossettes are mixed with hot water and sugar extracted by diffusion. Later goes forward for the production of sugar crystals, while the spent cossettes, known as wet pulp, form the basis of various valuable animal feeds. The wet pulp is either sold to the farmer or passed through heavy presses to squeeze out nearly all the surplus water to produce pressed pulp. The pressed pulp is also sold to the farmer for feeding fresh or ensiling, or is dried to produce dried plain sugar beet pulp. In some countries the pressed pulp is mixed with molasses and then dried to produce dried molassed sugar beet pulp (feed) (Harland et al., 2006).

The dried molasses feeds are popular in some industrialized countries. Dried sugar beet pulp is sold as shreds, or pelletized. Later is conducive for transporting greater distances from the beet sugar producing facility (Harland et al., 2006).

Sugar beet pulp is primarily used as an energy source in livestock rations. However, it has also been suggested that the digestibility of poor quality forage diets for ruminants, such as those based on straw or hay, could be enhanced by the inclusion of sugar beet pulp. The pulp increases the number and variety of fiber-digesting bacteria in the rumen, resulting in more extensive digestion and utilization of all fiber in the ration (Harland et al., 2006). This has enabled the intake of straw by sheep to be increased from 414 to 505 g DM/day (Silva and Orskov, 1985). Other studies have demonstrated that dried plain sugar beet pulp fed at high levels with ammonia-treated straw has a less depressing effect than barley on straw digestibility (Fahmy et al., 1984).

A recent suggestion has been inclusion of sugar beet pulp in cereal mixes fed with grass silage. These improve overall utilization of the diet. Resulting balance of carbohydrates leads to higher rumen microbial output and rumen pH is buffered leading to optimization of events in the rumen. Sugar beet pulp is able to improve the digestibility as well as intake of forage such as straw. Recent studies demonstrate that total intake and output of modest quality grass silage can also be enhanced (Humphries et al., 2003).

According to Harland et al. (2006) pet food manufacturers have also started including sugar beet pulp in their formulations as it provides the fibrous filler (bulking agent), plus pectin, for many of their formulations. Sunvold et al. (1995) has suggested that moderately fermentable dietary fiber sources like beet pulp, promotes excellent stool characteristics without compromising nutrient digestibility in dogs, and may promote gastrointestinal tract health by optimizing short-chain fatty acid production.

14.5 CONCLUSIONS

All over the world, research workers are attempting to improve the economics and sustainability of suagrbeet growing, minimize any threat posed to the environment and find other sources of income from the crop. A major research objective has been to improve profit by increasing yields, decreasing inputs, and minimizing root impurities, etc. Since sugar beet is the most effective scavenger nitrogen fertilizer it is receiving an environmental acceptability by all, as at harvest it leaves little in soil then move into groundwater. The tops, insoluble root material and molasses as by-products are extensively used as animal feeds as well as in human dietary fiber to very small extent. The principal by-product (bagasse) is used as fuel in the factory. Researchers are striving hard to find new uses for sugar beet. Most notable is ethanol production for use as fuel or as a feedstock for the chemical industry. Undoubtedly, the use of beet sugar is a well-established industry currently. It usually is not economical currently in developed countries at present energy costs. However, sugar is used in the manufacture of a range of potentially high-volume products like polyurethane foams as well as high-intensity sweeteners, vitamins, and antibiotics; which are of high-value but low-volume products (Draycott, 2006).

KEYWORDS

- alternative uses
- ecophysiology
- sugar beet
- taxonomy

REFERENCES

Abdollahian-Noghabi, M., & Froud-Williams, R. J., (1998). Effect of moisture stress and re-watering on growth and dry matter partitioning in three cultivars of sugar beet. *Aspects Appl. Biol., 52*, 71–78.

Abyaneh, H. Z., Farrokhi, E., Bayat Varkeshi, M., & Ahmadi, M., (2012). Determining water demand and effect of its variations on some quantitative and qualitative traits of sugar beet product. *J. Sugar Beet, 27* (2), 21–27.

AGBIOS, (2005). *Agriculture and Biotechnology Strategies Inc.* GM Database crop plant. Sugar beet. http://www.agbios.com/dbase.php.

Aksu, Z., & Işoğlu, I. A., (2005). Removal of copper (II) ions from aqueous solution by biosorption onto agricultural waste sugar beet pulp. *Process Biochemistry, 40* (9), 3031–3044.

Allen, R., Pereira, L. A., & Smith, M., (1998). Crop evapotranspiration: Guidelines for computing crop water requirement. *Irrig. Sci., 56*, 116–127.

Altıntaş A., (2004). The analysis of input and costs for some agricultural product in Tokat and Amasya and Yozgat provinces (Tokat, Amasya ve Yozgat yörelerinde yetiştirilen bazı tarımsal ürünlerin üretim girdileri ve maliyetleri). Ministry of Agriculture and Rural Affairs of Turkey, General Director of Rural Affairs, Tokat Research Institute. Tokat, Turkey: Tarım ve Köyişleri Bakanlığı Tokat Tarımsal Araştırma Enstitüsü. [in Turkish].

Andrieu, B., Allirand, J. M., & Jaggard, K. W., (1997). Ground cover and leaf area index of maize and sugar beet crops. *Agronomie, 17*, 315–321.

Aydeniz, A., (1985). Toprak amenajmani. [Soil Management]. A. U. Z. F. Publishing, *Textbook No. 263*. Ankara, Turkey.

Baclin, M., & Celik, S., (1994). Tokat-Kazova koşullarında kısıntılı su uygulamasında şeker pancarının su-verim ilişkisi. In: *Seker Pancarı Yetistirme Teknigi Sempozyumu, II* (pp. 92–103). Gubreleme ve Sulama, Konya, Mayıs 6–7.

Biala, J., & Wynen, W., (1998). *Is There a Market for Compost in Agriculture?* International Composting Conference, Melbourne.

Biancardi, E., McGrath, J. M., Panella, L. W., Lewellen, R. T., & Stevanato, P., (2010). Sugar beet. In: Bradshaw, J. E., (ed.), *Root, and Tuber Crops, Handbook of Plant Breeding 7* (pp. 173–219). Springer Science+Business Media, LLC.

Bloch, D., & Hoffman, C., (2005). Seasonal development of genotypic differences in sugar beet (*Beta vulgaris* L.) and their interaction with water supply. *J. Agron. Crop Sci., 191*, 263–272.

Bloch, D., Hoffman, C. M., & Märländer, B., (2006). Solute accumulation as a case for quality losses in sugar beet submitted to continuous and temporary drought stress. *J. Agron. Crop Sci., 192*, 17–24.

Çakmakçı, R., & Oral, E., (1998). Effect of different field emergence rates on the yield and quality of sugar beet (*Beta vulgaris* L.) grown with and without thinning. *Turkish Journal of Agriculture and Forestry, 22* (5), 451–462.

Çalışkan, M. E., Işler, N., Günel, E., & Güler, M. B., (1999). The effects of sowing date and row spacing on yield and quality of some sugar beet (*Beta vulgaris* L.) varieties in Hatay ecological conditions. *Turkish Journal of Agriculture and Forestry, 23* (EK5), 1155–1162.

Campbell, L. G., (2002). Sugar beet quality improvement. *J. Crop Prod., 5*, 395–413.

Chołuj, D., Karwowska, R., Ciszewska, A., & Jasińska, M., (2008). Influence of long-term drought stress on osmolyte accumulation in sugar beet (*Beta vulgaris* L.) plants. *Acta Physiologiae Plantarum, 30* (5), 679–687.

Chołuj, D., Karwowska, R., Jasi´nska, M., Heber, G., & Pietkiewicz, S., (2001). Mechanism of sugar beet acclimation to drought stress. *Buletinul USAMV-CV, A-H, 55*, 132–138.

Chołuj, D., Karwowska, R., Jasinska, M., & Haber, G., (2004). Growth and dry matter partitioning in sugar beet plants (*Beta vulgaris* L.) under moderate drought. *Plant Soil Environ., 50*, 265–272.

Chołuj, D., Wiśniewska, A., Szafrański, K. M., Cebula, J., Gozdowski, D., & Podlaski, S., (2014). Assessment of the physiological responses to drought in different sugar beet genotypes in connection with their genetic distance. *Journal of Plant Physiology, 171* (14), 1221–1230.

Clark, G. A., Stanley C. D., & Maynard, D. N., (2000). Municipal solid waste compost (MSWC) as a soil amendment in irrigated vegetable production. *Trans. ASAE, 43* (4), 847–853.

Clarke, N. A., Hetschkun, H., Jones, C., Boswell, E., & Marfaing, H., (1993). Identification of stress tolerance traits in sugar beet. In: Jackson, M. B., & Black, I. C. R., (eds.), *Interacting Stresses in Plants in Changing Climate* (pp. 511–524). Berlin: Springer-Verlag.

Clover, G. R. G., Jaggard, K. W., Smith, H. G., & Azam-Ali, S. N., (2001). The use of radiation interception and transpiration to predict the yield of healthy, droughted, and virus-infected sugar beet. *J. Agric. Sci. Camb., 136,* 169–178.

Clover, G. R. G., Smith, H. G., Azam-Ali, S. N., & Jaggard, K. W., (1999). The effect of drought on sugar beet growth in isolation and combination with beet yellow virus infection. *J. Agron. Sci., 133,* 251–261.

Coca, M., Garcia, M. T., Gonzalez, G., Pena, M., & Garcia, J. A., (2004). Study of colored components formed in sugar beet processing. *Food Chem., 86,* 421–433.

Craig, S. A., (2004). Betaine in human nutrition. *American Journal of Clinical Nutrition, 80,* 539–549.

De Bock, T. S., (1985). The genus *Beta*: Domestication, taxonomy, and interspecific hybridization for plant breeding. *International Symposium on Taxonomy of Cultivated Plants, 182,* pp. 335–344.

De Wit, M., & Faaij, A., (2010). European biomass resource potential and costs. *Biomass and Bioenergy, 34,* 188–202.

De Wit, M., Junginger, M., Lensink, S., Londo, M., & Faaij, A., (2010). Competition between biofuels: Modeling technological learning and cost reductions over time. *Biomass and Bioenergy, 34,* 203–217.

Demirer, T., Brohi, A. R., Koç, H., & Kahraman, M. R., (1994). Degisik azot ve fosfor dozlarının seker pancarı verim ve kalitesi üzerine etkisi. In: *Seker Pancarı Yetistirme Teknigi Sempozyumu, II.* Gubreleme ve Sulama, Konya, Mayıs 6–7.

Dodic, S., Popov, S., Dodic, J., Rankovic, J., Zavargo, Z., & Jevtic, M. R., (2009). Bioethanol production from thick juice as intermediate of sugar beet processing. *Biomass and Bioenergy, 33,* 822–827.

Draycott, A. P., (2006). Introduction. In: Draycott, A. P., (ed.), *Sugar Beet* (pp. 1–8). Blackwell Publishing Ltd.

Dronnet, V. M., Renard, C. M. G. C., Axelos, M. A. V., & Thibault, J. F., (1997). Binding of divalent metal cations by sugar-beet pulp. *Carbohydr. Polym., 34,* 73–82.

Durr, C., & Boiffin, J., (1995). Sugar beet seedling growth form germination to first leaf stage. *J. Agric. Sci. Camb., 124,* 427–435.

Durrant, M. J., Draycott, A. P., & Payne, P. A., (1974). Some effects of sodium chloride on germination and seedling growth of sugar beet. *Ann. Bot., 38,* 1045–1051.

Elverenli, M. A., (1985). Cesitli azotlu gubre seviyeleriyle sulamanın seker pancarı verimine ve kalitesine etkileri, Doktora Tezi, Ankara.

Erdal, G., Esengün, K., Erdal, H., & Gündüz, O., (2007). Energy use and economical analysis of sugar beet production in Tokat province of Turkey. *Energy, 32* (1), 35–41.

Esmaeili, M. A., (2010). Evaluation of the effects of water stress and different levels of nitrogen on sugar beet (*Beta vulgaris* L.). *Int. J. Biol., 3* (2), 89–93.

Evanylo, G., Sherony, C., Spargo, J., Starner, D., Brosius, M., & Haering, K., (2008). Soil and water environmental effects of fertilizer-, manure-, and compost-based fertility practices in an organic vegetable cropping system. *Agri. Ecosyst. Environ., 127* (1), 50–58.

Faberio, C., Martin de Santa Olalla, F., Lopez, R., & Dominguez, A., (2003). Production and quality of the sugar beet cultivated under controlled deficit irrigation conditions in a semi-arid climate. *Agric. Water Manage, 62*, 215–227.

Fahmy, S. T. M., Lee, N. H., & Orskov, E. R., (1984). Digestion and utilization of straw. 2, The effect of supplementing ammonia-treated straw with different nutrients. *Animal Production, 30*, 75–81.

FAO, (2005). *Food and Agriculture Organization of the United Nations.* Agriculture production. http://faostat. fao. org/faostat.

FAO, (2008). *The State of Food and Agriculture.* Biofuels: Prospects, risks, and opportunities. Rome, Italy: FAO Electronic Publishing Policy and Support Branch Communications Division. http://www.fao.org/publications/sofa-2008/en/.

FAO, (2011). *FAOSTAT Online Database*, Available from: http://faostat.fao.org.

Ford-Lloyd, B. V., & Williams, J. T., (1975). A revision of *Beta* section *Vulgares* (Chenopodiaceae), with new light on the originof cultivated beets. *Bot. J. Linn. Soc., 71*, 89–102.

Ford-Lloyd, B. V., (2005). Sources of genetic variation, Genus *Beta.* In: Biancardi, E., et al. (eds.), *Genetics*, and *Breeding of Sugar Beet* (pp. 25–33). Science Publishers Inc., Enfield, NH.

Freckleton, R. P., Watkinson, A. R., Webb, D. J., & Thomas, T. H., (1999). Yield of sugar beet in relation to weather and nutrients. *Agric. For. Meteorol., 93*, 39–51.

Gerbens-Leenes, W., Hoekstra, A. Y., & Van der Meer, T. H., (2009). The water footprint of bioenergy. *Proceedings of the National Academy of Sciences of USA, 106*, 10219–10223.

Gerente, C., Couespel, Du, Mesnil, P., Andres, Y., Thibault, J. F., & Le Cloirec, P., (2000). Removal of metal ions from aqueous solution on low-cost natural polysaccharides: Sorption mechanism approach. *React. Func. Polym., 46*, 135–144.

Ghoulam, Ch., Foursy, A., & Fares, K., (2002). Effects of salt stress on growth, inorganic ions and proline accumulation in relation to osmotic adjustment in five sugar beet cultivars. *Environ. Exp. Bot., 47*, 39–50.

Gzik, A., (1996). Accumulation of proline and pattern of a-amino acids in sugar beet plants in response to osmotic, water, and salt stress. *Environ. Exp. Bot., 36*, 29–38.

Hacıseferogulları, H., Acaroglu, M., & Gezer, I., (2003). Determination of the energy balance of the sugar beet plant. *Energy Sources, 25*, 15–22.

Hanson, A. D., & Rhodes, D., (1983). 14C tracer evidence for synthesis of choline and betaine via phosphoryl base intermediates in salinized sugar beets. *Plant Physiol., 71*, 692–700.

Hanson, A. D., & Wyse, R., (1982). Biosynthesis, translocation, and accumulation of betaine in sugar beet and its progenitors in relation to salinity. *Plant Physiol., 70*, 1191–1198.

Hargreaves, J. C., Adl, M. S., & Warman, P. R., (2008). A review of the use of composted municipal solid waste in agriculture. *Agri. Ecosyst. Environ., 123* (1), 1–14.

Harland, J. I., (1989). Beta fiber: A positive route to lower cholesterol. In: Proceedings of Food Ingredients Europe. *Porte de Versailles*, Paris, pp. 204–205.

Harland, J. I., (1993). Beta Fiber: A case history. *International Journal of Food Sciences and Nutrition, 44* (1), S87–S96.

Harland, J. I., Jones, C. K., & Hufford, C., (2006). Co-Products. In: Draycott, A. P., (ed.), *Sugar Beet* (pp. 443–463). Blackwell Publishing Ltd.

Hassanli, A. M., Ahmadirad, S., & Beecham, S., (2010). Evaluation of the influence of irrigation methods and water quality on sugar beet yield and water use efficiency. *Agric. Water Manage, 97*, 357–362.

Hatano, K. I., Kikuchi, S., Nakamura, Y., Sakamoto, H., Takigami, M., & Kojima, Y., (2009). Novel strategy using an adsorbent-column chromatography for effective ethanol production from sugarcane or sugar beet molasses. *Bioresource Technology, 100*, 4697–4703.

Hodanova, D., (1975). Specific leaf weight and photosynthetic rate in sugar beet leaves of different age. *Biol. Plant, 17*, 314–317.

Hoffmann, C. M., Huijbregts, T., van Swaaij, N., & Jansen, R., (2009). Impact of different environments in Europe on yield and quality of sugar beet genotypes. *European Journal of Agronomy, 30*, 17–26.

Hoogeveen, J., Faure's, J.-M., & Van de Giessen, N., (2009). Increased biofuel production in the coming decade: To what extent will it affect global freshwater resources? *Irrigation and Drainage, 58*, S160.

Hülsbergen, K. J., Feil, B., Biermann, S., Rathke, G. W., Kalk, W. D., & Diepenbrock, W. A., (2001). Method of energy balancing in crop production and its applications in a long-term fertilizer trial. *Agric. Ecosystem Environ., 86*, 303–321.

Humphries, D. J., Sutton, J. D., Cockman, D. M., Witt, M. W., & Beever, D. E., (2003). Pressed sugar beet pulp as an alternative fiber source for lactating dairy cows. *Proceedings British Society of Animal Science Annual Meeting*, New York, p. 9.

Içöz, E., Tugrul, K. M., Saral, A., & Içöz, E., (2009). Research on ethanol production and use from sugar beet in Turkey. *Biomass and Bioenergy, 33*, 1–7.

Jaggard, K. W., & Qi, A., (2006). Crop physiology and agronomy. In: Draycott, A. P., (ed.), *Sugar Beet* (pp. 134–168). Blackwell Publishing, Oxford.

Jaggard, K. W., Dewar, A. M., & Pidgeon, J. D., (1998). The relative effects of drought stress and virus yellows on the yield of sugar beet in the UK, 1980–1995. *J. Agric. Sci. Camb., 130*, 337–343.

Kaffka, S. R., & Hembree, K., (1999). The emergence of autumn-planted sugar beet seedlings under saline conditions. *Proceedings of the 62th IIRB Congress* (pp. 195–199). Sevilla, Spain.

Kaffka, S. R., (2009). Fertilizer N effects on yield and root quality for high-yielding, fall-planted sugar beets in the Imperial Valley, 37. *Denver, CO: ASSBT*.

Kandil, A. A., Badawi, M. A., El-Moursy, S. A., & Abdou, U. M. A., (2002). Effect of planting dates, nitrogen levels and biofertilization treatments on: II-yield, yield components and quality of sugar beet (*Beta vulgaris* L.). *J. Agric. Sci. Mansoura Univ., 27* (11), 7257–7266.

Kartel, M. T., Kupchik, L. A., & Veisov, B. K., (1999). Evaluation of pectin binding of heavy metal ions in aqueous solutions. *Chemosphere, 38*, 2591–2596.

Katerij, N., van Hoorn, J. W., Hamdy, A., Mastrorilli, M., & Mou Karezl, E., (1997). Osmotic adjustment of sugar beets in response to soil salinity and its influence on stomatal conductance, growth, and yield. *Agric. Water Manage, 34*, 57–69.

Kenter, C., & Hoffmann, C., (2003). Impact of weather on yield formation of sugar beet in Germany. In: *Advances in Sugar Beet Research* (pp. 19–32). 5. Institut International de Recherches Betteravieres.

Kenter, C., (2003). Ertragsbildung von Zuckerruben in Abhangigkeit von der Witterung. PhD Thesis, University of Gottingen, Cuvillier, Gottingen.

Kenter, C., Hoffmann, C. M., & Märländer, B., (2006). Effects of weather variables on sugar beet yield development (*Beta vulgaris* L.). *European Journal of Agronomy, 24* (1), 62–69.

Khalilian, A., Sulivan, M. J., Mueller, J. D., Shiralipour, A., Wolak, F. J., & Williamson, R. E., (2002). Effects of surface application of MSW compost on cotton production –soil properties, plant responses, and nematode management. *Comp. Sci. Util., 10* (3), 270–279.

Kirda, C., (2002). *Deficit Irrigation Scheduling Based on Plant Growth Stages Showing Water Stress Tolerance, Deficit Irrigation Practices*. Food and Agriculture Organization of the United Nations, Rome, pp. 3–10.

Kiymaz, S., & Ertek, A., (2015). Yield and quality of sugar beet (*Beta vulgaris* L.) at different water and nitrogen levels under the climatic conditions of Kırsehir, Turkey. *Agricultural Water Management, 158*, 156–165.

Klocke, M., Mahnert, P., Mundt, K., Souidi, K., & Linke, B., (2007). Microbial community analysis of a biogas-producing completely stirred tank reactor fed continuously with fodder beet silage as mono-substrate. *Systematic and Applied Microbiology, 30*, 139–151.

Knapp, E., (1958). Beta-Ruben, Bes. Zuckerruben. In: Kappert, H., & Rudorf, W., (eds.), *Handbuch der Pflanzenzüchtung* (Vol. 3, pp. 196–284). Berlin &Hamburg: Paul Parey.

Koga, N., (2008). An energy balance under a conventional crop rotation system in northern Japan: Perspectives on fuel ethanol production from sugar beet. *Agriculture, Ecosystems, and Environment, 125*, 101–110.

Koga, N., Takahshi, H., Okazaki, K., Kajiyama, T., & Kobayashi, S., (2009). Potential agronomic options for energy-efficient sugar beet-based bioethanol production in northern Japan. *Global Change Biology and Bioenergy, 1*, 220–229.

Kondili, E. M., & Kaldellis, J. K., (2007). Biofuel implementation in eastern Europe: Current status and future prospects. *Renewable and Sustainable Energy Reviews, 11*, 2137–2151.

Krajnc, D., Mele, M., & Glavic, P., (2007). Improving the economic and environmental performances of the beet sugar industry in Slovenia: Increasing fuel efficiency and using by-products for ethanol. *Journal of Cleaner Production, 15*, 1240–1252.

Kretschmer, H., & Hoffmann, F., (1985). Einfluß von Wasserdefizit auf Blattwachstum, Netto-CO_2-Assimilation und Ertragsbildung der Zuckerr"uben unter Feldbedingungen. *Arch. Acker- Pflanzenb. Bodenkd., 29*, 521–530.

Kuchaki, A., & Soltani, A., (1995). *Sugar Beet Agronomy*. Mashhad University Publisher.

Lange, W., Brandenburg, W. A., & De Bock, T. S. M., (1999). Taxonomy and cultonomy of beet (*Beta vulgaris* L.). *Bot. J. Linnean. Soc., 130*, 81–96.

Letschert, J. P. W., (1993). *Beta* section *Beta*, biogeographical patterns of variation, and taxonomy. PhD Thesis, Wageningen Agricultural University.

Letschert, J. P. W., Lange, W., Frese, L., & Van Der Berg, D., (1994). Taxonomy of the section *Beta. J. Sugar Beet Res., 31*, 69–85.

Liu, W., Pullammanappallil, P. C., Chynoweth, D. P., & Teixeira, A. A., (2008). Thermophilic anaerobic digestion of sugar beet tailings. *Trans. Asabe, 51* (2), 615–621.

Mack, G., & Hoffman, C. H. M., (2006). Organ-specific adaptation to low precipitation in solute concentration of sugar beet (*Beta vulgaris* L.). *Eur. J. Agron., 25* (3), 270–279.

Madejón, E., López, R., Murillo, J. M., & Cabrera, F., (2001). Agricultural use of three (sugar beet) vinasse composts: Effect on crops and chemical properties of a Cambisol soil in the Guadalquivir river valley (SW Spain). *Agri. Ecosyst. Environ., 84* (1), 55–65.

Mahmoud, E. A., & Hill, M. J., (1981). Salt tolerance of sugar beet at various temperatures. *NZ J. Agric. Res., 24*, 67–71.

Malnou, C. S., Jaggard, K. W., & Sparkes, D. L., (2006). A canopy approach to nitrogen fertilizer recommendation for the sugar beet crop. *Eur. J. Agron., 25*, 254–263.

Malnou, C. S., Jaggard, K. W., & Sparkes, D. L., (2008). Nitrogen fertilizer and the efficiency of the sugar beet crop in late summer. *European Journal of Agronomy, 28* (1), 47–56.

Manning, B. A., & Goldberg, S., (1996). Modeling competitive adsorption of arsenate with phosphate and molybdate on oxide minerals. *Soil Sci. Soc. Am. J., 60*, 121–131.

McGinnis, R. A., (1982). *Sugar-Beet Technology*. Beet Sugar Development Foundation, Fort Collins, CO.

Milford, G. F. J., & Lawlor, D. W., (1976). Water and physiology of sugar beet. In: *Proceedings 39th IIRB Winter Congress* (pp. 95–108). Brussels, Belgium.

Milford, G. F. J., & Riley, J., (1980). The effects of temperature on leaf growth of sugar beet varieties. *Ann. Appl. Biol., 94*, 431–443.

Milford, G. F. J., &Thorne, G. N., (1973). The effect of light and temperature late in the season on the growth of sugar beet. *Ann. Appl. Biol., 75*, 419–425.

Milford, G. F. J., Pocock, T. O., & Riley, J., (1985a). An analysis of leaf growth in sugar beet. I. Leaf appearance and expansion in relation to temperature under controlled conditions. *Ann. Appl. Biol., 106*, 163–172.

Milford, G. F. J., Pocock, T. O., Jaggard, K. W., Biscoe, P. V., Armstrong, M. J., Last, P. J., & Goodman, P. J., (1985b). An analysis of leaf growth in sugar beet. IV. The expansion of the leaf canopy in relation to temperature and nitrogen. *Ann. Appl. Biol., 107*, 335–347.

Mohammadian, R., Rahimian, H., Moghaddam, M., & Sadeghian, S. Y., (2003). The effect of early season drought on chlorophyll a fluorescence in sugar beet (*Beta vulgaris* L.). *Pak. J. Biol. Sci., 6*, 1763–1769.

Monreal, J. A., Jimenez, E. T., Remesal, E., Morillo-Velarde, R., García-Mauriño, S., & Echevarría, C., (2007). Proline content of sugar beet storage roots: Response to water deficit and nitrogen fertilization at field conditions. *Environmental and Experimental Botany, 60* (2), 257–267.

Morgan, L. M., Tredger, J. A., Williams, C. A., & Marks, V., (1988). Effects of sugar beet fibre on glucose tolerance and circulating cholesterol levels. *Proceedings of the Nutrition Society, 47*, 185A.

Morillo-Velarde, R., & Ober, E. S., (2006). Water use and irrigation. In: Draycott, A. P., (ed.), *Sugar Beet* (pp. 221–255). Oxford: Blackwell Publishing Ltd.

Morillo-Velarde, R., Cavazza, L., Cariolle, M., & Beckers, R., (2001). Irrigation de la betterave scurie`re en zone me´diterrane´enne. *Advances in Sugar Beet Research* (Vol. 3). Brussels, Belgium: IIRB.

Ober, E. S., & Rajabi, A., (2010). Abiotic stress in sugar beet. *Sugar Tech., 12*, 294–298.

Ober, E. S., Bloa, M. L., Clark, C. J. A., Royal, A., Jaggard, K. W., & Pidgeon, J. D., (2005). Evaluation of physiological traits as indirect selection criteria for drought tolerance in sugar beet. *Field Crops Res., 91*, 231–249.

Ober, E. S., Clark, C. J. A., Bloa, M. L., Royal, A., Jaggard, K. W., & Pidgeon, J. D., (2004). Assessing the genetic resources to improve drought tolerance in sugar beet: agronomic traits of diverse genotypes under drought and irrigated conditions. *Field Crops Res., 90*, 213–234.

Ouédraogo, E., Mando, A., & Zombré, N. P., (2001). Use of compost to improve soil properties and crop productivity under low input agricultural system in West Africa. *Agri. Ecosyst. Environ., 84* (3), 259–266.

Ozturk, M., Saba, N., Altay, V., Iqbal, R., Hakeem, K. R., Jawaid, M., & Ibrahim, F. H., (2017). Biomass and bioenergy: An overview of the development potential in Turkey and Malaysia. *Renewable and Sustainable Energy Reviews, 79*, 1285–1302.

Panella, L., & Kaffka, S. R., (2010). Sugar beet (*Beta vulgaris* L) as a biofuel feedstock in the United States. In: Eggleston, G., (ed.), *Sustainability of the Sugar and Sugar-Ethanol Industries* (pp. 163–175).

Panella, L., (2010). Sugar beet as an energy crop. *Sugar Tech., 12* (3–4), 288–293.

Paolo, S., (1996). The role of composting in sustainable agriculture. *The Science of Composting* (pp. 23–29). Springer Netherlands.

Pehlivan, E., Yanık, B. H., Ahmetli, G., & Pehlivan, M., (2008). Equilibrium isotherm studies for the uptake of cadmium and lead ions onto sugar beet pulp. *Bioresource technology, 99* (9), 3520–3527.

Pidgeon, J. D., Werker, A. R., Jaggard, K. W., Richter, G. M., Lister, D. H., & Jones, P. D., (2001). Climatic impact on the productivity of sugar beet in Europe (1961–1995). *Agric. For. Meteorol., 109* (1), 27–37.

Pilon-Smits, E. A., Terry, N., Sears, T., & Van Dun, K., (1999). Enhanced drought resistance in fructan-producing sugar beet. *Plant Physiology and Biochemistry, 37* (4), 313–317.

Polat, H., & Almaca, N. D., (2006). *The Application Compost and Green Fertilizer the Effect on Improvement of Leveled Soil Properties and Cotton Yield in Harran plain.* Soil and Water Resources Research Institute – Şanliurfa-Turkey.

Power, N., Murphy, J. D., & McKeogh, E., (2008). What crop rotation will provide optimal first-generation ethanol production in Ireland, from technical and economic perspectives? *Renewable Energy, 33*, 1444–1454.

Raviv, M., Zaidman, B. Z., & Kapulnik, Y., (1998). The use of compost as a peat substitute for organic vegetable transplants production. *Compost Science & Utilization, 6* (1), 46–52.

Reddad, Z., Gerente, C., Andres, Y., Ralet, M-C., Thibault, J-F., & Le Cloirec, P., (2002). Ni (II) and Cu (II) binding properties of native and modified sugar beet pulp. *Carbohydr. Polym., 49, 23–*31.

Roe, N. E., (1998). Compost utilization for vegetable and fruit crops. *Hort Science: A Publication of the American Society for Horticultural Science* (USA).

Romano, A., Sorgona, A., Lupini, A., Araniti, F., Stevanato, P., Cacco, G. et al. (2013). Morpho-physiological responses of sugar beet (*Beta vulgaris* L.) genotypes to drought stress. *Acta Physiol. Plant, 35*, 853–865.

Rover, A., & Buttner, G., (1999). Eidfluß von Trockenstreß auf die technische Qualita"t von Zuckerru"ben. In: *Proceedings 39th IIRB Congress* (pp. 97–109). Spain, Sevilla.

Sadeghian, S. Y., & Yavari, N., (2004). Effect of water-deficit stress on germination and early seedling growth in sugar beet. *Journal of Agronomy and Crop Science, 190* (2), 138–144.

Sadeghian, S. Y., Fasli, H., Taleghani, D. F., & Mesbah, M., (2000). Genetic variation for drought stress in sugar beet. *J. Sugar Beet Res., 37*, 55–77.

Sadeghian, S. Y., Mohammadian, R., Taleghani, D. F., & Abdollahian-Noghabi, M., (2004). Relation between sugar beet traits and water use efficiency in water stressed genotypes. *Pak. J. Biol. Sci., 7*, 1236–1241.

Sakellariou-Makrantonaki, M., Kalfountzos, D., & Vyrlas, P., (2002). Water saving and yield increase of sugar beet with subsurface drip irrigation. *Global Nest Int. J., 4* (2–3), 85–91.

Schafer, W., Dannowski, M., & Wurl, B., (1979). Bestimmung des Temperaturoptimums der CO_2-Aufnahmerate bei Zuckerruben. *Archiv fur Acker-und Pflanzenbau und Bodenkunde.*

Schittenhelm, S., (1999). Agronomic performance of root chicory, jerusalem artichoke, and sugar beet in stress and non-stress environments. *Crop Sci., 39*, 1815–1823.

Schmer, M. R., Vogel, K. P., Mitchell, R. B., & Perrin, R. K., (2008). Net energy of cellulosic ethanol from switchgrass. *Proceedings of the National Academy of Sciences of USA, 105*, 464–469.

Scott, R. K., & Jaggard, K. W., (1993). Crop physiology and agronomy. In: Cooke, D. A., & Scott, R. K., (eds.), *The Sugar Beet Crop: Science into Practice* (pp. 179–237). Chapman & Hall, London.

Scott, R. K., English, S. D., Wood, D. W., & Unsworth, M. H., (1973). The yield of sugar beet in relation to weather and length of growing season. *J. Agric. Sci. Camb., 81*, 339–347.

Shapouri, H., Salassi, M., & Fairbanks, J. N., (2006). *The Economic Feasibility of Ethanol Production From Sugar in the United States.* Joint publication of OEPNU, OCE, USDA, and LSU. www.usda.gov/oce/reports/energy/EthanolSugarFeasibilityReport3.pdf.

Sharmasarkar, F. C., Sharmasarkar, S., Miller, S. D., Vance, G. F., & Zhang, R., (2001). Assessment of drip and flood irrigation on water and fertilizer use efficiencies for sugar beets. *Agric. Water Manage, 46* (3), 241–251.

Shaw, B., Thomas, T. H., & Cooke, D. T., (2002). Responses of sugar beet (*Beta vulgaris* L.) to drought and nutrient deficiency stress. *Plant Growth Regul., 37,* 77–83.

Sibma, I., (1983). Effects of soil covers on air and soil temperature and on growth and yield of sugar beets. *Neth. J. Agric. Sci., 31,* 201–210.

Silva, A. T., & Orskov, E. R., (1985). Effect of unmolassed sugar beet pulp on the rate of straw degradation in the rumen of sheep given barley. *Proceedings of the Nutrition Society, 44,* 8A.

Srivastava, H. M., Sharma, V. K., & Bhargava, Y., (2008). Genetic potential of sugar beet genotypes for ethanol production under different agro-climatic conditions of India. *Proceedings of the International Institute of Beet Research.* 71st Congress, Brussels, Belgium, pp. 305–311.

Stephen, K., & Jackson, F. H., (1994). Sugar beet. *Encylopedia of Agriculture Science* (Vol. 4). New York: Academic Press Inc.

Stockfisch, N., Koch, H. J., & Marlander, B., (2002). Einfluss der Witterung auf die Trockenmassebildung von Zuckerruben. *Pflanzenbauwiss, 6,* 63–71.

Stout, B. A., (1999). *CIGR Handbook of Agricultural Engineering* (Vol. 5). Energy & biomass engineering. St. Joseph, MI: American Society of Agricultural Engineers.

Suheri, S., Topak, R., & Yavuz, D., (2008). Effect of different irrigation regimes applied in different growth stages on root and sugar yield of sugar beet. In: *Proceedings of the Conference on Groundwater and Drought in Konya Closed Basin* (pp. 342–354), Konya, Turkey.

Sunvold, G. D., Fahey, Jr. G. C., Merchen, N. R., Titgemeyer, E. C., Bourquin, L. D., Bauer, L. L., & Reinhart, G. A., (1995). Dietary fiber for dogs: IV. In vitro fermentation of selected fiber sources by dog fecal inoculum and in vivo digestion and metabolism of fiber-supplemented diets. *Journal of Animal Science, 73,* 1099–1109.

Terry, N., (1968). Developmental physiology of sugar beet. I. The influence of light and temperature on growth. *J. Exp. Bot., 19,* 795–811.

The Plant List, www.theplantlist.org.

Tian, Y., Zhao, L., Meng, H., Sun, L., & Yan, J., (2009). Estimation of un-used land potential for biofuels development in (the) People's Republic of China. *Applied Energy, 86,* S77–S85.

Tognetti, R., Palladino, M., Minnocci, A., Delfine, S., & Alvino, A., (2003). The response of sugar beet to drip and low-pressure sprinkler irrigation in southern Italy. *Agric. Water Manage, 1804,* 1–21.

Topak, R., Suheri, S., & Acar, B., (2010). Effect of different drip irrigation regimes on sugar beet (*Beta vulgaris* L.) yield, quality, and water use efficiency in Middle Anatolian, Turkey. *Irrig. Sci., 29,* 79–89.

Tugnoli, D. V., & Bettini, D. G., (2001). Verifying the germinability of commercial sugar beet seeds under laboratory conditions and from emergence in the field. *Proceedings of the 64th IIRB Congress* (pp. 333–340). Bruges.

Tzilivakis, J., Warner, D. J., May, M., Lewis, K. A., & Jaggard, K., (2005). An assessment of the energy inputs and greenhouse gas emissions in sugar beet (*Beta vulgaris*) production in the UK. *Agricultural Systems, 85,* 101–119.

Ucan, K., & Gencoglan, C., (2004). The effect of water deficit on yield and yield components of sugar beet. *Turk. J. Agric. For., 28,* 163–172.

USDA-ERS, (2010). Home/briefing rooms/sugar and sweeteners/recommended data. Table 14, http://www.ers.usda.gov/briefing/sugar/data.htm.

Viator, R. P., Kovar, J. L., & Hallmark, W. B., (2002). Gypsum and compost effects on sugarcane root growth, yield, and plant nutrients. *Agronomy Journal, 94* (6), 1332–1336.

Von Felde, A., (2008). Trends and developments in energy plant breeding-special features of sugar beet. *Zuckerindustrie, 133*, 342–345.

Watson, D. J., Motomatsu, T., Loach, K., & Milford, G. F. J., (1972). Effects of shading and seasonal differences in weather on the growth, sugar content and sugar yield of sugar-beet crops. *Ann. Appl. Biol., 71*, 159–185.

Werker, A. R., & Jaggard, K. W., (1997). Modeling asymmetrical growth curves that rise and then fall: applications to foliage dynamics of sugar beet (*Beta vulgaris* L). *Ann. Bot., 79*, 657–665.

Werker, A. R., & Jaggard, K. W., (1998). Dependence of sugar beet yield on light interception and evapotranspiration. *Agric. For. Meteorol., 89*, 229–240.

Windt, A., & Marlander, B., (1994). Wurzelwachstum von Zuckerr"uben unter besonderer Ber"ucksichtigung des Wasserhaushaltes. *Zuckerind. 119*, 659–663.

Winter, S. R., (1980). Suitability of Sugar Beet for limited irrigation in a semi-arid climate. *Agron. J., 72*, 118–123.

Wright, E., Carr, M. K. V., & Hamer, P. J. C., (1997). Crop production and water use. IV. Yield functions for sugar beet. *J. Agric. Sci. Camb., 129*, 33–42.

Yaldiz, O., Ozturk, H. H., Zeren, Y., & Bascetincelik, A., (1993). Energy usage in production of field crops in Turkey (Türkiye'de tarla bitkileri üretiminde enerji kullanımı), V. *International Congress on Mechanization and Energy in Agriculture*. Kusadası, Turkey: Ege Universitesi Ziraat Fakultesi Tarım Makinaları Bölümü, Izmir, pp. 527–536 [in Turkish].

Yao, Y., Gao, B., Inyang, M., Zimmerman, A. R., Cao, X., Pullammanappallil, P., & Yang, L., (2011a). Biochar derived from anacrobically digested sugar beet tailings: Characterization and phosphate removal potential. *Bioresource Technology, 102* (10), 6273–6278.

Yao, Y., Gao, B., Inyang, M., Zimmerman, A. R., Cao, X., Pullammanappallil, P., & Yang, L., (2011b). Removal of phosphate from aqueous solution by biochar derived from anaerobically digested sugar beet tailings. *Journal of Hazardous Materials, 190* (1), 501–507.

CHAPTER 15

AN OVERUTILIZED INDUSTRIAL CROP: TOBACCO: *NICOTIANA TABACUM* L. CASE STUDY FROM TURKEY

H. VAKIF MERCIMEK[1], VOLKAN ALTAY[2], MÜNIR ÖZTÜRK[3], EREN AKCICEK[4], and SALIH GUCEL[3]

[1]*Tobacco (TAPDK) Izmir Office Specialist, Bornova, Izmir, Turkey*

[2]*Biology Department, Faculty of Science and Arts, Hatay Mustafa Kemal University, Hatay, Turkey*

[3]*Botany Department and Center for Environmental Studies, Ege University, Izmir, Turkey, E-mail: munirozturk@gmail.com*

[4]*Ege University, Faculty of Medicine, Department of Gastroenterology, Izmir, Turkey*

ABSTRACT

The large-leaved and coarse textured tobacco plant was brought to Turkey by the British, Venetian, and Spanish sailors nearly 400 years back. It flourished well under the varying ecological conditions of the country with the help of skillful, meticulous, and determined Turkish farmers. The plant acquired very high-quality, being thin, fragrant, sweet tasting, and low in nicotine. Very different types of tobacco are cultivated around the world. Out of these, the oriental type is mainly produced in Turkey. This type of tobaccos has been in demand for many years because of its exquisite smell, taste, impressive color, good burning quality, and the blend. The reason why other types cannot keep the quality of tobacco grown here is the climate and soil conditions as well as intensive hand labor operations undertaken by Turkish farmers. This paper presents an overview of the production techniques of oriental tobacco together with information on the tobacco varieties at global level.

15.1   INTRODUCTION

Cristof Colomb and colleagues after landing in America around the later years of 15th century observed that local inhabitants emitted fumes through their mouths and noses. They learned that these fumes originated from a dried leaf named tobacco. This way they contributed to the recognition of tobacco plants in Europe. Today, tobacco occupies an important place in the agriculture, trade, and industry. The cigarette production has added to its distribution in the world, with monetary contributions from the budgets of the countries. Despite the measures and prohibitions taken in various forms, its use has steadily increased. Initially it was widely believed that this plant had both religious importance and had medicinally beneficial effects. However, today it is accused of being entirely harmful to health. Although its exact origin is not known but America is regarded as the homeland of this tobacco. The recognition of tobacco in the world came to the front after it was brought to Europe (Ozturk et al., 2014).

The patterns of Tobacco use and pipe shapes are frequently encountered in the pictures engraved on the Maya's historic stones in Yukataan and the historical monuments in northern Ohio area of America. When Christopher Columbus and his friends stepped on the island of San Salvador, they saw smoke coming out from the mouth and nose of the indigenous people living in the island. They learned that this smoke originated from dried leaves called tobacco. They also saw that indigenous people were chewing tobacco in the mouth and inhaled smoke through the nose with a pipe (Figure 15.1). The name used in the pipes was named as "tobacco." The botanical name "*Nicotiana tabacum*" was given after it reached Europe (Er, 1997).

Later on other people who traveled to America also report that people smoked tobacco like indigenous people, and they were addicted to it after using it for a long time. In 1518, bishop Romano Pane brought tobacco seeds to Europe and contributed to the start of tobacco production in Spain. It was used by the Europeans as an ornamental as well as in healing (Er, 1997). After Spain it reached Portugal from Brazil and was taken to France as well where French ambassador in Portugal; Jean Nicot; planted it in Lisbon in 1559. It was presented to the queen with the idea that the leaves were burned, smoke was smoked, and nose withdrawal applied with the crumbling of dried leaves. It was good for coughing, asthma, headache, stomach diseases and gynecological disorders (Er, 1997). The Queen showed great interest in tobacco and the herb was called "Queen herb." Jean Nicot's interest in this plant lead to its name as "Nicotiana." Later on the alkaloid found in tobacco in 1828 was named as "Nicotine" (Er, 1997).

The tobacco was used in both worship and as a healing drug. This was finally used as pleasure herb. This way the habit-forming substance spread widely. After its use as a pleasure herb consumption increased rapidly. In some sources this plant is said to have found a wide use around the churches. Almost all religious officials got addicted to this plant. In some countries its use was forbidden, because of its harmful effects on health. In 1575, the church especially in Spain and America, in 1603 in UK, and in 1620 in Japan its use was forbidden. In 1642, the Pope issued a circular prohibiting religious officials from smoking tobacco, both during the ceremony, and outside the church. The use of tobacco was prohibited in 1634 in Russia, in 1652 in Germany-Bavaria, in 1653 in Saxony-Austria, and in 1657 in Switzerland (Er, 1997). Penalties and death penalties were given to the people using tobacco, large number of people died but people continued to smoke (Er, 1997). Despite all the prohibitions and penalties, tobacco clubs and schools were opened in Europe's big cities, and the numbers steadily increased. The European sources mention that during 17th and 18th centuries tobacco use was used much by the adult men in the colonies of Europe and America, as well as children and even pregnant women. The European states, who understood that tobacco consumption cannot be avoided with these prohibitions, wanted to benefit financially from this end, and within a short time they put customs and tax as the first step in this connection. The use of tobacco and its agriculture received an encouragement because the states recognized the income they will earn from the tobacco sales. They established various taxes on tobacco production and sales to get concessions and monopolies.

Consumption as snuff gradually decreased and use in porcelain pipes increased. Later, the cut tobacco leaves were wrapped in a "Cigar" shape, or in paper. This started at the beginning of 18th century in Central and South America. First cigarettes were especially popular in Brazil and they were called "Papelitos." The use of tobacco in the form of cigarettes in Europe was first in Spain, followed by France. The first cigarettes were made in France in 1844, and paper cigars made in Italy in the same year attracted much interest. The spread of the cigarette making and use followed the Crimean War of 1856. The tobacco was wrapped in newspapers and smoked as such during the war. This became very popular among Turks, British, French, and other armies. The soldiers returning to their homelands after the war continued the smoking habit, which constituted the basis of the cigarette industry (Mercimek, 1997).

FIGURE 15.1   Pictures related to the history of tobacco.

15.2   TOBACCO IN TURKEY

The studies conducted in the Ottoman archives in Turkey has revealed that tobacco has entered the Ottoman geography in 1570s. It spread in the empire with the orders and instructions sent to the "officers" in different regions for the tobacco agriculture around 1570. All these decrees and their places of destination reveal that the use of tobacco agriculture may have begun simultaneously in Rumeli, Anatolia, and Middle East regions in early 17th century (Doğruel and Doğruel, 2000; Yılmaz, 2005). It seems more likely that tobacco entered the empire at an earlier date before 1598, and spread fast in 15–20 years. This can be named as the recognition and adaptation period, when small-scale production began (Savaş, 1969; Borio, 2001; Er et al., 2011; Şahin and Taşlıgil, 2013).

There are some clues that after the American continent, the first tobacco agriculture was started on the land of Ottoman empire as a commercial product. European traders (especially English, French, and Dutch) started bringing American tobacco to major cities in Turkey around 1598, especially Istanbul. Since the beginning of 17th century, tobacco-related consumption, production, and trade spread quickly, because of increase of tobacco use (Figure 15.1). The Islamic scholars argued that the practice of smoking is not appropriate on religious grounds. On top of that, a ban on smoking was brought by Sultan Ahmed I (Yılmaz, 2005). Despite this ban the state tried to apply against tobacco between 1609–1649, decrees were repeatedly sent. However, the use of tobacco increased in number everywhere when we look at the tobacco planted within the territory of Ottoman Empire. This expansion was an important indicator of the expansion of tobacco consumption throughout the empire. After the fire in Cibali, where tobacco production was intensive, severe punishments were applied by Sultan IV. Murad. The ban imposed on tobacco continued until it was removed by Sultan Mehmet the IVth. During the prohibition period, the use of tobacco increased due to the ban on smoking, but Hookah smoking improved in this period (Unal, 1997; Borio, 2001). On the contrary, the prohibition was mostly successful in not making tobacco farming in the vicinity of big cities and their surroundings. The production however continued at smaller scale in smaller settlements such as towns and villages where control was less (Yılmaz, 2005). This prohibition continued until 1649. Following this, tobacco consumption, trade, and production increased at a greater speed on economic grounds. Macedonia in Rumelia, Marmara in Anatolia, and Aleppo and Lazkiye in the Middle East emerged as places where intensive tobacco farming was done (Mercimek, 1999; Yılmaz, 2005). In the Middle East region, the number of tobacco farming areas increased around Jerusalem and Gaza (Yılmaz, 2005).

Until 1678, the tobacco was imported freely. During the time of II Sultan Süleyman, tobacco produced in Yenice and Kırcaali, and brought to Istanbul was permitted to enter only with "8–10 akçe" (Ottoman currency) custom tax. This tax was increased and applied to both the purchaser as well as the seller. Generally 8 akçe were charged from the purchaser and 12 akçe from the seller (Unal, 1997). Tobacco farming attracted great interest during these years and cultivation started in Anatolia as well. Large areas were allocated to the cultivation of tobacco in Anatolia and Thrace. A regulation was issued to improve tobacco production and to increase government revenues (Özdemir, 2010). Tax was collected from buyers, sellers, and tobacco producers according to the type of tobacco. The customs weight (oka) was up to the amount ranging from 20–50 akçe. The tobacco producers received two and a half cents per decar, while their tobacco was on the field. The application of this rule was continued until the establishment till recently. The tobaccos produced in Turkey was of high-quality due to climatic and soil conditions together with the skill of tobacco producers. Thus, the tobacco imported from Europe was replaced by the local production and was also exported. The tobacco export leads to the establishment of tobacco customs organizations in many cities in Turkey (Unal, 1997).

Tobacco agriculture in the first half of the 18th century showed the trend of expansion within the Asian part of Turkey, especially around Marmara and its surroundings, the expansion started especially around Bolu, Kastamonu, and their environs (Yılmaz, 2005). During 1750–1800, the number of tobacco cultivated districts in different areas of empire showed a greater increase compared to the earlier stages. In all 41 new places were added to the then existing tobacco cultivation districts. The activity was especially very high in the Marmara and Aegean regions. This was followed by a revival starting from Kocaeli and including İznik and Adapazarı, while the actual concentration was around the areas surrounding Bursa and Balıkesir. In the Aegean region, this movement followed the Marmara region, and went down to the south, with concentration mostly in the Aydın region. In the Black Sea area, except for the Sinop and Bafra, there was no tobacco farming. In the Central Anatolian region the cultivation and tobacco farming was done for the first time around Ankara, Çankırı and Haymana (Yılmaz, 2005). In the first half of the 19th century, the cultivation extended from Kocaeli to Eskişehir, and onwards to the Muğla region. In the Black Sea region, the tobacco farming was done around Sinop, Bafra, Samsun, Ordu, and Trabzon, as well as Bozok area of Canik near Samsun. At the same time, in East Anatolia Bitlis and Mus became important centers of tobacco cultivation.

At the end of the 17th century, almost 20% of the Ottoman Empire spread over three continents was practicing tobacco farming. The number of farmers engaged in tobacco farming is reported to lie then around 50,000, and the area of farming was approximately 50,000 acres. In 18th century period this development went on and till the second half of 19th century. The number of tobacco farming districts reached 38% in the whole empire, while the number of farmers working in this business reached a number of 150,000. The number of sowing areas was around one million acres (Yılmaz, 2005). The tobacco trade over time, merchant class only engaged in tobacco trade emerged, some cities fully got involved in the tobacco trade network and route became an important tobacco trade route. The most important development in tobacco farming in the Ottoman Empire took place in 1861, when tobacco imports were banned, with a regulation issued in 1862, tax were levied according to the quality of tobacco. Subsequently, this regulation was changed to a tax of 12 cents per oka regardless of the quality of tobacco. Such arrangements and changes took place until 1872 (Oktar, 1997). First State monopoly was established in 1872 and it was decided to make changes in the tobacco tax. Exclusive right to sell tobacco and its operation was sold to two Greek bankers at a price of 3,500 gold coins, but was terminated six months later. In 1873, some new regulations were made and an organization was established under the name of "Administration in Hisariye Dukhan" (Erdoğan, 2014). In 1874, cigarette, and packet tobacco manufacturing factories were established, while tobacco farming continued to be free, and tobacco sales prices were subject to records. With a specification made in 1883, the operating right of tobacco was granted to "Memaliki Osmaniye Dukhans Müşterekilmenfaa Reji Company" named French Joint Stock Company, for a period of 30 years (Thobie, 2000; Quataert, 2004). This company continued its activities until June 13, 1921; when the agreement between the company and the government passed completely to the state. During the Republic period, taxes were taken in various amounts, sometimes according to their quality, sometimes as a standard per kilogram, and sometimes according to the locality.

During the first Economy Congress held in Izmir in 1923, decision was taken to close the Reji Company. With the Law No. 558 dated February 26, 1923, it was ruled that tobacco purchase, operation, manufacture, and sale of cigarettes for domestic consumption, and tobacco-related works will be undertaken by the government. In the new law, the monopoly administration was like state monopoly starting from March 1, 1923. The period of authorization granted on February 26, 1926 was extended. The import of products such as leaf cigarettes, shredded tobacco, cigarettes, snuff, cigars; and Nationwide sale of these products was totally under the state monopoly. So Turkey had a state tobacco monopoly for long years in this trade after the republics was established. On 12 June, 1930, Law No. 557 was accepted with the consideration that the tobacco monopoly was not dependent on the duration (Ozavcı, 2007). After this date, various arrangements were made from time to time on the tobacco monopoly in the country. The law 1701 issued on June 5, 1930 on tobacco monopoly included the reorganization of tobacco agriculture, processing, transport, trade, and fabrication phases. These regulations were found to be inadequate and on 10 June 1938 the law 3437 as "Tobacco and Tobacco Monoly Law" was adopted, it entered into force on October 26, 1938 (Ozavcı, 2007). It remained in effect for more than 30 years and was abolished in 1969 and replaced with the law 1177 on "Tobacco and Tobacco Monopoly Law." This too could not respond to the developments in the tobacco sector over time, and was replaced with "The Tobacco and Tobacco Charter Regulation" with various principles related to this law, and issued in 1975 (Ozavcı, 2007). Recently, the conditions of the TEKEL privatization process and a new tobacco law has been established, which entered into force in January 9, 2002 under law 4733. The tobacco production was 150 million kilograms, with 450,000 producers, which dropped to 60,000 producers and 80 million kilograms in 2014 (TUIK, 2014).

15.3 TOBACCO TAXONOMY

The plant list includes 274 scientific plant names of species rank for the genus *Nicotiana* (family Solanaceae). Of these 55 are accepted as species (www.theplantlist.org). The four species of *Nicotiana* that are most widely used in the world nowadays are (Yılmaz and Katar, 1996):

1. *Nicotiana tabacum* L.
2. *Nicotiana rustica* L.
3. *Nicotiana paniculata* L.
4. *Nicotiana glutinosa* L.

Nicotiana tabacum (Figure 15.2) constitutes 80–90% of the tobacco varieties. At present the Virginia, smoking tobacco and oriental tobacco make up a large part of world tobacco consumption and all these belong to *Nicotiana tabacum*. Other *Nicotiana* species evaluated in the world is the yellow flowering *Nicotiana rustica*. Tobacco grown in some countries such as Russia, Poland, South America, Syria, Lebanon, and Libya are usually *Nicotiana rustica*. In Turkey Hasankeyf tobacco produced in Gaziantep region belongs to *Nicotiana rustica* (Şahin and Taşlıgil, 2013; Ozturk et al., 2014).

FIGURE 15.2   Flowering shoot of *Nicotiana tabacum*.

The origins of today's tobacco types is predominantly from *N. tabaccum* and partly *N. rustica*. The varieties of tobacco produced all over the world have in general originated from these two species. The type features, the effect of ecological conditions and the different methods of drying are of great importance in the "classification" of tobacco today. The grouping of tobaccos in the world according to the drying methods is as follows (Şahin and Taşlıgil, 2013):

1. **Flue-dried: Smoking tobacco–Virginia:** It is dried by using hot air obtained by using a heat source, being the most common type in the world and constitutes more than half of tobacco imports and exports. It is cultivated in about 75 countries, from New Zealand in the south to Germany in the north. China, Brazil, India, USA, and Zimbabwe top the list in production.
2. **Air-dried: Cigar, Burley, etc.:** Air-dried in partially or completely enclosed spaces, including Burley and Maryland as main ones. The

countries producing this type of Burley tobacco are; Malawi, Brazil, USA, Mozambique, and Argentina.

3. **Sun-dried: smoking of the Turkish type–Oriental tobacco:** This tobacco is dried in the sun, generally divided into different subgroups as Oriental, Semi-Oriental, light sun-dried and dark sun-dried. Mainly cultivated in Turkey as Oriental type. Although the ecological amplitude of this tobacco group is not as wide as that of other types, it is usually grown in countries like Turkey, Macedonia, Greece, and Bulgaria.

4. **Fire-dried: More chewing, snuff, pipe tobacco:** It is dried under fire, in closed hangars by exposure to the smoke of various aromatic fragrant trees that are burned in the fire. It is produced in America, India, Indonesia, and Italy.

There are some un-classified tobaccos groups, usually grown in countries like Morocco, Tunisia, Libya, Ethiopia, Iran, Jordan, and Lebanon. The tobacco generally produced is called Perique, Black-Fat, Hasankeyf, and Tömbeki. The nicotine content in this group ranges from 4–15% (Yılmaz and Katar, 1996). The production of these tobacco types is also important in Turkey. The share of Virginia tobacco has increased from 50 to 72% during the last 20 years (being around 37% during 1960–1965). Burley tobacco has a share of 14–15% (production rate was around 8% during 1960–1965). The share of oriental tobacco has decreased from 13–15% to 4% level (share was around 16% in 1960–1965). The share of other tobaccos has decreased from 25% to 9–10% (in 1960–1965 the share was around 38%).

15.4 DEVELOPMENTS IN TOBACCO CULTIVATION

Tobacco was something new to the Ottoman Empire. At the beginning of 17th century it was sown in the vineyards and gardens, and average was less than ¼ of the total area present. It was started probably as planting in furrows on trial basis, but prohibitions during 1609–1649 restricted any increase in the area. It was secretly cultivated between the mountains and hilly tracts. Some reports mention that first tobacco farming in Turkey started in Selçuk Ayasuluk site; however, many researchers report that tobacco production was undertaken over a wide range at the same time in Rumeli, Anatolia, and Middle East regions at the beginning 17th century. It is seen that 100 years after the entrance of tobacco in Ottoman Empire, 2/3 of the tobacco production was made in Western Thrace, Macedonia, and Rumeli; and the remaining 1/3 around Aleppo, Damascus, Bursa, Bandirma, and Bergama. Following next 100 years, reveals that there was an increase in the tobacco cultivation, extending in the Marmara, Aegean, and Black Sea regions. At the beginning 20th century, Tobacco farming seems to have spread over a very large area in Macedonia, Skopje, Edirne, Bursa, Manisa, Aydin, Çorum, Bandırma, Eskişehir, Kocaeli, Adana, Kars, Kütahya, Samsun, Trabzon, Konya, Maraş, Malatya, Antakya (Figure 15.3) (Yılmaz, 2005; Ozturk et al., 2014).

FIGURE 15.3   (See color insert.) Tobacco cultivation area in Turkey.

The tobacco production in the Ottoman Empire started from the seeds. All tobacco produced up to 1900s was evaluated with its qualitative classification. However, it was observed that weather conditions and soil characteristics are the priorities which lead to differences in the quality of uniform tobacco production. After 1950s, tobacco farming got concentrated in the Aegean region because of immigrants from Balkan countries and dense areas were in the Bursa region. In the Black Sea, this production too was supported by the immigrants. The tobacco farming increased in Eastern Anatolia due to political causes. In Central Anatolia, the production almost came to an end. The reasons for all these changes can be summarized as follows:

1. With the developments in cigarette making techniques together with the production and use of matches cigarette consumption in general increased. The diversification of filters, the differentiation

of permeability of the cigarette papers, possibilities of replacing the contents of cigarette tobacco on the basis of fires and smoke, all lead to an increase in the cigarette addiction, which further helped to develop more blended cigarettes. American Blend cigarettes, with a high degree of nicotine content, such as aroma and burning ability show extremely negative features. In order to ensure the regulation of smoking, decrease in the consumption of Oriental tobacco in Turkey into the blend inside dropped consumption in Turkey. This has been accelerated with legal changes affecting change in cigarette blends.

2. Decisions taken in 2001 and the International Monetary Fund (IMF) were responsible for the privatization of monopoly. It was one of the key ingredients in the agreement on loans. These changes experienced in the organization of the tobacco in the world and cigarette industry in Turkey step by step ended up in varying the structure of the tobacco produced in Turkey. Some of the effects of these changes in cigarette sector were less, while others had irreversible effects.

3. The support of tobacco production in Turkey started in 1947 and ended in 2001. In this process, the biggest effect in tobacco production was through politics.

4. In 2001, support for tobacco under 4733 law was abolished, monopoly privatization started, tobacco purchasing system changed, contract production started, cigarette factories opened and operating conditions were changed, and the amount of funds obtained from foreign tobacco imports decreased. The legislation which were used during the years are summarized as follows:

    •   At the beginning of 2000, first five production areas were Izmir, Samsun, Silvan, Adıyaman, Basma, while the overall production rate of was around 88%. Silvan and Adiyaman constituted 8%, not good for sale. 11 years later, with the influence of the new tobacco legislation, in Izmir and Samsun, Basma, Katerini, and Sun-Cured Virginia, were used and overall production increased to 98%. This production was entirely export-oriented. Only İzmir and Samsun remained as classical areas. Private firms do not produce varieties that are not worth export, 10% of these sown cultivars are newly adapted types.

    •   The productions of tobacco in Trabzon, Malatya, Bucak, Bursa, Edirne, Bahçe, Hendek had no chance to export and came to an end. The production of Katerini, Sun-Cured Virginia, Talgar, and Prilep was started. However, this does not mean that locally produced ones are of poor quality, their successors are of better quality (Mercimek, 2014).

- As the production of tobacco in Greece and Bulgaria decreased, the producibility of some of the local forms was investigated by the private sector in Turkey, and in 1990s Basma type tobacco originating from Gümüşhacıköy Basma or Tokat Topbas were found, but today. We see that Basma is called Greek Basma. The Gümüşhacıköy Basma, which is our classical type is no longer produced. Instead, the seeds of Greek Basma tobacco are used, which is sought in the world markets. The Greek Tobacco production started in 2002, but was different from the tobacco grown in Greece, till the seeds were taken from the plants grown in Turkey. The outward features and chemical content of these tobaccos vary much from the original taken from Turkey. It is estimated that its production will increase as long as it holds the ability to be sold at the world market.

- Katerini tobacco; with its origin from the seeds from Greece; is now rarely grown in Europe. The seeds were imported from Greece and adaptation started. It was observed that its production performance in various regions especially in Hatay was good. The result obtained in terms of quality/yield relation support this. The sales capacity of this type can go up to 4 million kilograms as of today and will increase in the following years.

- Semioriental tobacco is not used for domestic consumption, as the production is very low. Its production was 108 million kilograms in 1993 when the quota implementation decision was taken, and decreased to 25 million kilograms in 2001 and 1.5 million kilograms in 2012. The number of producers is around 1.500 people (Mercimek, 2014).

- Suppliers who want to produce tobaccos of different origin, related to the potential tobacco products are cultivating "Sun-Dried Virginia" tobacco, which is less costly than the original Virginia veneer seeds and sun-dried. The future of this tobacco seems to be optimistic, nearly 1.5 million kilograms are produced and sowing continues till it is seen who are concerned with the sustainability of demand. This tobacco type is dried on the place in short time, and are easily accepted by the buyers.

- The breeding studies have begun for "Dübek tobacco" originating from Kyrgyzstan, "Prilep tobacco" from Macedonia, and "Talgar tobacco" from Kazakhstan. The compatibility of these varieties with sauce blends is being tried. These are sown on irrigable land because of high market potential and no longer produced in the countries of origin. It is also seen that production has been continuing to diversify local types due to the possibility of selling "Talgar tobacco."

- In addition to these new tobacco types, we also see that the supplier companies are working to increase the appeal of producers on "İzmir tobacco," which is sought as an original brand abroad. To ensure the continuity of production, the equation is keep harvesting, drying screens, nets drying, hand in hand recruiting procedures.

- Training of producers who never cared about it in 90s and aware-ness of companies today, has pushed the village organizations that they unite under one village-village producer group. The producers are working and discussing, increasing production in the field, to reduce cost and to minimize the residue field studies seriously. Moreover, the "social responsibility project" scope for family producers is developing a variety of utility projects.

- Seed breeding trials have been initiated in order to ensure the sustainability of the elite and rootstock characteristics of Turkish tobacco, both by requesting to capture a certain standard in the blends of the foreign buyer and, if necessary, by seriously degrading the degraded tobacco seeds.

According to this, more than 80% of the production is based in Izmir tobacco (Table 15.1). The ecological area is very wide, and it is produced in Izmir, Manisa, Muğla, Aydın, Denizli, Balıkesir, Uşak and Kutahya. The Izmir tobacco is sown on a large area in the southwestern part of Anatolia and is the most uniform region as compared to other varieties, even though there are some differences in quality. In addition to Izmir tobacco, "Yenice Basma tobacco," too has responded well in the Aegean region to the soil and climatic conditions. The present Izmir tobacco has become more popular than the original Basma Tobacco withindustrial qualities (Incekara, 1979; Ekren and Sekin, 2008; Şahin and Taşlıgil, 2013). In Izmir tobacco, the leaves are usually small in size, large elliptical in form similar to Basma tobacco. Its color ranges from light yellow to yellow, with greenish and reddish interferences. Some of the top foliage leaves have dark brown specks, which are considered as a sign of maturity. Their tissues are moderately thick, flexible, resistant to hand and machine. Because of the thinness of the veins involved in shredded tobacco with the number of leaves, there is no negative effect both during the manufacturing of cigarettes. Cigarette yields are good. Hygroscopic properties are good because of its soil and climatic conditions, and the tissue structure is composed of small and plentiful cells arranged frequently. The most fragrant tobaccos are grown in many parts of Akhisar, Saruhanlı, Sındırgı, Soma Kirkagac and Manisa. In Denizli, Mugla, and their environs, production has increased lately, the colors of tobacco are more open, the texture is thinner and leaves are bigger. The smoking characteristics and physiological potency is too slow. It could be described as having very sweet and fragrant aromatic flavor, very slow and very sweet smoking quality, however, it holds true and is valid for Izmir tobacco, an unrivaled type among its counterparts in the world. It comes first in terms of its breeding characteristics. Their roles in the blends are multilateral, as they have hard-character, sweet, bitterness, dreary, smelly, and low-quality tobaccos that have no sweetness. Because of these attributes it is used to increase the quality of cigarette smoking.

TABLE 15.1   Tobacco Production in Turkey in 2014 (TAPDK, 2014)

Production Origin	Number of Farmers	Weight (kg)
Izmir	62,540	71,125,400
Samsun	9,900	8,127,500
Basma	4,140	4,266,200
Katerini	3,630	2,804,600
SC Virginya	550	1,168,600
Yayladağı	740	445,100
Adıyaman	150	133,700
Bitlis	70	50,900
Talgar	40	39,800
Hasankeyf	30	35,700
Muş	10	24,700
Total Origin	**81,800**	88,222,200

Izmir tobacco is useful for blend-type cigarette, because it has the characteristics of absorbing and protecting additives. In particular, to ensure color harmony in the blends of tobacco, colors which are light yellow and yellow are effective. In order to reduce the amount of nicotine in cigarettes and to slow the smoking, it is suitable to add the blends of Virginia, Virginia Blend and black cigarettes at high ratios. It is also a convenient blend of pipe tobacco (Incekara, 1979). The combustion conditions are of moderate degree, this is accepted as a positive quality for Izmir tobacco. The improvement qualities are more effective and useful in the inside of cigarettes. The chemical constituents of Izmir tobacco are known for their very low nicotine content, high protein and high sugar contents. The mean nicotine ratio is less than 0.70% and the amount of nicotine found is as low as 0.25%. The amount of protein nitrogen is 0.90 to 1.30%, amount of total reducing agents ranges from 15 to 20% (Sekin and Peksüslü, 1995; Peksüslü, 2005; Peksüslü et al., 2012). Generally 23–28 kg boxes in weight are put up for sale by manufacturers. Due to low-cost of labor in recent times, Izmir tobacco produced in different parts of the world does not reach the quality of the one grown in the Aegean region (Incekara, 1979; Sarıoğlu, 1980; Otan and Apti, 1989; Peksuslu, 2000).

15.5   TOBACCO FARMING

Tobacco is sown between 60° north and 40° south latitudes. The vegetative period lasts 80–100 days depending on climatic conditions. The leaves show different shapes. The form of leaves, their connection to the stem, their posture and size in the plant are regarded as the physical characteristics of the variety or types (Incekara, 1979; Şahin and Taşlıgil, 2013). The seeds are too small, weight of one seed varies between 0.07–0.09 grams. The average temperature for seed germination is 12–14°C, and the optimum growth temperature for the plant is 25–30°C (Incekara, 1979). Tobacco production begins in a nursery, which should be well chosen. It is appropriate to select the area closed to the north winds. The nurseries with south-facing slopes have good aeration characteristics and areas can be kept under constant surveillance. The plants are sown at least 10 cm apart in the nursery, the nursery should be changed every year or soil disinfect done. Oriental tobaccos prefers organic matter-poor soils, while other tobaccos prefer deeply profiled, organic matter-rich, irrigable soils (Savaş, 1969; Taşlıgil, 1990; Er et al., 2011; Doğanay and Çoşkun, 2012; Şahin and Taşlıgil, 2013). The superficial release prior to sowing is discarded or dragged, a type of treatment applied to the soil type, depending on the time of sowing and planting (Esendal, 2007).

Tobacco planting furrow opening is made in the field mechanically or with hand.

- In the Aegean region, 30–40 cm between rows, 5–10 cm above the row are enough.

- In Marmara and Black Sea regions, distance between the rows is 40–50 cm, rows are 15–20 cm.
- In the East and Southeast, distance is 30–70 cm between rows, 10–25 cm above the row.
- In Burley, Virginia, and Tömbeki tobacco, the distance is 80–100 cm and the rows are 60–80 cm.

The tobaccos are dried in the Aegean by kırmandal, in the Marmara by mirror and partly in the wagon, in the Black Sea by mirror or on fixed scaffold. In Eastern and Southeastern regions, tobacco is dried by different shapes on the "gam," arbor, and ayvan" (Incekara, 1979; Şahin and Taşlıgil, 2013).

In some tobacco types, the leaves are hand-unbreakable. In these types, when majority of the leaves reach a general level, the plant is cut from the bottom. In this way, the tobacco leaves are dried altogether without tearing. This is called "Sak ile kurutma" in Turkish. Tobacco, such as smoking tobacco and Burley tobacco, is dried by hanging in closed areas. It is ensured that good drying is done and ventilation is good through the window. Foreign type tobaccos are dried in private cabins with Flue-Cured (FC). The mature FC tobacco leaves are usually hand-knotted and passed through swollen-like mechanism. The tobacco here is housed in specially manufactured cabins. The hot air obtained by using a heat source is routed between the leaves with hanging-stacked lines to dry the tobacco. Costly facilities are needed to invest in FC and Burley and smoking tobacco drying. Oriental type tobacco can be dried with modest methods at a low-cost that cannot be compared with such facilities. In Tobacco nursery and field periods, infection of nearly 40 diseases is possible. In the nursery period, black root decay, collapsing, and blue mold are important, while slug insect, mole, and earthworms are harmful ones. In the field period, blue mold and virus diseases and nematodes are important. Disease and pest control pesticides are often seen because of ignorance or improper use (type of drug to be administered, dosage, number of applications, style, and period errors made). These adversely affect the yield and quality of tobacco. The biggest problem of pesticide residues in tobacco leaves is caused by the incorrect use of drugs (residual), which is an emerging problem.

15.6 PROCESSING AND MAINTENANCE OF TOBACCO

One of the most important stages in the Oriental type tobacco production is the fermentation of tobacco and processing. At the end of the fermentation; a kind of activity of the microorganisms; the fermentation of tobacco occurs and leads to chemical changes in tobacco. These changes are acquisition of tobacco and smoking characteristics which represent the way of changes. If tobacco does not undergo a healthy fermentation, a heavy smell in tobacco is seen, as well as substances change, all resulting in undesirable qualities. The fermentation process is passed in a controlled manner (mandatory) during the processing period, thus accelerating the fermentation process of the tobacco and accelerates the ripening process.

Tobacco is purchased from the farmers sowing it, but they are taken into operation within the framework of a program, and consideration given to the conditions. They are first subjected to a preliminary examination, by dividing into groups according to their type, color, and fault conditions and they are marked. Thus, processing parties are prepared. During processing; size, color, texture, smell, leaf integrity, physical features, such as malfunction and disease are noted and all divided into categories based on appearance quality. Tobaccos in American Grad classification are classified as AG, BG, KP, CU, DKP, AB, BKP, KDKP, XK, Thin Broken, Broken, Thick Broken. These separate tobaccos are subjected to guided fermentation under certain heat and humidity conditions by passing through a so-called soft-driver drying cylinder and the moisture is fixed. Tobaccos are classified according to this evaluation and marked separately. The tobacco placed in the box in a package varies in weight according to the characteristics of the origin. As the durability of leaves increases, the amount of tobacco added to the box increases. This weight ranges from 230 to 110 kilograms.

15.7 ORIENTAL TOBACCO PRODUCTION

Cultivation of Seedlings: High efficiency and high-quality products in tobacco farming start with a robust and disease-free seedling. The sturdy, puffy, and time-related fidelities of tobacco effect the maturation on time in the field. These are broken in time, dried, and high-quality products obtained. The seeds of the farmer must be of a certified tobacco variety with qualities that are congruent with the soil structure and climate of the region following the adaptation trials. Care must be taken to ensure that the seeds of the cultivars to be cultivated are absolutely free from disease, and high in germination ability. An acre tobacco field is obtained from the seedlings of 5–6 m². To grow healthy seedlings; good selection of seedlings, proper seedlings with good maintenance are required (Dimon, 2007).

Selection of the fields for seedlings: The best fields for the seedlings are: the north side must be closed, south, and southwest facing areas. It must be as far as possible from vegetable gardens and tobacco fields to protect from disease and harm. It should have plenty of sunshine at all hours of the day. Seedlings should be close to the village in order to be constantly checked. Water needs should be readily procured. It should not be installed in shade and under humid places. If any tobacco-infected disease or harmful organism is found, the seedlings must be replaced or the soils of seedlings must be disinfected. The diseases and harmful factors which can harm tobacco should be destroyed.

Preparation of the seedbed compost: The soil to be used should be taken from moor or a non-agricultural area. The animal feces to be used (sheep, goat) must be very well burned. The accumulation of animal manure should be started at least 6 months before the preparation of the seedling pillows. The fertilizer should be irrigated at certain intervals, so as to protect from unwanted weed seeds germination in the manure. Fresh and unburned fertilizer should not be used. Before preparing the area, soil, and fertilizer are sieved separately using 1 cm diameter sieves. One part is burned, sifted sheep or goat dung and two parts of raw soil are mixed and the seedling cushions are left until preparation. Part of the manure is passed through the smaller mesh screen and cover separated for fertilizer (Dimon, 2007).

Establishment of the seedling fields: According to the climatic features of the area, seedling should be sown in the field with preparations started in January and March Soils are first plowed, 10–15 days before the seedlings are sown in the field, at a depth of 15–20 cm. Pillows 1 to 1.20 meters wide are set with 50 to 60 cm path, parceled. The soil in the seedling section is excavated to a depth of 20–30 cm and soil removed. If the soil is not filtered, gravel or bushes are laid on the bottom. 20 cm of rough soil is put on it, and then 10 cm thick mortar is spread on it.

Plantation of seedlings: Plantation is calculated as 2 g/1 m². (Technically, 0.6 gr seeds per square meter are recommended, but the producers use one seed in one square meter area). In sprinkling by hand, the seed itself is mixed with 30–40 times fine sand or wood ash and is thrown into the seedling bed and seed dropped regularly. Seeds can be planted dry as such, 1–2 days before sowing the seeds are dipped in a wet bag so that the germination of the seeds is accelerated (Dimon, 2007). Seed planting should be done in a calm atmosphere without wind. After the seeds are laid on the pillows, a 1 cm layer of very well burnt, sifted sheep or goat manure is sprinkled over and pressed with wood. This is called Cover Fertilizer. This manure also feeds seedlings as well as protects the seeds and the moisture. After Cover Fertilizer, the seed pillow is irrigated with a strainer bucket. It is covered with nylon to control the moisture and heat of the pillow, to protect from the external environmental conditions and to ensure more homogenous germination and development of the seedling.

Preparation of Plastic Tunnels: For nylon-covered tunnel, pike or bending iron bars, at intervals should be 50 cm high over the pillow, the seedbed is immersed in soil on both. Nylon is stretched over the hollows. One side of the nylon is covered with soil, and the other with large stones to open and close. Pillows are constantly lightly watered with a strainer bucket until the seeds germinate, ensuring that the soil does not lose its wetness. The watering of the pillows until the seed germinates is done in the morning or at noon, due to the cooling of the night air. Seeds germinate between one week to fifteen days depending on seedling condition and weather conditions. Irrigation is done once or twice a day depending on the weather condition after seeds germinate. If weather is cold, single irrigation may suffice. The increase in air temperature and with the development of seedling the number of irrigation should be increased. Irrigation should be done in the morning as much as possible. If watering is done on hot days at noon, wet leaves can be damaged by the sun. When seedlings begin to grow, it is necessary to give abundant but infrequent water in order to obtain seedlings and protect them from diseases. Irrigation water must be clean and clear, and water should not be too cold. One week before taking the seedlings the irrigation is stopped and thus the mature seedlings are obtained (Dimon, 2007).

Seedlings should be covered at night to maintain the temperature inside the tunnel on very cold nights, but must remain off during the day. On luke-warm days, seedling cover should be entirely open at noon. If there is no ventilation on hot days, water vapor accumulates with the increase of heat in the tunnel and the environment suitable for the development of diseases is prepared. It also causes the seedlings to burn. Especially 6–7 days before planting, the seedling covers must be removed and the seedlings should be kept fully open day and night. This allows the seedlings to acclimatize and mature into outdoor conditions (Dimon, 2007).

Before weed cleaning, the pillows are watered so that roots of seedling are not damaged. In frequent seed throwing cushions, seeds should not be thrown too often as seedlings will be thin, long, and weak. If too much seed is thrown, it must be diluted. During the seedling period, the fast-growing seedling should have long uniform seedling growth. After seedlings and weed cleaning is done. In order to prevent seedling roots from opening, mixture of fine sand, shaft, and cover grains are spread over the seedlings and watering is done. The root parts are filled. The same process is repeated when the soil is sealed due to irrigation. Under normal climatic conditions, about 1.5–2.5 months after sowing, the seedlings get ready for planting. A good and mature seedling with 10–15 cm length and 4–5 leaves is suitable for planting. To understand that the seedlings are matured, one or two seedlings are pulled to the finger, the seedling is not broken, and if it shows a flexible state, the seedlings are matured.

Tobacco planting, according to the region begins in April and May. Seedling should be done early in the morning. Dismantling should be done as soon as the seedlings are planned to be planted. It must be watered first with plenty of water in order to be able to uptake the water from the soil without disturbing the structure of the roots. The seedlings are disassembled into bundles, and roots should be arranged in large basket or box inside. For the seedlings roots to remain wet, they should be covered with wet cloth. Yet undeveloped seedlings are left on the pillow. In order to close the root collar of the seedlings which are left after each seedling sown, mixture of fine sand, silt, cover fertilizer are put and irrigation is done.

15.8 FIELD OPERATIONS

Field selection and preparation: The quality "Turkish Tobacco," with good reputation in the world, is sown on moderate-fertile lands. Leaves grow too much at the water-holding capacity. The type characteristic of tobacco breaking down by leaf texture disappears. The field is plowed at a depth of 20–25 cm in the fall, so that rainfall gets stored in the soil in winter and the weeds in the field are cleaned. In the spring, fields are cleaned one or two times. The disc harrow or dredge slider are pulled in such a way that the clods are broken up well, without losing the humidity (Dimon, 2007).

Planting of Seedling Field: In Aegean tobacco, planting is done according to the climatic conditions and tobacco variety is generally sown by hand or by machine in April and May.

Hand planting: The time of planting is adjusted according to the wetness of soil. Before planting, the furrows are opened. The direction of the furrows is determined according to the maximum benefit from sun and the dominant winds. The seedling is erected by placing on the previously opened furrows. Immediately after planting, life water is given. Planting frequency in the rows is 5–10, the distance between the rows is 35–40 cm. In general, tobacco is often planted under good conditions.

Machine planting: In successful machine planting, very good preparation of the fields is a must and the seedlings should be longer. The distance between the rows is adjusted according to the soil adherence, while the order depends on the hand habits of the planters on the machine and the speed of the tractor. After both types of planting, the field is checked for 10 days. Replacement is done for the non-holding seedlings (Dimon, 2007).

Field care: First hoeing is done 10–15 days after planting, the second approximately 15 days after the first one, the second hoeing takes care of filling the tobacco root collars. Anchor is required to break the slippy layer formed after the rain. Otherwise, the soil will be in a bad condition, preventing the plant from air effect and development. During the hoeing process, the bottom leaves are cut down because they have no commercial value, and are removed from the field.

Irrigation and fertilization: No irrigation and fertilization is generally needed in Oriental type tobacco, because irregular irrigation and fertilization effect the leaf growth and the quality declines. However, in dry years it can be irrigated once, provided that it is not extreme between the first and second anchor. If fertilizer use is mandatory, the soil should be analyzed to determine what nutrients the soil needs. The type and amount of fertilizer to be applied needs to be determined before hand by the authorized person (Dimon, 2007).

Crop rotation: It is beneficial to apply crop rotation in order to protect it against diseases and prevent exploitation of soil with the same plant for a longer duration. After 3–5 years of tobacco farming, one year of grain (rye, barley) is recommended.

Harvesting: The commercial value of tobacco leaf is to a large extent related to the maturity, which should be well detected. It depends on the time of harvesting the crop. Only mature leaves should be cut to obtain higher and uniform quality. Usually 2–2.5 months after planting and maturity, the crop harvesting can begin. Same happens in the beginning of June. Since bottom leaves are not of commercial value, they are stripped and removed from the field. The maturity in the tobacco is understood by the start of yellowing from the end of the leaf towards the edges, and yellow bubbles appear on the leaf face. The best time for harvesting of crop is early in the morning. Tobacco leaves cut at these times are easier to harvest because they are perfect with full life activity. The leaves are kept in bunches after harvesting in such a way that they do not stick to each other, the arrangement is easier. The tobacco plants which get dew and get water after rain should not be harvested until they become dry. A fully matured leaf should break in every harvested crop. The cut leaves are put on top and bundle made. These are placed in the molds, and moved to a shadow place where they are to be placed longer. The number of harvesting the crop in Izmir tobacco varies from 3–4 times (Dimon, 2007).

Aligning: Leaves are inserted 1–2 cm below the site where they are connected to the plant stem, and needles are passed one by one through the middle vein of the leaf. Hand aligning should not be preferred as many leaves will be damaged (5–6 leaves). When the needle is full, a rope is put in the hole at the tip of the needle and leaves transferred to the ropes. These transferred leaves to ropes should not be frequent; otherwise loss of the leaves will occur. Nylon ropes are never used in tobacco aligning and baling. The tobacco leaves are left to fade in shady places for 24–48 hours, when these leaves lose 5–10% of water. With the chlorophyll fragmentation, the leaf color begins to turn from green to yellow. These leaves, which become pale in color are placed on the grill for complete drying.

Grill drying (Kirmandal): A good drying space should be as close to the village as possible and must be far from threshing field and roads. It must always be installed in places where there is plenty of sunshine, and areas where there is no wind. The ground must be dry and clean, and soil surface must be cleaned from waste grass cover. The grill is made by two wires parallel to each other, ranging in height from 40 to 60 cm from the floor, to the extent that tobacco sequences can be exceeded. The sequences completing turn pale and are placed on the grille to allow air to flow side by side. Depending on the weather conditions, the tobacco is dried for 5–10 days on the grill, and then transferred to the main exhibit for 5–7 days. Drying times in the first and second main exhibition should not exceed 1 day. At this stage the leaves turn from light green to yellow. At the exhibition, the lines are left on the sun for a day or two, then other side is turned and the color of leaves followed. At the same time, the internal parts of the aligning, and middle veins dry up. After exhibition, the sequences are collected in such a way that the head portions of the leaf face upwards, and the vein drying is completed for two days. In order that tobacco lines are not damaged by dew, rain or sun rays, a roof is built on the grill and covered with nylon. It is better to dry in the tunnel greenhouse (Dimon, 2007).

Issues in drying tunnel greenhouse: Under normal conditions, the temperature inside the tunnel should not exceed 50–55°C, and ventilation should be present, so that the relative humidity is at the most 70%. Drying is very important in the first 1–2 days, temperature, and humidity inside the tunnel should be monitored closely. Not the early harvested crop, only mature leaves should be dried in the tunnel. Tobacco leaves waiting to turn pale should not be carried to the tunnel greenhouse immediately, in particular when sun is effective and weather is not windy.

Boxing: Tobacco is tempered before they packing, in order to soften it. Pesticides or coloring materials are not added to the tempering water. It is expected that the given tempered water will be absorbed by the leaf. Defective, diseased, burned, faded, dark green tobacco leaves must be removed from series. Tobacco should be chosen and balance kept that all are of the same stage and good quality. Before boxing, the lacing ropes are taken, boxes are stored in clean, dry, and airy environment. No foreign material should be present in the vicinity of tobacco boxes. Box weights should not exceed 25 kilograms depending on the origin and type of tobacco. In Turkey, information about tobacco processing both in the field and in the industry is done as visualized in Figures 15.4 and 15.5.

15.9 GLOBAL CLIMATE CHANGE SCENARIOS AND TOBACCO PRODUCTION IN TURKEY

Tobacco is a subtropical climate plant with good adaptation ability. The duration of vegetative period varies from 60 to 150 days, depending on the ecological differences in its environment. Total temperature demand varies between 18 and 32°C, depending on the varieties. The minimum temperature required for germination of tobacco seeds is 12–14°C and the optimum temperature needed is 27–28°C. For successful plant growth, a minimum temperature of 10°C and an optimum temperature of 25–30°C are needed. The growth of tobacco is affected adversely after 35–40°C and it withstands temperatures below 0°C for a very short time (Incekara, 1979; Taşlıgil, 1992; Yılmaz and Katar, 1996; Esendal, 2007; Doğanay and Çoşkun, 2012; Şahin and Taşlıgil, 2013).

Annual precipitation requirement for tobacco plants varies between 300–2000 mm, depending on the type. In the case of Flue-cured (Virginia) and Air-cured (Burley) type tobacco annual rainfall needed should be about 1300 mm, with 400–800 mm during vegetation period and 65–75% moisture content. Cigar tobacco needs 2000 mm or more of rainfall per year and can grow well in places with 62–83% humidity. Oriental (Turkish tobacco) type tobaccos requires 300–600 mm of rainfall per year and quality increases due to an increase in the relative humidity. Especially, if precipitation takes place in the month following planting, Turkish type tobacco accepts it as a desirable condition, lack of accurate rainfall during autumn allows it to retain the quality to a desirable extent. In Turkish tobacco, depending on the amount of rainfall, watering can be done 1–2 times a year, depending on the regime (Incekara, 1979; Taşlıgil, 1990; Yilmaz and Katar, 1996; Esendal, 2007; Şahin and Taşlıgil, 2013).

FIGURE 15.4 Collection and drying methods in tobacco cultivation in Turkey.
(Row 1: Preparations of fields for tobacco farming; Row 2: Crop Harvesting; Row 3: Tobacco Aligning; Row 4: Tobacco drying and packaging).

FIGURE 15.5   Processing of harvested tobacco in the industry in Turkey.

The variations and increase in the temperatures is a common phenomenon in Turkey as in other countries of the world. Many different factors are responsible for this. The ratio of greenhouse gases effects the temperature regime, rainfall, and other changes add to the temperature vulnerability. The temperatures in Turkey are said to have increased by about 1–1.5 degrees compared to 1961–2000. It is predicted that these increases may go up to

2–6 degrees in the years between 2070–2100. The global climate change has produced no change in the annual precipitation in Central Anatolia, Eastern Anatolia and Marmara Regions, while increase in Black Sea Region and decrease in Aegean, Mediterranean, and Southeastern Anatolia regions has been recorded. The temperatures are increasing in the winter months in the eastern parts and in the western regions during summer months (Anonymous, 2008; Ozturk et al., 2014). This is leading to a decrease in the level of groundwater, decrease in the yield values of crops grown in arid areas, and changes in disease-damaging factors.

Tobacco farming is carried out in all regions of Turkey except Central Anatolia (where it was done in the past). The most of the tobacco farming in the country is observed in the Aegean Region, followed by the Black Sea, the South Eastern Anatolia, the Marmara, the Eastern Anatolia and the Mediterranean region, respectively. More than half of Turkey's tobacco production is carried out in the Aegean region. In recent years there have been large reductions in the quantity of tobacco produced (Table 15.2). Most important reason for this is the changes in the country's tobacco policy. In addition to this, decrease in yield values due to changing climate factors has occurred.

TABLE 15.2   Regional Annual Tobacco Production in Turkey (1000 tons) (Anonymous, 2013)

Years	Regions					Total
	Aegean	Black Sea	Marmara	Eastern Anatolia	Southeastern Anatolia + Mediterranean	
2003	104	24	9	4	18	160
2004	66	15	6	5	20	112
2005	82	16	9	5	22	134
2006	83	18	8	6	20	135
2007	54	15	6	5	19	98
2008	39	13	4	4	15	75
2009	60	12	2	4	16	93
2010	59	15	5	0	2	81
2011	38	7	3	1	5	53
2012	32	5	2	1	4	45
2013	51	9	3	0	6	69

15.10 CONCLUSION

Tobacco is a subtropical climate loving plant. It is adversely affected by changes in the amount, regime, and duration of precipitation, while being affected positively (increase in quality) by temperature increases. In the Aegean region where tobacco is most cultivated in Turkey, increase in temperature is within acceptable limits and a decreasing tendency of rainfall proves fruitful for growth and development of this plant. Too many rainy days decrease the efficiency of tobacco, which in turn effects the production negatively. The increase of harmful populations of disease causing organisms, both in number as well as their spread prevalence end up with diseases in the plant and cause loss in yields. Similarly sudden air changes such as rapid warming of the air after torrential rainfall, especially in the Black Sea region, fungal diseases (blue mold) are become more effective. We can conclude that, tobacco production in Turkey will be affected negatively by certain level of climate change expected as a result of global warming.

KEYWORDS

- **production techniques**
- **tobacco**
- **turkey**
- **types**

REFERENCES

Anonymous, (2008). Küresel Isınmanın Etkileri ve Su Kaynaklarının Sürdürülebilir Yönetimi Konusunda Kurulan (10/1, 4, 5, 7, 9, 10, 11, 13, 14, 15, 16, 17) Esas Numaralı Meclis Araştırması Komisyonu Raporu. Ankara.

Anonymous, (2013). Tütün ve Alkol Piyasası Düzenleme Kurumu, Tütün Piyasası, Tütün Alım Verileri. www.tapdk.gov.tr. Başbakanlık Osmanlı Arşivlerini

Borio, G., (2001). The Tobacco Timeline. http://www.tobacco.org/resources/history/Tobacco_History21.html.

Dimon, (2007). Türk Tütün A. Ş., Tütün Tarımı. İzmir.

Doğanay, H., & Coşkun, O., (2012). *Tarım Coğrafyası. Güncellenmiş II. Baskı, Pegem Akademi, s., 488*, Ankara.

Doğruel, F., & Doğruel, A. S., (2000). *Osmanlı'dan Günümüze TEKEL. Türkiye Ekonomik ve Toplumsal Tarih Vakfı Yayınları,* s. 398, İstanbul.

Ekren, S., & Sekin, S., (2008). Ege bölgesi tütünlerinin verim ve bitkisel özellikleri ile aralarındaki ilişkilerin saptanması. *Ege Üniversitesi Ziraat Fakültesi Dergisi, 45* (2), 77–84.

Er, C., (1997). *Tütünün Hikâyesi.* Ankara Üniversitesi Ziraat Fakültesi Yay., Ankara.

Er, C., Başalma, D., Ekiz, H., & Sancak, C., (2011). Tarla Bitkileri II. T. C. *Anadolu Üniversitesi Yayın No: 2254,* s. 235, Eskişehir.

Erdoğan, A., (2014). Düyun-I Umumiye'nin Kurulmasının Siyasi Yönü ve Reji İdaresi İle İlişkisi. In: Mercimek, H. V., & Akçiçek, E., (eds.), *Mucizeden Belaya Yolculuk – Tütün.* Tarihçi Kitabevi, İstanbul-Türkiye.

Esendal, E., (2007). *Tütün Tarımı.* NKÜ Tekirdağ Ziraat Fakültesi Yayınları. Tekirdağ.

İncekara, F., (1979). *Endüstri Bitkileri.* 4. Cilt (Keyf Bitkileri). E. Ü. Z. F. Yay. No: 84, Bornova-Izmir.

Mercimek, H. V., (1997). *Türkiye'de Tütün.* In: Naskalli, E. G. (ed.), Tütün. İstanbul.

Mercimek, H. V., (2014). Yeni tütün yasasının işlevlik kazanmasıyla oluşan tütün profilindeki değişmeler. In: Mercimek, H. V., & Akçiçek, E., (eds.), *Mucizeden Belaya Yolculuk – Tütün.* Tarihçi Kitabevi, İstanbul-Türkiye.

Mercimek, V., (1999). *Tarihi süreç içerisinde Türkiye'de yetiştirilen tütün miktarı ve değişim sebepleri.* Tütün Eksperler Derneği Bülteni, Sayı: 42, Mart-Nisan 1999.

Oktar, T., (1997). Osmanlı döneminde Reji şirketinin kurulmasından sonraki gelişmeler. In: Naskalli, E. G., (ed.), Tütün. İstanbul.

Otan, H., & Apti, R., (1989). *Tütün.* T. C. T. O. K. İ. B. Ege T. A. E. Yay. No: 83, Menemen-Izmir.

Özavcı, Ö., (2007). *Cumhuriyet'ten Günümüze Türkiye Tütün Ekonomisi ve Politikaları: AB Ülkeleri ile Bir Karşılaştırma.* Osmangazi Üniversitesi Sosyal Bilimler Enstitüsü, Yüksek Lisans Tezi, Eskişehir.

Özdemir, M., (2010). *Türkiye'de Tütün Sektörünün Tarihi ve Ekonomik Yapısı.* Gaziosmanpaşa Üniversitesi Sosyal Bilimler Enstitüsü, Yüksek Lisans Tezi, Tokat.

Öztürk, M., Güvensen, A., Altay, V., & Altundağ, E., (2014). Tütünün botaniksel özellikleri hakkında genel değerlendirme. In: Mercimek, H. V., & Akçiçek, E., (eds.), *Mucizeden Belaya Yolculuk – Tütün.* Tarihçi Kitabevi, İstanbul-Türkiye, pp. 17–51.

Peksüslü, A., (2000). *Bazı Türk tütün çeşitlerinin İzmir-Bornova koşullarında morfolojik fizyolojik ve agronomik özellikler.* Tyuap, 23–25 Mayıs 2000, ETAE. Yayın No: 98, pp. 249–268, Menemen-Izmir.

Peksüslü, A., (2005). *Ege bölgesine uygun yeni tütün çeşitleri ve bölgedeki performansları.* TAYEK 06–08 Eylül 2005, E. T. A. E. Yay. No: 120. pp. 217–231, Menemen-İzmir.

Peksüslü, A., Yılmaz, I., Inal, A., & Kartal, H., (2012). Türkiye tütün genotipleri. *Anadolu Ege Tarımsal Araştırma Enstitüsü Dergisi, 22* (2), 82–90.

Quataert, D., (2004). *Osmanlı İmparatorluğu'nun Ekonomik ve Sosyal Tarihi.* İstanbul.

Şahin, G., & Taşlıgil, N., (2013). Türkiye'de tütün (*Nicotiana tabacum* L.) yetiştiriciliğinin tarihsel gelişimi ve coğrafi dağılımı. *Doğu Coğrafya Dergisi, 18* (30), 71–101.

Sarıoğlu, M., (1980). Türk Tütünleri. Tekel Genel Müd. Yay. 199 EAG/DKY 61., İstanbul.

Savaş, R., (1969). *Ticaret ve Endüstri Bitkileri (Özel Tarla Ziraati).* Kardeş Matbaası, Ankara.

Sekin, S., & Peksüslü, A., (1995). *Ege tütün genotiplerinin agronomik özellikleri.* I. Milli Tütün Komitesi. Bilimsel Araştırma Alt Komitesi 13. Toplantısı. 25–27 Ekim 1995. Istanbul.

TAPDK, (2014). *Tütün ve Alkol Piyasası Düzenleme Kurumu 2013 Faaliyet Raporu.* T. C. Tütün ve Alkol Piyasası Düzenleme Kurumu Yay, Ankara.

Taşlıgil, N., (1990). *Manisa Ovaları ve Çevresi*. İstanbul Üniversitesi Deniz Bilimleri ve Coğrafya Enstitüsü, Doktora Tezi, İstanbul.

Taşlıgil, N., (1992). Türkiye'de Tütün Ziraati. *Türk Coğrafya Dergisi, 27*, 129–138.

The Plant List, www.theplantlist.org.

Thobie, J., (2000). *Osmanlı Devletinin Son Yıllarına Bakış*. Muğla Üniversitesi SBE Dergisi Cilt: 1/2.

TUIK, (2014). *Türkiye İstatistik Kurumu Bitkisel Üretim İstatistikleri*. http://tuikapp.tuik.gov. tr/bitkiselapp/bitkisel.

Unal, M., (1997). Tütünün dört yüz yılı. In: Naskalli, E. G., (ed.), *Tütün*. İstanbul.

Yılmaz, F., (2005). *Osmanlı İmparatorluğu'nda Tütün: Sosyal, Siyasî ve Ekonomik Tahlili*. İstanbul.

Yılmaz, G., & Katar D., (1996). *Keyf Bitkileri Üretimi*. GOÜ Ziraat Fak. Yayınları No: 11, Tokat.

CHAPTER 16

WILD RELATIVES AND THEIR ROLE IN BUILDING RESISTANCE IN BREAD WHEAT: AT A GLANCE

SAMEEN RUQIA IMADI and ALVINA GUL

Atta-ur-Rahman School of Applied Biosciences, National University of Sciences and Technology, Islamabad, Pakistan, E-mail: alvina_gul@yahoo.com

ABSTRACT

Wheat is the domesticated form of many of the grasses from Poaceae and tribe Triticeae. These annual and perennial grasses comprising of approximately 325 species also include the diploid ancestors of wheat known as its wild relatives. These relatives exhibit more genetic diversity as compared to their cultivated form, hence are a potent means of allelic enrichment of wheat via intergeneric, interspecific or intraspecific hybridization. Wheat relatives include many grasses like *Aegilops, Thinopyrum, Triticum, Elymus, Leymus, Elytrigia, Lophopyrum, Secale, Heteranthelium, Taeniantherum, Haynaldia, and Hordeum* from the tribe Hordeae that are distributed across the three gene pools (primary, secondary, and tertiary) based upon their genetic distance attribute. Due to the allelic variation that is present, these relatives have been utilized in wheat research and production for over a century commencing in 1904 with wheat/barley combinations or as early as 1872 if X Triticale is considered. During these many decades via basic, strategic, and applied research efforts genes have been exploited for intro-gressing resistances/tolerances to several biotic and abiotic stresses of global significance. The trait coverage has increased progressively and currently targets include insect/virus resistance, micronutrients with growing focus on cereal quality parameters towards which these alien resources are contributing new diversity. Wild relative variation is not restricted to bread wheats alone and the tetraploid durums are also recipients of their practical benefits for crop yield maximization. Potential of the wild resource also exists for inducing sterility; crucial for hybrid wheat development. In addition, within the strategic research scenario wild species have emerged as a significant base for producing user-friendly wheat genomic genetic stocks for future usage practicality. We in this review have targeted major facets of wild relative utilization for wheat improvement since 1904 and in addition have provided an in-depth overview of the germplasm holdings that global researchers could tap upon.

## 16.1	INTRODUCTION

The term wild relatives refer to the plants, which are ancestors as well as other species that are closely related to cultivated crops. Wild relatives of a plant are a source of genes resistant to pests and different stresses therefore are used for the improvement of cultivated plants. They are also used to raise the nutritional value of crops like protein content in durum wheat. Crop wild relatives are excellent tools, which are used to adapt to new environmental conditions and human needs.

Currently the cultivated wheat is a domesticated plant with agronomically improved traits. The original primitive wheat was produced by natural spontaneous crossing of wild grasses. The process of domestication of wheat is extended over centuries. First criterion for the domestication of wheat is selection of mutants with non-shattering ears. Archaeologically their domestication appeared 9,500 years ago (Tanno and Willcox, 2006). Domestication of wheat was followed by selection. The plants showing high adaptation and best characters were selected and cultivated. Modern hexaploid and tetraploid wheat are originated by the progenitor *Triticum dicoccoides,* which is a diploid (Dubcovsky and Dvorak, 2007). A study performed by Giles and Brown suggested that the hexaploid wheat was firstly originated in Southeast Turkey and northern Syria (Giles and Brown, 2006).

Wheat wild relatives have been largely used for the improvement of crop. To improve the salt tolerance in wheat, Colmer, and colleagues used its wild relatives. They observed that there is a variation in salt tolerance in different species of Triticeae and some Triticeae species are halophytes. It is also seen that halophyte species display sodium exclusion whereas other species display chloride exclusion. It is also predicted that halophyte species can be hybridized with bread wheat and durum wheat to produce salt tolerant wheat variety (Colmer et al., 2006; Nevo and Chen, 2010).

Wild relatives of wheat are used in improvement of characters like drought tolerance (Nevo and Chen, 2010), ozone sensitivity (Biswas et al., 2008), stability, disease resistance (Warburton et al., 2006; Zhang et al., 2009), snow mold resistance, freezing tolerance (Iriki et al., 2001), stress tolerance (Farooq and Azam, 2001), heat tolerance (Zaharieva et al., 2001a, b), mildew resistance (Jarve et al., 2000), stem rust resistance (Rouse and Jin, 2011) and enhancing zinc and iron content (Cakmak et al., 2004; Rawat et al., 2009) in cultivated crops.

16.2   WILD RELATIVES OF TRITICUM

16.2.1   AEGILOPS SPP.

Aegilops is a plant genus, which is commonly known as goat grasses. It belongs to Poacaea family. It contains 23 known species. They are most closely related to winter wheat, which is *Triticum aestivum;* therefore they are also classified as agricultural weeds. The hexaploid genome of *Triticaea* (BAD) when studied phylogenetically showed that D genome is derived from diploid *Aegilops tauschii* and B genome is derived from diploid *Aegilops speltoides*. This shows that *Triticum* and *Aegilops* are monophyletic (Peterson et al., 2006).

16.2.2   AEGILOPS TAUSCHII

Aegilops tauschii is the most primitive relative of wheat grown nowadays. Bread wheat has obtained its D genome from *Ae. tauschii* (Guyomarc'h et al., 2002). Hybridization of *Ae. tauschii* with durum wheat produces F_1 and F_2 hybrids, which are morphologically similar to bread wheat and have vigorous growth habits (Matsuoka and Nasuda, 2004). A set of 84 bread wheat lines containing single homozygous introgression of *Ae. tauschii* was developed. This development was accompanied by microsatellite marker-assisted selection. It is observed that genome of *Ae. tauschii* is fully represented in these lines except three telomeric regions at chromosomes 1DL, 4DL and 7DS (Pestsova et al., 2006).

There is relatively low level of polymorphism in the genomes of *Aegilops tauschii* and wheat, as shown by microsatellite analysis because D genome of wheat is relatively young (Bossolini et al., 2006; Pestsova et al., 2000; Lelley et al., 2000). On capillary electrophoretic examination of high molecular weight glutenin subunits in *Aegilops tauschii,* it is observed that there are 42 high molecular weight glutenin subunit alleles in *Ae. tauschii,* which are similar to that of primitive wheat. But most of these alleles are eliminated from present bread wheat. Hence *Ae. tauschii* can be used as a resource of quality improvement of bread wheat and its bread-making properties (Yan et al., 2003). Genes for low molecular weight glutenins when sequenced from *Aegilops tauschii* showed that they are very similar to those isolated from bread wheat (Johal et al., 2004).

Aegilops tauschii is a potential source for the improvement of *Triticum aestivum. Pm34,* a powdery mildew resistance gene found in *Ae. tauschii* can be transferred to common wheat to avoid the loss of wheat by mildew (Miranda et al., 2006). Miranda and colleagues also worked for isolation of *Pm35* powdery mildew resistance gene from *Aegilops tauschii* to administer in wheat (Miranda et al., 2007). Leaf rust is one of the major diseases of wheat. *Lr40* and *Lr21*; leaf rust resistance genes are isolated from *Aegilops tauschii* and are administered to *Triticum* to gain resistance against leaf rust (Huang and Gill, 2001). Synthetic hexaploid wheat showed that 18 accessions of *Ae. tauschii* are resistant to leaf rust. Out of these 18 accessions, one is immune, 13 are highly resistant and 4 are moderately resistant. Six of the synthetic hexaploid wheat showed resistance while 4 were susceptible. This shows that in inheritance, resistance is dominant over susceptibility (Assefa and Fehrmann, 2000).

Glutenin genes of *Ae. tauschii* significantly affect the bread-making properties in hexaploid wheat. Synthetic hexaploid wheat produced by durum wheat and *Ae. tauschii* showed that prolamin alleles of parents are additively expressed. Bread loaf volume, gluten index and dough surface are highly influenced (Hsam et al., 2001). Iron and zinc content in cultivated wheat can be increased by the use of *Aegilops tauschii* as it contains high level of iron and zinc. Fortification of wheat with micronutrient contents may help reduce malnutrition in many countries as wheat is used as a staple food by a large section of world population (Chhuneja et al., 2006).

Synthetic hexaploid wheat was experimented for Hessian fly resistance. It is found that genome donated by *Ae. tauschii* parents contain *SW8* gene, which is resistant to Hessian fly biotype Great Plains (GP) and strain vH13, whereas another gene *SW34* is resistant to biotype GP. Thus *Ae. tauschii* can be used for increasing the resistance of wheat against Hessian fly (Wang et al., 2006). W7984, a synthetic hexaploid wheat line is resistant to green bug. This resistance is due to a single dominant gene placed on chromosome arm 7DL, which is inherited by *Aegilops tauschii* (Weng et al., 2005).

16.2.3 AEGILOPS UMBELLULATA

Aegilops umbellulata is a diploid species belonging to Poacaea family. *Ae. umbellulata* is a primitive relative of wheat and is found to be phylogenetically related to wheat (Molnar et al., 2001). Short interspersed nuclear element (SINE) is found in second intron of acetyl CoA carboxylase gene of *Aegilops umbellulata*. This SINE is designated as 'Au.' Au elements are also discovered in genomes of *Triticum* species. This shows a phylogenetic relationship among *Triticum* and *Aegilops umbellulata* (Yasui et al., 2001).

Aegilops umbellulata derived leaf rust resistant gene *Lr9* can be used for the improvement of wheat as leaf rust resistant marker (Gupta et al., 2005). *Ae. umbellulata* accession 3732 is an excellent source of resistance of wheat to major diseases like leaf rust resistance. Chhuneja and colleagues conducted an experiment in which synthetic bread wheat was introgressed with *Ae. umbellulata* accession 3732. It was observed that at least three putatively new genes, out of which two are for leaf rust resistance (*LrU1* and *LrU2*) and one for stripe rust resistance (*YrU1*) are introgressed into *Triticum aestivum* from *Aegilops tauschii* (Chhuneja et al., 2008). *Triticum carthlicum* accession PS5 and *Aegilops umbellulata* accession Y39 were crossed to produce amphiploid line Am9. Am9 is shown to be resistant to powdery mildew. This resistance is transferred to common wheat cultivar by continuous backcrossing. Thus powdery mildew resistance can be gained in wheat by *Ae. umbellulata* species (Zhu et al., 2006).

Genome of *Aegilops umbellulata* codes for two high molecular weight glutenin subunits, which are phylogenetically linked to high molecular weight glutenin subunits of D genome of *Triticum aestivum* (Liu et al., 2003).

16.2.4 AEGILOPS BIUNCIALIS

Aegilops biuncialis is allo-tetraploid relative of wheat belonging to Poaceae family. *Aegilops biuncialis* lines, which are grown in drought condition, can be used for improving the drought tolerance of *Triticum aestivum* (Molnar et al., 2002). Polyethylene glycol (PEG) induces drought stress in *Aegilops biuncialis* in nutrient solution. Osmotic stress when applied to *Ae. biuncialis* and *Triticum aestivum* shows that *Ae. biuncialis* has high adaptability and increased ability to tolerate the stress as compared to common wheat. Under high osmotic stress *Aegilops biuncialis* shows significant water loss, low degree of stomatal closure and decrease in inter-cellular carbon dioxide concentration (Molnar et al., 2004).

A synthetic amphiploid wheat line made by *Ae. biuncialis* germplasm 15–3–2 showed that the line had spikes intermediate to both of the parents. It displays stable immunity and fertility to stripe rust and wheat powdery mildew (Tan et al., 2009). Therefore, this wheat line 15–3–2 can be further studied for production of new synthetic wheat.

16.2.5   AEGILOPS GENICULATA

Aegilops geniculata is another allo-tetraploid relative of wheat. It is originated in Meditteranean region, as predicted by phylogenetic AFLP analysis (Arrigo et al., 2010). *Ae. geniculata* can be used to produce leaf rust and stripe rust resistant introgression lines. These lines were developed by induced homoeologous chromosome pairing between wheat chromosomes and *Ae. geniculata*. These lines showed three resistant introgressions. All of the three introgressions are highly resistant against most prevalent and most virulent races of leaf rust and stripe rust (Kuraparthy et al., 2006). *Lr57* and *Yr40* are designated as new genes for leaf rust resistance and stripe rust resistance (Kuraparthy et al., 2007). *Aegilops geniculata* can be used for the improvement of *Triticum aestivum* as its accessions when studied from different eco-geographical regions showed that they are resistant to drought and heat stress. Some of the accessions are also resistant to Hessian fly and cereal cyst nematodes, whereas some are resistant to barley yellow dwarf virus and rusts (Zaharieva et al., 2001a, b).

Chromosome 5M^g isolated from *Aegilops geniculata* confers resistance to both leaf rust and stripe rust. This chromosome has been transferred to bread wheat in the form of DS5M^g chromosome substitution line. After the transfer it is observed that wheat gains resistance against both leaf rust and stripe rust. But this germplasm needs further studies and engineering (Aghaee-Sarbarzeh et al., 2002). *Sr53* is a stem rust resistant gene found in *Aegilops geniculata*. This gene is transferred to synthetic bread wheat and it is observed that bread wheat gains the properties of *Aegilops* parents and become resistant to stem rust. Thus this gene can be used to improve the quality of cultivated wheat (Liu et al., 2011). Homoeologous metaphase I pairing of *Triticum aestivum* cross with *Aegilops geniculata* when examined by *in situ* hybridization explores that certain genes of wheat are transferred by *Ae. geniculata*. 3DL and 5DL are wheat genome locations that highly resemble *Aegilops geniculata* (Cifuenetes and Benavente, 2009). Herbicide tolerant *Triticum aestivum* can be produced with the use of *Aegilops geniculata*. This experiment requires further studies to be implemented commercially to cultivated crop (Loureiro et al., 2008).

Excess copper was applied to hybrid seedlings of *Triticum aestivum* and *Aegilops geniculata*. It was observed that hybrid lines are more tolerant towards copper excess as compared to *Triticum aestivum* lines. This shows that *Ae. geniculata* plays an important role in copper tolerance and hence can be used to improve wheat tolerance for copper (Landjeva et al., 2004).

16.2.6   AEGILOPS TRIUNCIALIS

Aegilops triuncialis also known as barb goat-grass belongs to Poaceae family of grasses. It is tetraploid, which is derived from multiple origins with minimal genome changes after its formation (Vanichanon et al., 2003). Its genome formula is UUCC. This plant is primitive relative of wheat. *Aegilops triuncialis* has functions in nutrient recycling and decomposition, which brings changes in plant community tissue chemistry and biomass production (Drenovsky and Batten, 2007). It is an excellent source of resistance to many wheat diseases.

Transfer line TR–3531 is a result of cross between *Triticum turgidum* and *Aegilops triuncialis,* the F_1 hybrids are crossed with *Triticum aestivum.* These transfer lines contain resistance for nematode, Heterodera avenae (causative agent of cereal cyst) due to presence of gene *Cre7*. This gene is highly resistant to the prevailing SW biotype. Further a single dominant gene of *H30* is found to be highly resistant against Hessian fly in the introgression line (Martin-Sanchez et al., 2003). Leaf rust is a foliar disease of bread wheat throughout the world. *Aegilops triuncialis* confers resistance to this disease due to its chromosome $5U^t$. This chromosome when transferred to *Triticum aestivum* in the form of disomic $DS5U^t$, showed an increased resistance in bread wheat for this disease. But this germplasm needs further engineering to be applied commercially (Aghaee Sarbarzeh et al., 2002).

Synthetic substitution line 4V of *Triticum aestivum* was crossed to *Aegilops triuncialis* 3C addition line. F_1 hybrids were then backcrossed to common bread wheat to induce changes of 4V. Chromosome C banding and genomic *in situ* hybridization (GISH) was applied to find out the genomic variations. Results indicated that gametocidal chromosome 3C from *Aegilops triuncialis* can induce effective structural changes in chromosome of wheat (Chen et al., 2002). Study on root of introgression line TR–3531 produced by crossing *Aegilops triuncialis* and *Triticum aestivum,* showed that on exposure to nematode Heterodera avenae these plants show resistance. This resistance is due to the presence of *Cre7* resistant gene. It is also observed that after four to six days exposure of the nematode, the levels of enzymes like peroxidase, esterase, and superoxide dismutase significantly increases in the roots of introgression lines. The results predict that enhanced peroxidase, esterase, and superoxide dismutase activities may have an important role in cell wall lignifications which in turn increases the resistance of plants towards cereal cyst nematode (Montes et al., 2004).

Wheat can gain leaf rust resistance by *Aegilops triuncialis*. An experiment conducted by crossing highly susceptible wheat cultivar WL711 with recurrent parent *Aegilops triuncialis* accession 3549, proved that resistance to leaf rust can be achieved by use of *Ae. triuncialis* hence this plant can be used to improve commercial wheat (Aghaee-Sarbarzeh et al., 2001).

16.2.7 THINOPYRUM SPP.

Thinopyrum spp. are a primitive relative of wheat and are also known as wheat grasses. They are pentaploid and their genome formula is written as SPEEE (Refoufi et al., 2001). These grasses are found in semi arid temperate region as they are cool season grasses (Wang et al., 2003). A set of 42 microsatellites of wheat is experimented for their cross amplification on DNA of different *Thinopyrum* species. It is found that the number of microsatellites that amplified *Thinopyrum ponticum* is 33, *Thinopyrum intermedium* is 28, *Thinopyrum elongatum* is 24 and *Thinopyrum bessarabicum* is 24. 24 pairs of microsatellites are seen to be successfully amplified by all of the experimented species (Kroupin et al., 2013).

Thinopyrum spp. particularly *Thinopyrum ponticum* can be used to increase resistance in *Triticum aestivum* against various diseases. These species contain genes for resistance of leaf rust, stem rust, powdery mildew, barley yellow dwarf virus (BYDV), wheat streak mosaic virus (WSMV) and wheat curl mite (vector for WSMV). Many of these genes have been transferred in wheat for experimentation and many experiments are successful in transferring the resistance to wheat. Thus *Thinopyrum spp.* can be used as an excellent alien source for the improvement of wheat (Li and Wang et al., 2009).

Leaf rust resistant gene *Lr19* is transferred from *Thinopyrum elongatum* to *Triticum aestivum* cultivars. This transfer of gene expresses resistance to most of the pathotypes of *Puccinia triticina*. SCAR marker co-segregating with *Lr19* is found. This marker is identified in 120 varieties of *Triticum aestivum*, which indicates that this marker can be used for the marker-assisted selection (MAS) in breeding of resistance to leaf rust in bread wheat and durum wheat (Hong Fei et al., 2009). Introduction of *Thinopyrum interme-dium* derived cereal yellow dwarf virus resistance gene in *Triticum aestivum* substitution line P29 results in dose-dependent resistance in wheat (Wiangjun and Anderson, 2004).

Chromosome number 4 of *Thinopyrum* species has an ability to re-grow in hexaploid wheat *Triticum aestivum* after sexual cycle and senescence. This chromosome is associated with resistance to necrotrophic eyespot pathogen, the *Oculimacula yallundae*. It is observed that after the transfer of chromosome 4 to wheat, wheat gains resistance against necrotrophic eyespot disease. It is also observed that wheat genome and *Thinopyrum* genome do not recombine in meiosis but chromosome engineering and genetic engi-neering techniques can be applied to them to transfer these novel sources of resistance to wheat (Okubara and Jones, 2011). Wheat introgression lines P1583794 and P1611939 contain a fragment of *Thinopyrum* species. These lines were tested for resistance against *Oculimacula, Fusarium culmorum, Puccina triticina* and *Barley yellow dwarf virus*. Three wheat cultivars Boomer, Esket, and Mirage were used for the marker-assisted backcross experiment. The BC_1F_1 and BC_2F_1 hybrids of introgression lines were seen to be resistant to all of these species despite of the presence of only a small fragment of *Thinopyrum* genome.

Thinopyrum elongatum contains disease-resistant genes, which are found in its 7E chromosome. It is thought that 7E chromosome is associ-ated with resistance to *Fusarium* head blight and wheat rust. Due to this hypothesis, 7E chromosome specific molecular marker linked to resistant genes is formed. This marker has an ability to detect 7E chromosome of *Thinopyrum elongatum* and provide an important theoretical and practical basis for breeding by marker-assisted selection (MAS) (Chen et al., 2013).

16.2.8 ELYTRIGIA SPP.

Elytrigia is a wild genus, which is related to common bread wheat (*Triticum aestivum*). *Elytrigia* belongs to the family Poaceae. It contains many wild grasses, which are commonly known as quack grasses or couch grasses. Many of its genes and traits are used for yield and quality improvement in *Triticum aestivum*. This genus is characterized by species with various levels of ploidy including diploids, tetraplods, hexaploids, octaploids, and decaploids (Mao et al., 2010). *Elytrigia* species can be used as forage for cattle. *Elytrigia elongata* is one of the diploid species of this genus. *E. elongata* contains many genes for resistance to many diseases. These genes can be incorporated to the genome of bread wheat to enhance its ability for the resistance towards diseases. Some experiments on transfer of genes for increased resistance to barley yellow dwarf virus and *Puccinia striiformis* are successful up till now. *Elytrigia* species can also be used to improve resistance against stresses like salt stress and drought stress (Wei-Dong et al., 2007).

Resistance against *Blumeria graminis* (causative agent of powdery mildew) can be gained in common wheat by hybridizing it with *Elytrigia intermedium* by crossing and backcrossing. After a long period of back-crossing a line GRY19 is found to have single dominant gene for resistance against powdery mildew. Resistant gene, which is transferred from *Elytrigia intermedium* to *Triticum aestivum* is named as *Pm40* (Luo et al., 2009). Resistance to salt stress in *Elytrigia* species can be transferred to *Triticum aestivum* by breeding between both the species and backcrossing the hybrid plants with *Triticum* as recurrent parent (Geng et al., 2007).

Eleven alien disomic addition lines of *Elytrigia intermedium* and with *Triticum aestivum* were isolated from hybrid generations. Mitosis of root tips is observed, and results indicate that these lines are stable for heredity (Ping-Li et al., 2002). As the lines produced by hybridization of *Triticum aestivum* and *Elytrigia* species are stable, they can be used to improve the quality of bread wheat (Hong- Gang et al., 2000).

16.2.9   LOPHOPYRUM ELONGATUM

Lophopyrum elongatum is a close relative of wheat and is commonly known as wheat grass. It provides gene sources for the improvement of qualities of bread wheat. *L. elongatum* chromosome segment can be introgressed in hexaploid wheat. Simple sequence repeats can be used to identify the locus of *Lophopyrum elongatum* genome. 41 SSRs are proved to locate the loci of *L. elongatum* in wheat (Mullan et al., 2005).

Waterlogging tolerance in *Triticum aestivum* can be increased by the use of *Lophopyrum elongatum*. It is found that *L. elongatum* is more tolerant to de-oxygenated stagnant nutrient solution and waterlogged soil than wheat. Amphiploids and addition line of hybrids of *L. elongatum* and *Triticum aestivum* are found to be tolerant to waterlogging. Roots of *L. elongatum* are more porous than wheat, which results in greater internal movement of oxygen in roots, which keep the plant intact. This trait helps the plant to survive efficiently in waterlogged soil. This trait can be transferred to wheat by crossing and backcrossing of hybrid plants (McDonald et al., 2001).

7DL.7Ag fragment of *Lophopyrum elongatum* contains *Lr19* gene, which is leaf rust resistant gene. This fragment is also associated with significant increase in grain yield under irrigated and disease free conditions. When 7DL.7Ag was translocated to *Triticum aestivum,* it is observed that yield is increased which is linked with increased rate of biomass production (Monneveux et al., 2003). Amphiploid cross of *Triticum aestivum* and *Lophopyrum elongatum* is subjected to salt stress to observe the influx of sodium and potassium ions inside the cell membrane. It is found that under first 24 hours of exposure to stress, plants absorb and translocate the ions to youngest leaves. During this period the amphiploids retain more potassium ions as compared to their parent wheat. Amphiploids exert a fast acting mechanism therefore they are sensitive to sodium and potassium concentration. In growing tissue of leaf, substitution of sodium ions by potassium ions is also observed (Santa-Maria and Epstein, 2001).

Stripe rust resistance in wheat can be achieved by the use of *Lophopyrum elongatum.* Addition and substitution lines of wheat cultivar Chinese spring and *L. elongatum* were inoculated with pathotypes CYR–30 and CYR–31 of stripe rust. It is observed that chromosome 7E1 from *L. elongatum* is responsible for resistance to stripe rust in adult plants but the expression of resistance by 7E1 chromosome is dependent on genotype of wheat (Zu-Jun and Zheng-Long, 2001). Genetic lines (substitution lines and addition lines) of *Triticum aestivum-L. elongatum* complement fusarium head blight resistance in common bread wheat (Shen et al., 2004). This resistance is also due to chromosome 7E of *L. elongatum.* Gene for resistance to fusarium head blight is located on the long arm of chromosome 7E (Shen and Ohm, 2006).

Triticum aestivum-Lophopyrum elongatum chromosome substitution lines are tested for resistance to yellow rust at seedling stage. The results explored that *L. elongatum* contains a gene on 3E chromosome, which is linked to resistance to yellow rust. This gene can be translocated into wheat as an alien chromosome (Ma et al., 2000).

16.2.10 HORDEUM SPP.

Hordeum genus is closely related to wheat. It belongs to the Triticeae family of plants. This genus contains many commercially available species like *Hordeum vulgare* (barley), *Hordeum marinum* (sea barley). Specie specific DNA markers of wheat can be amplified in *Hordeum* species, which shows their ability of natural cross-hybridization (Guadagnuolo et al., 2001). *Hordeum species* specifically *Hordeum bulbosum* can be used to increase the resistance against drought and salt stress in wheat (Nevo and Chen, 2010). Waterlogging tolerance in wheat can be achieved by crossing and backcrossing *Triticum aestivum* with *Hordeum marinum* (sea barley). Amphiploids of *Hordeum marinum-Triticum aestivum* showed increased porosity of roots which result in movement of water and the root remains intact. Barrier to radial oxygen loss (ROL) is also observed in these amphiploids. Thus it can be deduced that *Hordeum* species can be used for improving the quality of wheat (Malik et al., 2011).

It is found that *Hordeum marinum* is more tolerant to salt stress than wheat. Amphiploids of *Hordeum marinum-Triticum aestivum* are shown to be intermediately tolerant to the stress. But salt tolerance of *Triticum* can be enhanced by producing the hybrid lines of *Hordeum marinum- Triticum aestivum,* and backcrossing them keeping wheat as recurrent parent (Islam et al., 2007). Genetic map of *Hordeum chilense* (wild barley) shows that it represents 13% of wheat simple sequence repeat primers. Overall structure of the genome linkage group is seen to be similar to B genome and D genome of *Triticum aestivum* (Hernandez et al., 2001).

16.2.11   TRITICUM DICOCCOIDES

Triticum dicoccoides is commonly known as wild emmer wheat. It is allo-tetraploid wheat, which is the progenitor of modern tetraploid (durum) and hexaploid (bread) cultivated wheat. Hexaploid wheat inherited its A-genome from *Triticum dicoccoides*. Cross between *T. dicoccoides* and *T. durum* revealed approximately 70 domestication quantitative trait loci (QTL) (Peng et al., 2003). Genetic differentiation in *Triticum dicoccoides* and *Triticum durum* is observed only in B genome as revealed by Amplified Fragment Length Polymorphism (AFLP) (Peng et al., 2000a).

Resistance to powdery mildew has been transferred to bread wheat by wild emmer accession G- 305-M by crossing and backcrossing. It is found that resistance to powdery mildew is controlled by single dominant gene at seedling stage. G–305-M derived powdery mildew gene is a novel gene and is temporarily designated as MIG (Xie et al., 2003). Two stripe rust resistant genes (*YrH52* and *Yr15*) derived from wild emmer wheat have been located on 1B genome of bread wheat. Large number of molecular markers is explored for these genes. These markers can help in marker-assisted intro-gression of *T. dicoccoides* derived *YrH52* and *Yr15* stripe rust resistant genes into elite cultivars of wheat (Peng et al., 2000b).

Powdery mildew is a very disastrous disease of wheat worldwide and is caused by *Blumeria graminis*. Wild emmer is a very potent donor of resistance towards disease and other important traits to bread wheat. Wild emmer accession G–303–1M is used to transfer disease resistance to bread wheat by crossing and backcrossing. This crossing and backcrossing result into new inbred lines of bread wheat, which are P63. When F_2 population of these inbred lines was analyzed genetically, it is observed that resistance to powdery mildew is controlled by a single recessive gene. This resistance gene is designated as *Pm42*. *Pm42* is very potent against many of the pathotypes of *Blumeria graminis*. Closely linked molecular markers can be used for the rapid transfer of *Pm42* to wheat breeding populations (Hua et al., 2009).

Wild emmer wheat can also be used to understand the response of bread wheat towards drought stress. microRNA expression patterns of drought resistance in wild emmer wheat is studied. It is observed that 438 microRNAs are expressed in drought stress in leaf and root tissues (Kantar et al., 2011). *Triticum dicoccoides* derived wheat line known as Zecoi–1 is resistant to powdery mildew effectively. F_3 segregation analysis of hybrids of Chinese spring and Zecoi–1 showed that resistance in Zecoi–1 is due to the presence of a single dominant gene. AFLP analysis helped in identification of eight markers, in which four were associated with resistance allele in repulsion phase. SSR markers showed that powdery mildew resistance gene is distal of breakpoint 0.89 in deletion line 2BL–6. Resistance gene in Zecoi–1 is temporarily designated as MIZec1 (Mohler et al., 2005).

Head blight of wheat is associated with yield and quality losses in wheat worldwide. 151 *T. dicoccoides* genotypes were subjected to fungus and responses were evaluated. Heritability for type II fusarium head blight resistance was 0.71. Tetraploid species are considered to be highly susceptible to disease. Resistance to head blight is the highest as observed in wheat cultivar Sumai3 (Buerstmayr et al., 2003). Linked genes for leaf rust resistance and stripe rust resistance are transferred from *Triticum dicoccoides* to *Triticum aestivum*. These genes are shown to provide resistance to bread wheat. The gene is found to be located on chromosome arm 6BS. The introgressed gene does not pair with Chinese spring wheat arm 6BS during meiosis which is probably due to lack of homology between the two chromosomes. But this 6BS appears to get paired with W84–17 and Avocet S. respective genes are designated as *Lr53* (for leaf rust resistance) and *Yr35* (Stripe rust resistance) (Marais et al., 2005).

Triticum dicoccoides can be used to transfer genes encoding resistance to powdery mildew, resistance to yellow (stripe) rust, high protein content and improved baking quality (Nevo, 2001). Wild emmer wheat accession

IW72 is found to be resistant to powdery mildew at seedling stage as well as adult stage. Its genetic analysis revealed that the resistance is due to a single dominant gene M11W72. Chromosome location and genetic mapping showed that the resistance gene might be a new allele at locus *Pm1* or a novel locus near *Pm1* (Ji et al., 2008).

16.2.12   TRITICUM URARTU

Triticum urartu is a close relative and A-genome donor of *Triticum aestivum* and *Triticum durum*. It has a diploid genome and its genome formula is AA. It is commonly known as wild einkorn wheat. It belongs to Triticeae family of plants. Omega gliadins which are encoded on chromosome 1 of A genome were purified from bread wheat and subjected to reverse phase high-performance liquid chromatography and sodium dodecyl sulfate polyacrylamide gel electrophoresis confirmed that omega gliadins extracted from A-genome are smaller as compared to B genome and D genome. The data also suggest that the donor of A-genome is *Triticum urartu* (DuPont et al., 2004). It can be used as genetic resource for wheat quality breeding.

Powdery mildew resistant gene from *Triticum urartu* accession UR206 is successfully transferred to hexaploid bread wheat by crossing and back-crossing. This gene is designated as *PmU* (Qiu et al., 2005). An amphiploid is produced from the cross between accession of *Triticum urartu* high molecular weight glutenin subunit and *Triticum durum* (durum wheat). The hybrid amphiploids produce high molecular weight glutenin subunits (Alvarez et al., 2009).

16.2.13   TRITICUM MONOCOCCUM

Triticum monococcum is one of the close relatives of wheat. It is diploid and belongs to Triticeae family. It is commonly known as einkorn wheat. Genome analysis of *T. monococcum* showed that about 70% of coting is composed of different classes of transposable elements. Different types of retrotransposons are also identified. There is evidence of major insertion, duplication, and deletion events in the genome, which suggest that genome is evolved by multiple mechanisms. Seven types of foldback transposons are also identified (foldback transposons were previously not expected to be present in Triticeae family) (Wicker et al., 2001).

Leaf rust resistance in *Triticum aestivum* can be improved by the use of *Triticum monococcum*. *T. monococcum* contains *Lr10* gene which is effective against leaf rust. This gene can be transferred to *T. aestivum* by cross-hybridization of bread wheat and einkorn wheat. The hybrids are then crossed and backcrossed by keeping bread wheat as recurrent parent (Stein et al., 2000).

16.3   CONCLUSION AND FUTURE PROSPECTS

Wild relatives of particular specied provide useful information about history, genome, life cycle, natural selection, characteristics, traits, breeding, and domestication of that particular species. Similarly, wheat wild relatives provide an evidence for the domestication of wheat and characterization of its genome. These wild relatives are used for the improvement of many traits in domesticated and cultivated variety of wheat. Most of the research has been performed to improve disease resistance traits, osmotic resistance, stress resistance, salt resistance, drought resistance, vigor, taste, and yield of domesticated wheat (Nevo and Chen, 2010). Major diseases which infect wheat include powdery mildew, head blight diseases, leaf rust, barley yellow dwarf and stem rust. Many of the genes from closed wheat relatives are characterized as resistant genes against these diseases and the insertion of these genes in cultivated variety of wheat has led to the production of disease resistance wheat varieties. Improvement of these characteristics has enabled the farmers and scientists to artificially select the wheat with advanced characters. Many genes have been inserted into modern cultivated wheat by its primitive relatives. Wild wheat relatives have been extensively used for studying the phylogenetic relationships of wheat with closely related species including *Aegilops* and *Hordeum spp.*

   In future research on wild relatives of wheat can be employed to increase the superior characteristics of cultivated wheat variety. Wheat varieties produced by hybridization, breeding, and gene insertion from wild relatives will be efficient enough for combating wheat diseases and stresses. Wheat forms the predominant basis of human nutrition and is a staple crop for many of the countries so it is mandatory to increase the yield, vigor, taste, resistance, and nutrition of this crop. Due to these reasons, research on wild wheat relatives and domesticated wheat improvement will bring a high impact on world's economy and global food security in near future (Table 16.1).

TABLE 16.1　General Information on Wild Relatives of Wheat

Name of relative	Ploidy Level	Family
Aegilops tauschii	Diploid	Poaceae
Aegilops umbellulata	Diploid	Poaceae
Aegilops biuncialis	Allo-tetraploid	Poaceae
Aegilops geniculata	Allo-tetraploid	Poaceae
Aegilops triuncialis	Tetraploid	Poaceae
Thinopyrum spp.	Pentaploid	Poaceae
Elytrigia spp.	Diploid, tetraploid, hexaploid, octaploid, decaploid	Poaceae
Lophopyrum elongatum	Diploid	Triticaea
Hordeum spp.	Diploid	Triticaea
Triticum dicoccoides	Allo-tetraploid	Triticaea
Triticum uratu	Diploid	Triticaea
Triticum monococcum	Diploid	Triticaea

KEYWORDS

- *Aegilops*
- *Hordeum*
- thinopyrum
- triticum aestivum
- *Triticum durum*

REFERENCES

Aghaee-Sarbarzeh, M., Ferrahi, M., Singh, S., Singh, H., Friebe, B., Gill, B. S., & Dhaliwal, H. S., (2002). Ph I-induced transfer of leaf and stripe rust-resistance genes from *Aegilops triuncialis* and *Ae. geniculata* to bread wheat. *Euphytica, 127* (3), 377–382.

Aghaee-Sarbarzeh, M., Singh, H., & Dhaliwal, H. S., (2001). A microsatellite marker linked to leaf rust resistance transferred from *Aegilops triuncialis* into hexaploid wheat. *Plant Breeding, 120* (3), 259–261.

Alvarez, J. B., Caballero, L., Nadal, S., Ramirez, M. C., & Martin, A., (2009). Development and gluten strength evaluation of introgression lines of *Triticum urartu* in durum wheat. *Cereal Res Comm. 37* (2), 243–248.

Arrigo, N., Felber, F., Parisod, C., Buerki, S., Alvarez, N., David, J., & Guadagnuolo, R., (2010). Origin and expansion of the allotetraploid *Aegilops geniculata*, a wild relative of wheat. *New Phytol., 187* (4), 1170–1180.

Assefa, S., & Fehrmann, H., (2000). Resistance to wheat leaf rust in *Aegilops tauschii* Coss. and inheritance of resistance in hexaploid wheat. *Genet Res Crop Evol., 47* (2), 135–140.

Biswas, D. K., Xu, H., Li, Y. G., Liu, M. Z., Chen, Y H., Sun, J. Z., & Jiang, G. M., (2008). Assessing the genetic relatedness of higher ozone sensitivity of modern wheat to its wild and cultivated progenitors/relatives. *J. Exp. Bot., 59* (4), 951–963.

Bossolini, E., Krattinger, S. G., & Keller, B., (2006) Development of simple sequence repeat markers specific for the *Lr34* resistance region of wheat using sequence information from rice and *Aegilops tauschii. Theor. App. Genetic., 113* (6), 1049–1062.

Buerstmayr, H., Stiersschneider, M., Steiner, B., Lemmens, M., Griesser, M., Nevo, E., & Fahima, T., (2003). Variation for resistance to head blight caused by Fusarium graminearum in wild emmer (*Triticum dicoccoides*) originating from Israel. *Euphytica, 130* (1), 17–23.

Cakmak, I., Torun, A., Millet, E., Feldman, M., Fahima, T., Korol, A., Nevo, E., Braun, H. J., & Ozkan, H., (2004). *Triticum dicoccoides*: An important genetic resource for increasing zinc and iron concentration in modern cultivated wheat. *Soil Sci. Plant Nutr., 50* (07), 1047–1054.

Chen, Q. Z., Qi, Z. J., Feng, Y. G., Wang, S. L., & Chen, P. D., (2002). [Structural changes of 4V chromosome of *Haynaldia villosa* induced by gametocidal chromosome 3C of Aegilops triuncialis]. *Yi Chuan Xue Bao., 29* (4), 355–358.

Chhuneja, P., Dhaliwal, H. S., Bains, N S., & Singh, K., (2006). *Aegilops kotschyi* and *Aegilops tauschii* as sources for higher levels of grain Iron and Zinc. *Plant Breed., 125* (5), 529–531.

Chhuneja, P., Kaur, S., Goel, R. K., Aghaee-Sarbarzeh, M., Prashar, M., & Dhaliwal, H. S., (2008). Transfer of leaf rust and stripe rust resistance from *Aegilops umbellulata* Zhuk. to bread wheat (*Triticum aestivum* L.). *Genet Res Crop Evol., 55* (6), 849–859.

Cifuentes, M., & Benavente, E., (2009). Wheat-alien metaphase I pairing of individual wheat genomes and D genome chromosomes in interspecific hybrids between *Triticum aestivum* L. and *Aegilops geniculata* Roth. *Theor App Genetic., 119* (5), 805–813.

Colmer, T. D., Flowers, T. J., & Munns, R., (2006). Use of wild relatives to improve salt tolerance in wheat. *J. Exp. Bot., 57* (5), 1059–1078.

Drenovsky, R. E., & Batten, K. M., (2007). Invasion by *Aegilops triuncialis* (Barb Goatgrass) Slows Carbon and Nutrient Cycling in a Serpentine Grassland. *Biol Invas., 9* (2), 107–116.

Dubcovsky, J., & Dvorak, J., (2007). Genome Plasticity a key factor in the success of polyploid wheat under domestication. *Science, 316* (5833), 1862–1866.

DuPont, F. M., Vensel, W., Encarnacao, T., Chan, R., & Kasarda, D. D., (2004). Similarities of omega gliadins from *Triticum urartu* to those encoded on chromosome 1A of hexaploid wheat and evidence for their post-translational processing. *Theor. App. Genetic., 108* (7), 1299–1308.

Farooq, S., & Azam, F., (2001). Production of low input and stress tolerant wheat germplasm through the use of biodiversity residing in the wild relatives. *Hereditas, 135* (2–3), 211–215.

Geng, Z., Hong-Wen, G., Zan, W., Xi, Y., Gui-Zhi, S., & Ning, G., (2007). Studies on screening identification indexes of salt tolerance and comprehensive evaluation at seedling stage of Elytrigia. *Acta Paracult Sinica.*

Giles, R. J., & Brown, T. A., (2006). GluDy allele variations in *Aegilops tauschii* and *Triticum aestivum*: implications for the origins of hexaploid wheats. *Theor. App. Genetic., 112* (8), 1563–1572.

Guadagnuolo, R., Savova-Bianchi, D., Keller-Senften, J., & Felber, F., (2001). Search for evidence of introgression of wheat (*Triticum aestivum* L.) traits into sea barley (*Hordeum marinum* s. str. Huds.) and bearded wheatgrass (*Elymus caninus* L.) in central and northern Europe, using isozymes, RAPD, and microsatellite markers. *Theor. App. Genetic., 103* (2–3), 191–196.

Gupta, S. K., Charpe, A., Koul, S., Prabhu, K. V., & Haq, Q. M. R., (2005). Development and validation of molecular markers linked to an *Aegilops umbellulata*–derived leaf-rust-resistance gene, *Lr9*, for the marker-assisted selection in bread wheat. *Genome., 48* (5), 823–830.

Guyomarch, H., Sourdille, P., Charmet, G., & Bernard, E. M., (2002). Characterization of polymorphic microsatellite markers from *Aegilops tauschii* and transferability to the D-genome of bread wheat. *Theor. App. Genetic., 104* (6–7), 1164–1172.

Hernandez, P., Dorado, G., Prieto, P., Gimenez, M. J., Ramirez, M. C., Laurie, D. A., Snape, J W., & Martin, A., (2001). A core genetic map of *Hordeum chilense* and comparisons with maps of barley (*Hordeum vulgare*) and wheat (*Triticum aestivum*). *Theor. App. Genetic., 102* (8), 1259–1264.

Hongfei, Y., WenXiang, Y., YunFang, C., QingFang, M., & DaQun, L., (2009). Specificity and stability of E-chromosome specific SCAR marker from *Thinopyrum* spp. for *Lr19*. *Acta Phytopathalogica Sinica., 39* (1), 76–89.

Hong-Gang, W., Shu-Bing, L., Zeng-Jun, Q. I., Fan-Jing, K., & Ju-Rong, G., (2000). Application studies of *Elytrigia intermedium* in hereditary improvement of wheat. *J. Shandong. Agricult. Univ. 31*, 333–336.

Hsam, S. L. K., Kieffer, R., & Zeller, F. J., (2001). Significance of *Aegilops Tauschii* Glutenin Genes on Breadmaking Properties of Wheat, 78 (5), 5210525.

Hua, W., Liu, Z., Zhu. J., Xie, C., Yang, T., Zhou, Y., Duan, X., Sun, Q., & Liu, Z., (2009). Identification and genetic mapping of *Pm42*, a new recessive wheat powdery mildew resistance gene derived from wild emmer (*Triticum turgidum* var. dicoccoides). *Theor. App. Genetic., 119* (2), 223–230.

Huang, L., & Gill, B. S., (2001). An RGA – like marker detects all known *Lr21* leaf rust resistance gene family members in *Aegilops tauschii* and wheat. *Theor. App. Genetic., 103* (6–7), 1007–1013.

Iriki, N., Kawakami, A., Takata, K., Kuwabara, T., & Ban, T., (2001). Screening relatives of wheat for snow mold resistance and freezing tolerance. *Euphytica., 122* (2), 335–341.

Islam, S., Malik, A. I., Islam, A. K. M. R., & Colmer, T. D., (2007). Salt tolerance in a *Hordeum marinum–Triticum aestivum* amphiploid, and its parents. *J. Exp. Bot., 58* (5), 1219–1229.

Jarve, K., Peusha, H. O., Tsymbalova, J., Tamm, S., Devos, K. M., & Enno, T. M., (2000). Chromosomal location of a *Triticum timopheevii* – derived powdery mildew resistance gene transferred to common wheat. *Genome, 43* (2), 377–381.

Ji, X., Xie, C., Ni, Z., Yang, T., Nevo, E., Fahima, T., Liu, Z., & Sun, Q., (2008). Identification and genetic mapping of a powdery mildew resistance gene in wild emmer (*Triticum dicoccoides*) accession IW72 from Israel. *Euphytica., 159* (3), 385–390.

Johal, J., Gianibelli, M. C., Rahman, S., Morell, M. K., & Gale, K. R., (2004). Characterization of low-molecular-weight glutenin genes in *Aegilops tauschii*. *Theor. App. Genetic., 109* (5), 1028–1040.

Kantar, M., Lucas, S. J., & Budak, H., (2011). miRNA expression patterns of *Triticum dicoccoides* in response to shock drought stress. *Planta., 233* (3), 471–484.

Kroupin, P. Y., Divashuk, M G., Fesenko, I. A., & Karlov, G. I., (2013). Evaluating wheat microsatellite markers for the use in genetic analysis of Thinopyrum, Dasypyrum, and Pseudoroegneria Species. *Dataset Papers in Biol.* 3 (Article ID 949637), https://doi.org/10.7167/2013/949637.

Kuraparthy, V., Chhuneja, P., Dhaliwal, H. S., Kaur, S., & Gill, B. S., (2006). Characterization and Mapping of *Aegilops geniculata* introgressions with novel leaf rust and stripe rust resistance genes *Lr57* and *Yr40* in wheat. *Am. Soc. Agronom., 246* (3).

Kuraparthy, V., Chhuneja, P., Dhaliwal, H. S., Kaur, S., Bowden, R. L., & Gill, B. S., (2007). Characterization and mapping of cryptic alien introgression from *Aegilops geniculata* with new leaf rust and stripe rust resistance genes *Lr57* and *Yr40* in wheat. *Theor. App. Genetic., 114* (8), 1379–1389.

Landjeva, S., Angelov, G., Nenova, V., Merakchijska, M., & Ganeva, G., (2004). Seedling growth and peroxidase responses to excess copper in wheat-*Aegilops geniculata* chromosome addition and substitution lines. *Genet Breed., 33* (3–4), 19–26.

Lelley, T., Stachel, M., Grausgruber, H., & Vollmann, J., (2000). Analysis of relationships between *Aegilops tauschii* and the D genome of wheat utilizing microsatellites. *Genome., 43* (4), 661–668.

Li, H., & Wang, X., (2009). *Thinopyrum ponticum* and *Th. intermedium*: the promising source of resistance to fungal and viral diseases of wheat. *J. Genet. Genom., 36* (9), 557–565.

Liu, W., Rouse, M., Friebe, B., Jin, Y., Gill, B., & Pumphrey, M. O., (2011). Discovery and molecular mapping of a new gene conferring resistance to stem rust, *Sr53*, derived from *Aegilops geniculata* and characterization of spontaneous translocation stocks with reduced alien chromatin. *Chromosome Res., 19* (5), 669–682.

Liu, Z., Yan, Z., Wan, Y., Liu, K., Zheng, Y., & Wang, D., (2003). Analysis of HMW glutenin subunits and their coding sequences in two diploid Aegilops species. *Theor. App. Genetic., 106* (8), 1368–1378.

Loureiro, I., Escorial, E., Baudin, J. M. C., & Chueca, M. C., (2008). Hybridization between wheat and *Aegilops geniculata* and hybrid fertility for potential herbicide resistance transfer. *Weed Res., 48* (6), 561–570.

Luo, P. G., Luo, H. Y., Chang, Z. J., Zhang, H. Y., Zhang, M., & Ren, Z. L., (2009) Characterization and chromosomal location of *Pm40* in common wheat: a new gene for resistance to powdery mildew derived from *Elytrigia intermedium*. *Theor. App. Genetic., 118* (6), 1059–1064.

Ma, J., Zhou, R., Dong, Y., & Jia, J., (2000). Control and inheritance of resistance to yellow rust in *Triticum aestivum-Lophopyrum elongatum* chromosome substitution lines. *Euphytica., 111* (1), 57–60.

Malik, A. I., Islam, A. K. M. R., & Colmer, T. D., (2011). Transfer of the barrier to radial oxygen loss in roots of *Hordeum marinum* to wheat (*Triticum aestivum*): evaluation of four *H. marinum*–wheat amphiploids. *New Phytol., 190* (2), 499–508.

Martin-Sanchez, J. A., Gomez-Colmenarejo, M., Moral, J. D., Sin, E., Montes, M. J., Gonzalez-Belinchon, C., et al. (2003). A new Hessian fly resistance gene (*H30*) transferred from the wild grass *Aegilops triuncialis* to hexaploid wheat. *Theor. App. Genetic., 106* (7), 1248–1255.

Matsuoka, Y., & Nasuda, S., (2004). Durum wheat as a candidate for the unknown female progenitor of bread wheat: an empirical study with a highly fertile F_1 hybrid with *Aegilops tauschii* Coss. *Theor App Genetic., 109* (8), 1710–1717.

McDonald, M. P., Galwey, N. W., Ellneskog-Staam, P., & Colmer, T. D., (2001). Evaluation of *Lophopyrum elongatum* as a source of genetic diversity to increase the waterlogging tolerance of hexaploid wheat (*Triticum aestivum*). *New Phytol., 151* (2), 369–380.

Miranda, L M., Murphy, J. P., Marshall, D., Cowger, C., & Leath, S., (2007). Chromosomal location of *Pm35*, a novel Aegilops tauschii derived powdery mildew resistance gene introgressed into common wheat (*Triticum aestivum* L.). *Theor App Genetic., 114* (8), 1451–1456.

Miranda, L. M., Murphy, J P., Marshall, D., & Leath, S., (2006). Pm34, a new powdery mildew resistance gene transferred from *Aegilops tauschii* Coss. to common wheat (*Triticum aestivum* L.). *Theor App Genetic., 113* (8), 1497–1504.

Molnar, I., Gaspar, L., Sarvari, E., Dulai, S., Hoffmann, B., Molnar-Lang, M., & Galiba, G., (2004). Physiological and morphological responses to water stress in *Aegilops biuncialis* and *Triticum aestivum* genotypes with differing tolerance to drought. *Func Plant Biol., 31* (12), 1149–1159.

Molnar, I., Gaspar, L., Stehli, L., Dulai, S., Sarvari, E., Kiraly, I., Galiba, G., & Molnar-Lang, M., (2002). The effects of drought stress on the photosynthetic processes of wheat and *Aegilops biuncialis* genotyptes originating from various habitats. *Acta Biologica Szegendiensis., 46* (3–4), 115–116.

Molnar, I., Kubalakova, M., Simkova, H., Cseh, A., Molnar-Lang, M., & Dolezel, J., (2001). Chromosome Isolation by Flow Sorting in *Aegilops umbellulata* and *Ae. comosa* and their allotetraploid hybrids *Ae. biuncialis* and *Ae. geniculata*. *PLoS ONE., 6* (11), e27708.

Monneveux, P., Reynolds, M. P., Gonzalez Aguilar, J., Singh, R. P., & Weber, W. E., (2003). Effects of the 7DL. 7Ag translocation from *Lophopyrum elongatum* on wheat yield and related morphophysiological traits under different environments. *Plant Breeding, 122* (5), 379–384.

Montes, M. J., Lopez-Brana, I., & Delibes, A., (2004). Root enzyme activities associated with resistance to Heterodera avenae conferred by gene *Cre7* in a wheat/*Aegilops triuncialis* introgression line. *J. Plant Physiol., 161* (4), 493–495.

Mullan, D. J., Platteter, A., Teakle, N. L., Appels, R., Colmer, T. D., Anderson, J. M., & Francki, M. G., (2005). EST-derived SSR markers from defined regions of the wheat genome to identify *Lophopyrum elongatum* specific loci. *Genome., 48* (5), 811–822.

Nevo, E., & Chen, G., (2010). Drought and salt tolerances in wild relatives for wheat and barley improvement. *Plant Cell Environ., 33* (4), 670–685.

Nevo, E., (2001). Genetic resources of wild emmer, emmer, *Triticum dicoccoides* for wheat improvement in the third millenium. *Isr J Plant Sci., 49,* 77–92.

Okubara, P. A., & Jones, S. S., (2011). Seedling resistance to Rhizoctonia and Pythium spp. in wheat chromosome group 4 addition lines from *Thinopyrum* spp. *Can. J. Plant Pathol., 33* (3), 416–423.

Peng, J. H., Fahima, T., Roder, M. S., Huang, Q. Y., Dahan, A., Li, Y C., Grama, A., & Nevo, E., (2000b). High-density molecular map of chromosome region harboring stripe-rust resistance genes *YrH52* and *Yr15* derived from wild emmer wheat, *Triticum dicoccoides*. *Genetica., 109* (3), 199–210.

Peng, J., Korol, A. B., Fahima, T., Roder, M. S., Ronin, Y. I., Li, Y. C., & Nevo, E., (2000a). Molecular Genetic Maps in Wild Emmer Wheat, *Triticum dicoccoides*: Genome-Wide Coverage, Massive Negative Interference, and Putative Quasi-Linkage. *Genome Res., 10,* 1509–1531.

Peng, J., Ronin, Y., Fahima, T., Roder, M. S., Li, Y., Nevo, E., & Korol, A., (2003). Domestication quantitative trait loci in *Triticum dicoccoides*, the progenitor of wheat. *PNAS, 100* (5), 2489–2494.

Pestsova, E. G., Borner, A., & Roder, M. S., (2006). Development and QTL assessment of *Triticum aestivum–Aegilops tauschii* introgression lines. *Theor. App. Genetic., 112* (4), 634–647.

Pestsova, E., Korzun, V., Goncharov, N. P., Hammer, K., Ganal, M. W., & Roder, M. S., (2000). Microsatellite analysis of *Aegilops tauschii* germplasm. *Theor. App. Genetic., 101* (1–2), 100–106.

Peterson, G., Seberg, O., Yde, M., & Berthelsen, K., (2006). Phylogenetic relationships of Triticum and Aegilops and evidence for the origin of the A, B, and D genomes of wheat (*Triticum aestivum*). *Mol. Phylogen. Evol., 39*, 70–82.

Ping-Li, L., & Hong-Gang, W., (2002). Studies on the cultivation and morphological and cytological identification of wheat-*Elytrigia intermedium* alien disomic addition lines. *Acta Botanica Boreali-occidentalia Sinica.*

Qiu, Y. C., Zhou, R. H., Kong, X. Y., Zhang, S. S., & Jia, J. Z., (2005). Microsatellite mapping of a *Triticum urartu* Tum. derived powdery mildew resistance gene transferred to common wheat (*Triticum aestivum* L.). *Theor. App. Genetic., 111* (8), 1524–1531.

Rawat, N., Tiwari, V. K., Singh, N., Randhawa, G. S., Singh, K., Chhuneja, P., & Dhaliwal, H. S., (2009). Evaluation and utilization of Aegilops and wild Triticum species for enhancing iron and zinc content in wheat. *Genet. Res. Crop. Evol., 56* (01), 53–64.

Refoufi, A., Jahier, J., & Esnault, M. A., (2001). Genome analysis of *Elytrigia pycnantha* and *Thinopyrum junceiforme* and of their putative natural hybrid using the GISH technique. *Genome., 44* (4), 708–715.

Rouse, M. N., & Jin, Y., (2011). Stem rust resistance in A-genome diploid relatives of Wheat. *Am Phytopathol Soc., 95* (08), 941–944.

Santa-Maria, G E., & Epstein, E., (2001). Potassium/sodium selectivity in wheat and the amphiploid cross wheat x *Lophopyrum elongatum. Plant Sci., 160* (3), 523–534.

Shen, X., & Ohm, H., (2006). Fusarium head blight resistance derived from *Lophopyrum elongatum* chromosome 7E and its augmentation with *Fhb1* in wheat. *Plant Breeding, 125* (5), 424–429.

Shen, X., Kong, L., & Ohm, H., (2004). Fusarium head blight resistance in hexaploid wheat (*Triticum aestivum*)-Lophopyrum genetic lines and tagging of the alien chromatin by PCR markers. *Theor. App. Genetic., 108* (5), 808–813.

Stein, N., Feuillet, C., Wicker, T., Schlagenhauf, E., & Keller, B., (2000). Subgenome chromosome walking in wheat: A 450-kb physical contig in *Triticum monococcum* L. spans the *Lr10* resistance locus in hexaploid wheat (*Triticum aestivum* L.). *PNAS, 97* (24), 13436–13441.

Tan, F., Zhou, J., Yang, Z., Zhang, Y., Pan, L., & Ren, Z., (2009). Characterization of a new synthetic wheat – *Aegilops biuncialis* partial amphiploid. *Af J Biotech., 8* (14).

Tanno, K., & Willcox, G., (2006). How fast was wild wheat domesticated? *Science, 311* (5769), 1886.

Vanichanon, A., Blake, N., Sherman, J., & Talbert, L., (2003). Multiple origins of allopolyploid *Aegilops triuncialis. Theor. App. Genetic., 106* (5), 804–810.

Wang, T., Xu, S. S., Harris, M. O., Hu, J., Liu, L., & Cai, X., (2006). Genetic characterization and molecular mapping of Hessian fly resistance genes derived from *Aegilops tauschii* in synthetic wheat. *Theor. App. Genetic., 113* (4), 611–618.

Wang, Z. Y., Bell, J., & Hopkins, A., (2003). Establishment of a plant regeneration system for wheatgrasses (*Thinopyrum, Agropyron,* and *Pascopyrum*). *Plant Cell, Tissue Organ Cult., 73* (3), 265–273.

Wartburton, M. L., Crossa, J., Franco, J., Kazi, M., Trethowan, R., Rajaram, S., et al. (2006). Bringing wild relatives back into the family: Recovering genetic diversity in CIMMYT improved wheat germplasm. *Euphytica., 149* (03), 289–301.

Wei-Dong, L., Peng-Bin, X. U., & Xun, P. U., (2007). Summary of the situation for applying genetic resources from Elytrigia in *Triticum aestivum* breeding. *Acta Paracult Sinica.*

Weng, Y., Li, W., Devkota, R N., & Rudd, J. C., (2005). Microsatellite markers associated with two *Aegilops tauschii*-derived greenbug resistance loci in wheat. *Theor. App. Genetic., 110* (3), 462–469.

Wiangjun, H., & Anderson, J. M., (2004). The basis for *Thinopyrum*-derived resistance to cereal yellow dwarf virus. *Genet. Resist., 94* (10), 1102–1106.

Wicker, T., Stein, N., Albar, L., Feuillet, C., Schlagenhauf, E., & Keller, B., (2001). Analysis of a contiguous 211 kb sequence in diploid wheat (*Triticum monococcum* L.) reveals multiple mechanisms of genome evolution. *Plant J., 26* (3), 307–316.

Xie, C., Sun, Q., Ni, Z., Yang, T., Nevo, E., & Fahima, T., (2003). Chromosomal location of a *Triticum dicoccoides*-derived powdery mildew resistance gene in common wheat by using microsatellite markers. *Theor. App. Genetic., 106* (2), 341–345.

Yan, Y., Hsam, S. L. K., Yu, J., Jiang, Y., & Zeller, F. J., (2003). Allelic variation of the HMW glutenin subunits in *Aegilops tauschii* accessions detected by sodium dodecyl sulfate (SDS-PAGE), acid polyacrylamide gel (A-PAGE) and capillary electrophoresis. *Euphytica., 130* (3), 377–385.

Yasui, Y., Nasuda, S., Matsuoka, Y., & Kawahara, T., (2001). The Au family, a novel short interspersed element (SINE) from *Aegilops umbellulata. Theor. App. Genetic., 102* (4), 463–470.

Zaharieva, M., Gaulin, E., Havaux, M., Acevedo, E., & Monneveux, P., (2001a). Drought and heat responses in the wild wheat relative *Aegilops geniculata* Roth. *Am. Soc. Agronom., 41* (04), 1321–1329.

Zaharieva, M., Monneveux, P., Henry, M., Rivoal, J., Valkoun, J., & Nachit, M. M., (2001b). Evaluation of a collection of wild wheat relative *Aegilops geniculata* Roth and identification of potential sources for useful traits. *Euphytica., 119* (1–2), 33–38.

Zhang, Z., Lin, Z., & Xin, Z., (2009). Research progress in BYDV resistance genes derived from wheat and its wild relatives. *J. Genet. Genom., 36* (09), 567–573.

Zhu, Z., Zhou, R., Kong, X., Dong, Y., & Jia, J., (2006). Microsatellite marker identification of a *Triticum aestivum-Aegilops umbellulata* Substitution Line with powdery mildew Resistance. *Euphytica. 150* (1–2), 149–153.

Zu-Jun, Y., & Zheng-Long, R., (2001). Chromosomal distribution and genetic expression of *Lophopyrum elongatum* (Host) A. Löve genes for adult plant resistance to stripe rust in wheat background. *Genet Res Crop Evol., 48* (2), 183–187.

CHAPTER 17

STRUCTURAL AND FUNCTIONAL GENOMICS OF POTATO (*SOLANUM TUBEROSUM*)

AQSA ABBASI, AYESHA SAJID, and ALVINA GUL

Atta-ur-Rahman School of Applied Biosciences, National University of Sciences and Technology, Islamabad, Pakistan

ABSTRACT

Potato (*Solanum tuberosum*), an important food crop is heterozygous, tetraploid, having inbreeding depressions and twelve chromosomes in its genome. The genome encodes many different proteins for tuber quality, tuber development, and disease and pest resistance. The genome sequencing of this crop helps in understanding different structural and functional features of potato. Recent development in genomic resources such as virus-induced gene silencing, expressed sequence tags, genome editing tools and genomic sequences libraries make it easy to study structural and functional genomics of potato. The knowledge obtained from structural genomics laid the foundation for functional genomics of potato by identifying different genes, proteins, and their function.

17.1 INTRODUCTION

All the information necessary for the species' morphology and characteristics resides in its genome in the form of DNA. Winkler in 1920 was the first person who used this term "genome" which was modified later. It is usually seen that genome of eukaryotic organisms are much more complex than prokaryotes in their functioning, organization, and structure but genomes of plants are more complex in structure than eukaryotes. Plants have nuclear genomes as well as extranuclear genomes that include chloroplasts and mitochondrial genomes. In plant species chloroplasts genome is usually conserved but there is a huge diversity among the sizes of plant genomes.

In 1986 term 'genomics' was used which was derived from 'genome.' It is the detailed study of genome including sequences of RNA, DNA, and protein using advanced methodologies and then genetic maps are generated which help in the determination of structural and functional features. This genomics has two main elements 'structural genomics' that involves structural elements, genomic map, gene sequencing, cloned fragments of DNA, etc. and the other is 'functional genomics' that involves functions related to different genes (Varma & Shrivastava, 2011).

Structural Genomics deals with the identification of three-dimensional structures of genes and proteins. By identifying sequences and structure of proteins, molecular function can also be determined effectively (Zhang & Kim, 2003). After structural genomic study, focus on the functional genomics begins. Functional genomics helps in exploring genes, RNA transcripts, proteins, and specifically interactions between different proteins can be determined through it. Both structural and functional genomics follow some common approaches in identifying function of genes and proteins as shown in the Figure 17.1.

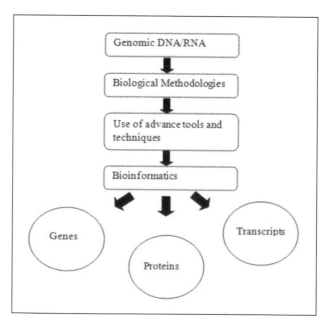

FIGURE 17.1 Pathways leading from genome sequencing to the identification of structure and function in genes and proteins.

17.2 POTATO STRUCTURAL AND FUNCTIONAL GENOMICS

Solanum tuberosum commonly known as potato belongs to the family Solanaceae which also includes pepper, tomato (*Solanum lycopersicum*), tobacco (*Nicotiana tabacum*) and petunia, etc. It is further divided into two types on the basis of their occurrences: *tuberosum* and *andigena* but *tuberosum* sub-species is studied in detail as compared to other one (Hannemann, 1994; Machida-Hirano, 2015). Potato is unique as it spreads to vast eco geographical region and it produces stolons or underground stems. When environmental conditions are favorable, these stems grow and develop tubers which are an important source of vitamins, starch, and proteins to consumers. It has two forms one is diploid and other is autopolyploid form. It can be vegetatively propagated and is heterozygous which means having multiple alleles at one or more loci and covers significant aspects of plant biology (Baker et al., 2000).

Potato is at fourth position among other food crops and it is both economically and agriculturally important as indicated by FAO statistics of 2014 which shows that 380 million tonnes of potatoes were produced in that particular year (FAO, 2016). Its genome is mapped at the diploid stage with the help of heterozygous and homozygous clones however researchers have also tried to work at tetraploid level. This helps in the development of molecular markers which help to study and compare different maps of potato genome with itself as well as with other species of Solanaceae family. It also leads to the identification of many other aspects of potato plant such as pathogen resistance genes and genes conferring important characteristics to potato. Genetic functioning of potato can be easily studied by different technologies such as by over expressing the genes and through antisense technology (Bryan & Hein, 2008). This review article provides information on structural and functional genomics of *Solanum tuberosum*.

17.3 STRUCTURAL FEATURES OF POTATO DETERMINED THROUGH GENOME SEQUENCING

The genetic study of potato is not well described because of its polyploidy nature. Regardless of this, global Potato Genome Sequencing Consortium (PGSC) has sequenced its genome in 2011 using a whole-genome shotgun method. PGSC sequenced homozygous doubled-monoploid potato clone and heterozygous diploid clone and determine genetic characteristics of these clones. This effort leads to significant improvement in studying the structural and functional characteristics of this vegetable (Sharma, 2010). Genetic map of potato was developed within the past decade and this could not be possible without the following two developments in this field:

- Reduction in the ploidy level, i.e., from tetraploid (4n) to diploid (2n).
- Development of DNA based markers (Gebhardt, 2007).

Genome sequencing shows that majority of the potato cultivars are auto-tetraploid, i.e., 2n=4x=48), have inbreeding depressions and genome size is 844 Mb (PGSC, 2011). It is considered that its genome size is double than the rice genomic size and six times as compared to Arabidopsis plant. Its basic chromosome number is twelve. In order to study those factors that are involved in the pests' resistance and tuber characteristics, restriction fragment length polymorphism or RFLP linkage maps have been developed comprising whole genome (Ballvora et al., 2007). There are very less number of repetitive sequences in the potato genome, i.e., 4 to 7%. There are more than 39,000 protein-coding genes encoded by its genome (Mclvor, 2011) and location of majority of genes (approximately 95%) is known. Apparently genomic differences between other potato varieties are contributed by different proteins. Within the asterid specific gene set, there are many genes for transcription factors, defense related proteins against many pathogens and they also impart uniqueness to asterids. Its genome also has Kunitz protease inhibitor genes (KTIs) which are double in number compared to tomato genome. They get activated when any foreign pathogen attacks and inhibit proteinases. It has also been noticed that genes for resistance against diseases encode nucleotide-binding site (NBS) and leucine-rich-repeat (LRR) domains. Proteins consisting of these domains are responsible for resistance against pathogen infection (Bendahmane et al., 2002). Most of the proteins have significant role in disease and defense responses against pathogens, carbohydrate metabolism and proteins storage (Li et al., 2005).

17.3.1 METHODOLOGIES FOR STRUCTURAL GENOMIC STUDY OF POTATO

17.3.1.1 EXPRESSED SEQUENCE TAGS (EST)

These are incomplete sequences of DNA which are constructed from cDNA clones from either of its ends. This type of sequencing has many advantages as compared to sequencing of whole genome. It is usually considered as a first step in the discovery and characterization of genes. Large number of ESTs of potato is available and is ranked at the second position among Solanaceae family. With the help of ESTs, genetic markers in potato and its candidate genes have been identified. Researchers have developed the EST sequences of a potato tuber complementary DNA library which helps in determining the expression of genes positioned at the tuber region (Cormack, 2004). It was found that genes important for protein formation and defensive purpose are present in this region of potato genome. Another study shows that large number of ESTs approximately 62000 was constructed from nine different cDNA libraries, which provides information on developmental stages of tuber, potato leaves, roots, and pathogen susceptible leaves (Flinn et al., 2005). Furthermore, microarrays have been developed using EST sequence data from EST libraries to analyze the gene expression profile of potato under biotic and abiotic stress (Tai et al., 2013; Bengtsson et al., 2014; Hancock et al., 2014).

17.3.1.2   BACTERIAL ARTIFICIAL CHROMOSOME (BAC) LIBRARIES

These are the large libraries of DNA which are helpful in sequencing the genome of any organism (Farrar & Donnison, 2007). In studying the genome of potato, BAC libraries are used for physical mapping and for gene cloning. There are large numbers of libraries developed from *Solanum tuberosum* which are the source of genome organization. For identification, specific DNA markers can also be developed with the help of BAC clones and fluorescence in situ hybridization (FISH) technique. By using such BAC libraries, it was observed that heterochromatin is densely located in the centromeric region and it has many rDNA-Related tandem repeats and is known as pericentromeric heterochromatin (Stupar et al., 2002).

17.3.1.3   IDENTIFICATION OF TANDEM REPEATS

Few studies show that out of the twelve, five centromeres of potato are having single or low copy number sequences of DNA and are without any repeats. So they have resemblance with the neo or newly formed centromeres which are devoid of tandem repeats. In contrast, six centromeres of potato contain large sized tandem or satellite repeats which are different from each other and one of the centromere have both single or low copy number sequences of DNA and satellite repeats (Mach, 2012). There is also a continuous evolutionary process in these repeats in all the members of Solanaceae family. These repeats have very significant role in the organization and stabilization of such chromosomal characteristics which are important for dividing cell (Presting et al., 2013).

17.3.1.4   ISOLATION OF RESISTANCE GENES FROM MOLECULAR MAPPING OF POTATO

Plants evoke complex immune responses against variety of pathogens, such as bacteria, fungi, virus, insects, and nematodes for their survival (Piquerez et al., 2014). This may lead to compromised growth and production which can be avoided by enhancing resistance mechanisms in plants (Atkinson & Urwin, 2012). It can be done efficiently with better understanding of molecular mechanisms involved in immunity against pathogens (Nicaise, 2014; Das & Rao, 2015). Mapping of the potato genome helps to study its different characteristics such as quality, disease, and pathogen resistance genes, etc. The genes of *S. tuberosum* which provide resistance against many pathogens such as *Phytophthora infestans,* potato cyst nematodes and potato virus X strains, are located at specific loci known as resistance gene cluster. Resistance (R) genes have very important role in providing qualitative and quantitative resistance (Mazourek, 2009). On chromosome 5 of potato genome there are five R genes loci which are described below:

i.   *R1* locus presents resistance to *Phytophthora infestans.*
ii. *Resistance against potato virus X is due to Rx2 and Nb loci.*
iii. *Gpa and Grp1* gives resistance to the nematodes residing in the roots of potato.

On chromosome 11, R genes are located which provide resistance to potato virus Y, potato virus A and potato root nematode. R3a and R3b genes on R1 locus offers resistance against *Phytophthora infestans* (Kanyuka, 2000) and hypersensitivity response genes namely *Ny–1* and *Ny-2* confer resistance to potato virus Y (Szajko et al., 2008, 2014). Another locus known as Gpa2 contains four homologous resistance genes which occupy 115 kilobases of genome. Out of these four, two are active genes. One is at distal end of centromere while other is proximal to centromere. Former resembles *Rx1 gene* which determine resistance to potato virus X while other correlates with *Gpa2* gene conferring resistance to *Globodera pallida*, a potato cyst nematode. On chromosome 5 there is Rx2 locus encoding a protein having strong homology with Gpa2 protein. The three proteins Gpa2, R11 and Rx2 have evolutionary relationship to each other because they have highly conserved sequences (Van der Vossen et al., 2000).

17.3.1.5   IDENTIFICATION OF CANDIDATE GENES THROUGH CANDIDATE GENE APPROACH

Genes conferring special traits to *S. tuberosum* can be isolated through candidate gene approach. It is based on the principle that if a structural related gene is present in a close vicinity of a known trait locus or its function is same to that of the target trait then this gene is considered as a candidate gene for that trait (Bryan, 2007). With the help of this method several genes from the potato genome have been isolated and studied which are described below:

*   Gro1 is present as a single allele dominant in nature and encodes resistance against two pathotypes of *Globodera rostochiensis* namely Ro1 and Ro5. Until now, 14 such R genes have been mapped on potato genome at 8 linkage groups. Out of these, four provide complete resistance to nematodes while other confers partial resistance (Finkers-Tomczak et al., 2011). Both the genes H1 on chromosome V and Gro1 positioned on the chromosome VII provides protection against potato root nematode *Globodera rostochiensis*. H1 gene specifically targets Ro1 and Ro4 pathotypes of *G. rostochiensis* (Galek et al., 2011).
*   Color of the potato is determined by many loci such as R, P, and I in diploid potatoes (in tetraploid potatoes it is also known as D locus) which are studied and mapped in detail. Potato has been classified according to color, appearance of tuber, pigmentation, and taste, etc. The R locus in any tissues of the potato encodes the formation of red anthocyanin pigments. Red color imparted by the R locus also requires some other regulatory genes which are tissue specific such as D or developer gene is required for the production of anthocyanin in the skin of tuber and F codes for the expression of this pigment in the flowers. While P locus codes for the purple pigments and it stimulates the production of anthocyanin biosynthetic enzyme flavonoid 3,5-hydroxylase (f 3,5 h). P locus also needs regulatory genes for its function (Cheng et al., 2005). Researchers have used candidate gene approach to get information about the genes residing in those loci. According to them genes in these loci are involved in the biosynthesis

of anthocyanin pigment. When compared with the tomato mapped genome, researchers came to know that on chromosome 2, R locus is located on which dihydroflavonol 4-reductase (*dfr*) gene is present. On P locus, flavanoid 3,'5'-hydroxylase (*f3'5'h*) gene is linked on chromosome 11 and on chromosome 10, *an2* gene which is a regulator of transcription of this pathway has association with I/D locus (Flinn et al., 2005).

- Another study was carried out in which variations at the DNA sequence level have been analyzed and molecular basis for the quantitative characters have been studied at quantitative trait loci (QTL). For the good quality of potatoes, carbohydrates such as starch and sugar amount is important and therefore it is considered as a quantitative trait. Low-quality potato relates with the high quantity of glucose and fructose which are reducing sugars. So these features were used as a model to identify more about QTL by candidate gene approach. The gene that was considered as a candidate is invertase genes, i.e., *InvGE* and *InvGF* located at invGE/GF locus having size of 806 kilobases (Draffehn, 2012). These genes are located on the chromosome IX which is adjacent to cold sweetening or cold stress quantitative trait loci and these were identified by studying tetraploid potato cultivars. It was observed that for good quality of potato chips, two alleles of invertase, *invGE-f*, and *invGF-d* play important role which are closely related to each other and another allele known as invGF-*b* is associated with low content of starch in the tuber. So this study showed that different functional variants of invertase gene control the quantity of sugar in potato tuber (Brigneti, 2004). If the expression of invertase gene is silenced using gene knock out approach, *the* cold induced-sweetening is inhibited in potato resulting in an improved quality as shown in another study (Clasen et al., 2016).

- Recently, it has been demonstrated that on chromosome 5 another candidate gene known as StCDF1 which belongs to the CYCLING DOF (transcription factor) family is present. The function of this gene is to induce maturation in the plant and thus affecting the maturity phenotype of potato (Kloosterman et al., 2013). These are some of the examples that explain the application of candidate gene approach in identifying different genes of potato. With the help of this approach, more resistance genes and traits of potato tuber can be studied and mapped for future understanding of potato uses in biotechnological field.

17.3.1.6 DIVERSITY ARRAYS TECHNOLOGY (DART)

It is the microarray based hybridization technology which develops sequence makers to identify the presence or absence of different gene fragments in the genome of an organism (Wenzl et al., 2008). In past few years DArT array was developed for potato. The DArT linkage maps by using DArT markers were constructed for *S. bulbocastanum, S. michoacanum* and *Solanum phureja* diploid map which comprises of linkage map of two solanum species (Mann et al., 2011; Sliwka et al., 2012). By using this technology *Rpi-rzc1* gene was mapped to chromosome 10 of potato and it prevents the infection of *P.infestans* in tuber and foliage region of potato. It has also been determined that *Rpi-rzc1* gene has close association with the gene encoded by F locus and this gene imparts violet color to flowers and it can be exploited as a phenotypic marker (Sliwka et al., 2012). Recently DArT marker sequences shows heterogeneity among wild species of potato. All these information is available due to DArT array methodology and in future it will assist in developing gene pyramids of resistant genes (R) and for cloning R genes in cultivated potato to improve its quantity and quality (Traini et al., 2013).

17.4 POTATO GENOME PROJECT

National Science Foundation (USA) in collaboration with other laboratories is involved in studying the structural and functional genomics of many plants including potato. Following are the achievements of this project related to potato:

- RB gene confers resistance to *P. infestans,* has been identified.
- There are approximately 180 resistance genes candidates have been identified conferring resistance to pathogens. Their sequence study help in the identification of several traits related to disease resistance.
- On chromosome 5 and 10, there are 'hot spots' where resistance genes are localized and their physical as well as genetic maps have been studied.
- 13000 Expressed Sequence Tags have been constructed from potato cDNA libraries. Potato microarray has been developed using these ESTs.
- Researchers have also analyzed the expression level of potato genome in order to recognize many resistance genes and to get knowledge about the biology of potato (Baker et al., 2000).

17.5   FUNCTIONAL GENOMICS OF POTATO

Functional genomics deals with the study of the function of specific genes. There are several techniques in order to focus the functional aspects of genes, including RNA interference (RNAi), virus-induced gene silencing (VIGS) and microarray technology. By combining structural and functional genomics tools, different traits of potato can be studied. Heterozygosity is very high in potato cultivars which also complicated the study of genes. With the advancement in structural genomics of potato, researchers have also focused in determining functional aspects of genes (Barone et al., 2008).

17.5.1   VIRUS-INDUCED GENE SILENCING

This method is used to study functional aspect of potato genes. It utilizes viral vectors to carry the target genes and direct them to plant mRNAs. By using VIGS vectors such as potato virus X (PVX) and tobacco rattle virus (TRV), tuber trait's genes and pathogen resistance genes can be identified (Brigneti, 2004). The potato virus X vector has ability to induce gene silencing in two species of *solanum*, one is wild diploid potato and other is cultivated tetraploid potato specie. It was shown that virus silences the phytoenedesaturase (pds) gene in the foliar tissues as well as in the tuber of the potato. In this way genes involved in development of tuber, metabolism process and defensive mechanism can be studied. This gene (pds) is involved in synthesizing carotenoids pigments in potato and have role in many physiological processes in plants (Lopez et al., 2008). TRV also induced silencing in different species of potato. Researchers have used pds gene as a silencing marker to suppress three more genes in order to find out function of other important resistance genes. The other three genes include R1, Rx, and RB that confer resistance to *P. infestans* and potato virus X. It has been observed in a study that the defense mechanism of potato is weakened when these genes are silenced and it results in a disease susceptible phenotype (Brigneti, 2004). Recently, function of exocyst subunits of Solanaceous family was determined through TRV mediated VIGS and it was observed that *Solanaceous exocysts* provide immunity against plant pathogens (Du et al., 2018).

17.5.2   SERIAL ANALYSIS OF GENE EXPRESSION (SAGE)

This method is used for analyzing expression patterns of genes. In this method, small cDNA tags are produced which are linked together to determine expression patterns of genes (Saha et al., 2002). In potato this method has been used which generates many thousands of such tags of which few tags were unique and have similarity with the potato EST sequences. By using this technology, many genes were identified that participate in biological, physiological, and metabolic processes (Crookshanks et al., 2001; Ronning et al., 2003).

17.5.3   AMPLIFIEDFRAGMENTLENGTHPOLYMORPHISM(AFLP)

AFLP is used to study stages of potato tuber life cycle. Out of 18000 transcripts derived fragments (TDFs), 200 are specific and they belong to different stages of tuber life cycle. The similarities between these fragments and known genes sequences give details about different processes occurring in potato life cycle such as proteins regulating dormancy and sprouting. This facilitate in the recognition of cis acting elements which are twelve in number and they are activated towards stimulus like sugars, hormones or light and they are playing their role during the life cycle of tuber (Bryan & Hein, 2008).

17.5.4   RNA SEQUENCING

RNA sequencing (RNA-Seq) is a valuable tool for studying detailed expression of genes associated with plant's molecular processes. It involves use of RNA molecules to identify functional elements of genome involved in diseases and developmental processes (Kukurba & Montgomery, 2015). Recently this molecular technique has been employed in understanding structural and functional genomics of potato. Gene expression of multiple responses of *S. tuberosum* has been quantified using RNA-Seq including defense mechanism against *P. infestans*, photoperiod controlled tuberization, pigment regulation, resistance to potato virus Y, response to drought stress and nitrogen fertilizers (Shan et al., 2013; Zhang et al., 2014; Frades et al., 2015; Goyer et al., 2015; Liu et al., 2015; Cho et al., 2016; Gálvez et al., 2016). Furthermore, genes anticipating cold induced sweetening in tubers were also determined through RNA sequencing (Neilson et al., 2016).

17.6   IDENTIFICATION OF DIFFERENT PROTEINS THROUGH TRANSGENIC APPROACH

In studying functional genomics of potato, transgenics play an important part in understanding different pathways of sugar and starch metabolism.

Transgenics rely on the recombinant DNA technology. There are two genes whose overexpression can lead to increased quantity of starch which are glucose–6-phosphate and adenylate translocator. In the metabolism of carbohydrates, more than eighty proteins are involved and all this information is due to transgenic biology (Zhang et al., 2008).

• In addition to this, effect of down regulation of another enzyme was demonstrated using this approach. The down regulation of zeaxanthine poxidase affects the carotenoid phenotype of potato. With the advancements in genetics, more alleles of potato have been identified. In some potatoes, Y locus is considered as a major quantitative trait locus for carotenoids pigments and candidate gene β-carotene hydroxylase has been recognized for this locus (Brown et al., 2006).

• Homeobox gene comprises the family of transcription factors and it is known as POTH1 gene which is present in variety of cell and tissues such as meristems, primordium of leaf and vascular tissues of stem. Through functional studies, it has been determined that this gene is responsible for vegetative development in potato and is associated with decreased levels of gibberellins (GA). It has been also noticed that overexpression of this gene alters the phenotype, leaf morphology and also results in small-sized plant. Therefore, POTH1 negatively regulates GA biosynthesis to mediate the development of potato (Rosin et al., 2003).

• MADS box genes activates the transcription factors involve in the developmental process of potato (Rosin et al., 2003b) book. MADS box protein contains two domains, MAD box and K box. Former is the DNA binding domain while later is the protein interaction domain. POTM1 is the potato MAD box gene and is expressed in roots, leaves, and tuber. During vegetative meristematic tissue development, its transcripts are increased when shoot formation begins but reduces during tuber development. Development of axillary bud is also due to POTM1 gene that regulates cell growth in meristems (Kang, 2002).

17.7 PROTEINS IN TUBER LIFE CYCLE

Functional genomics helps in understanding phases of potato tuber life cycle and gives information about the proteins or genes involved in these stages. The main stages include tuber development also known as tuberization, dormancy, and sprouting (Bachem et al., 2000). Little description of these stages is given in the following subsections.

17.7.1   TUBERIZATION

It is a developmental process of tuber in potato. Through past studies, it has become clear that some genes or proteins involved in photoperiod dependent tuberization and flowering have similarity in their function (Rodriguez-Falcon et al., 2006). In this perspective, phytochrome B is the main enzyme responsible for tuber growth but along with this, there are certain homologs of GIGANTEA, flowering locus T and CONSTANS which also participates in development process of tuber. CONSTANS (CO) is a transcription factor in many plants including Arabidopsis and is translated into CO protein which further encodes transcription of another protein, flowering locus T (Kim et al., 2008). Both these proteins interact with each other and induces flowering in plants and in the same way they interact in the potato but increased expression of CO gene delay the tuberization (Davies et al., 2008).

StBEL5, POTH1 are the transcription factors involved in regulating GA levels. It was observed that as tuberization is reached, GA level decreases and it ultimately induces changes in the cell division process. Another protein GA2 oxidase is also involved in the mechanism of decreasing GA level in stolon so the two mechanisms involved in tuberization are:

- Photoperiod, short day or long night dependent pathway; and
- Gibberellin dependent pathway.

It is still unknown which genes are involved in the interaction between these pathways and research is needed to identify those proteins or genes on those specific quantitative trait loci. But environmental and signaling factors involved in tuberization along with their favoring condition are known. According to Fernie and Willmitzer (2001), the favoring conditions are high with cytokinin, sucrose, light, and jasmonic acid but low with gibberellin and phytochrome, short days favor this process.

17.7.2   DORMANCY AND SPROUTING

The final stage of tuber life cycle is dormancy. It provides protection to the tuber tissues during unfavorable environmental conditions. During this period sprouting starts and roots begin to grow rapidly. Tuber provides nourishment to the sprouts so that new organism can develop. To induce and extend this period, several chemicals are being used such as Maleic Hydrazide, chlorpropham, and dimethyl naphthalene, etc. It has been observed that during dormancy, cell cycle stops at G1 phase where cell division occurs and it does not enter into S phase to begin DNA synthesis. To break this period, several environmental and genetic factors including D-cyclins (CYCD), histones 3 and 4 protein and cyclin dependent kinases are involved (Sonnewald & Sonnewald, 2014). Cyclin D3 gene stimulates the sprouting outgrowth so all these genes shifts dormancy towards sprouting (Aksenova, 2013).

The phytohormones such as gibberellins and abscisic acid (ABA) regulate growth of potato tubers. ABA induces and maintains dormancy period and in the presence of ethylene, it delays dormancy break (Foukaraki et al., 2016b). On the contrary, gibberellins–1 is associated with bud, sprout outgrowth and enzymatic activity of α- amylases. Activity of both α- and β- amylases is enhanced during sprouting in potato tuber (Rentzsch, 2012).

17.8   ROLE OF GENOMICS FOR THE IMPROVEMENT OF POTATO

The different approaches of functional genomics can provide insights into characteristics feature of potato such as quality, taste, and texture, etc. Besides this; the genome sequence of potato can be used as a reference to explore structural variations and diversity in different species of *Solanum* (Gálvez et al., 2017). In this perspective, a study has reported the draft genome sequence of *Solanum commersonii*, which is the wild species of potato using the reference genome sequence of *S. tuberosum*. This advancement provides better understanding of genes responsible for stability of crop to endure severe environmental and climatic variations (Aversano et al., 2015). Furthermore, information from genome sequencing of related species of potato such as tomato (The Tomato Genome Consortium, 2012), chili pepper (Kim et al., 2014), tobacco (Sierro et al., 2014) and petunia (Bombarely et al., 2016) is also significant in overcoming challenges associated with the improvement of potato (Gálvez et al., 2017). Here are the points which should be given priority to improve potato cultivation.

- Enhancing potato use for the development of processed products with improved quality, taste, and should have low glycemic index content in them.
- Develop or design tools for the identification of important components present in the potato for production of healthy diets.
- There should be a need of assessing the impact of climatic and other factors in the quality of potatoes.

- Reference genomic sequence of potato can be used to modify the potato traits so that new varieties are resistant to diseases and pests (Davies et al., 2008).
- In order to study complex structural features of potato, association mapping should be employed instead of quantitative trait locus (QTL) mapping. Association mapping will help in analyzing more complex potato traits and it also reduces the time required to produce crosses for QTL mapping approach (D'hoop et al., 2007; Morrell et al., 2012).

17.8.1 GENOME EDITING

It is necessary to ensure sufficient food supply to meet the increasing nutritional demand across the world. For this purpose, there must be increased production of food crops using both conventional and modern techniques. Plants are susceptible to pathogens and severe environmental conditions which can affect their growth and yield. However, in the recent years, new methods have been introduced for the improvement of crops and to protect them from harmful conditions. Genome editing is one such tool that enables mutation at a highly specific and targeted position for the enhancement of genetic traits (Georges & Ray, 2017). It involves endonucleases such as sequence-specific nuclease (SSN), transcription activator-like effector nucleases (TALENs), zinc finger nucleases, and clustered regularly interspaced short palindromic repeats (CRISPR) associated nucleases. These nucleases activate DNA repair pathways and homologous recombination by introducing DNA double strand breaks at a specific sites (Palmgren et al., 2014; Cermak et al., 2015; Alamar et al., 2017). This has been demonstrated in many food crops to enhance their characteristics including tomato, potato, maize, eggplant and pepper, etc. (Shi et al., 2017; Van Eck, 2017). A study showed that *S. tuberosum* displayed decreased susceptibility to herbicide when treated with Geminivirus mediated genome editing (Butler et al., 2016).

To improve the quality of vegetables, mutations at multiple alleles have been incorporated using CRISPR associated 9 (CRISPR/Cas9) system (Ma et al., 2016; Karkute et al., 2017). The quality of starch in *S. tuberosum* is important factor in food. High amylopectin potatoes also known as waxy potatoes were developed by inducing multiple allelic mutations in the granule *bound starch synthase* (*GBSS*) gene using CRSIPR/Cas9 technology (Andersson et al., 2017). This research has proved the efficiency of CRISPR/Cas9 for targeting multiple alleles at a time. In addition to this, the function of *S. tuberosum* transcription factor gene *MYB44* was also determined with CRSIPR/Cas9, thus presenting it as a perfect tool to be exploited in reverse genetics (Zhou et al., 2017). It can also be used to introduce mutations in the regulatory regions of the gene such as cis-regulatory elements (CREs) located in the promoter regions. These elements controls expression of genes with great specificity so mutations in these regions will slightly affect the level of gene expression thus changes the traits quantitatively (Swinnen et al., 2016; Li et al., 2017).

17.9 CONCLUSION AND PROSPECTIVES

In this review different methodologies have been discussed for studying structural and functional genomics of *Solanum Tuberosum* such as expressed sequence tags (EST), Bacterial Artificial Chromosome (BAC) Libraries, molecular mapping, Candidate Gene Approach, Virus-induced gene Silencing (VIGS), Serial analysis of gene expression (SAGE) and transgenic approach. National Science Foundation along with other laboratories has studied structural and functional characteristics of this crop and identified pest and disease resistance genes. Functional genomics approach helps in better understanding of potato tuber life phases such as tuberization, dormancy, and sprouting.

In the scientific community it is considered that new era for the potato has began after the completion of its genomic sequencing. The development of large potato EST collections and microarrays and most importantly genome editing has helped in better understanding of genome of Solanaceae family. Due to the increasing demand of potato as a food, it is the responsibility of researchers to get insights into potato genome sequencing in order to improve qualitative as well as quantitative characteristics of this vegetable. There is a need to study this plant in detail in order to produce new and better varieties that are able to produce higher yield even in stressed and unfavorable conditions. The sequencing of the potato genome was a scientific milestone that led to the discoveries of its different genes, their encoded proteins along with their structure and function. For genetic characterization and determining gene function, more complex computational approaches are likely to develop in near future.

The rapid and swift increase in the world population has resulted in the increase in the food demand. Thus the complete insight on potato structural and functional genomics can help in advanced and novel plant breeding techniques to meet the terms of today's growing population. Moreover, this information can serve as a basis for other ongoing and future projects. Rapid generation of breeding markers, agronomically significant genes identification and capacity building in developing countries can be accomplished through these studies. However, there is a need in the advancement and improvement of the molecular tools so that progress can be made.

KEYWORDS

- functional genomics
- potato (*Solanum tuberosum*)
- structural genomics

REFERENCES

Aksenova, N. P., Sergeeva, L. I., Konstantinova, T. N., Golyanovskaya, S. A., Kolachevskaya, O. O., & Romanov, G. A., (2013). Regulation of potato tuber dormancy and sprouting. *Russian Journal of Plant Physiology, 60* (3), 301–31.

Alamar, M. C., Tosetti, R., Landahl, S., Bermejo, A., & Terry, L. A., (2017). Assuring future potato tuber quality during storage. *Frontiers in Plant Science, 8,* 2034.

Andersson, M., Turesson, H., Nicolia, A., Fält, A. S., Samuelsson, M., & Hofvander, P., (2017). Efficient targeted multiallelic mutagenesis in tetraploid potato (*Solanum tuberosum*) by transient CRISPR-Cas9 expression in protoplasts. *Plant Cell Rep., 36*, 117–128.

Atkinson, N. J., & Urwin, P. E., (2012). The interaction of plant biotic and abiotic stresses: From genes to the field. *J. Exp. Bot., 63*, 3523–3544.

Aversano, R., Contaldi, F., Ercolano, M. R., Grosso, V., Iorizzo, M., Tatino, F., et al. (2015). The Solanum commersonii genome sequence provides insights into adaptation to stress conditions and genome evolution of wild potato relatives. *The Plant Cell., 27* (4), pp. 954–968.

Bachem, C., Van der Hoeven, R., Lucker, J., Oomen, R., Casarini, E., Jacobsen, E., & Visser, R., (2000). Functional genomics analysis of potato tuber life-cycle. *Potato Research, 43* (4), 297–312.

Baker, B., Brown, C., Buell, R., May, G., Jiang, J., Cornell, M. J., et al. (2000). *Potato Genome Project*. National Science Foundation.

Ballvora, A., Jöcker, A., Viehover, P., Ishihara, H., Paal, J., Meksem, K., et al. (2007). Comparative sequence analysis of Solanum and Arabidopsis in a hot spot for pathogen resistance on potato chromosome V reveals a patchwork of conserved and rapidly evolving genome segment. *BMC Genomics 8*, 112.

Barone, A., Chiusano, M. L., Ercolano, M. R., Giuliano, G., Grandillo, S., & Frusciante, L., (2008). Structural and Functional Genomics of Tomato. *International Journal of Plant Genomics, 2008*, p. 12.

Bendahmane, A., Farnham, G., Moffett, P., & Baulcombe, D. C., (2002). Constitutive gain-of-function mutants in a nucleotide binding site-leucine rich repeat protein encoded at the Rx locus of potato. *Plant J., 32* (2), 195–204.

Bengtsson, T., Weighill, D., Proux-Wéra, E., Levander, F., Resjö, S., Burra, D. D., et al., & Alexandersson, E., (2014). Proteomics and transcriptomics of the BABA-induced resistance response in potato using a novel functional annotation approach. *BMC Genomics, 15*, 315.

Bombarely, A., Moser, M., Amrad, A., Bao, M., Bapaume, L., Barry, C. S., et al. (2016). Insight into the evolution of the Solanaceae from the parental genomes of Petunia hybrida. *Nat Plants, 2*, 16074.

Brigneti, G., Martín-Hernández, A. M., Jin, H., Chen, J., Baulcombe, D. C., Baker, B., & Jones, J. D. G., (2004). Virus-induced gene silencing in *Solanum* species. *The Plant Journal, 39* (2), 264–272.

Brown, C. R., Kim, T. S., Ganga, Z., et al. (2006). Segregation of total carotenoid in high level potato germplasm and its relationship to beta-carotene hydroxylase. *Am. J. Potato Res., 83*, 365–372.

Bryan, G. J., & Hein, I., (2008). Genomic resources and tools for gene function analysis in potato. *International Journal of Plant Genomics, 9.*

Bryan, G. J., (2007). Genomics. *Potato Biology and Biotechnology: Advances and Perspectives*, 1–857.

Butler, N. M., Baltes, N. J., Voytas, D. F., & Douches, D. S., (2016). Geminivirus-mediated genome editing in potato (*Solanum tuberosum* L.) using sequence-specific nucleases. *Frontiers in Plant Science, 7*, 1045.

Čermák, T., Baltes, N. J., Čegan, R., Zhang, Y., & Voytas, D. F., (2015). High-frequency, precise modification of the tomato genome. *Genome Biology, 16* (1), 232.

Cho, K., Cho, K. S., Sohn, H. B., Ha, I. J., Hong, S. Y., Lee, H., et al. (2016). Network analysis of the metabolome and transcriptome reveals novel regulation of potato pigmentation. *J. Exp. Bot., 67*, 1519–1533.

Clasen, B. M., Stoddard, T. J., Luo, S., Demorest, Z. L., Li, J., Cedrone, F., et al. (2016). Improving cold storage and processing traits in potato through targeted gene knockout. *Plant Biotechnol. J., 14*, 169–176.

Crookshanks, M., Emmersen, J., Welinder, K. G., & Nielsen, K. L., (2001). The potato tuber transcriptome: analysis of 6077 expressed sequence tags. *FEBS Letter, 506* (2), 123–126.

Das, G., & Rao, G. J. N., (2015). Molecular marker-assisted gene stacking for biotic and abiotic stress resistance genes in an elite rice cultivar. *Front. Plant Sci., 6*, 698.

Davies, H., Bryan, G. J., & Mark, T., (2008). Advances in functional genomics and genetic modification of potato. *Potato Research, 51*, 283–299.

Draffehn, A. M., (2010). Structural and functional characterization of natural alleles of potato (*Solanum tuberosum*L.) invertases associated with tuber quality traits Inaugural – Dissertation. *Max Planck Institute for Plant Breeding Research*, 1–338.

Du, Y., Overdijk, E. J., Berg, J. A., Govers, F., & Bouwmeester, K., (2018). Solanaceous exocyst subunits are involved in immunity to diverse plant pathogens. *Journal of Experimental Botany*.

Farrar, K., & Donnison, I. S., (2007). Construction and screening of BAC libraries made from Brachypodium genomic DNA. *Nat Protoc., 2* (7), 1661–74.

Felcher, K. J., (2013). Candidate Gene Approach to Identify an Economically Important Gene in Cultivated Potato. *Plant Breeding and Genomics*.

Fernie, A. R., & Willmitzer, L., (2001). Molecular and Biochemical Triggers of Potato Tuber Development. *Plant Physiol., 127*, 1459–1465.

Finkers-Tomczak, A., Bakker, E., De Boer, J., Van der Vossen, E., Achenbach, U., Golas, T., et al. (2011). Comparative sequence analysis of the potato cyst nematode resistance locus H1 reveals a major lack of co-linearity between three haplotypes in potato (*Solanum tuberosum* ssp.). *Theor. Appl. Genet., 122,* 595–608.

Flinn, B., Rothwell, C., Griffiths, R., Lägue, M., DeKoeyer, D., Sardana, R., et al. (2005). Potato expressed sequence tag generation and analysis using standard and unique cDNA librarie s. *Plant Mol. Bio., 59* (3), 407–33.

Food and Agriculture Organization, (2016). Food and Agricultural commodities production/ Commodities by regions. FAOSTAT. Available from: http://faostat3. fao. org/browse/ rankings/commodities_by_regions/E

Foukaraki, S. G., Cools, K., & Terry, L. A., (2016b). Differential effect of ethylene supplementation and inhibition on abscisic acid metabolism of potato (*Solanum tuberosum* L.) tubers during storage. *Postharvest Biol. Tec., 112,* 87–94.

Frades, I., Abreha, K. B., Proux-Wéra, E., Lankinen, Å., Andreasson, E., & Alexandersson, E., (2015). A novel workflow correlating RNA-seq data to Phythophthora infestans resistance levels in wild Solanum species and potato clones. *Front Plant Sci., 6,* 718.

Galek, R., Rurek, M. S. D., Jong, W., Pietkiewicz, G., Augustyniak, H., & Sawicka-Sienkiewicz, E., (2011). Application of DNA markers linked to the potato H1 gene conferring resistance to pathotype Ro1 of Globoderarostochiensis. *J. Appl. Genetics., 52,* 407–411.

Gálvez, J. H., Tai, H. H., Barkley, N. A., Gardner, K., Ellis, D., & Strömvik, M. V., (2017). Understanding potato with the help of genomics. *AIMS Agriculture and Food, 2* (1), 16–39.

Gálvez, J. H., Tai, H. H., Lagüe, M., Zebarth, B. J., & Strömvik, M. V., (2016). The nitrogen responsive transcriptome in potato (*Solanum tuberosum* L.) reveals significant gene regulatory motifs. *Sci. Rep., 6,* 26090.

Gao, L., Tu, Z. J., Millett, B. P., & Bradeen, J. M., (2013). Insights into organ-specific pathogen defense responses in plants: RNA-seq analysis of potato tuber-Phytophthora infestans interactions. *BMC Genomics, 14,* 340.

Gebhardt, C., (2007). Molecular Markers, Maps, and Population Genetics. Potato Biology and Biotechnology. *Advances and Perspectives, 857.*

Georges, F., & Ray, H., (2017). Genome editing of crops: a renewed opportunity for food security. *GM Crops Food, 8,* 1–12.

Gong, Z., Wu, Y., Koblížková, A., Torres, G. A., Wang, K., Iovene, M., et al. (2012). Repeatless and repeat-based centromeres in potato: Implications for centromere evolution. *American Society of Plant Biologists, 24* (9), 3559–3574.

Goyer, A., Hamlin, L., Crosslin, J. M., Buchanan, A., & Chang, J. H., (2015). RNA-Seq analysis of resistant and susceptible potato varieties during the early stages of potato virus Y infection. *BMC Genomics, 16,* 472.

Gregory, C., (2004). An EST based Genomics Project in Potato Solanum Tuberosum. *Department of Biology, 85.*

Hancock, R. D., Morris, W. L., Ducreux, L. J., Morris, J. A., Usman, M., Verrall, S. R., et al. (2014). Physiological, biochemical, and molecular responses of the potato (*Solanum tuberosum* L.) plant to moderately elevated temperature. *Plant Cell Environ., 37,* 439–450.

Jung, C. S., Griffiths, H. M., De Jong, D. M., Cheng, S., Bodis, M., & De Jong, W. S., (2005). The potato P locus codes for flavonoid 3,' 5'-hydroxylase. *Theor. Appl. Genet., 110* (2), 269–75.

Kang, S. G., & Kang, H., (2002) Characterization of Potato Vegetative MADS Box Gene, POTMI–1, in Response to Hormone Applications. *Journal of Plant Biology, 45* (4), 196–200.

Karkute, S. G., Singh, A. K., Gupta, O. P., Singh, P. M., & Singh, B., (2017). CRISPR/Cas9 Mediated Genome Engineering for the improvement of Horticultural Crops. *Frontiers in Plant Science, 8,* 1635.

Kim, S. Y., Yu, X., & Michaels, S. D., (2008). Regulation of Constans and Flowering Locus T Expression in Response to Changing Light Quality. *Plant Physiology, 148,* 269–279.

Kim, S., Park, M., Yeom, S. I., Kim, Y. M., Lee, J. M., Lee, H. A., et al. (2014). Genome sequence of the hot pepper provides insights into the evolution of pungency in Capsicum species. *Nature Genetics, 46* (3), pp. 270–278.

Kloosterman, B., Abelenda, J. A., Del Mar, C. G. M., Oortwijn, M., De Boer, J. M., Kowitwanich, K., et al. (2013). Naturally occurring allele diversity allows potato cultivation in northern latitudes. *Nature, 495,* 246–250.

Kukurba, K. R., & Montgomery, S. B., (2015). RNA Sequencing and Analysis. *Cold Spring Harbor Protocols, 11,* 951–969.

Li, L., Strahwald, J., Hofferbert, H. R., Lu Beck, J., Tacke, E., Junghans, H., Wunder, J., & Gebhardt, C., (2005). DNA Variation at the Invertase Locus *invGE/GF* Is Associated with Tuber Quality Traits in Populations of Potato Breeding Clones. *Genetics, 170,* 813–821.

Li, X., Xie, Y., Zhu, Q., & Liu, Y. G., (2017). Targeted Genome Editing in Genes and cis-Regulatory Regions Improves Qualitative and Quantitative Traits in Crops. *Molecular Plant, 10* (11), 1368–1370.

Liu, B., Zhang, N., Wen, Y., Jin, X., Yang, J., Si, H., & Wang, D., (2015). Transcriptomic changes during tuber dormancy release process revealed by RNA sequencing in potato. *J. Biotechnol., 198,* 17–30.

Lopez, A. B., Yang, Y., Thannhauser, T. W., & Li, L., (2008). Phytoenedesaturase is present in a large protein complex in the plastid membrane. *Physiologia Plantarum., 133,* 190–98

Ma, X., Zhu, Q., Chen, Y., & Liu, Y. G., (2016). CRISPR/Cas9 platforms for genome editing in plants: developments and applications. *Mol. Plant., 9,* 961–974.

Mach, J., (2012). Rapid Centromere Evolution in Potato: Invasion of the Satellite Repeats. *American Society of Plant Biologists, 24* (9), 3487.

Machida-Hirano, R., (2015). Diversity of potato genetic resources. *Breed Sci., 65,* 26–40.

Mann, H., Iorizzo, M., Gao, L., D'Agostino, N., Carputo, D., Chiusano, M. L., & Bradeen, J. M., (2011). Molecular linkage maps: Strategies, resources, and achievements. In: Bradeen, J. M., (ed.), *Genetics, Genomics, and Breeding of Crop Plants: Potato* (pp. 68–89). Enfield KC.

Mazourek, M., Cirulli, E. T., Collier, S. M., Landry, L. G., Kang, B. C., Quirin, E. A., et al. (2009). The Fractionated Orthology of Bs2 and Rx/Gpa2 Supports Shared Synteny of Disease Resistance in the Solanaceae. *Genetics, 182,* 1351–1364.

McIvor, C., (2011). All eyes on the potato genome: Cracking of tricky genetic code may offer clues to fighting blight. *Nature.*

Neilson, J., Lagüe, M., Thomson, S., Aurousseau, F., Murphy, A. M., Bizimungu, B., et al. (2017). Gene expression profiles predictive of cold-induced sweetening in potato. *Functional & Integrative Genomics.* pp. 1–18.

Nicaise, V., (2014). Crop immunity against viruses: Outcomes and future challenges. *Front. Plant Sci., 5,* 660.

Palmgren, M. G., Edenbrandt, A. K., Vedel, S. E., Andersen, M. M., Landes, X., Østerberg, J. T., et al. (2015). Are we ready for back-to-nature crop breeding?. *Trends in Plant Science, 20* (3), 155–64.

Piquerez, S. J., Harvey, S. E., Beynon, J. L., & Ntoukakis, V., (2014). Improving crop disease resistance: Lessons from research on Arabidopsis and tomato. *Front. Plant Sci., 5,* 671.

Rentzsch, S., Podzimska, D., Voegele, A., Imbeck, M., Muller, K., Linkies, A., & Leubner-Metzger, G., (2012). Dose- and tissue-specific interaction of monoterpenes with the gibberellin-mediated release of potato tuber bud dormancy, sprout growth and induction of a-amylases and b-amylases. *Planta., 235,* 137–151.

Rodriguez-Falcon, M., Bou, J., & Prat, S., (2006). Seasonal control of tuberisation in potato: conserved elements with the flowering response. *Annu. Rev. Plant Bio., 57,* 151–180.

Ronning, C. M., Stegalkina, S. S., & Ascenzi, R. A., (2003). Comparative analyzes of potato expressed sequence tag libraries. *Plant Physiology, 131* (2), 419–429.

Rosin, F. M., Hart, J. K., Horner, H. T., Davies, P. J., & Hannapel, D. J., (2003a). Overexpression of a *Knotted*-Like Homeobox Gene of Potato Alters Vegetative Development by Decreasing Gibberellin Accumulation. *Plant Physiology, 132,* 106–117.

Rosin, F. M., Hart, J. K., Horner, H. T., Davies, P. J., & Hannapel, D. J., (2003b). Overexpression of a *Knotted*-Like Homeobox Gene of Potato Alters Vegetative Development by Decreasing Gibberellin Accumulation. *Plant Physiology, 131,* 1613.

Saha, S., Sparks, A. B., Rago, C., et al. (2002). Using the transcriptome to annotate the genome. *Nature Biotechnology, 20* (5), 508–512.

Sharma, A., Wolfgruber, T. K., & Presting, G. G., (2013). Tandem repeats derived from centromeric retrotransposons. *BMC Genomics, 14,* 1471–2164.

Sharma, S. K., (2010). The Potato Genome Sequencing Initiative. *The Potato Genome Sequencing Consortium, 1.*

Shi, J., Gao, H., Wang, H., Lafitte, H. R., Archibald, R. L., Yang, M., et al. (2017). ARGOS8 variants generated by CRISPR-Cas9 improve maize grain yield under field drought stress conditions. *Plant Biotechnology Journal, 15* (2), 207–216.

Sierro, N., Battey, J. N., Ouadi, S., Bakaher, N., Bovet, L., Willig, A., et al. (2014). The tobacco genome sequence and its comparison with those of tomato and potato. *Nat Commun., 5,* 3833.

Śliwka, J., Jakuczun, H., Chmielarz, M., Hara-Skrzypiec, A., Tomczyńska, I., Kilian, A., et al. (2012). Late blight resistance gene from Solanum ruiz-ceballosii is located on potato chromosome X and linked to violet flower color. *BMC Genetics, 13,* 12.

Sonnewald, S., & Sonnewald, U., (2014). Regulation of potato tuber sprouting. *Planta., 239* 27–38.

Stupar, R. M., Song, J., Tek, A. L., Cheng, Z., Dong, F., & Jiang, J., (2002). Highly condensed potato pericentromeric heterochromatin contains rDNA-related tandem repeats. *Genetics, 162* (3), 1435–1444.

Swinnen, G., Goossens, A., & Pauwels, L., (2016). Lessons from domestication: Targeting Cis regulatory elements for crop improvement. *Trends in Plant Science, 21,* 506–515.

Szajko, K., Chrzanowska, M., Witek, K., Strzelczyk-Zyta, D., Zagórska, H., Gebhardt, C., et al. (2008). The novel gene *Ny*–1 on potato chromosome IX confers hypersensitive resistance to potato virus Y and is an alternative to *Ry* genes in potato breeding for PVY resistance. *Theor. Appl. Genet., 116,* 297–303.

Szajko, K., Strzelczyk-Żyta, D., & Marczewski, W., (2014). *Ny*–1 and *Ny*–2 genes conferring hypersensitive response to potato virus Y (PVY) in cultivated potatoes: mapping and marker-assisted selection validation for PVY resistance in potato breeding. *Mol. Breed.,* *34*, 267–271.

Tai, H. H., Goyer, C., De Koeyer, D., Murphy, A., Uribe, P., & Halterman, D., (2013). Decreased defense gene expression in tolerance versus resistance to Verticillium dahliae in potato. *Funct Integr. Genomics., 13*, 367–378.

The Potato Genome Sequencing Consortium, (2011). Genome sequence and analysis of the tuber crop potato. *Nature, 475*, 189–195.

The Tomato Genome Consortium, (2012). The tomato genome sequence provides insights into fleshy fruit evolution. *Nature, 485*, 635–641.

Traini, A., Iorizzo, M., Mann, H., Bradeen, J. M., Carputo, D., Frusciante, L., & Chiusano1, M. L., (2013). Genome microscale heterogeneity among wild potatoes revealed by diversity arrays technology marker sequences. *International Journal of Genomics,* p. 9.

Van Der Vossen, E. A. G., Rouppe Van Der, V. J. N. A. M., Kanyuka, K., Bendahmane, A., Sandbrink, H., Baulcombe, D. C., et al. (2000). Homologs of a single resistance-gene cluster in potato confer resistance to distinct pathogens: A virus and a nematode. *The Plant J., 23* (5), 567–576.

Van Eck, J., (2017). Genome editing and plant transformation of solanaceous food crops. *Curr. Opin. Biotechnol., 49*, 35–41.

Varma, A., & Shrivastava, N., (2011). The role of plant genomics in biotechnology. *Biotechnology, 8*, 1–6.

Wenzl, P., Huttner, E., Carling, J., Xia, L., Blois, H., Caig, V., & Kilian, A., (2008). Diversity Arrays Technology (DArT): A generic high-density genotyping platform. In: *Safflower: Unexploited Potential and World Adaptability* (pp. 1–7). 7th International Safflower Conference.

Wisser, R. J., Balint-Kurti, P. J., & Nelson, R. J., (2006). The genetic architecture of disease resistance in maize: a synthesis of published studies. *Phytopathology, 96*, 120–129.

Zhang, C., & Kim, S. H., (2003). Overview of structural genomics: from structure to function. *Curr. Opin. Chem. Biol., 7* (1), 28–32.

Zhang, N., Yang, J., Wang, Z., Wen, Y., Wang, J., He, W., Liu, B., Si, H., & Wang, D., (2014). Identification of novel and conserved microRNAs related to drought stress in potato by deep sequencing. *PLoS One, 9*, e95489.

Zhou, X., Zha, M., Huang, J., Li, L., Imran, M., & Zhang, C., (2017). StMYB44 negatively regulates phosphate transport by suppressing expression of PHOSPHATE1 in potato. *J. Exp. Bot., 68*, 1265–1281.

INDEX